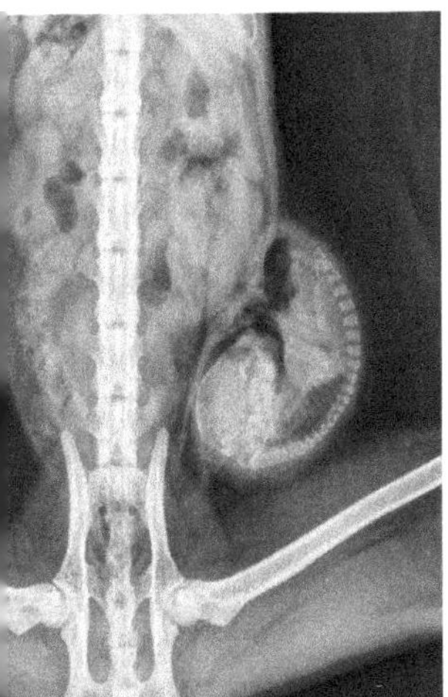
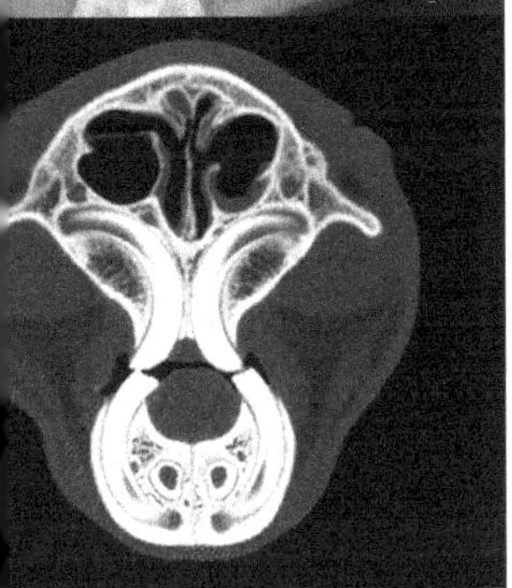
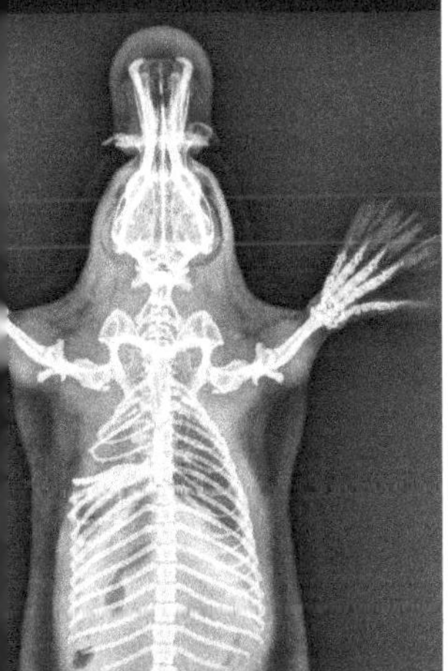
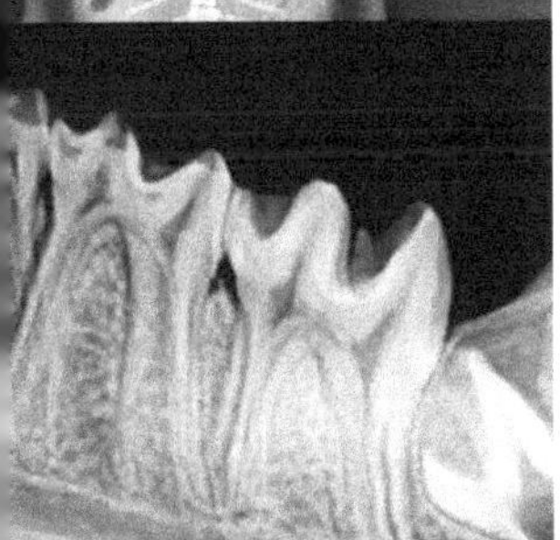
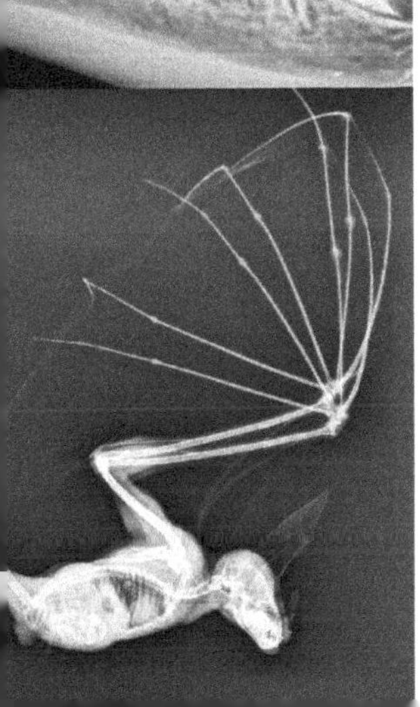

RADIOLOGY OF AUSTRALIAN MAMMALS

LARRY VOGELNEST AND GRAEME ALLAN

CSIRO PUBLISHING

National Library of Australia Cataloguing-in-Publication entry

Vogelnest, L. (Larry), author.

Radiology of Australian mammals / Larry Vogelnest and Graeme Allan.

9780643108646 (hardback)
9780643108653 (epdf)
9780643108660 (epub)

Includes bibliographical references and index.

Mammals – Australia.
Veterinary radiology.

Allan, Graeme Sutcliffe, author.

599.0994

Published by

CSIRO Publishing
36 Gardiner Road, Clayton VIC 3168
Private Bag 10, Clayton South VIC 3169
Australia

Telephone: +61 3 9662 7666
Local call: 1300 788 000 (Australia only)
Fax: +61 3 9662 7555
Email: publishing.sales@csiro.au
Website: www.publish.csiro.au

Front cover (main image): tammar wallaby (*Macropus eugenii*)
Front cover (top to bottom): yellow-bellied glider (*Petaurus australis*), common wombat (*Vombatus ursinus*), platypus (*Ornithorhynchus anatinus*), sub-adult eastern grey kangaroo (*Macropus giganteus*), ghost bat (*Macroderma gigas*)
Back cover (top to bottom): squirrel glider (*Petaurus norfolcensis*), common wombat (*Vombatus ursinus*), spot-tailed quoll (*Dasyurus maculatus*), spot-tailed quoll (*Dasyurus maculatus*), greater bilby (*Macrotis lagotis*)

Set in 10.5/14 Palatino & Optima
Edited by Kerry Brown
Cover design by James Kelly
Typeset by Thomson Digital
Index by Bruce Gillespie
Printed by Ingram Lightning Source

Oct25_RP_ILS

CONTENTS

PREFACE

This text has been compiled to fill a void in the current literature. We recall on many occasions, while interpreting radiographs of various Australian mammals, saying 'We need to do a text on the normal radioanatomy of Australian mammals'. We have acted on it and produced *Radiology of Australian Mammals.*

The recognition, diagnosis and treatment of injury and disease in wildlife species present unique challenges for the veterinarian. This is particularly true for many species of Australian native mammals. Many are cryptic in nature and frequently mask their clinical signs when injured or suffering from disease – the 'preservation response'. When animals can no longer compensate and appear clinically affected, the effects of their injuries or diseases are advanced and efforts to treat these animals are often fruitless. Veterinarians dealing with wildlife must draw on all their skills and powers of observation to recognise the signs of injury and disease. In the majority of cases, further examination and diagnostic workup will be required to confirm a diagnosis. As with domestic species, a suite of diagnostic modalities can be used to further define the nature and extent of injury or disease, guide therapeutic decisions and determine prognosis. Radiology is a fundamental diagnostic modality that is generally readily available and, with the advent of digital radiology, provides high-quality images that can be viewed almost instantly. These can also be easily electronically transferred for referral to a specialist wildlife veterinarian or radiologist for interpretation and opinion if required.

Two important aspects of radiology are the recognition and description of abnormal findings and the interpretation of these findings. In order to recognise abnormalities, knowledge of normal radioanatomy is required. This is particularly challenging with wildlife because of the frequent lack of familiarity with a species and anatomic descriptions, and because radiographic images of normal animals are not readily available for comparison. A further challenge is the variation in anatomy among the different wildlife species; in some instances, even those within the same taxonomic group. One option that can increase one's confidence in distinguishing normal from abnormal is consultation of reference sources on normal radioanatomy. This, until now, has not been available for Australian mammals. *Radiology of Australian Mammals* is the first comprehensive resource on the normal radioanatomy of Australian mammals.

The primary aim of this text is to provide a detailed reference on the normal radioanatomy of Australian mammals. As there are over 380 native mammal species in 48 families and 150 genera in Australia, to represent all of these radiographically would be impossible. We have used selected, readily available species to represent a typical species within each of the major taxonomic groups. We have not included rodents, as Australian native rodents are not anatomically sufficiently different from exotic rodent species described in other texts. We have not included marine mammals, as acquiring complete image sets for available species was not possible. We have also not included the dingo, as this species has similar radioanatomy to that of domestic canids. In most cases clinically normal, healthy animals were used. In some instances, radiographs were opportunistically obtained from animals undergoing investigative procedures for injury or disease. After careful scrutiny by the authors, only radiographs representing normal radioanatomy have been used in Chapters 2 to 9. In some cases, however, radiographs with abnormalities were selected if the abnormality did not detract from the intent to demonstrate other normal radiographic features. Radiographic abnormalities in these images are indicated by annotations on the images and described in the caption. Variations of normal radioanatomy, for example between sexes or ages, have been indicated where relevant. The animals used were primarily sourced from the animal population at Taronga Zoo or wildlife admitted to the Taronga Wildlife Hospital, Sydney, Australia. Animal use was approved by the Zoo's Animal Ethics Committee (approval numbers 4a/11/10 and 4a/10/13). All animals were anaesthetised for radiographic studies and strict radiation safety protocols were adhered to for personnel. A small number of cadaver specimens were also used. If species or specific images were not available from Taronga, these were sourced from other institutions and have been acknowledged appropriately. The majority of images are conventional digital radiographs. A small selection of computed tomography images is included. No magnetic resonance images have been used. A chapter on radiographic technique is included and provides useful information on radiography of wildlife species, highlighting the unique challenges inherent in working with wildlife. There are nine chapters covering the major Australian mammal taxonomic groups. Each chapter covers radioanatomy of the skeletal system and major soft tissue structures and systems. A chapter on dental radiology discusses and demonstrates normal dental radioanatomy of Australian marsupials and monotremes. The final chapter includes selected radiographic pathology case studies, providing an appreciation of radiographic findings seen in some common or noteworthy diseases of Australian mammals. Anatomical descriptions are based on available literature (some of which is antiquated), examination of skeletal and cadaver specimens and comparison with domestic species anatomy. There were many gaps in anatomical descriptions and in some instances skeletal structures identified on radiographs had not previously been described.

Images at Taronga were taken using a GE AMX 4 or GE Optima X-ray generator, Canon CXDI-55G digital flat-panel detector and Sound-Eklin Mark V Portable Digital Radiography System. To ensure the uniformity of all the images presented throughout the book, positioning and side markers are not shown. Instead, we followed the convention that the head or cranial is always to the left in lateral images and the right side is on the left side of all dorsoventral/ventrodorsal, craniocaudal/caudocranial, dorsopalmar/palmarodorsal, dorsoplantar/plantarodorsal and rostrocaudal images. The radiographic position in each image is indicated in the caption. All lateral images are right lateral unless otherwise indicated.

The scientific name for species listed in the text is included in parentheses after the common name when first mentioned in the text for each chapter. The common names (and their construction) and scientific names have been standardised throughout. A complete checklist of the mammals of Australia and its territories is provided in Appendix 1. Abbreviations used to annotate images are included in Appendix 2. Wherever possible we have used terminology that is in common usage and for anatomical terms we have used *Nomina Anatomica Veterinaria* (http://www.wava-amav.org).

Finally, *Radiology of Australian Mammals* complements and utilises the resources of *Pathology of Australian Native Wildlife* (Ladds 2009), *Medicine of Australian Mammals* (Vogelnest and Woods 2008), *Haematology of Australian Mammals* (Clark 2004) and *Australian Mammals: Biology and Captive Management* (Jackson 2003), all CSIRO Publishing publications.

Larry Vogelnest and Graeme Allan

ACKNOWLEDGEMENTS

Compiling a text such as this is an enormous task and could not possibly be completed without the assistance of many others.

This project would not have been possible without the animals we used. We thank the Taronga Zoo Australian mammal supervisors and keepers: Vanessa Stebbings, Nick de Vos, Sam Bennet, Tony Britt-Lewis, Jodie Carr, Paul Davies, Robert Dockerill, Brian Gee, Wendy Gleen, Tim Grul, Nat Holdsworth, Laura Jones, Elli Todd, Rebecca Usmar and Evelyn Weston, for the dedication and care they provide for these animals and for their enthusiasm to assist us with this project. Maintaining an often small animal under anaesthesia for a prolonged period in order to obtain perfectly positioned and high quality radiographs requires the dedication, patience and skill of an exceptional veterinary nurse. We thank the Taronga Wildlife Hospital nursing team: Liz Arthur, Tammy Hadenham, Libby Hall, Liz McConnell, Debbie Pritchard, Annabelle Sehlmeier, Amy Twentyman and Gemma Watkinson.

We have produced a text, which required accurate reproduction of radiographic images. Although digital radiography can often simplify acquisition of wonderful images, there were times when we reached the limits of our equipment to resolve small parts of tiny animals. We were fortunate to be able to call on the services of Jon and Alex Marriott (Radincon), who revised the algorithms used by the Sound-Eklin equipment and so enhanced our image clarity beyond that which is normally available in commercial digital radiography. We thank CSIRO Publishing, who have been supportive, helpful and great to work with during the course of the project. We also thank the following people:

Dr Susan Hemsley for proof reading Chapters 2–9.

Taronga veterinarians, Drs Frances Hulst, Kimberly Vinette Herrin, Gabi Tobias, Benn Bryant and Michelle Campbell for assistance with acquiring some of the images.

Vanessa Di Giglio for assistance with records management.

Dr Sarah Davies (Veterinary Imaging Associates, Sydney) for assistance with CT reconstructions.

Alastair Lyon (Senior Radiographer, Amberley RAAF Base) for Figures 2.2a and b, 2.17b, 2.23a–c, and 2.25a–c (short-beaked echidna CT images) that were acquired for a collaborative project with Taronga during his radiology training at the University of Sydney.

Dr Margaret Stewart (Royal Prince Alfred Hospital) for Figures 5.11a, c and e (koala CT images).

Dr Amber Gillett (Australia Zoo Wildlife Hospital) for Figures 8.22d and 8.23d (greater glider images).

UQ Veterinary Medical Centre and the Native Wildlife Teaching and Research Facility, The University of Queensland, Gatton for Figure 5.1c (fat-tailed dunnart image).

North Shore Veterinary Hospital for use of their CT scanner.

The veterinarians that submitted cases for consideration in the pathology chapter. Unfortunately not all cases could be included; however, we appreciate you sending them to us. Those that were accepted have been individually acknowledged in the chapter.

Alena Felkai for assistance with sourcing some of the reference material.

Dan Vogelnest for technical assistance with Chapter 1.

Dr Gabi Tobias for proof reading Chapter 1.

Sandy Ingelby (Australian Museum, Sydney) for providing access to skeletal specimens.

Small Animal Specialist Hospital (SASH) for use of dental radiography equipment.

Dr Cath Herbert for providing cadaver specimens.

The Taronga Conservation Society Australia for supporting this project.

BIOGRAPHIES

Larry Vogelnest BVSc (Hons) (Syd), MVS (Syd), MACVSc (Zoo Animal Medicine), PSM
Senior Veterinarian, Taronga Zoo, Taronga Conservation Society Australia
Sydney, New South Wales, Australia

Dr Vogelnest graduated with a BVSc degree from the University of Sydney in 1984. He has been employed as a veterinarian with the Taronga Conservation Society Australia since January 1990. He has been Senior Veterinarian at Taronga Zoo since July 1994 and is responsible for the health and welfare of the zoo's animal population and wildlife admitted to the Taronga Wildlife Hospital. In 1990 Dr Vogelnest completed a Masters in Veterinary Studies in Wildlife Medicine and Husbandry. In 1996 he gained membership to the Australian College of Veterinary Scientists by examination in Zoo Animal Medicine. Dr Vogelnest is the Principal Editor of "Medicine of Australian Mammals", CSIRO Publishing (2008) and author and co-author on several chapters. He has published widely in the veterinary literature, including as principal author and co-author on several peer-reviewed journal articles, conference and course proceedings and book chapters. In 2009 he was awarded an Australian Public Service Medal for his contribution to conservation medicine and service to the Taronga Conservation Society Australia. Dr Vogelnest specialises in the health and reproductive management of small populations including ex situ breeding programs for endangered species. He has participated in and advised on both in situ and ex situ components of numerous conservation projects.

Graeme Allan DVSc (Syd), MVSc (Syd), FACVSc, DACVR
Veterinary Imaging Associates
Sydney, New South Wales, Australia

Dr Allan graduated with a BVSc degree from the University of Sydney in 1966. He worked in private practice for 11 years, during which time he was awarded an MVSc degree from the University of Sydney for his research into radiography of the biliary system of dogs published by thesis as 'Cholecystography in the dog' in 1971. In 1973 he gained membership of the Australian College of Veterinary Scientists by examination. In 1977 he commenced a training program in Veterinary Radiology at Cornell University in New York, USA, which culminated in successfully completing the examinations in 1979 that permitted him to qualify as a Diplomate of the American College of Veterinary Radiology. He returned to Australia to commence practice as a specialist veterinary radiologist, in which capacity he has continued to practice to this day. In 1982 he was admitted to Fellowship of the Australian College of Veterinary Scientists by examination. During the last 20 years Dr Allan has taught small animal radiology to undergraduates at the University of Sydney and to graduate veterinarians through the Centre for Veterinary Education at the University of Sydney. He was awarded the TG Hungerford Award for Excellence in Post Graduate Education by the Post Graduate Foundation in Veterinary Science at the University of Sydney in 1993 and the Seddon Memorial Prize by the NSW Division of the AVA in 1999. Dr Allan is published widely in the veterinary literature and has published chapters in veterinary textbooks on 12 occasions. He has been a faculty member at the Veterinary Faculty at the University of Sydney and in 1999 was appointed Adjunct Professor of Veterinary Science within the Faculty. In 2012 he was awarded a DVSc by the University of Sydney after examination of his thesis entitled 'Radiological Studies of Diseases in Companion and Zoo Animals'.

Susan Hemsley BVSc (Hons) (Syd), MVSc (Syd), PhD, Grad Cert Ed Stud
Education Program Developer, Centre for Veterinary Education
University of Sydney, Sydney, New South Wales, Australia

Dr Hemsley graduated with a BVSc degree from the University of Sydney in 1986. She then worked in private practice and completed a MVSc exploring diseases and injuries of possums before returning to the University of Sydney in 1993 to commence a PhD investigating the mucosal immune system of possums and koalas and aspects of koala chlamydial disease. From 1996 to 2012 she was a teaching/research academic in the Faculty of Veterinary Science at the University of Sydney, with a particular interest in histology, anatomy, immunology and disease of marsupials. She is currently an education program developer with the Centre for Veterinary Education, University of Sydney – a provider of continuing professional development programs for veterinarians. She has co-authored 22 scientific journal articles, one book chapter and 29 conference presentations.

Frances Hulst BVSc (Hons) (Syd), MVS (Syd)
Veterinary Officer, Taronga Zoo, Taronga Conservation Society Australia
Sydney, New South Wales, Australia

Dr Hulst graduated from the University of Sydney in 1983. She worked in private practice before commencing a veterinary internship at Taronga Zoo in 1990 and completing a Masters in Veterinary Studies in Wildlife Medicine and Husbandry in 1991. Dr Hulst then worked as a locum at several zoos in Australia and in small animal and exotic practice in England. She has been employed by the Taronga Conservation Society Australia as a veterinary officer since 1995. Dr Hulst's duties include maintaining the health and welfare of the zoo's animals together with the veterinary care, rehabilitation and release of wildlife admitted to the Taronga Wildlife Hospital. She has wide ranging experience with aspects of in situ and ex situ conservation of mammals, birds, reptiles and amphibians. Dr Hulst is a co-author of several papers and one book chapter.

Nadine Fiani BVSc (Hons) (Syd), DAVDC
Clinical Assistant Professor in the Section of Dentistry and Oral Surgery
Cornell University, Ithaca NY, USA

Dr Fiani graduated with a BVSc degree from the University of Sydney in 2005. She then went on to complete an internship in small animal medicine and surgery at the University of Sydney Veterinary Teaching Hospital, Camden. It was there that she became interested in advanced veterinary dentistry and oral surgery. In 2011 Nadine completed a three-year residency training program in Dentistry and Oral Surgery at the University of California, Davis, USA. This culminated in the successful completion of her specialist examinations allowing her to qualify as a Diplomate of the American Veterinary Dental Collage. She returned to Sydney, Australia, in 2011 and took up the post of Specialist Veterinary Dentist at the Small Animal Specialist Hospital. During her time in Sydney, Nadine developed an interested in zoo dentistry and worked extensively with both Taronga Zoo and Mogo Zoo. She also coordinated and led several continuing education programs for general practitioners as well as teaching the 4th year veterinary students at the University of Sydney. In 2014, Nadine accepted a clinical position at Cornell University, New York, USA. She is hoping to expand her interests in the fields of zoo dentistry, endodontics and orthodontics.

Paul Andrew BSc (Hons), MA (Philosophy)
Curator, Taronga Zoo, Taronga Conservation Society Australia
Sydney, New South Wales, Australia

Mr Andrew graduated with a BSc degree from the University of East Anglia in 1977. He has been employed as a curator with the Taronga

Conservation Society Australia since 1993. He spent the intervening years bird watching. In this period he also published the first checklist of Indonesian birds and was second author of the first global assessment of the threat status of birds. The latter became the first iteration of the Red List for birds and included the development of protocols for the construction of common names for birds now adopted by Bird Life International. In 1996 Mr Andrew took on the role of taxonomic advisor to the Australasian Zoo and Aquarium Association and extended the protocols developed for birds to other taxa in regional zoo collections. In 2011 he completed a Masters degree at the University of Melbourne and advocates for the ethical treatment of animals.

1

Radiographic technique

Regardless of the reason for undertaking any radiographic study, the primary objective is to obtain high-quality diagnostic images, while ensuring the safety of the patient and personnel. This can be accomplished through preparation before the procedure and proper radiographic technique: using suitable radiographic equipment (X-ray generation and image formation), patient preparation and restraint, positioning and positioning aids, radiation safety equipment and protective barriers, and post-image capture quality assurance and processing.

The principles of X-ray generation and image production for wildlife species are no different than those for domestic species and are not discussed in detail here. There are several excellent radiology texts that provide this information (Capello and Lennox 2008; Silverman and Tell 2010; Thrall 2013). There are, however, some special considerations inherent in radiographing wildlife species. These primarily relate to the need for safe and humane restraint, and to radiographing small species, which are frequently encountered in zoo and wildlife practice.

There are several fundamental tenets of radiographic technique that are worth reinforcing.

- Take two views at 90° (orthogonal views) whenever possible. This is useful for thorough evaluation of both soft tissue and skeletal anatomy and pathology.
- One joint should be included in every long bone image. As long bone joints are usually characteristic, they assist with identification of the bone adjacent to the joint. In addition, the joint provides a reference point, allows consistency in serial imaging, may guide surgical decisions when located near pathology and may assist in assessing the maturity of the patient through the degree of physeal closure.
- The comparative view. This is mainly relevant to the appendicular skeleton, where imaging the contralateral side may be useful when the clinician is unfamiliar with normal anatomy or a comparison view may assist in confirming a diagnosis.
- Image labelling. All images must be labelled with patient information (at a minimum, the species and individual animal identifier such as a transponder, band, tag or accession number) and side markers (left or right). Labels indicating the view may be useful.
- Review previous images. This will allow the clinician to more adequately plan a radiographic study, become familiar with previous or pre-existing pathology and review radiographic anatomy.
- Peer or specialist review of images. This is particularly valuable if the clinician is less familiar with a species, or with the subtle radiographic changes that may be associated with particular disease processes (e.g. pulmonary disease). It will result in improved patient care and assist in the development of radiographic technique and interpretation skills, which is particularly valuable for wildlife because information on many species and disease processes is limited. The review process is now relatively straightforward with PACS

(picture archiving and communication systems) whereby digital images can be readily transferred using various web-based platforms. The universal image format for PACS is DICOM (digital imaging and communications in medicine). Digital images can also be saved in different formats (jpeg, tiff) and sent via e-mail.

Digital radiography

Digital radiography has revolutionised radiology because of improved image quality; superior image contrast; greater exposure latitude; immediate viewing; reduced need for repeat exposures caused by incorrect exposure factors and film processing errors; the ability to manipulate images; the ability to electronically transmit data to an appropriate storage medium from which it can be electronically retrieved; reduced physical storage space of images; and ease of electronically transferring images.

However, despite these advantages over conventional screen-film (SF) systems, getting the most out of a digital system requires an understanding of the principles of digital imaging. SF technique is not necessarily applicable to digital radiography and may be counter-intuitive. A common misconception is that digital systems correct for exposure errors, when in fact they do not. During the analysis of the image data, the potential exists for the digital system to make adjustments to the image data so that the image has an acceptable level of brightness in the presence of underexposure and overexposure. However, the exposure error remains regardless of what occurs during image processing by the system's software (Herrmann *et al.* 2012).

Two digital radiographic systems are currently available: **computed radiography** (CR) and direct **digital radiography** (DR). The primary difference between the two systems is the method by which the X-ray photons are captured and converted to a digital image. The component of these systems where this occurs is in the **digital receptor** (digital detector, flat-panel detector, sensor plate), which is the plate that receives the X-ray photons after they pass through the patient. In the case of CR, it is a flexible plate coated with a photostimulable phosphor (PSP) that temporarily stores a latent image of the X-ray attenuation pattern of the patient. The latent image is read optically as photostimulated luminescence when the PSP plate is subsequently stimulated using a scanning laser. The CR cassette must be processed in a plate reader following exposure (Robertson and Thrall 2013). In the case of DR, the transmitted X-ray signal is read immediately after exposure and converted to an electronic file within the receptor plate and therefore does not use a cassette (Williams *et al.* 2007; Robertson and Thrall 2013).

After the X-ray photons pass through the patient, digital receptors sense their intensity variation over an area, from which a matrix of many individual picture elements or **pixels** is formed. Each pixel is represented by an intensity value that equates to a shade of grey. This value ranges from black (0) to white (2^x-1), where x is the **bit-depth** and equates to the number of possible grey shades in a pixel. A greater bit-depth means a higher number of independent shades of grey that can be represented by each pixel. For example, these intensity values will be 0–255 (2^8), 0–1023 (2^{10}) or 0–4095 (2^{12}) for bit-depths of 8, 10 and 12, respectively. The number of available grey shades for each pixel will influence the contrast of the image. If a pixel has only a few grey shades, the image can only have a very narrow range of contrast because there will be only black pixels, white pixels and pixels of a few grey shades. If, conversely, there are many grey shade options for each pixel, the scale of contrast can be long. A typical digital receptor has a linear relationship between exposure and the resulting pixel value (Fig. 1.1). Low exposures result in low pixel values and higher exposures give greater pixel values. The usefulness of bit-depth becomes apparent in the image post processing. A high bit-depth provides much greater latitude for adjusting image contrast and brightness. A low bit-depth results in very limited range of adjustment, where the difference between dark and bright in the grey scale is narrow. Achieving the optimal image therefore requires the selection

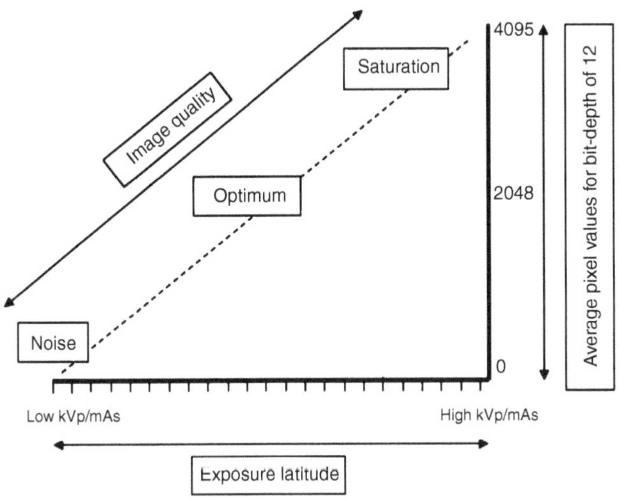

Figure 1.1: Relationship between exposure, average pixel values and image quality.

of appropriate exposure factors for a particular body part or patient size, as well as familiarity with the capability of the system being used (Robertson and Thrall 2013; Sprawls 2014).

The number of pixels per inch or centimetre (PPI, PPCM) is a measurement of the pixel density in a system and determines the **spatial resolution**. This relates to the ability of an imaging system to allow two adjacent structures to be visualised as separate (the distinctness or sharpness of an edge in the image). The greater the PPI/PPCM, the greater the spatial resolution, which in turn gives the impression of greater detail. Other factors contributing to spatial resolution are X-ray tube focal spot size, collimation, scatter, grid efficiency and motion of the patient.

The **focal spot** of the X-ray machine is the area on the anode of the X-ray tube struck by electrons from which the resulting X-ray photons are emitted. The smaller the focal spot the sharper the image. A large focal spot reduces image quality by reducing the ability to define small structures, and the edges of anatomical structures are less sharp, which is called 'penumbra' (Fig. 1.2) (Robertson and Thrall 2013). The focal spot size cannot be altered; however, some X-ray machines provide two filaments of different sizes to make a dual-focus tube, adding flexibility to the system.

Collimation of the beam to the appropriate region of interest is equally as important with digital radiography as it is with SF radiography. Limiting the area that receives radiation reduces the patient's dose and minimises scatter radiation to the digital receptor. Digital receptors are more sensitive to low levels of radiation and the resulting

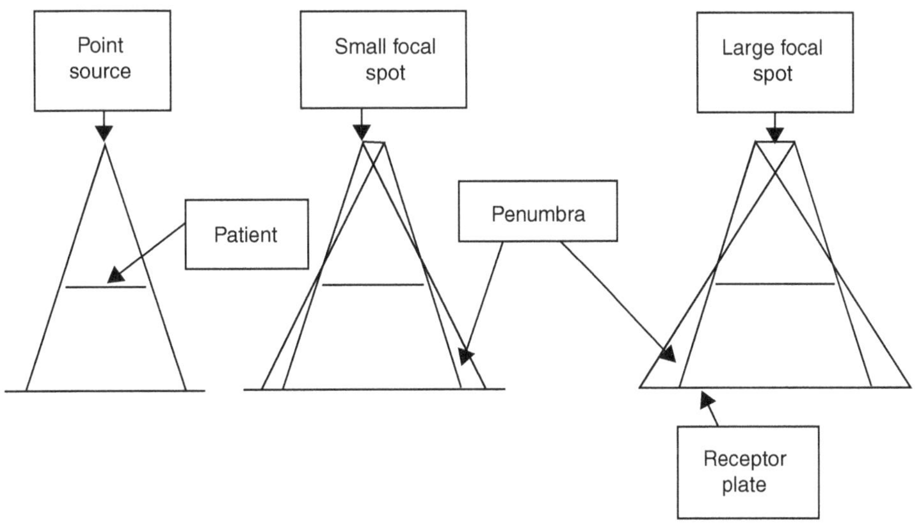

Figure 1.2: Diagram illustrating the effect of focal spot size on image quality. Large focal spot size results in loss of detail, whereas a small focal spot produces sharper images.

digital image might have reduced contrast because of excess scatter radiation striking the receptor. Close collimation will improve contrast and wide collimation may result in more grey and less detail. However, care must be taken with close collimation to ensure annotation of the image does not obscure the structures of interest. Some digital receptors also require a minimum area of the receptor to be exposed to enable effective image formation. Post-processing cropping should not be used as a replacement for beam restriction achieved through physical collimation before capturing the image (Herrmann *et al.* 2012). The primary beam must be centred on the centre of the digital receptor.

Although antiscatter grids can be used in digital radiography, some manufacturers have incorporated a scatter-correction algorithm into the image-processing software, which makes grid use unnecessary. Antiscatter grids may be recommended for large patients. If an antiscatter grid is to be used, it must be specific for the digital receptor being used (Robertson and Thrall 2013).

Patient motion because of poor patient restraint or long exposure time (s) may result in blurred images. Short exposure times minimise motion artefact associated with rapid respiratory rates or muscular tremors (Silverman and Tell 2010).

A characteristic of digital systems is the ability to correct for exposure deficiencies to a much greater degree than with SF systems and thus create an image of suitable quality that is relatively independent of the kVp and mAs values selected. This is because of the greater **dynamic range** of digital receptors and the software capabilities of digital systems. The dynamic range of a digital receptor is the ratio between the maximum and minimum X-ray intensities that can be detected. For all systems, the smallest useful intensity is determined by the intrinsic system noise. The largest intensity is determined by receptor saturation (Williams *et al.* 2007). For example, if the minimum exposure that the receptor can detect is low and the maximum is high, then the system has a wide dynamic range. SF systems have a narrow dynamic range, whereas digital receptors can detect a greater range of X-ray intensities. The magnitude of the range of exposures that can be used to create diagnostic images is called the **exposure latitude** (Robertson and Thrall 2013) (Fig. 1.1). As a result of these properties of digital systems, incorrect exposures are more likely to go unnoticed. However, gross underexposure (too few X-ray photons reaching the receptor plate) will result in image **noise** and gross overexposure will result in **saturation** (the detector becomes saturated with too many X-ray photons and cannot respond to additional exposure) (Williams *et al.* 2007; Herrmann *et al.* 2012). Noise on digital radiographic images manifests as graininess or mottling and obscures detail, particularly in low-contrast images and when images are magnified (Williams *et al.* 2007) (Figs 1.3 and 1.4). Underexposure in digital radiography does not result in 'light' images; rather it results in images that are 'noisy'. Overexposure does not result in 'dark' images; rather it may result in saturated or 'light' images with loss of contrast. Although still useful, exposure guides or technique charts are less important with digital systems than they are with SF radiography and can even be misleading. The basic principles of SF radiography do not apply to digital radiography. Image quality is dependent on a sufficient number of X-ray photons of appropriate energy reaching the image receptor plate. Radiographing small species poses a challenge because the tendency is to choose low exposure factors to accommodate small body part thickness and tissue density. However, this is more likely to result in image noise and a low average pixel value, thus reducing image quality with a narrow range of contrast (because of a very narrow range of shades of grey, the background will be light with little contrast between it and the image). Although counter-intuitive, with small species it is beneficial to increase the exposure factors. This may be kVp and/or mAs, although increasing the mAs can result in motion artefact, particularly in small species where heart and respiration rates may be high (Figs 1.3 and 1.4). The same applies to large species or dense body parts where enough X-ray photons must reach the receptor plate to provide an image without noise.

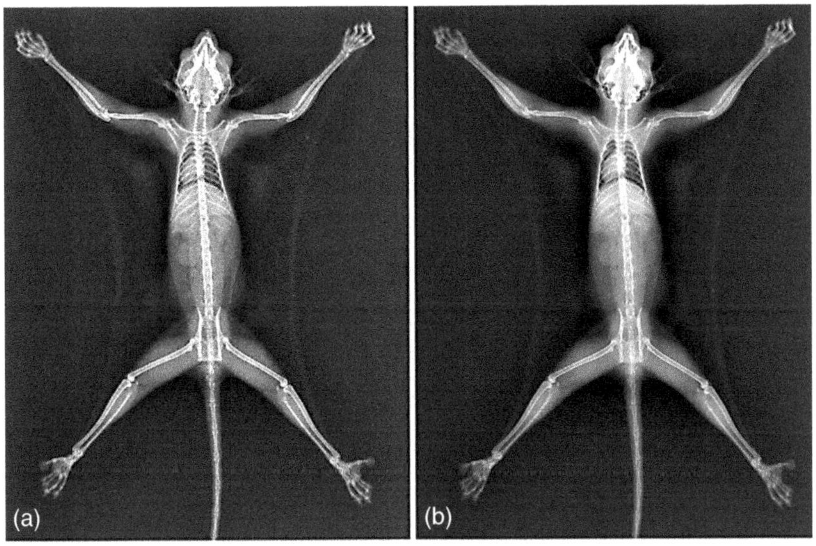

Figure 1.3: Dorsoventral radiograph of a 133 g sugar glider (*Petaurus breviceps*). (a) Image taken using 50 kVp and 0.4 mAs. This gross underexposure has resulted in a grainy, mottled image, an effect known as 'noise'. (b) Image taken using 85 kVp and 2 mAs. Despite this gross overexposure, the image is not saturated and is of good quality, a demonstration of the wide exposure latitude and dynamic range of the digital system used (Canon CXDI-55G digital flat-panel detector and Sound-Eklin Mark V portable digital radiography system).

Bit-depth, dynamic range and PPI/PPCM are intrinsic features of the receptor system being used. When choosing a digital system, digital receptors with a bit-depth of 12 or greater are ideal.

Until excessively high exposures are used, images usually appear ideal and without noise. There can therefore be a tendency, over time, to increase the exposure factors unnecessarily to create images without noise. This is known as **exposure creep**, whereby unnecessarily high exposure factors are not recognised as they would be with SF systems (i.e. 'dark' images). The disadvantage of exposure creep is unnecessary patient and personnel exposure to radiation. Most digital radiography manufacturers have incorporated exposure indicators (EIs), a metric that enables the user to evaluate the amount of incident plate exposure, and this can be used to judge whether exposure factors are too high or too low (Robertson and Thrall 2013). Exposure information is usually displayed on the image, although EIs vary among manufacturers, having different names, symbols and units. Examples include S (Fuji, Konica), REX (Canon), IgM (Agfa) or EI [exposure index] (Carestream, Philips, Siemens). To confuse the issue further, some EIs increase with increasing exposure and others decrease with increasing exposure. Operators should become familiar with the range of values that equate to optimum exposures for their system. These values, however, will vary between digital systems, algorithms used, patient size and the body part being examined.

The key is to choose factors that provide optimum exposure and produce a good balance between image noise and patient exposure. One of the challenges with digital radiography is knowing when the receptor is optimally exposed. It is not like film radiography where under- and overexposures are obvious. The differences in the quality of images within the extremes of under- and overexposure may not be obvious, but they do become apparent when images are magnified (Fig. 1.3). Contrast and definition therefore depend on having adequate exposure (either kVp or mAs or both) for the size of the patient and density of the tissue being imaged. It is difficult to make specific recommendations on exposure factors, because they need to be tailored to each digital system and receptor combination.

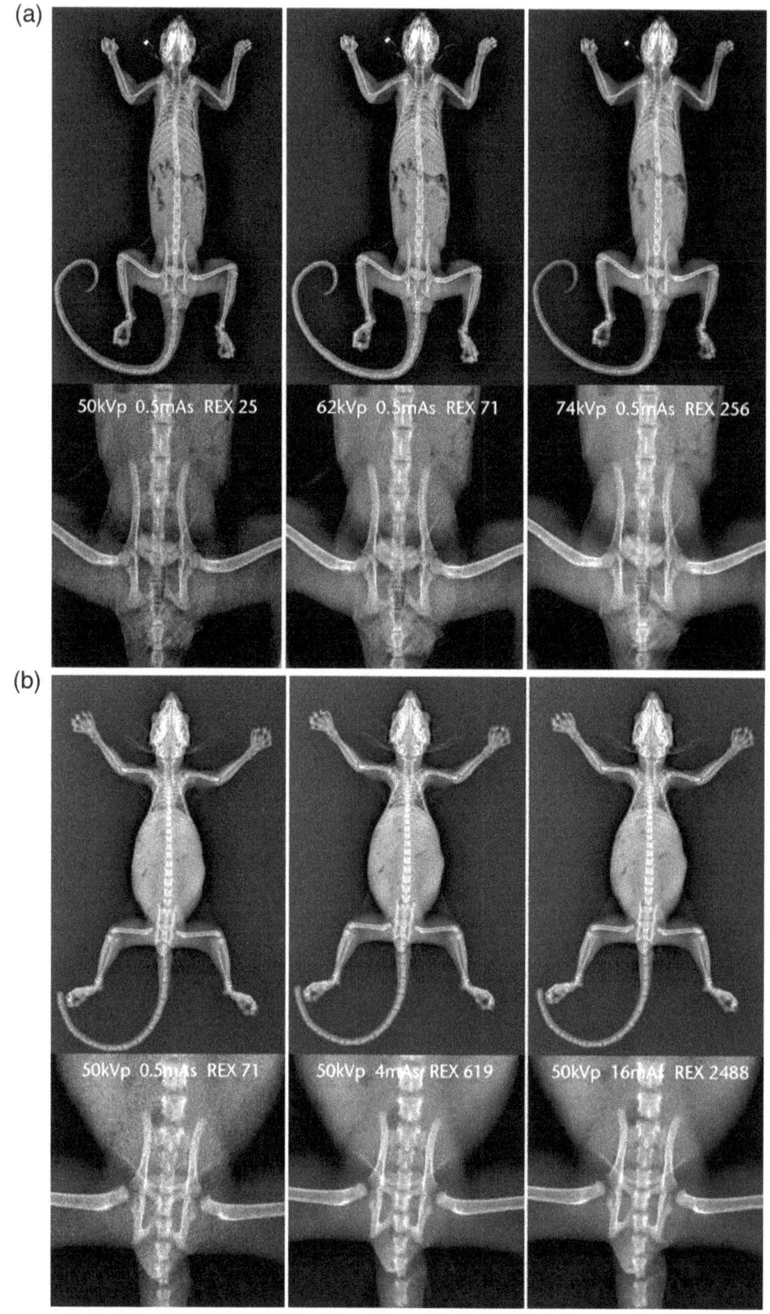

Figure 1.4: (a) Whole body, common ring-tailed possum (*Pseudocheirus peregrinus*), dorsoventral view, abdominal thickness 3.5 cm, bodyweight 456.8 g, variable kVp. (b) Whole body, common ring-tailed possum, ventrodorsal view, abdominal thickness 2.5 cm, bodyweight 131 g, variable mAs. These images demonstrate two important principles of digital radiography. The first is that using a constant kVp and variable mAs or vice versa results in little variation in the overall brightness of the image because the image processing software compensates for variation in overall exposure. The second is that at low mAs or kVp (low REX values [the exposure indicator for the Canon flat-panel detector used]) whereby fewer X-ray photons reach the receptor, detail is lost because of noise. As the exposure increases, noise has less influence and image clarity improves, despite excessively increased exposures in the right-hand images. In these sequences you can also see that overall image resolution is related to noise. At low magnification, there appears to be little difference between the images, other than the left-hand image in which exposure is too low, resulting in noise. It is when the images are magnified that a difference can be seen. Increasing exposure provides better definition in the magnified images.

Most digital systems have pre-set **algorithms** configured by the manufacturer for specific body parts or studies and are selected before image acquisition. Algorithms optimise for the range of tissue densities, from bone to soft tissue, in a particular body part. They optimise contrast and other aspects of image presentation for the body part being radiographed. Most systems allow customisation of the algorithms by expert technicians, including installation of high-contrast algorithms that can further improve image quality. As a user, exposure variables are limited to kVp, mAs and the algorithm selected for the exposure.

Digital systems allow for **post-processing** of images between acquisition and saving (e.g. cropping, rotating, adjusting brightness and contrast, applying markers). This is designed to convert the raw image data that has been corrected to a degree by the system software to a useable radiographic image. Once post-processing adjustments have been made, the image is ready for saving, viewing and interpretation. The image file is usually converted to DICOM format and sent through a local network to the viewing computer(s) and an appropriate archive. The DICOM image can be further manipulated to assist with interpretation. The intrinsic features of the system (bit-depth, PPI/PPCM, dynamic range) and the exposure factors used will influence the ability to improve the image during post-processing and once converted to DICOM format. For example, a small bit-depth will give fewer shades of grey and therefore narrower range in contrast. This will also be affected by exposure factors. The resolution of magnified images will be affected by the PPI/PPCM and exposure factors. The flexibility provided by post-processing maximises the viewing options for the clinician (Robertson and Thrall 2013).

Radiographic projections

The radiographic projections (or views) chosen for a specific study will depend on the intention of the study and the desired outcomes. At a minimum, two orthogonal projections should be taken. Correct use of directional terms in naming radiographic projections is important because it assists in identifying the exact location of structures and lesions seen on radiographs. In this text, we have used the standard method of naming radiographic projections approved by the American College of Veterinary Radiography (Smallwood *et al.* 1985). This system is based on two rules: (1) the correct use of veterinary anatomical directional terms and (2) naming radiographic projections or views according to the entry and exit points of the beam. The *Nomina Anatomica Veterinaria* (ICVGAN 2012) provides correct anatomical directional terms and has been used in this text for all anatomical terms.

Directional terms are listed in Table 1.1. The correct directional terms used for describing radiographic projections are illustrated in Fig. 1.5. Standard terms and abbreviations for radiographic projections as they apply to various parts of the body are listed in Table 1.2.

The naming of oblique projections is more complex, but the same system applies (i.e. by anatomically designating the entrance and exit points of the beam). Oblique projections are primarily used for the head, specifically for extraoral dental radiography, and for the limbs, particularly the distal limbs. For example, to evaluate maxillary or mandibular molar teeth, ventrolateral-dorsomedial oblique or dorsolateral-ventromedial oblique projections are useful (Table 1.3 and Fig. 1.6). For the distal limbs, dorsolateral-palmaromedial oblique, dorsomedial-palmarolateral oblique, dorsolateral-plantaromedial

Table 1.1. Directional terms used to describe radiographic projections (Smallwood *et al.* 1985)

Left	Medial
Right	Lateral
Dorsal	Proximal
Ventral	Distal
Cranial	Palmar
Caudal	Plantar
Rostral	Oblique

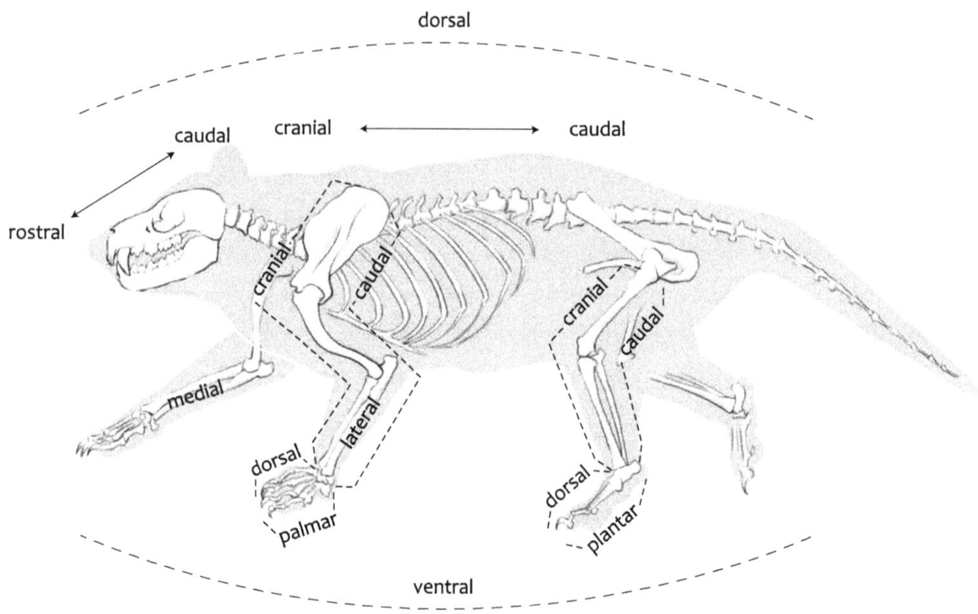

Figure 1.5: Anatomical directional terms for radiographic projections as they apply to various parts of the body. © Beth Croce, Bioperspective.com.

oblique or dorsomedial-plantarolateral oblique projections can be used (Smallwood *et al.* 1985). Designating the angle of obliquity is not usually necessary and only included when considered important for the exact reproduction of a desired image (e.g. having the head positioned at 45° is most commonly used for extraoral projections when evaluating the maxillary molar teeth in macropods).

Additional terms may be applied to further describe projections. These include combinations of terms, angles of obliquity, positions of the body (recumbent, standing, erect), orientation of the central beam relative to space (horizontal, vertical), position of joints (flexed, extended, hyperextended, frog leg) and position of the head (open-mouth). Terms describing body planes are less applicable to conventional radiography than with CT, MRI and sonography; for example, transverse (a plane perpendicular to the long axis of the body or part), median or midsagittal (a plane that bisects the body vertically through the midline), sagittal or parasagittal (refers to any plane parallel to the median

Table 1.2. Standard terms and abbreviations for radiographic projections as they apply to various parts of the body (Smallwood *et al.* 1985)

Directional Term	Abbreviation	Body Part
Dorsoventral/ventrodorsal	DV/VD	Head, neck, trunk, tail
Right or left lateral (right or left denotes the side adjacent to the sensor plate)	RtL/LeL	Head, neck, trunk, tail
Craniocaudal/caudocranial	CrCd/CdCr	Limbs proximal to the antebrachiocarpal joint and tarsocrural joint
Lateromedial/mediolateral	LM/ML	Limbs
Dorsopalmar/palmarodorsal	DPa/PaD	Forelimb distal to the antebrachiocarpal joint
Dorsoplantar/plantarodorsal	DPl/PlD	Hindlimb distal to the tarsocrural joint
Rostrocaudal	RCd	Head

Table 1.3. Recommended oblique extraoral projections for the evaluation of maxillary or mandibular molar teeth: dorsolateral-ventromedial oblique (DLVMO), ventrolateral-dorsomedial oblique (VLDMO)

Teeth to be Examined	Position	Recommended Projection
Right maxillary	Right lateral	DLVMO 45° (uppermost maxillary teeth on the image is the right arcade)
Left maxillary	Left lateral	DLVMO 45° (uppermost maxillary teeth on the image is the left arcade)
Right mandibular	Left lateral Right lateral	DLVMO 45° (uppermost mandibular teeth on the image is the left arcade) VLDMO 45° (lowermost mandibular teeth on the image is the right arcade)
Left mandibular	Right lateral Left lateral	DLVMO 45° (uppermost mandibular teeth on the image is the right arcade) VLDMO 45° (lowermost mandibular teeth on image is the left arcade)

plane) and dorsal (parallel to the dorsal surface and perpendicular to the median and transverse planes).

Patient positioning

Correct positioning of the patient is critical to obtaining diagnostic images. All wildlife patients, regardless of size and temperament, should be chemically restrained (sedated or anaesthetised) for radiography. This ensures the safety of the patient and personnel, reduces stress on the patient and the need for repeat exposures. The anaesthetised patient can be more easily manipulated into the correct position. Various positioning aids such as tape, ties, sandbags or foam pads can be used, minimising the need for personnel to physically hold an animal in position. Even for critically ill patients, the benefits of chemical restraint may outweigh the risks and stress associated with physical restraint alone.

Table 1.4 provides some basic protocols that can be used for chemical restraint of Australian mammal species. For more detail on this topic, readers are referred to Vogelnest and Woods (2008) and Vogelnest (2015).

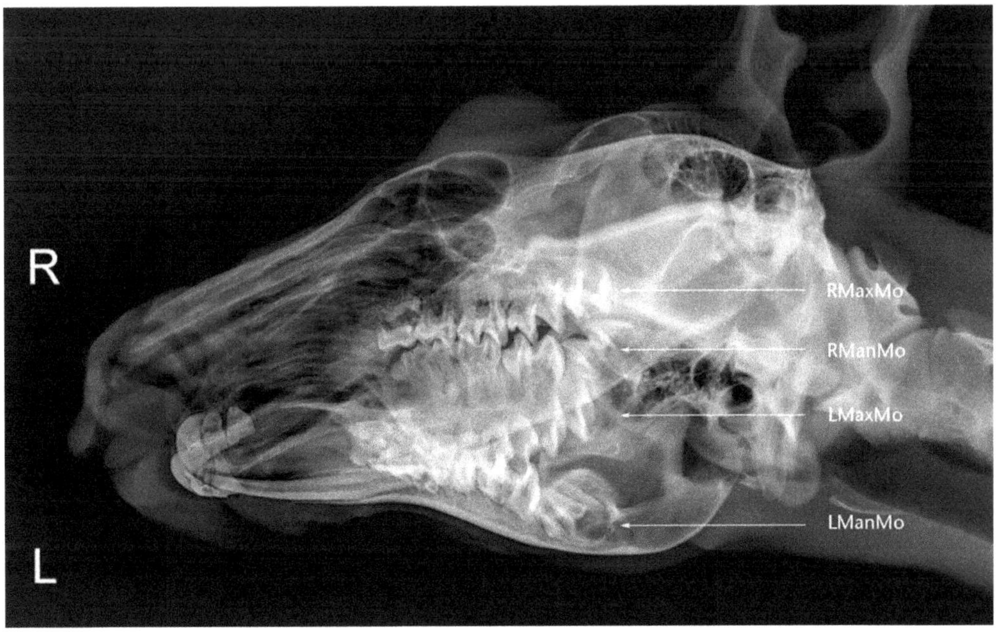

Figure 1.6: Skull, red kangaroo (*Macropus rufus*), right dorsolateral-ventromedial oblique taken at 45°. In this view, the right maxillary molars (RMaxMo) and right mandibular molars (RManMo) are uppermost in the image, while the left maxillary molars (LMaxMo) and left mandibular molars (LManMo) are towards the bottom of the image. To assist with ease of identification of dental arcades, markers can be used as indicators of which arcades are uppermost and lowermost.

Table 1.4. Suggested chemical restraint agents, regimens and doses for Australian mammals (Holz 2007; Vogelnest and Woods 2008; Vogelnest 2015)

Echidna

Sedation
- Diazepam 1–5 mg/kg IM

Anaesthesia
- Alfaxalone 3–8 mg/kg IM
- Tiletamine/zolazepam 3–10 mg/kg IM
- Ketamine 5 mg/kg + medetomidine 300 µg/kg IM, reversal atipamezole (5 × medetomidine dose IM)
- Isoflurane in oxygen (induction and/or maintenance) via face mask or induction chamber

Platypus

Sedation
- Diazepam 0.5–1 mg/kg IM

Anaesthesia
- Isoflurane in oxygen via face mask or induction chamber

Macropods

Sedation
- Diazepam 0.1–2 mg/kg IM, 0.1–1 mg/kg IV
- Midazolam 0.3 mg/kg IM, 0.2 mg/kg IV

Anaesthesia
- Tiletamine/zolazepam 4–10 mg/kg IM, 1–3 mg/kg IV. Lower doses for tractable animals, higher for excited animals and western grey kangaroos
- Tiletamine/zolazepam 2–3 mg/kg IM + medetomidine 40 µg/kg IM, reversal atipamezole (4 × medetomidine dose IM)
- Tiletamine/zolazepam 3–4 mg/kg IM + acepromazine 0.3 mg/kg IM
- Ketamine 3–5 mg/kg IM + medetomidine 40–125 µg/kg IM, reversal atipamezole (4–5 × medetomidine dose IM). Use higher doses for nervous or excited animals
- Ketamine 3–4 mg/kg IM + medetomidine 50 µg/kg IM + midazolam 0.02–0.1 mg/kg IM, reversal atipamezole 0.25 mg/kg IM
- Alfaxalone 5–8 mg/kg IM, 1.5–3 mg/kg IV. Very short duration of action
- Isoflurane in oxygen (induction and/or maintenance) via face mask or endotracheal tube

Koalas

Sedation
- Diazepam 0.5–1 mg/kg IM, 0.5 mg/kg IV

Anaesthesia
- Alfaxalone 3 mg/kg IM, 1.5 mg/kg IV
- Tiletamine/zolazepam 5–10 mg/kg IM, 2.5 mg/kg IV
- Tiletamine/zolazepam 3.5 mg/kg IM + medetomidine 55 µg/kg IM, reversal atipamezole 2 mg IV
- Isoflurane in oxygen (induction and/or maintenance) via face mask or endotracheal tube

Wombats

Sedation
- Diazepam 0.5–1 mg/kg IM
- Midazolam at 0.5 mg/kg IM

Anaesthesia
- Tiletamine/zolazepam 2–9 mg/kg. Effect and duration are dose dependent
- Ketamine 2 mg/kg IM + medetomidine 125 µg/kg IM, reversal atipamezole (4 × medetomidine dose IM)
- Tiletamine/zolazepam 3 mg/kg IM + medetomidine 40 µg/kg IM, reversal atipamezole (5 × medetomidine dose IM)
- Alfaxalone 3–5 mg/kg IM. Effect is dose dependent
- Isoflurane in oxygen (induction and/or maintenance) via face mask or endotracheal tube

Table 1.4. *(Continued).*

Dasyurids

Sedation
- Diazepam 1–2 mg/kg IM

Anaesthesia
- Isoflurane in oxygen (induction and/or maintenance) via face mask or endotracheal tube
- Tiletamine/zolazepam 7–10 mg/kg IM

Bandicoots and bilby

Sedation
- Diazepam 0.5–1 mg/kg IM
- Midazolam 0.1–0.2 mg/kg IM

Anaesthesia
- Isoflurane in oxygen (induction and/or maintenance) via face mask or endotracheal tube

Possums and gliders

Sedation
- Diazepam 0.5–1 mg/kg IM
- Butorphanol tartrate 0.4–1 mg/kg SC or IM

Anaesthesia
- Isoflurane in oxygen (induction and/or maintenance) via face mask or endotracheal tube
- Ketamine 1–3 mg/kg IM + medetomidine 20–100 µg/kg IM, reversal atipamezole 0.05–0.4 mg/kg IV
- Alfaxalone 5–8 mg/kg IM, 5 mg/kg IV. Rapid short-acting anaesthesia in common brush-tailed and common ring-tailed possums

Bats

Anaesthesia
- Isoflurane in oxygen (induction and/or maintenance) via induction chamber, face mask or endotracheal tube

For the majority of standard projections, symmetry is important in patient positioning. Tape is useful for securing the patient to the sensor plate and can be used in combination with pieces of foam to position the patient. Sand bags can be used to secure limbs to hold them in an extended and symmetrical position; however, having the sandbag within the primary beam should be avoided because the presence of a radiodense object on the sensor plate will affect image processing by the system software. The positioning for various projections is no different to those used in domestic species and readers are referred to standard radiology texts (Fig. 1.7).

Contrast radiography

Contrast radiography in wildlife practice is primarily used for the diagnosis of gastrointestinal and urogenital disorders, although it may also be used in myelography and angiography. Both positive (barium sulfate, organic iodine compounds) and negative (air, oxygen, carbon dioxide) contrast agents can be used. The techniques used are no different to those used for domestic species; however, consideration must be given to differences in the anatomy and physiology of wildlife species when choosing to utilise contrast studies. As chemical restraint is required in most cases, serial functional studies, particularly those of the gastrointestinal tract, may be difficult, if not impossible.

Radiation safety

The ionising properties of X-rays are harmful to living tissues. Precautions must be taken at all times when undertaking radiographic studies of animals to minimise exposure of personnel to

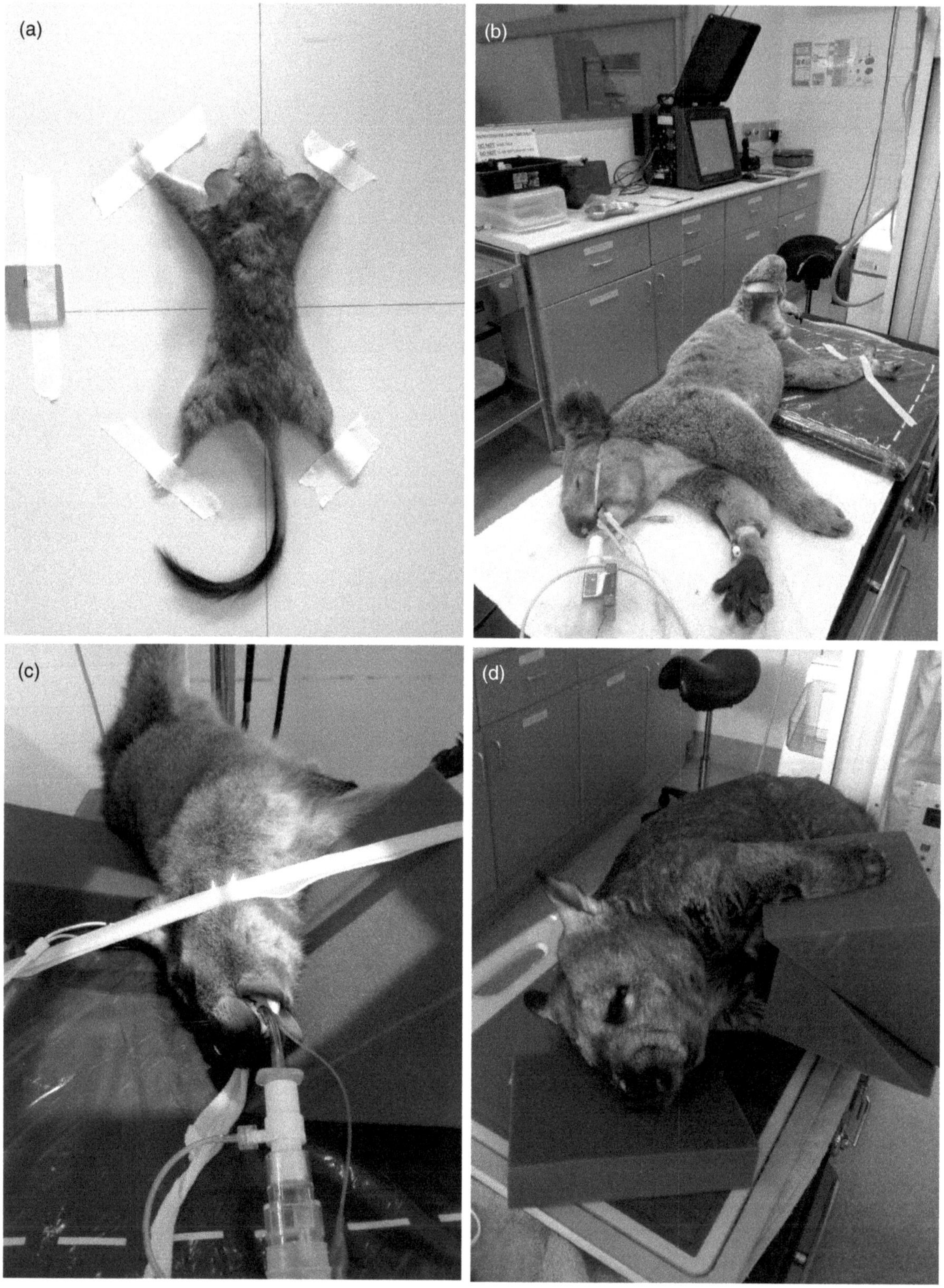

Figure 1.7: Positioning the anaesthetised patient with aids such as foam wedges, tape and sandbags.

radiation. Only properly trained, authorised and appropriately licensed personnel should perform radiographic procedures.

Sources of radiation exposure include the primary beam and radiation scatter. Personnel should never be exposed to the primary beam, so safety measures are aimed at limiting exposure to scattered radiation. Scatter radiation is the main source of radiation hazard and is produced when the primary X-ray beam interacts with the patient. The following measures must be used to minimise exposure to scatter radiation (Capello and Lennox 2008).

- Use chemical restraint and positioning techniques that do not require personnel to hold the patient or be in the room.
- If personnel are required to be in the room or hold the patient, protective lead shielding such as aprons, gloves and thyroid shields must be used. Note: lead gloves do not protect from radiation from the primary beam.
- Radiation monitors must be worn by all personnel performing radiography.
- Distance from the radiation source is important. X-rays diverge as they travel and their intensity falls rapidly the further away from their source. Maximising the distance from the patient during an exposure is an important safety measure.
- Use shortest possible exposure times when selecting exposure factors.
- Minimise the number of exposures through careful planning, preparation and use of correct radiographic technique.

References

Capello V, Lennox AM (2008) *Clinical Radiology of Exotic Companion Mammals.* Wiley-Blackwell, Iowa.

Herrmann TL, Fauber TL, Gill J, Hoffman C, Orth DK, Peterson PA, Prouty RR, Woodward AP, Odle TG (2012) *Best Practices in Digital Radiography.* American Society of Radiologic Technologists, Albuquerque.

Holz P (2007) Marsupials. In *Zoo Animal and Wildlife Immobilization and Anesthesia.* (Eds G West, D Heard and N Caulkett) pp. 341–346. Blackwell Publishing, Ames.

ICVGAN (2012) International Committee on Veterinary Gross Anatomical Nomenclature. *Nomina Anatomica Veterinaria.* 5th ed (revised). <http://www.wava-amav.org/> Accessed April 2014.

Robertson ID, Thrall DE (2013) Digital radiographic imaging. In *Textbook of Veterinary Diagnostic Radiology.* 6th edn. (Ed DE Thrall) pp. 22–37. Elsevier-Saunders, Missouri.

Silverman S, Tell LA (2010) *Radiology of Birds and Atlas of Normal Anatomy and Positioning.* Saunders-Elsevier, Missouri.

Smallwood JE, Shively MJ, Rendano VT, Habel RE (1985) A standardized nomenclature for radiographic projections used in veterinary medicine. *Veterinary Radiology* **26**, 2–9. doi:10.1111/j.1740-8261.1985.tb01105.x

Sprawls P (2014) *Digital Radiography: The Physical Principles of Medical Imaging.* Sprawls Educational Foundation http://www.sprawls.org/resources/DIGRAD/module.htm

Thrall DE (2013) *Textbook of Veterinary Diagnostic Radiology.* 6th edn. Elsevier-Saunders, Missouri.

Vogelnest L (2015) Marsupialia (Marsupials). In *Fowler's Zoo and Wild Animal Medicine,* Vol. 8. (Eds RE Miller and ME Fowler) pp. 255–274. Elsevier-Saunders, Missouri.

Vogelnest L, Woods R (eds) (2008) *Medicine of Australian Mammals.* CSIRO Publishing, Melbourne.

Williams MB, Krupinski EA, Strauss KJ, Breeden WK, Rzeszotarski MS, Applegate K, Wyattg M, Bjork S, Seibert JA (2007) Digital radiography image quality: image acquisition. *Journal of the American College of Radiology* **4**, 371–388. doi:10.1016/j.jacr.2007.02.002

2

Short-beaked echidna

The short-beaked echidna (*Tachyglossus aculeatus*) is native to Australia and New Guinea and is the only extant species within its genus. The three extant species of long-beaked echidnas (*Zaglossus* spp.) are confined to New Guinea. There are five recognised subspecies of *T. aculeatus*: *T. aculeatus acanthion* from the Northern Territory, northern Queensland, inland Australia and Western Australia; *T. aculeatus aculeatus* from eastern New South Wales, Victoria and southern Queensland; *T. aculeatus setosus* from Tasmania; *T. aculeatus multiaculeatus* from South Australia; and *T. aculeatus lawesii* from New Guinea (Augee *et al.* 2006).

Echidnas, along with the platypus (*Ornithorhynchus anatinus*), are monotremes and the only egg-laying mammals. They are thought to represent the most ancient mammalian lineage surviving to the present day. Echidnas probably evolved from the platypus during the Cretaceous period (Augee and Gooden 1993) and it is possible that monotremes have existed for more than 50 million years. Monotremes exhibit reptilian, avian and mammalian characteristics in their physiology, reproduction and anatomy. They possess unique bones in their pectoral girdles (the epicoracoids and the interclavicle) and therefore have similarities to the extinct cyanodonts and to living crocodilians. Paired strips of vestigial bone (the *pilae antoticae*) arise from the base of the chondrocranium of embryonic monotremes. These are not present in any other extant mammals, but are found in reptiles and birds, where they form a large part of the cranium. A cartilaginous scleral ring is found in the monotreme eye, which is also a structure not found in any other extant mammals but occurs in birds and reptiles. Embryonic monotremes possess an 'egg tooth', a structure also found in birds and reptiles to facilitate the emergence of hatchlings from the egg. Echidnas lack teeth and, unlike the platypus, there is no evidence of vestigial dentition. They have the most reduced mandible of any mammal, represented only by fine, rod-like dentaries (Augee *et al.* 2006). Monotremes have cervical ribs, another reptilian feature. Adult male short-beaked echidnas have well-developed hindlimb spurs that are evident externally, but most females do not (some females have small spurs) (Johnston *et al.* 2006; Middleton 2008).

Wild short-beaked echidnas have been known to live for approximately 20 years, whereas captive echidnas have lived to over 50 years of age (Augee *et al.* 2006). In the wild, adult short-beaked echidnas are myrmecophagous; they feed almost exclusively on ants and termites. However, some prefer ants and they will also eat other invertebrates such as earthworms and the larvae of beetles and moths (Booth 1999).

The dorsum and flanks of the short-beaked echidna are covered by fine hair and keratinised spines. The spines are moderately radiopaque and obscure much of the tissue detail on radiographs (Fig. 2.1). The use of CT imaging overcomes some

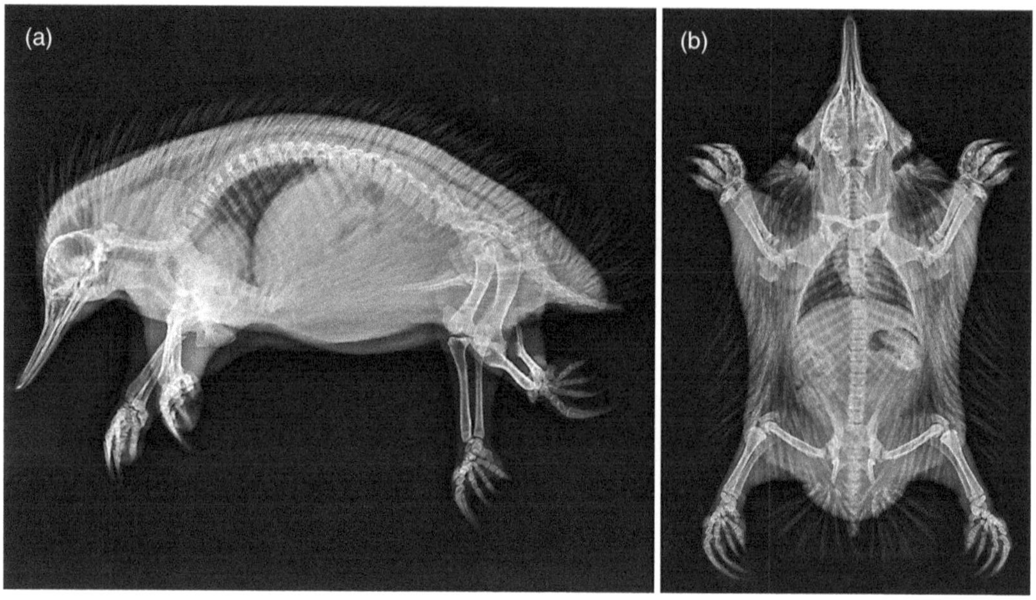

Figure 2.1: Whole body: (a) lateral view, (b) dorsoventral view.

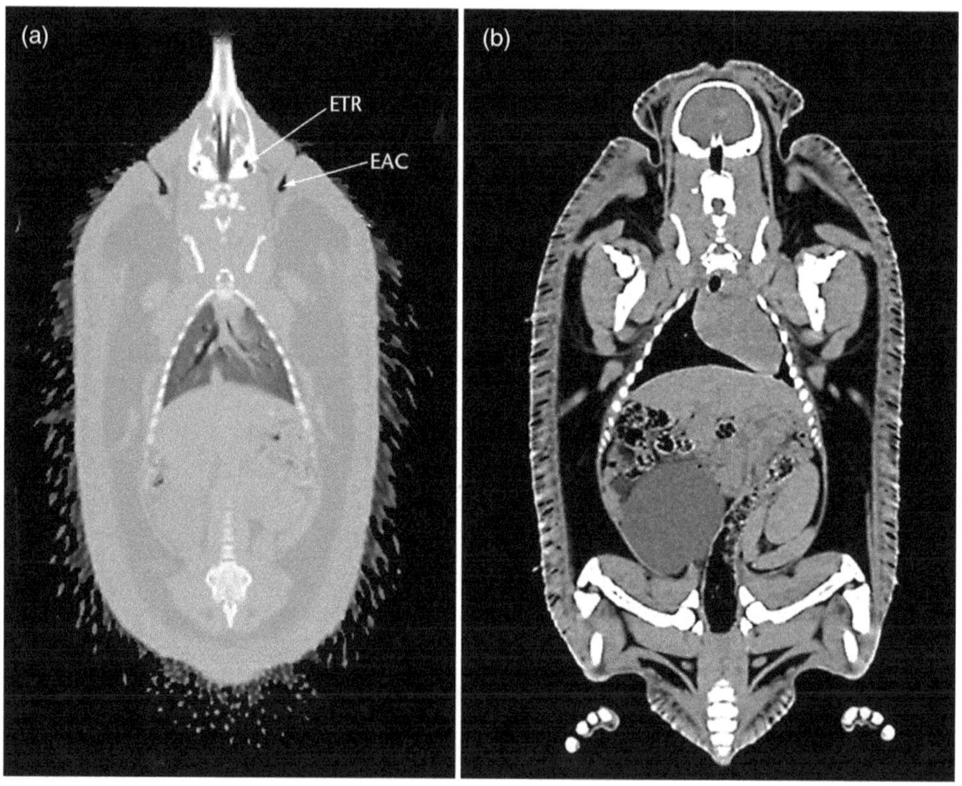

Figure 2.2: Whole body, dorsal CT reconstruction: (a) lung window, (b) soft tissue window. (a) The lung window optimises pulmonary opacities while facilitating detail such as air in the external auricular canal (EAC) and epitympanic recess (ETR). (b) The soft tissue window optimises detail of abdominal viscera and muscles.

of the problems created by the keratinised spines and permits internal anatomic detail to be recognised (Fig. 2.2).

Skeletal system

Appendicular skeleton

Pectoral girdle and forelimb

The pectoral girdle of monotremes is unique and has no parallel among living mammals. The pectoral girdle in the echidna has two clavicles, two scapulae (each with a well developed acromion with which the clavicle articulates at the cranial border), two short and broad coracoids, two epicoracoids and a median T-shaped interclavicle (Figs 2.3–2.5). The scapula is expanded dorsally, has a shallow infraspinous fossa and lacks a cranial recess for the attachment of the supraspinatus muscle. The scapula is large and projects cranially (Figs 2.3 and 2.4). It forms a unit on one side with the ventral coracoid and epicoracoid. This unit is fixed dorsally to the ribs and ventrally to the interclavicle, which is united caudally to the sternum. The transverse portion of the interclavicle is

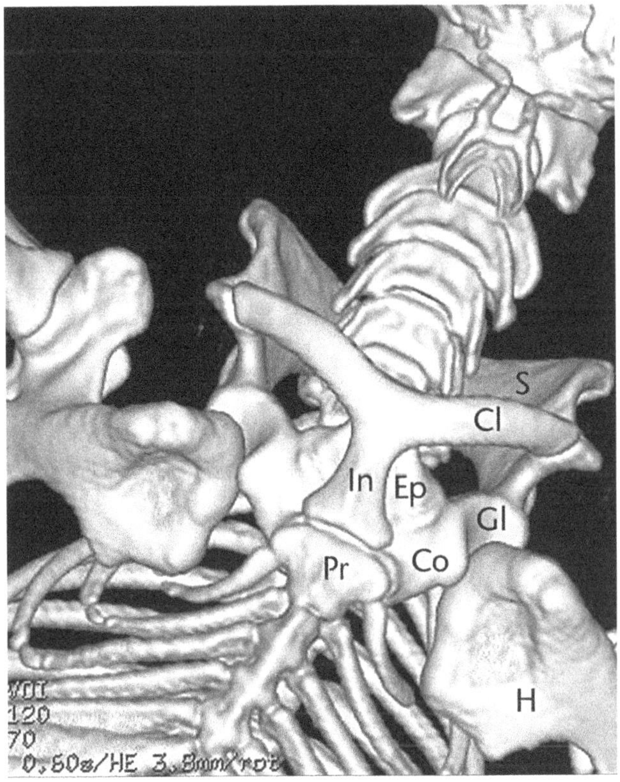

Figure 2.4: Pectoral girdle: three-dimensional volume-rendered CT reconstruction. Presternum (Pr), interclavicle (In), coracoid (Co), epicoracoid (Ep), clavicle (Cl), scapula (S), glenoid cavity (Gl), humerus (H).

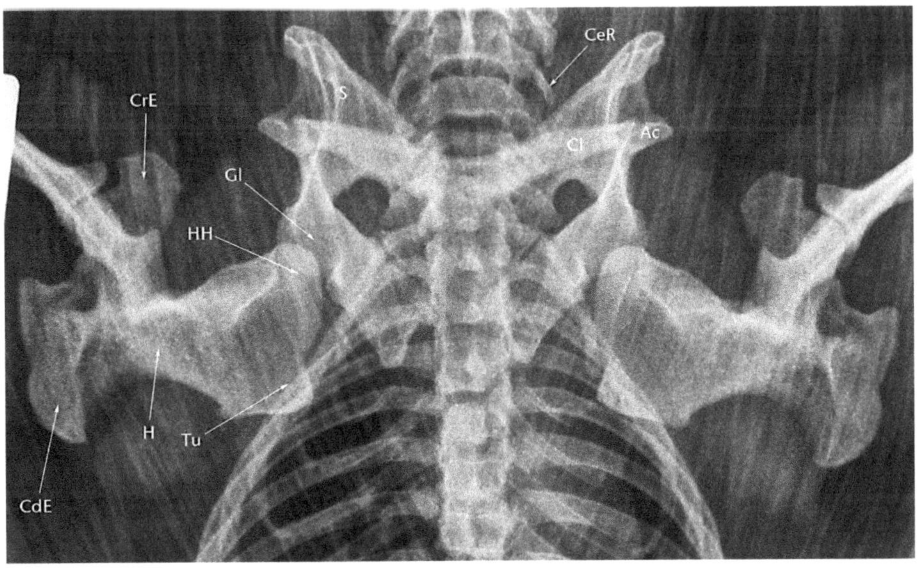

Figure 2.3: Pectoral girdle and brachium, ventrodorsal view. Note the complex pectoral girdle, the shape of the humerus (H) and the relationship of the clavicle (Cl) and acromion (Ac). There are vestigial cervical ribs (CeR) on the cervical vertebrae. The scapula (S), glenoid cavity (Gl) and humeral head (HH) are visible. The humerus has a prominent proximal tubercle (Tu), a diaphysis and two large distal epicondyles, one directed cranially (CrE) and the other caudally (CdE).

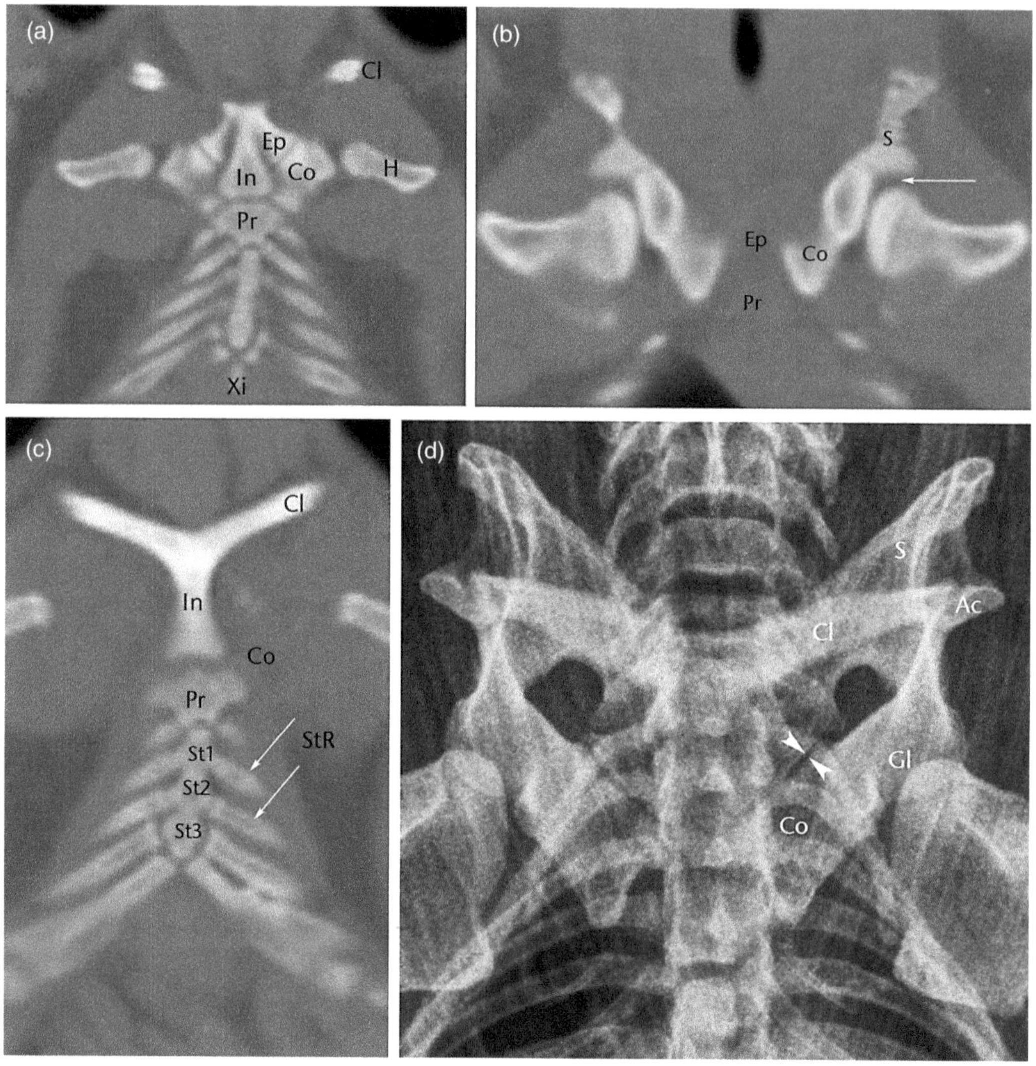

Figure 2.5: Pectoral girdle: (a–c) CT reconstructions in the dorsal plane, (d) dorsoventral radiograph illustrating the complex structure of the rigid pectoral girdle. (a) The complex relationships of the presternum (Pr), interclavicle (In), coracoid (Co) and epicoracoid (Ep) are demonstrated. The humerus (H), clavicle (Cl) and xiphisternum (Xi) are labelled. (b) The cranially directed scapula (S) and the coracoid (Co) share the glenoid cavity (arrow). The coracoid articulates with the epicoracoid (Ep) and the presternum (Pr), but neither is visible in this plane. (c) The clavicles (Cl) are continuous with the interclavicle (In), which articulates with the presternum (Pr). The relative location of the coracoid (Co) is indicated. There are three sternal elements (St1, St2, St3) and sternal ribs (StR) arising from the presternum and the sternal elements. Radiographically (d), the epicoracoid, interclavicle and presternum are obscured by midline structure superimposition. The articulation between the epicoracoid and the coracoid (Co) is visible however (arrowheads). The clavicular (Cl) articulation with the acromion (Ac) of the scapula (S) and the glenoid cavity (Gl) are visible in this view.

clamped rigidly to the scapulae by the clavicles. The glenoid cavity is a deep and laterally oriented (to receive the horizontally oriented humerus), wedge-shaped depression at the union of scapula and coracoids (Griffiths 1989). The epicoracoids and T-shaped interclavicle are not found in any mammals other than monotremes. The bones of the pectoral girdle are firmly connected to the body, giving the echidna a sturdy base for digging action of the forelimbs.

The echidna and platypus are humeral rotation diggers and/or swimmers (the humerus rotates about its axis rather than moving back and forward as in other mammals). This adaptation is reflected in the characteristic 'hourglass' shape (twisted on its long axis) of the humerus, in which both epicondyles are extremely broad, a shape that occurs relatively rarely within mammals (Weisbecker 2011). The greatly expanded extremities are particularly evident in the echidna, in which there are prominent tuberosities and epicondyles (Figs 2.3 and 2.6). The expanded and wide distal humerus serves as an area of attachment for the extensive forelimb musculature (Augee *et al.* 2006). The radius and ulna are stocky and have similar dimensions. The ulna extends proximal to the elbow joint with an elongated olecranon (Fig. 2.6). In the carpus, the radial and intermediate carpal bones are united and there is no separate central carpal bone. There are a radial sesamoid bone and two very large palmar sesamoid bones, which are sometimes united (Parker and Haswell 1921). The radial sesamoid bone may not be visible radiographically. The manus has five digits with wide, spade-like claws (Fig. 2.7).

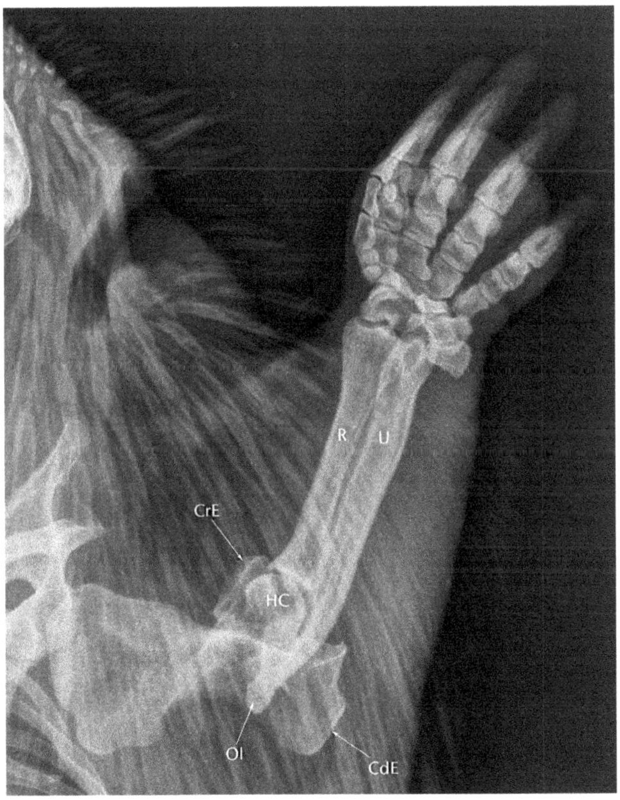

Figure 2.6: Antebrachium, craniocaudal view, and manus, dorsopalmar view. The humeral condyle (HC) articulates with the proximal radius (R) and the ulna (U). The ulna extends proximal to the elbow joint with an elongated olecranon (Ol). The cranial epicondyle (CrE) and the caudal epicondyle (CdE) of the humerus complete the complex of structures surrounding the elbow.

Pelvis and hindlimb

The pelvis is short and broad and consists of the usual three elements: cranial ilia, caudal ischia and ventral pubic bones. The ischia and pubic bones are united ventrally at a symphysis, but are separated dorsally by the obturator foramen. The symphysis of the three bones at the acetabulum is incomplete and the latter exhibits a central foramen, as in birds. Epipubic bones project cranially from the cranial margin of each pubis. Their articulation with the pubis is mobile. The cranial portions of the ilia are elongated, narrow and extend forward and dorsally to unite with the sacrum (Parker and Haswell 1921; Griffiths 1989) (Fig. 2.8).

The process of endochondral ossification in juvenile echidnas is interesting because separate centres of ossification, such as the femoral head and the femoral condyle, appear later than in the normal bone development of most species. Physeal cartilage also ossifies late, creating an image radiographically of a large space between the metaphysis and the epiphysis of some bones of the appendicular skeleton, such as the tibia (Fig. 2.9). The sequence of ossification of the appendicular skeleton of monotremes

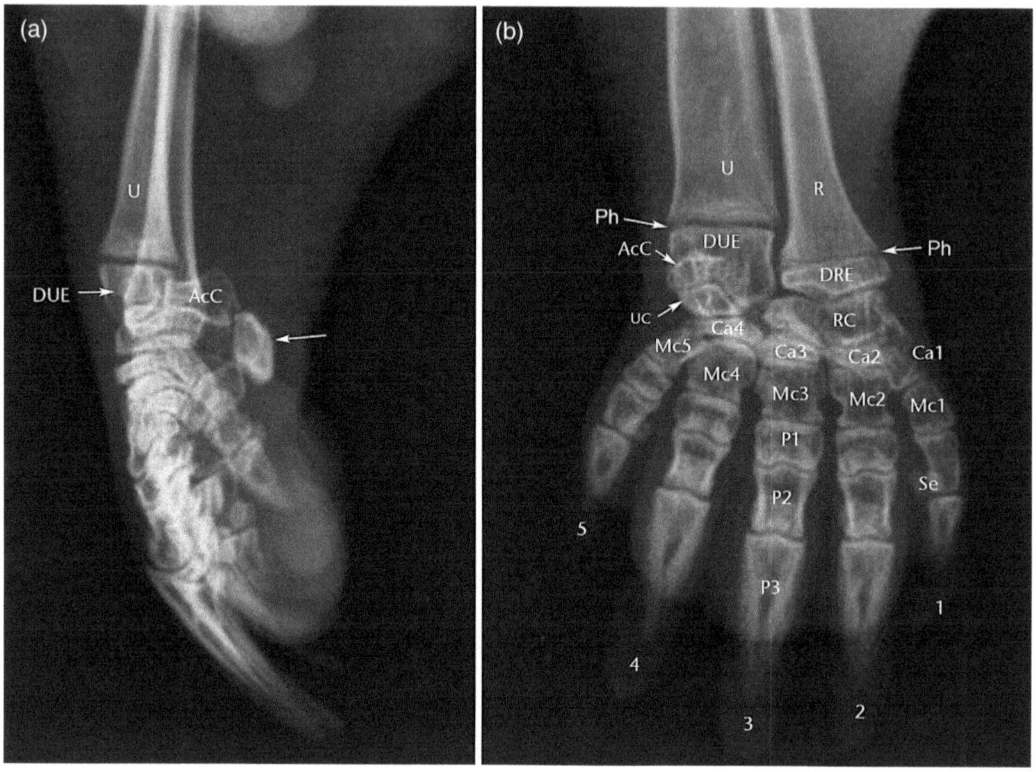

Figure 2.7: Carpus and manus, skeletally immature short-beaked echidna (*Tachyglossus aculeatus*): (a) mediolateral view, (b) dorsopalmar view. There are five digits (1, 2, 3, 4, 5). Each digit, except the 1st, consists of three phalangeal bones (P1, P2, P3). Each digit has an associated metacarpal bone (Mc1–Mc5) and a sesamoid bone (Se) on the palmar surface of the distal interphalangeal articulation. The proximal row of carpal bones consists of a radial (RC), ulnar (UC) and accessory (AcC) carpal bone. The distal row of four carpal bones (Ca1–Ca4) articulates with the metacarpal bones. The distal radial (DRE) and ulnar (DUE) epiphyses are separated from the adjacent metaphysis of the radius (R) and ulna (U) by a lucent cartilaginous physis (Ph). The large palmar sesamoid (arrow) is visible in (a).

differs from other mammals; the more distal bones (radius/ulna, tibia/fibula) ossify earlier that the humerus and femur and this can be recognised radiographically (Weisbecker 2011). The patella develops from two separate ossification centres, thus appearing bipartite in juvenile echidnas. These fuse to form a single bone as the skeleton matures (Fig. 2.9b and 2.10).

The femur is short (relative to the humerus), stout and flattened with expanded extremities and prominent external and internal trochanters (Parker and Haswell 1921). The head of the femur is inserted horizontally into the acetabulum, projecting at right angles to the body. The head is not inflected medially, the trochanters diverge laterally and the neck is short. The femur projects laterally from the body and rotates during locomotion (Augee *et al.* 2006) (Fig. 2.11). It has a broad groove for the large ossified patella and flat, subequal, separated condyles.

The tibia and fibula are transposed and of equivalent size, and distally can be confused with one another radiographically (Griffiths 1989). They are rotated caudally so that the pes is rotated outwards and backwards. The flattened proximal end of the monotreme fibula has an elongated proximal process that resembles the olecranon of the ulna (Figs 2.11 and 2.12). In the tarsus, there are two

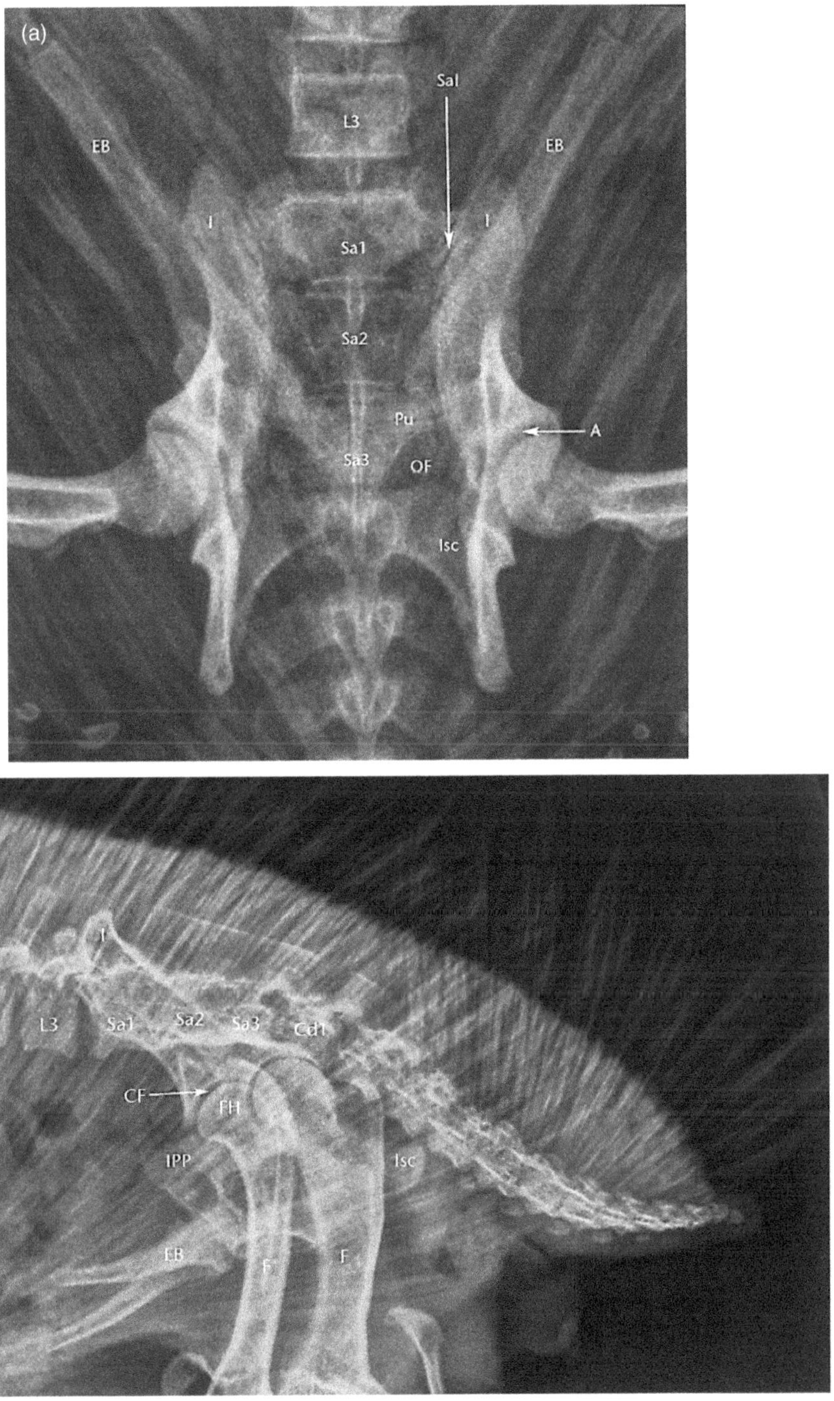

Figure 2.8: Pelvis: (a) ventrodorsal view, (b) lateral view. The epipubic bones (EB) are partly obscured by opaque spines in the ventrodorsal image. Paired iliopectineal processes (IPP) project cranially from the pubis. The sacrum has three elements (Sa1, Sa2, Sa3). The ilia (I) and sacral elements are separated by the sacroiliac (SaI) joint. The pubis (Pu) and ischium (Isc) are located on the cranial and caudal sides of the obturator foramen (OF). The acetabulum (A) contributes the pelvic component and the femoral head (FH) the femoral component of the coxofemoral joint (CF). The caudal vertebrae (Cd1 labelled) bear large transverse processes that are visible on the ventrodorsal image. The 3rd lumbar vertebra (L3) and femurs (F) are labelled.

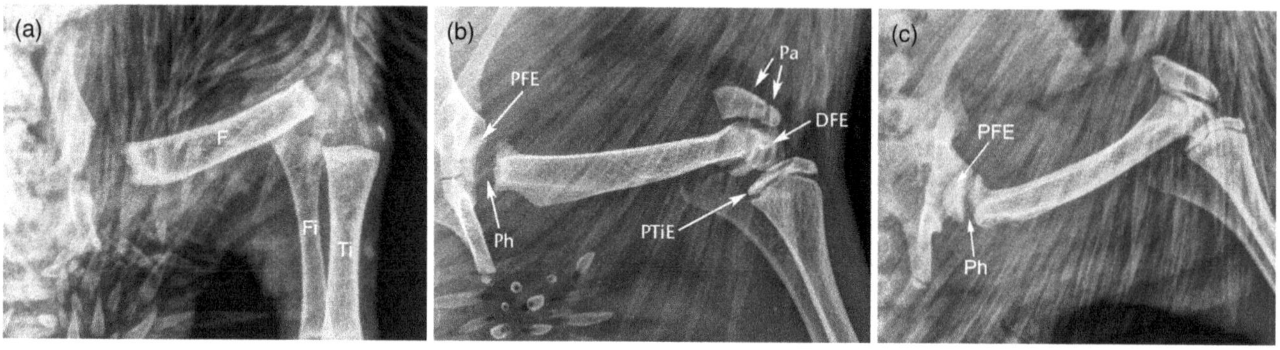

Figure 2.9: Hindlimb, three juvenile short-beaked echidnas (*Tachyglossus aculeatus*) of increasing maturity from left to right (a–c), dorsoventral/lateromedial view. The diaphyses of the femur (F), fibula (Fi) and tibia (Ti) are visible radiographically before their respective epiphyses: the proximal femoral epiphysis (PFE), distal femoral epiphysis (DFE), the patella (Pa) and proximal tibial epiphysis (PTiE). The two ossification centres of the patella are evident in (b). The physes, such as the femoral capital physis (Ph), are last to ossify.

rows of tarsal bones that articulate with the metatarsal bones distally (Parker and Haswell 1921).

There are five digits on the pes. Digit 1 is greatly reduced and has a short claw; digit 2 is elongated with a long claw. Digit 1 has two phalanges, whereas digits 2–5 have three (Figs 2.12 and 2.13) (Augee *et al.* 2006).

All male and some female echidnas possess spurs on the caudomedial aspect of the tarsus (Griffiths 1989; Johnston *et al.* 2006). When present, the spur is housed within a sheath of skin. The spur is present in juveniles of both sexes and appears as a soft tissue structure on radiographs. Most females lose their spurs as they age. In adult

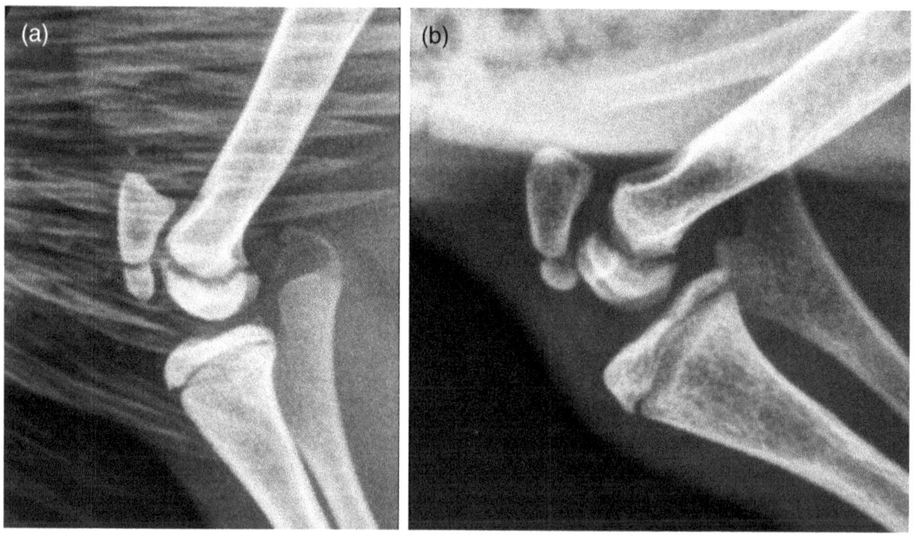

Figure 2.10: Stifle, two juvenile short-beaked echidnas (*Tachyglossus aculeatus*): (a) lateromedial view, (b) mediolateral view. The two ossification centres of the patella are demonstrated. (b) Note the Salter fracture of the distal femoral physis and displacement of the distal patella ossification centre by the physeal fracture.

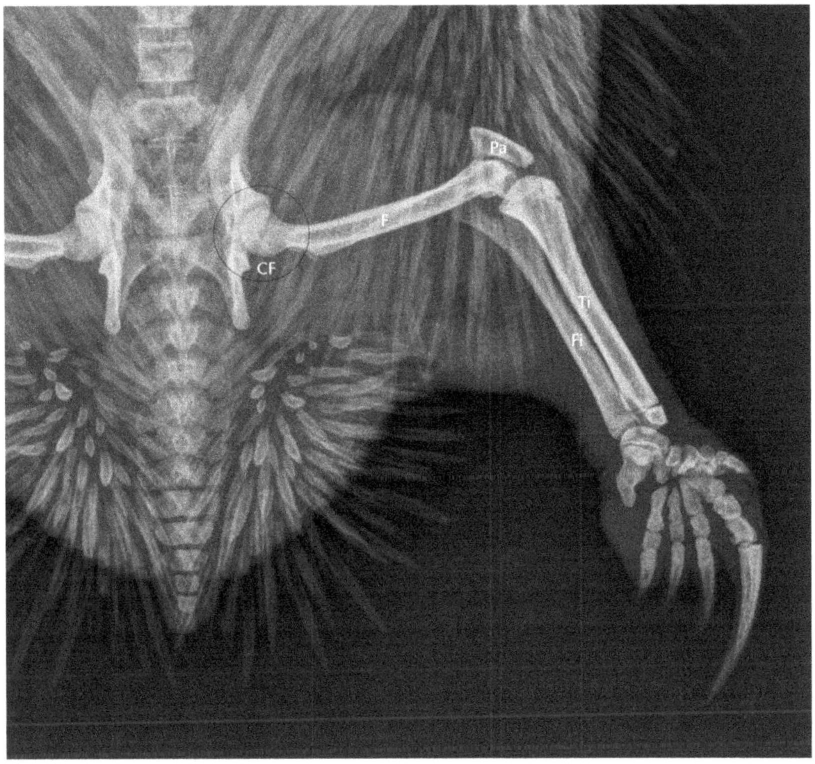

Figure 2.11: Hindlimb, dorsoventral/lateromedial view. The coxofemoral joint (CF, circle) houses the femoral head. At the distal end of the femur (F), a large cuboidal patella (Pa) is located on the cranial surface of the femorotibial articulation. The proximal end of the fibula (Fi) possesses a proximally directed process. Both the tibia (Ti) and fibula contribute to the tarsocrural articulation.

males, the spur sits on a nub of bone (the *os calcaris*) that is not present in females (Figs 2.13 and 2.14). Male echidnas possess a crural gland that is located within the musculature of the hindlimb, just proximal and caudal to the stifle. There is significant seasonal variation in the size of the gland, which does not produce venom, unlike the gland of platypus. It is not visible radiographically, although it is easily visualised on ultrasound examination.

Axial skeleton

Skull

The skull of monotremes differs greatly from that of other mammals. The bones become fused together early in life, so it is difficult to trace suture lines (Fig. 2.15). The cranium is larger and more rounded in the echidna than in the platypus, in accordance with the larger size of the brain in the former. On the floor of the cranial compartment there is a large, horizontal cribiform plate, perforated for the passage of olfactory nerves from olfactory epithelium of the ethmoid turbinates (Augee *et al.* 2006) (Fig. 2.16). There is a pterygoid bone similar to that of lower vertebrates. Perforating the rostral root of the zygomatic arch is the temporal canal, a structure not present in higher mammals. It apparently represents the post-temporal fossa (canal) of reptiles (Parker and Haswell 1921). The zygomatic arches are thin (Fig. 2.16) and the jugal, the rostral component of the zygomatic arch of most mammals, is absent, although it is present in the platypus. The base of the skull is flat (Augee *et al.* 2006).

In adult echidnas, the tympanic cavities remain open, so the tympanic membrane is visible, as is

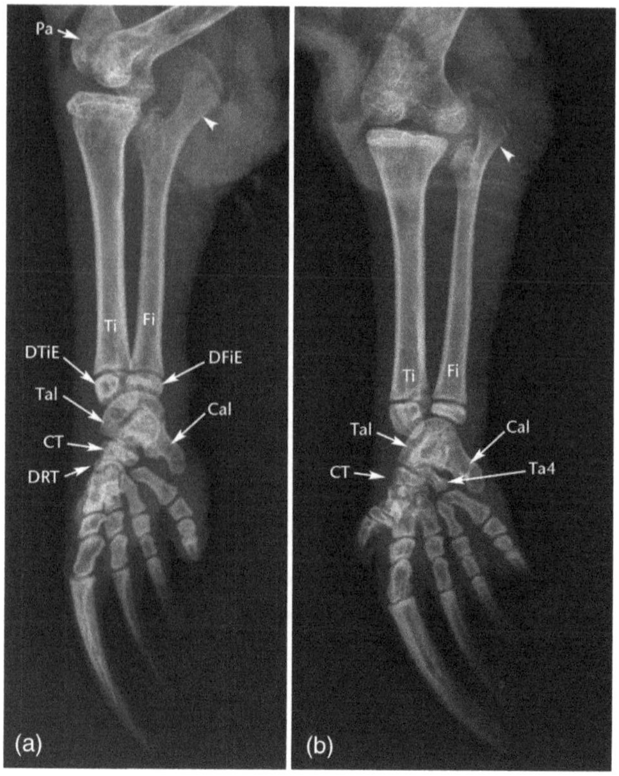

Figure 2.12: Crus and pes, skeletally immature short-beaked echidna (*Tachyglossus aculeatus*): (a) mediolateral and slightly oblique dorsoplantar view, (b) mediolateral and dorsoplantar view. The patella (Pa) is visible on the cranial aspect of the distal femur. The elongated process of the proximal fibula (arrowhead) is visible in both views. The crus comprises the tibia (Ti) and fibula (Fi). The distal tibial (DTiE) and fibular (DFiE) epiphyses are separated from their respective metaphyses by a lucent cartilaginous physis. The tarsal bones consist of the talus (Tal), calcaneus (Cal), central tarsal bone (CT) and a distal row of small tarsal bones (DRT). In (b), the 4th tarsal bone (Ta4) is separate from the remaining distal tarsal bones.

the manubrium of the malleus (Griffiths 1989). There are no tympanic bullae in monotremes (Augee *et al.* 2006).

The beak comprises long, thin extensions of the nasal, premaxillary, septomaxillary and maxillary bones, together with the rostral process of the vomer within the septal basement (Fig. 2.16). Except for the vomer, prominent syndesmoses link the

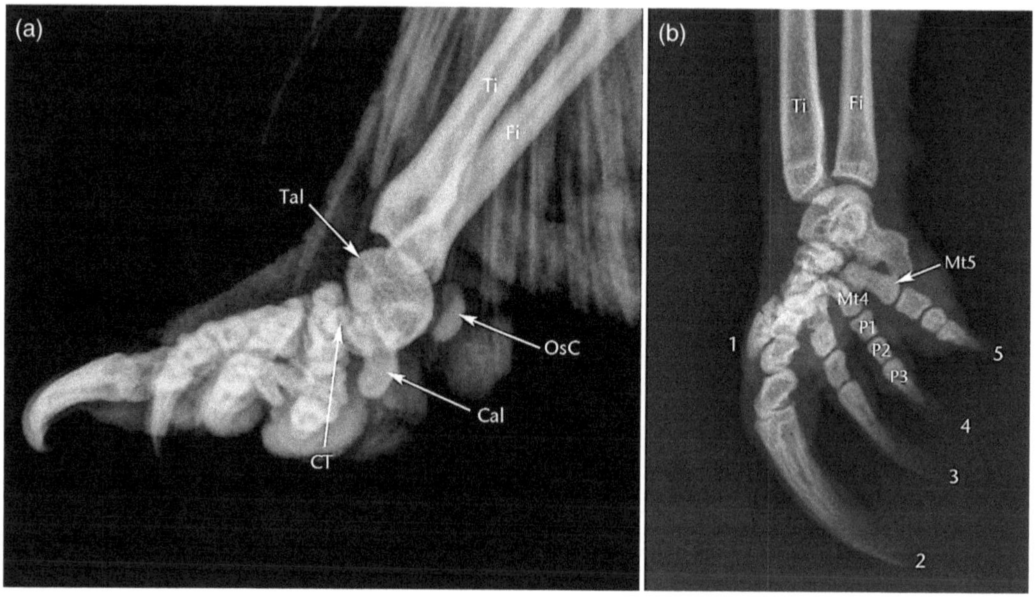

Figure 2.13: Pes, male short-beaked echidna (*Tachyglossus aculeatus*): (a) mediolateral view, (b) dorsoplantar view. There are four main digits (2, 3, 4, 5) each comprising three phalanges (P1, P2, P3). The 1st digit (1) is abbreviated. There are five metatarsal bones (Mt). The calcaneus (Cal) is directed plantarodistally. The bone on the plantar aspect of the talus is the *os calcaris* (OsC). The tibia (Ti), fibula (Fi), talus (Tal) and central tarsal bone (CT) are labelled.

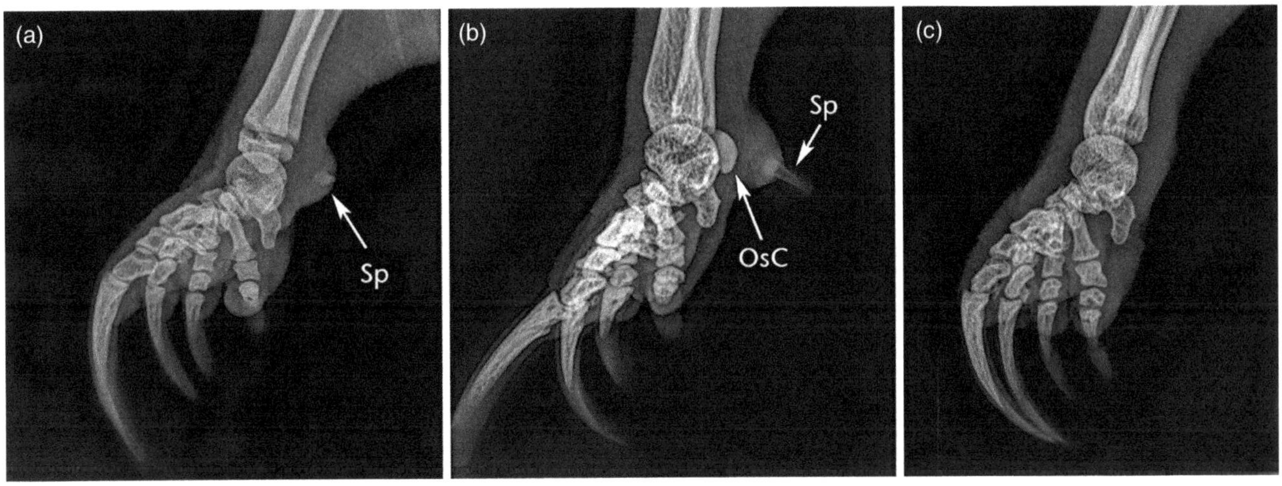

Figure 2.14: Tarsus, mediolateral view: (a) juvenile with a clearly defined spur (Sp), (b) adult male, in which the *os calcaris* (OsC) and spur are present, (c) adult female in which both the *os calcaris* and spur are absent.

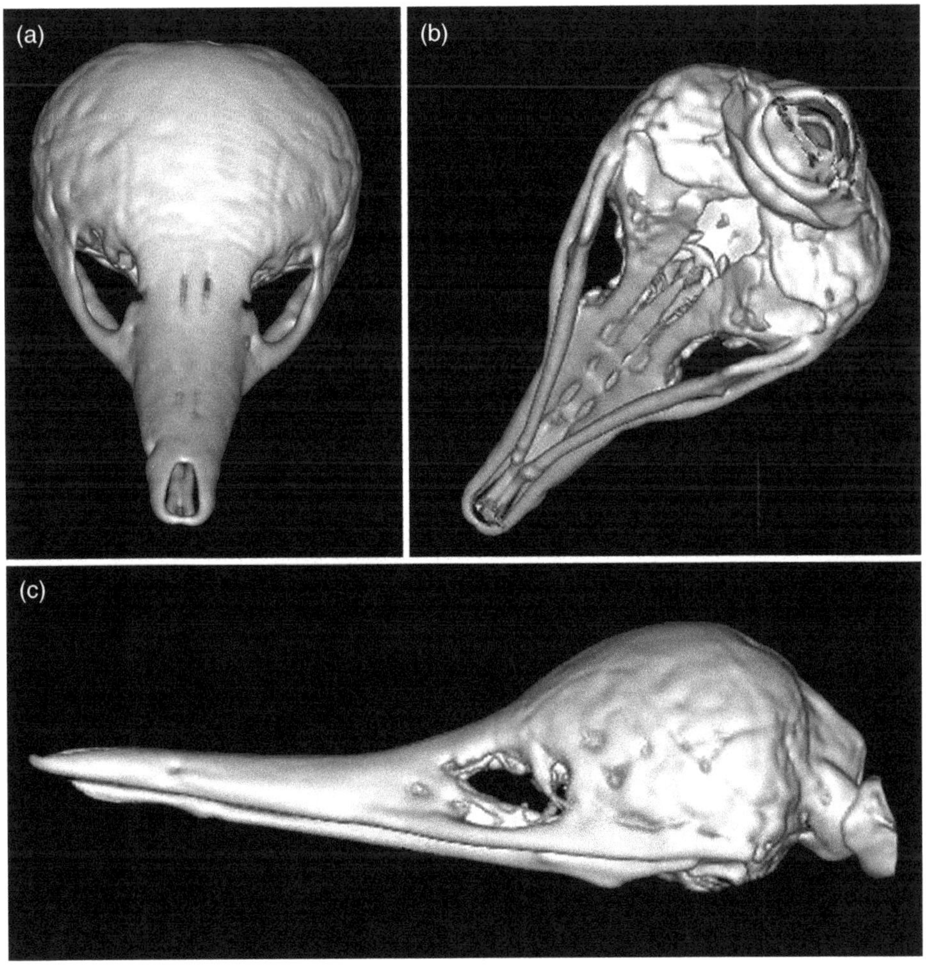

Figure 2.15: Skull, volume-rendered three-dimensional CT reconstruction: (a) rostrocaudal dorsoventral view, (b) ventrodorsal view, (c) lateral view. The contribution of the maxilla and mandible to form the beak is demonstrated.

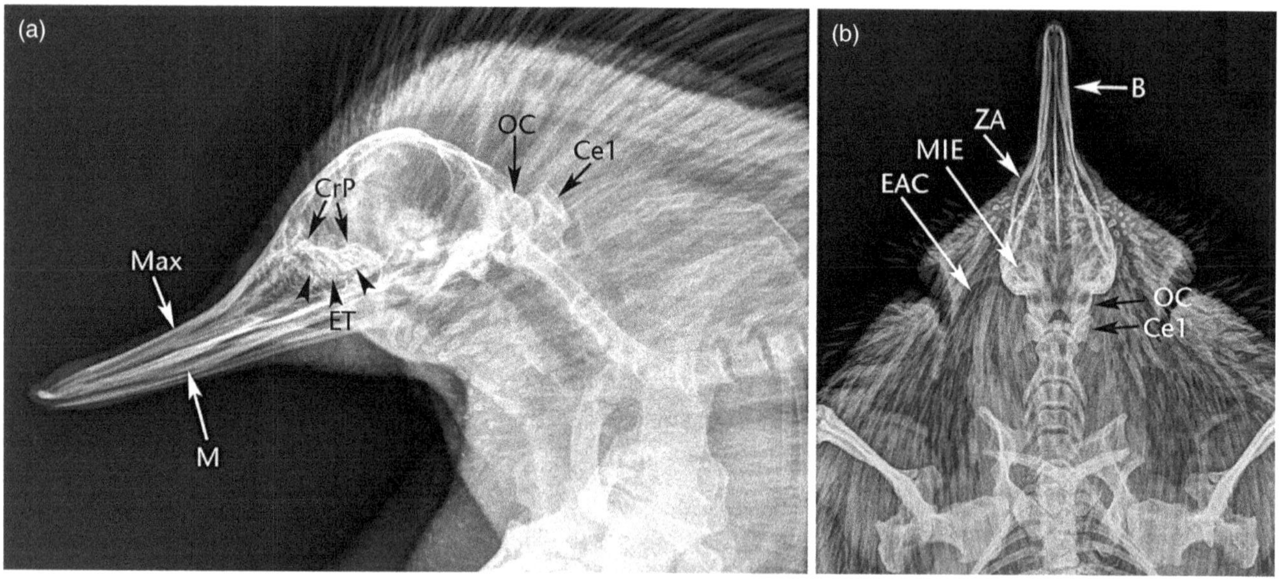

Figure 2.16: Head and neck: (a) lateral view, (b) ventrodorsal view. The maxilla (Max) and mandible (M) form the beak (B). The ethmoid turbinates (ET) are visible extending ventrally from the cribiform plate (CrP). The occipital condyles (OC) of the skull articulate with the 1st cervical vertebra (Ce1). The zygomatic arches (ZA) are fine. The middle and internal ear (MIE) structures and the external auricular canal (EAC) are visible. Note the non-displaced rostral maxillary fracture in (a).

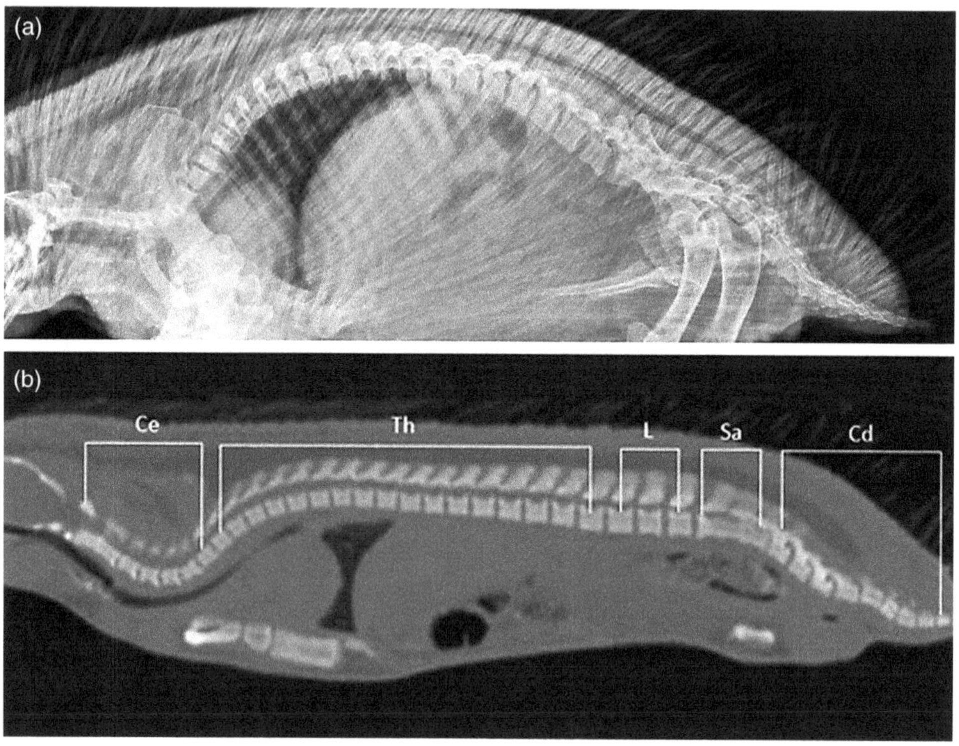

Figure 2.17: Vertebral column: (a) lateral radiograph, (b) lateral reconstructed CT image. There are 7 cervical (Ce), 17 (in this individual) thoracic (Th) and 3 lumbar (L) vertebrae, 3 sacral elements (Sa) and 12 (only 9 visible in this image) caudal vertebrae (Cd).

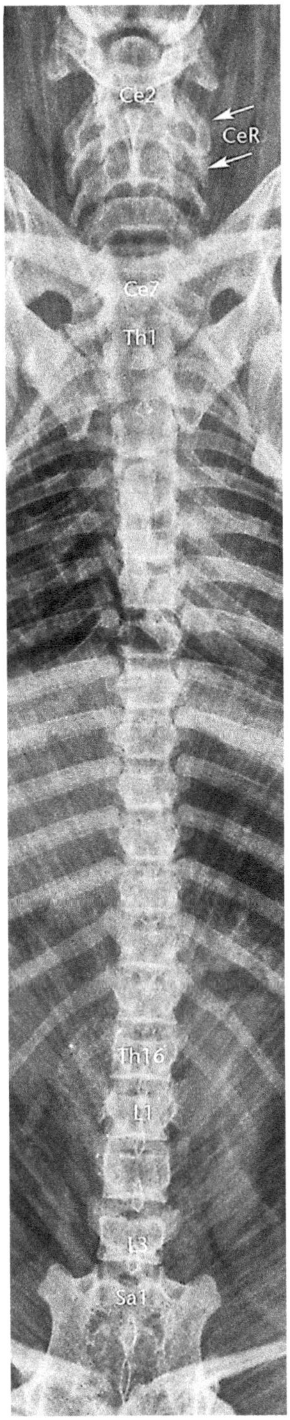

Figure 2.18: Vertebral column: ventrodorsal view. There are 7 cervical vertebrae (Ce2, Ce7 are labelled), 16 (in this individual) thoracic (Th1, Th16 are labelled) and 3 lumbar (L1, L3 are labelled) vertebrae. There are 3 sacral elements (Sa1 is labelled). The cervical vertebrae have rudimentary cervical ribs (CeR).

bones. The nasal cartilages comprise the hard palate and nasal septum. Several small holes for branches of the trigeminal nerve are found along the maxilla (Augee *et al.* 2006).

The mandibles are delicate and consist of very fine styliform rami (dentaries), weakly fused at the symphysis and articulated loosely at the mandibular fossa, which is a shallow depression on the underside of the skull (Augee *et al.* 2006) (Figs 2.15 and 2.16). The condyle is narrow and more elongated rostrocaudally than transversely. There are very slight rudiments of the angle of the mandible and the coronoid and angular processes (Parker and Haswell 1921). The paired mandibles (dentaries) assist in the opening of the mouth by rotating medially.

Echidnas do not have teeth and, unlike the platypus, lack keratinous grinding plates and vestigial juvenile dentition.

Vertebrae

Short-beaked echidnas have 7 cervical, 16–17 thoracic, 3 lumbar, 3 sacral and 12 caudal vertebrae (Figs 2.17 and 2.18) (Griffiths 1989). There are small cervical ribs (Fig. 2.18). The cervical–thoracic spine junction is lordotic. The dorsal spinous processes between the first thoracic and last pre-sacral vertebrae are directed caudally (Augee *et al.* 2006). Vertebral epiphyses are absent.

The atlas has the form of a ring. Its cranial ventral surface is concave and articulates with the laterally placed occipital condyles of the skull. The axis bears an odontoid process, which projects cranially into the centre of the atlas. The atlanto-occipital joint permits flexion and extension, whereas the atlanto-axial joint allows rotation (Jenkins 1969).

The echidna spinal cord is one of the shortest recorded for any mammal, ending at the mid-thoracic level (Ashwell and Zhang 1997). The tail is very short and the vertebrae are flattened and, unlike those of the platypus, do not have ventral spines. There are about five subvertebral bones, which differ from ordinary chevron bones in being

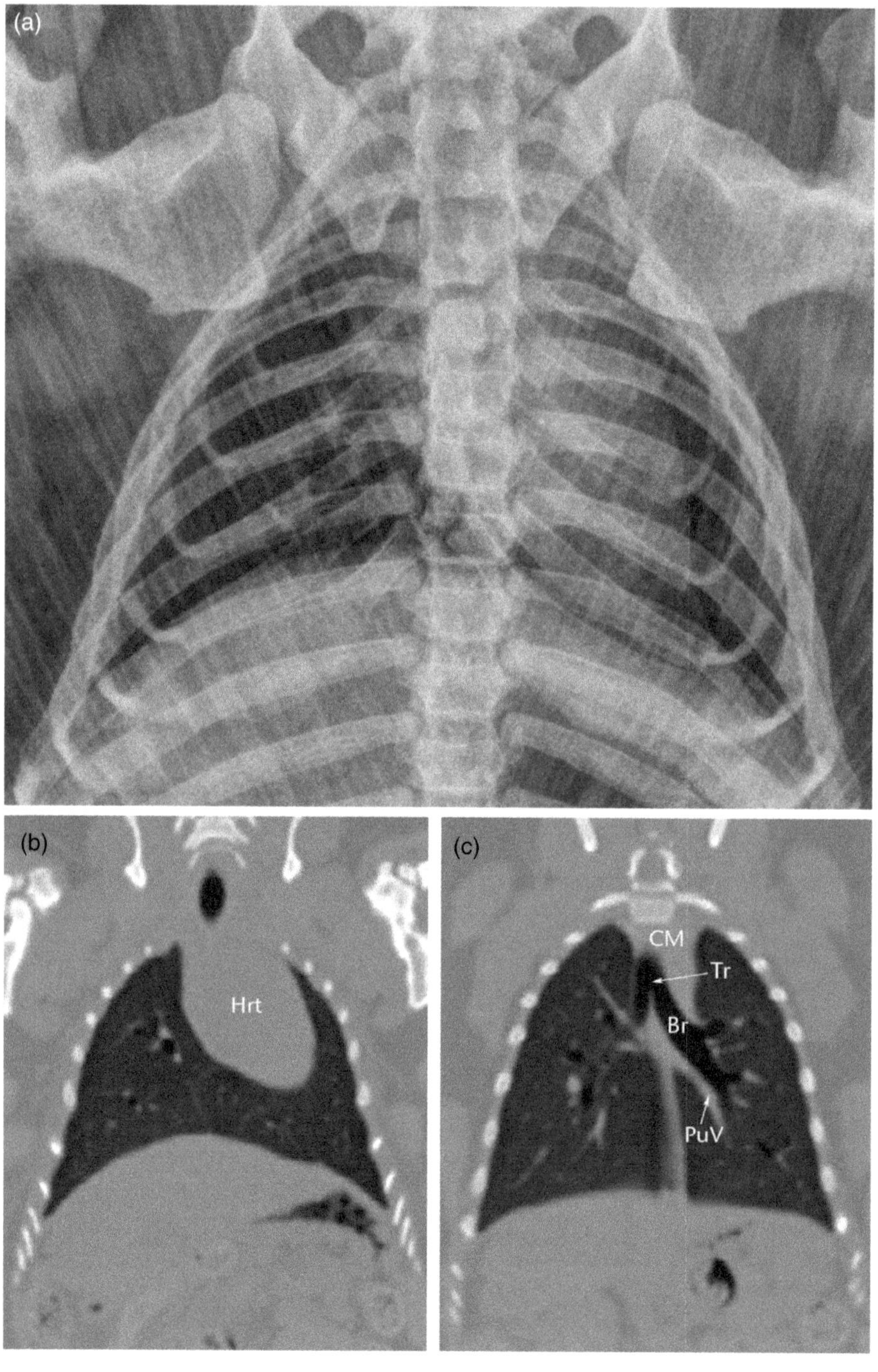

Figure 2.19: Thorax: (a) ventrodorsal radiograph, (b, c) dorsal planar CT reconstruction, lung window. In conventional radiographs, detail of the pulmonary parenchyma is obscured by opaque spines. The heart (Hrt) is located cranially within the thorax and appears to be more left-sided than midline. The pulmonary parenchyma, airways and pulmonary vasculature are clearly visible using CT, which enables removal of the superimposed obstruction of the spines. The trachea and its bifurcation (Tr) are superimposed on the cranial mediastinum (CM). A lobar bronchus (Br) and a pulmonary vein (PuV) can be seen.

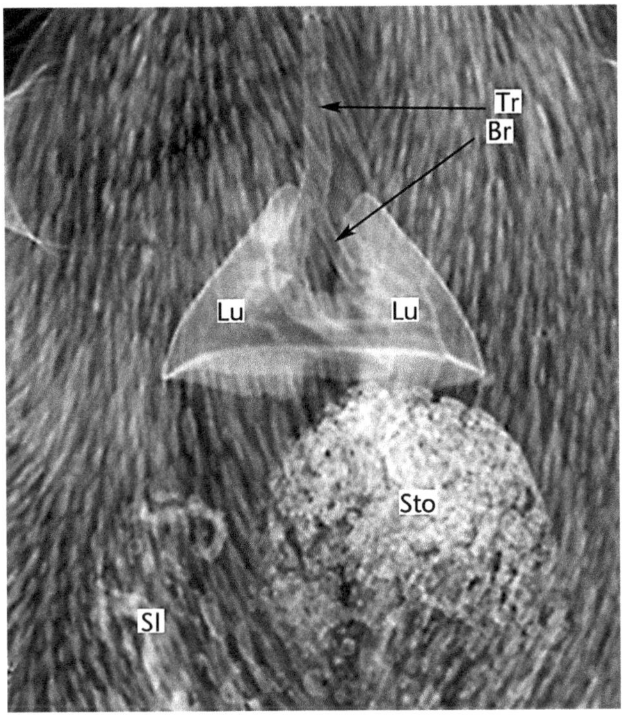

Figure 2.20: Thorax and cranial abdomen, three-dimensional rendered CT image optimised for gas-containing structures. The airways, lungs and gas in the gastrointestinal tract are demonstrated. The trachea (Tr), a bronchus (Br), lungs (Lu), stomach (Sto) and small intestine (SI) are labelled.

mere flat nodules (Parker and Haswell 1921). The caudal vertebrae have large lateral processes (Figs 2.8 and 2.11).

Ribs and sternum

Monotremes have both vertebral and sternal ribs, which are joined by cartilage. The broad, overlapping sternal ribs are strongly ossified and attach directly to the sternum, rather than the cartilaginous connection typical of other mammals. They are connected with the vertebral ribs by imperfectly ossified intermediate ribs (Parker and Haswell 1921). Ossified sternal ribs are a unique feature of monotremes. The vertebral ribs of monotremes are not divided into a head and tubercle, as in other mammals, and attach directly to the sides of the vertebral bodies (Augee *et al.* 2006). The sternum consists of a presternum, three sternebrae and a xiphisternum (Fig. 2.5).

Soft tissue

Head and neck

The paired submandibular salivary glands are palpable as two distinct firm, flattened, mobile subcutaneous masses on the ventral neck. Despite their prominence, they are not readily identifiable radiographically. The external ear canals are well disguised by hair and open as vertical slits protected by a ridge of muscle and cartilage. They are clearly visible on radiographs and CT images (Figs 2.2 and 2.16).

Thorax

The lungs are asymmetrical, the right lung making up approximately 65% of the total lung volume (Figs 2.19 and 2.20). The right lung consists of three parts: a large dorsal lobe, a ventrolateral lobe and a ventral (cardiac) lobe. Each lobe is supplied by a separate bronchus. The left lung is unilobate (Perry *et al.* 2000). When viewed on a lateral radiograph, the cranial part of the heart and the cranial lung lobes are obscured by superimposition of the bulky pectoral girdle. As the pectoral girdle is rigid, the cranial thorax will always be difficult to image using a lateral projection (Fig. 2.21). A ventrodorsal projection is more useful, although the pectoral girdle is also superimposed on some of the cranial lung field in this view (Fig. 2.19).

Abdomen

The digestive tract of the echidna is a simple monogastric arrangement (Figs 2.22–2.25). The intestine is 3.5 m long and exhibits a slow rate of throughput, with a clearance of 2 days for a termite- and ant-based meal (Griffiths 1989). Soil and sand are ingested during feeding

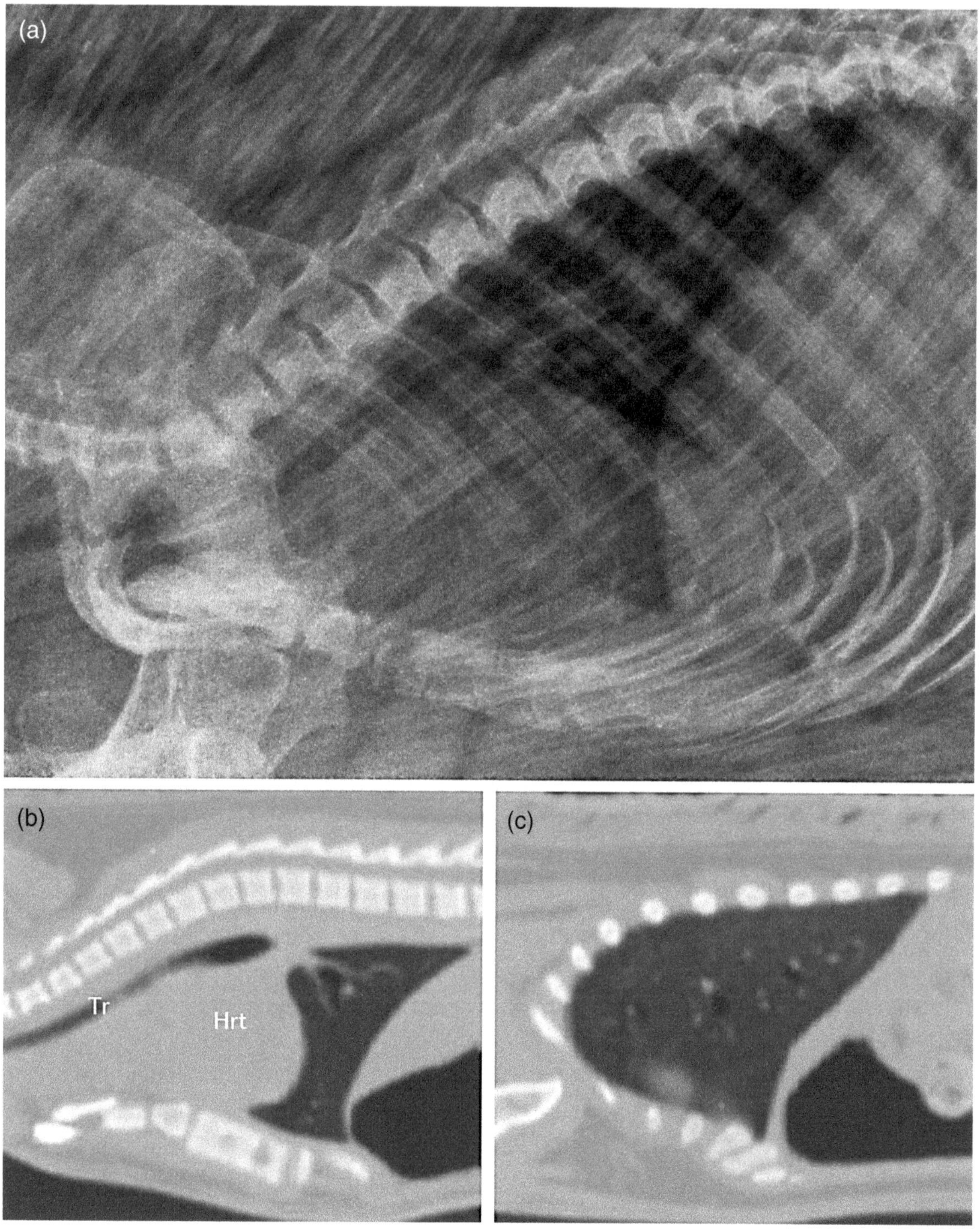

Figure 2.21: Thorax: (a) lateral view, (b, c) reconstructed sagittal plane CT images, lung window. (a) The rigid pectoral girdle and cranially located cardiac silhouette obscure cranial thoracic detail. (b) Thoracic detail such as the heart (Hrt) and trachea (Tr) are more obvious. Pulmonary detail that cannot be appreciated radiographically on conventional images is better visualised, particularly in the parasagittal plane (c).

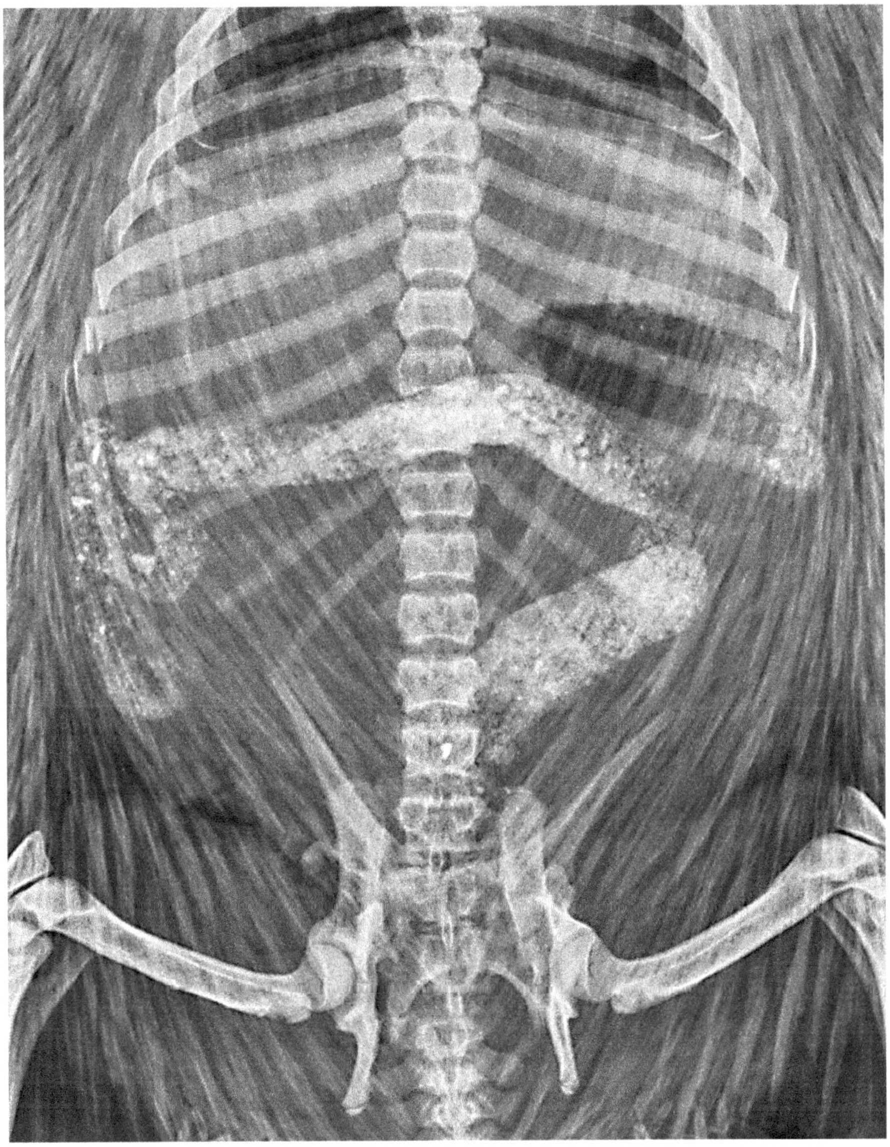

Figure 2.22: Abdomen: ventrodorsal view. Soft tissue detail of abdominal viscera is obscured by spines. Sand is visible in the colon.

and, consequently, are seen on radiographs (fine, radiopaque particles within any part of the gastrointestinal tract) (Figs 2.22 and 2.24). The spleen is strap-like and T-shaped. The two ovoid testes are suspended within the abdominal cavity caudal to the kidneys. There are no seminal vesicles and no discrete prostate gland. The penis is securely retracted into the preputial sac inside the cloaca when not erect. The penis in adult males is approximately 7 cm long (Johnston *et al.* 2006).

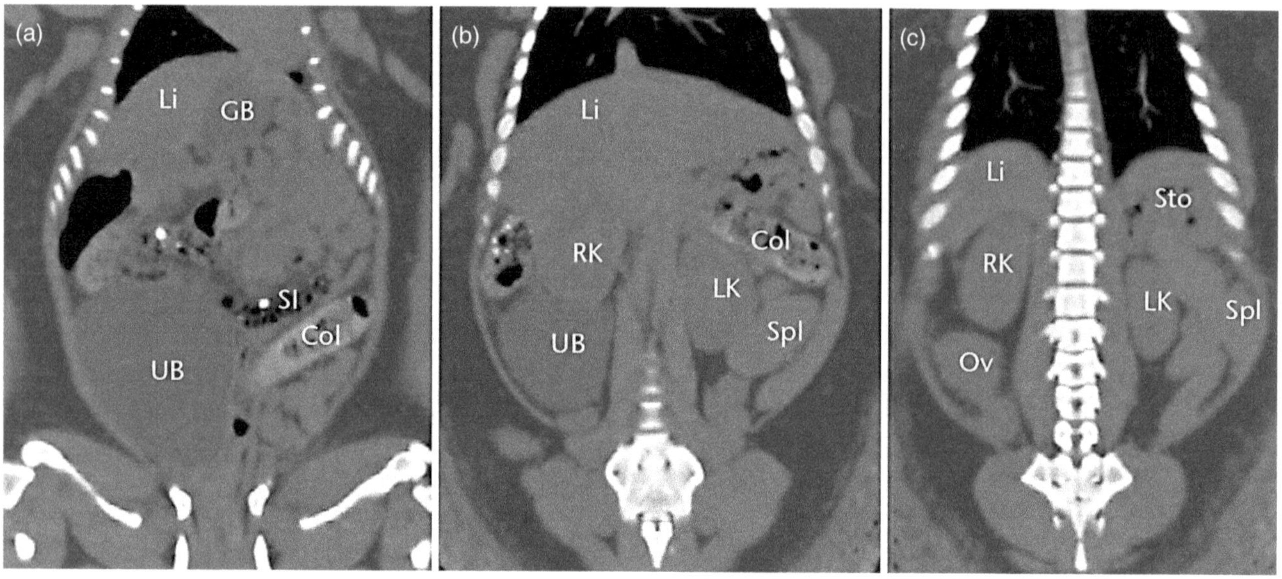

Figure 2.23: Abdomen. Using reconstructed CT images in the dorsal plane (soft tissue window) the effect of spines has been eliminated by reconstruction of the abdominal image into three separate planes: (a) ventral, (b) mid-abdominal, (c) dorsal. The two kidneys (RK, LK), spleen (Spl), liver (Li), gall bladder (GB) stomach (Sto), small intestine (SI), colon (Col) and urinary bladder (UB) can each be seen as separate structures. An ovary (Ov) is visible, superimposed on the urinary bladder on the right side.

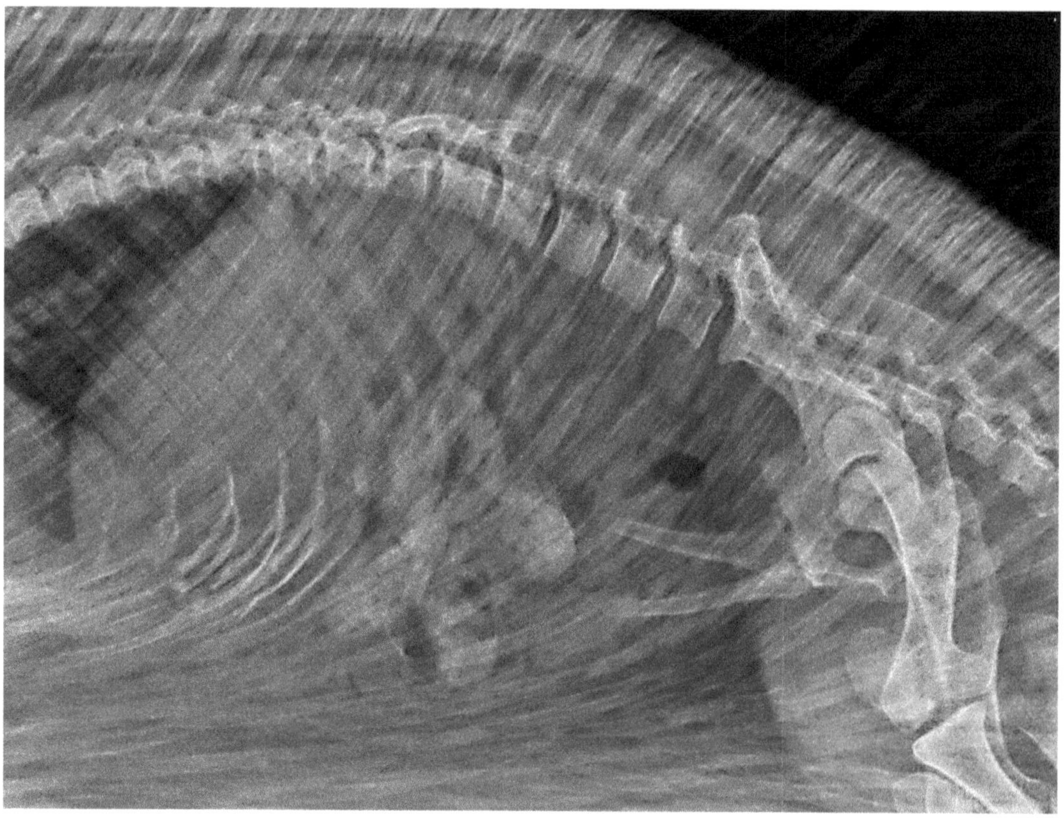

Figure 2.24: Abdomen: lateral view. The conventional radiograph provides poor detail of abdominal viscera.

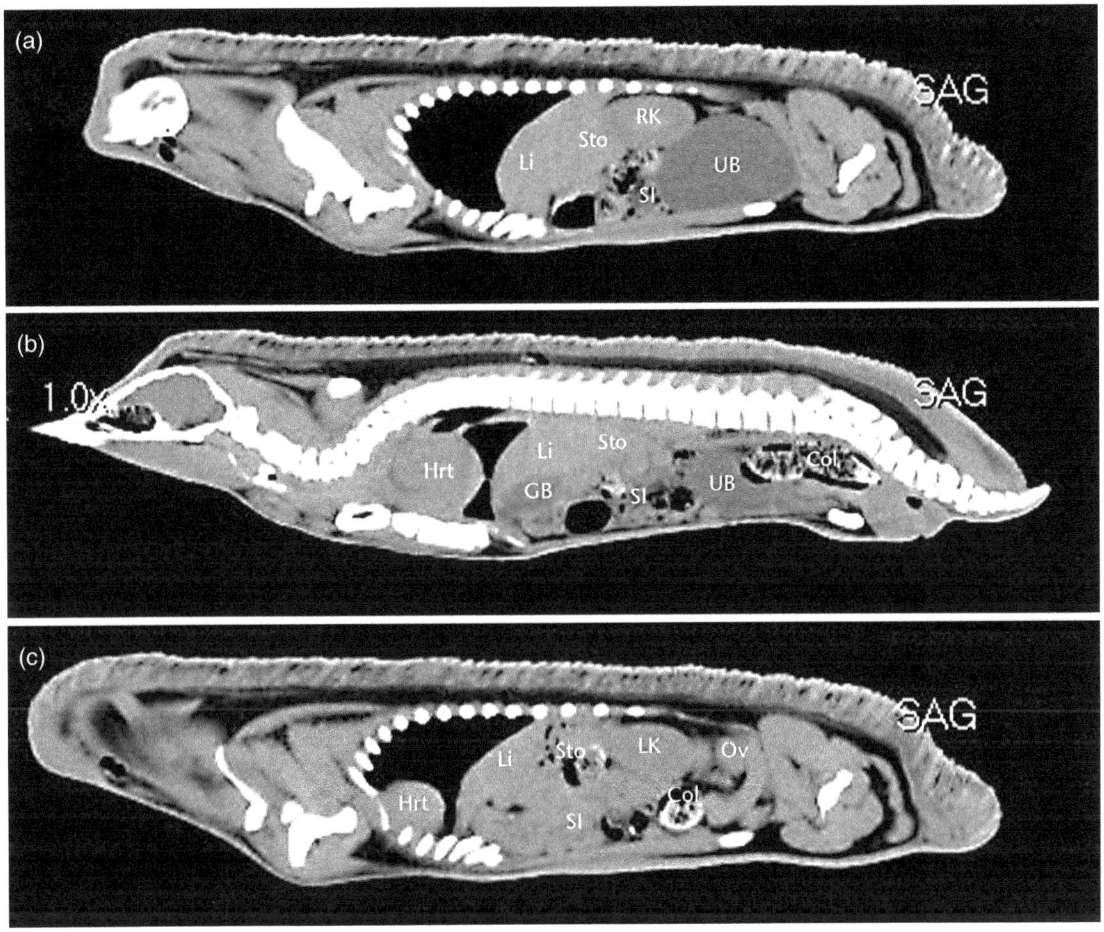

Figure 2.25: Thorax and abdomen, sagittal plane reconstructed CT images, soft tissue window: (a) right parasagittal plane, (b) mid-sagittal plane, (c) left parasagittal plane. CT reconstructed images permit separation of individual intra-abdominal structures: liver (Li), gall bladder (GB), stomach (Sto), left kidney (LK), right kidney (RK), small intestine (SI), colon (Col), urinary bladder (UB), ovary (Ov). The heart (Hrt) is visible in the thorax.

References

Ashwell KWS, Zhang LL (1997) Cyto- and myeloarchitectonic organization of the spinal cord of an echidna (*Tachyglossus aculeatus*). *Brain, Behavior and Evolution* **49**, 276–294. doi:10.1159/000112998

Augee M, Gooden B (1993) *Echidnas of Australia and New Guinea.* University of New South Wales Press, Sydney.

Augee M, Gooden B, Musser A (2006) *Echidna: Extraordinary Egg-laying Mammal.* CSIRO Publishing, Melbourne.

Booth R (1999) Care and medical management of monotremes. In *Wildlife in Australia: Healthcare and Management.* Proceedings 327, pp. 41–52. Post Graduate Foundation in Veterinary Science, Sydney.

Griffiths M (1989) Tachyglossidae. In *Fauna of Australia, Vol. 1B, Mammalia.* (Eds DW Walton and BJ Richardson) pp. 407–435. Australian Government Publishing Service, Canberra.

Jenkins FA (1969) The evolution and development of the dens of the mammalian axis. *The*

Anatomical Record **164**, 173–184. doi:10.1002/ar.1091640205

Johnston SD, Madden C, Nicholson V, Pyne M (2006) Identifying the sex of short-beaked echidnas. *Australian Veterinary Journal* **84**, 63–65. doi:10.1111/j.1751-0813.2006.tb13131.x

Middleton D (2008) Echidnas. In *Medicine of Australian Mammals.* (Eds L Vogelnest and R Woods) pp. 77–101. CSIRO Publishing, Melbourne.

Parker JT, Haswell WA (1921) Skeleton of prototheria. In *A Text-book of Zoology, Volume II.* pp. 501–506. Macmillan and Co. Ltd, London.

Perry SF, Schmitz A, Andersen NA, Wallau BR, Nicol S (2000) Descriptive study of the diaphragm and lungs in the short-nosed echidna, *Tachyglossus aculeatus* (Mammalia: Monotremata). *Journal of Morphology* **243**, 247–255. doi:10.1002/(SICI)1097-4687(200003)243:3<247::AID-JMOR2>3.0.CO;2-S

Weisbecker V (2011) Monotreme ossification sequences and the riddle of mammalian skeletal development. *Evolution* **65**, 1323–1335. doi:10.1111/j.1558-5646.2011.01234.x

3

Platypus

The platypus (*Ornithorhynchus anatinus*) is the only extant species in the family Ornithorhynchidae. Together with the echidnas, the platypus belongs to the order Monotremata – the egg-laying mammals. The monotremes ('one hole') possess a common opening, the cloaca, for the reproductive, urinary and digestive systems (Booth and Connolly 2008). Platypuses are highly specialised, semi-aquatic, semi-fossorial, egg laying mammals that are so well adapted that they have survived relatively unchanged since the time of the dinosaurs. Like echidnas (see Chapter 2), they exhibit reptilian, avian and mammalian characteristics in their physiology, reproduction and anatomy. The platypus is one of a small number of venomous mammals, the males having a venom gland connected by a duct to a hollow keratinous spur on the medial tarsus.

Platypuses are sexually dimorphic, with males larger than females. Males are 45–63 cm long and weigh 1000–3000 g, while females are 39–55 cm long and weigh 700–1750 g. Males can also be distinguished from adult females by the presence of spurs (Grant 1989; Connolly and Obendorf 1998). Platypuses are opportunistic benthic carnivores, feeding mainly on aquatic insect larvae and small, soft-shelled crustaceans, but small fish, frogs and worms are also eaten (Booth and Connolly 2008).

The platypus has a streamlined body that is compressed dorsoventrally (Fig. 3.1) and is covered by fur except for the bill and feet. There are two pairs of pentadactyl limbs, with claws and webbing on the digits of the manus and pes. The pes is turned caudolaterally. Webbing on the manus extends beyond the claws to form fan-shaped paddles for swimming (Booth and Connolly 2008).

Skeletal system

Appendicular skeleton

Pectoral girdle and forelimb

The pectoral girdle of monotremes is unique among living mammals. There are two clavicles, a T-shaped interclavicle (similar to that of reptiles) and two short, broad coracoids. The coracoids articulate medially and caudally with the presternum and unite laterally with their respective scapula to form the glenoid cavity (Fig. 3.2). The glenoid cavity is deep and laterally oriented (to accommodate the horizontally oriented humerus). Cranial to each coracoid is a flat plate, the epicoracoid. The scapula of monotremes is unlike that of other mammals. It has a well-developed acromion that articulates with the clavicle. The acromion and spine of the scapula essentially form the cranial margin of the scapula. The cranial part of the medial surface is actually the supraspinous fossa. The majority of the lateral surface is the infraspinous fossa and the part caudal to this, which is separated from it by a slight ridge, together with the caudal portion of the inner surface, is the subscapular fossa (Parker and Haswell 1921; Griffiths 1989). The scapula is large, expanded dorsally and projects cranially. The pectoral girdle of the platypus provides a sturdy base for the digging action of the forelimbs and is more like that of lizards than that of mammals (Griffiths 1978).

The platypus and echidna are humeral rotation diggers and/or swimmers (the humerus rotates

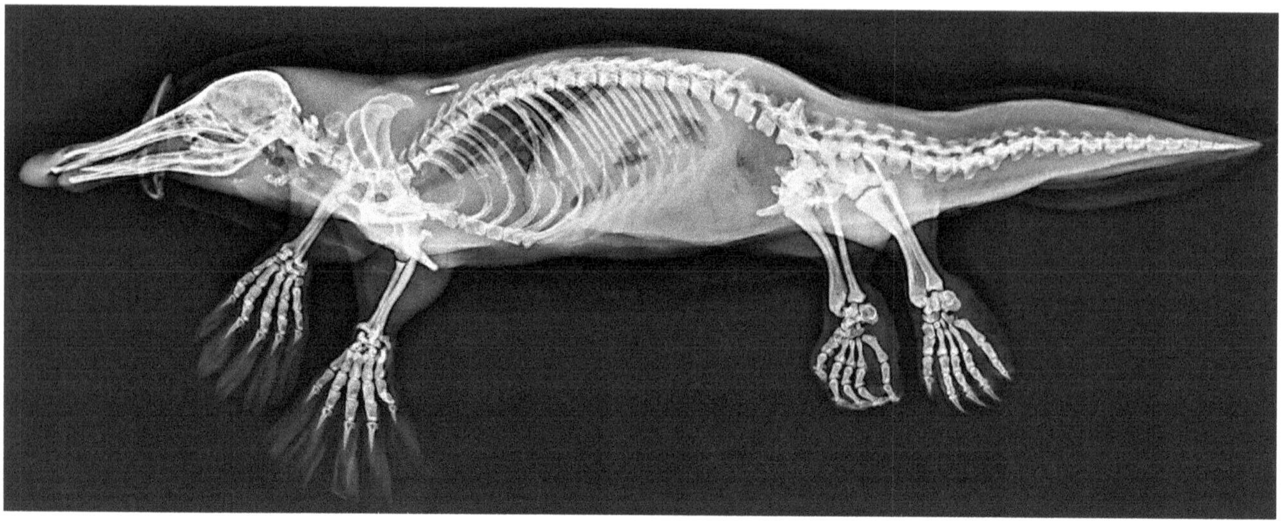

Figure 3.1: Whole body, female platypus (*Ornithorhynchus anatinus*), lateral view. Note: one of the fourth pair of ribs is displaced as a result of trauma, and a transponder can be seen.

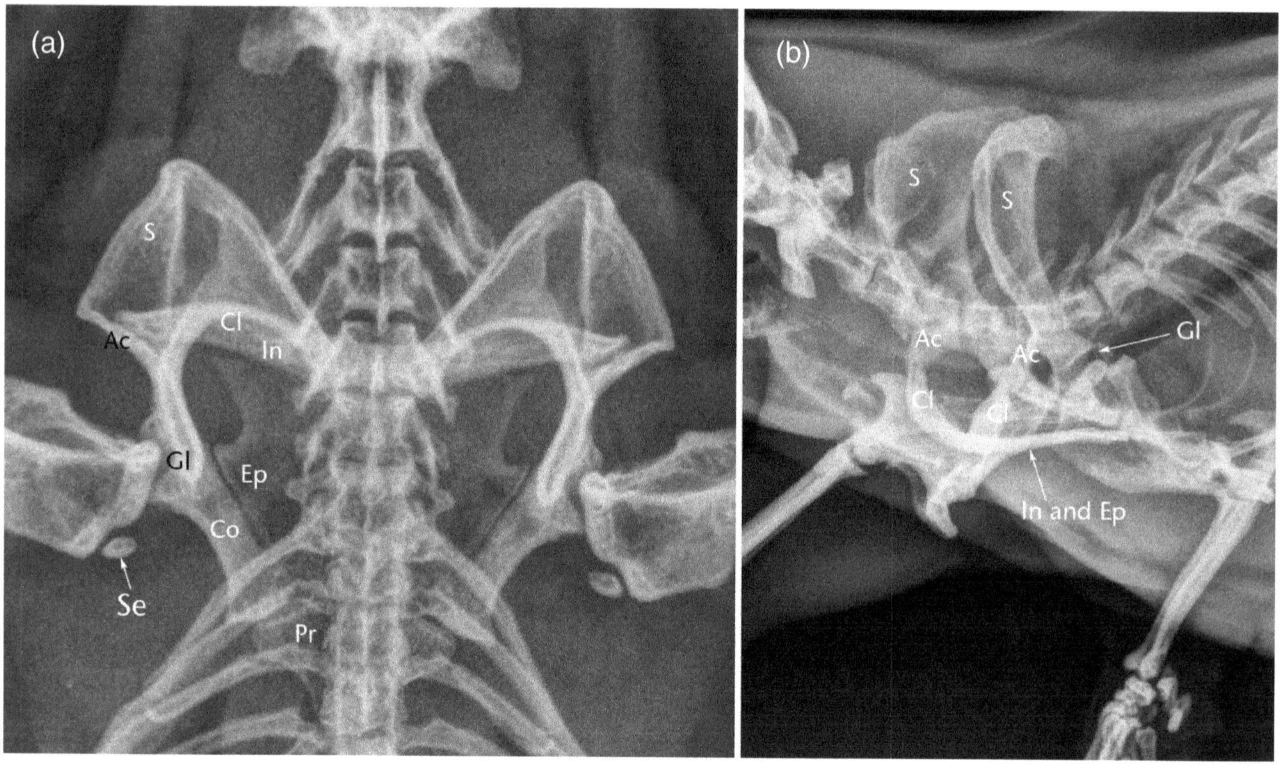

Figure 3.2: Pectoral girdle: (a) ventrodorsal view, (b) lateral view. The scapula (S), clavicle (Cl), acromion (Ac), interclavicle (In), glenoid cavity (Gl), epicoracoid (Ep), coracoid (Co) and presternum (Pr) are labelled. The interclavicle and the epicoracoid appear as a merged bone in the lateral view (In and Ep). There is a large sesamoid bone (Se) adjacent to the glenoid cavity and its humeral articulation.

about its axis, rather than moving back and forth as in other mammals). This adaptation is reflected in the characteristic 'hourglass' (twisted on its long axis) shape of the humerus (Weisbecker 2011). The humerus of the platypus has greatly expanded extremities, although less so than that of the echidna, and prominent tuberosities and condyles (Figs 3.3 and 3.4). The radius is short relative to the ulna and slender with slightly expanded extremities. The ulna extends proximal to the elbow joint with an elongated olecranon, the proximal end of which has a characteristic spade-handle shape (Fig. 3.3). In the carpus, the radial carpal bone and intermediate carpal bone are united and there is not a separate central carpal bone. There is a radial sesamoid bone and two very large palmar sesamoid bones, which are sometimes united (Parker and Haswell 1921) (Figs 3.3 and 3.4).

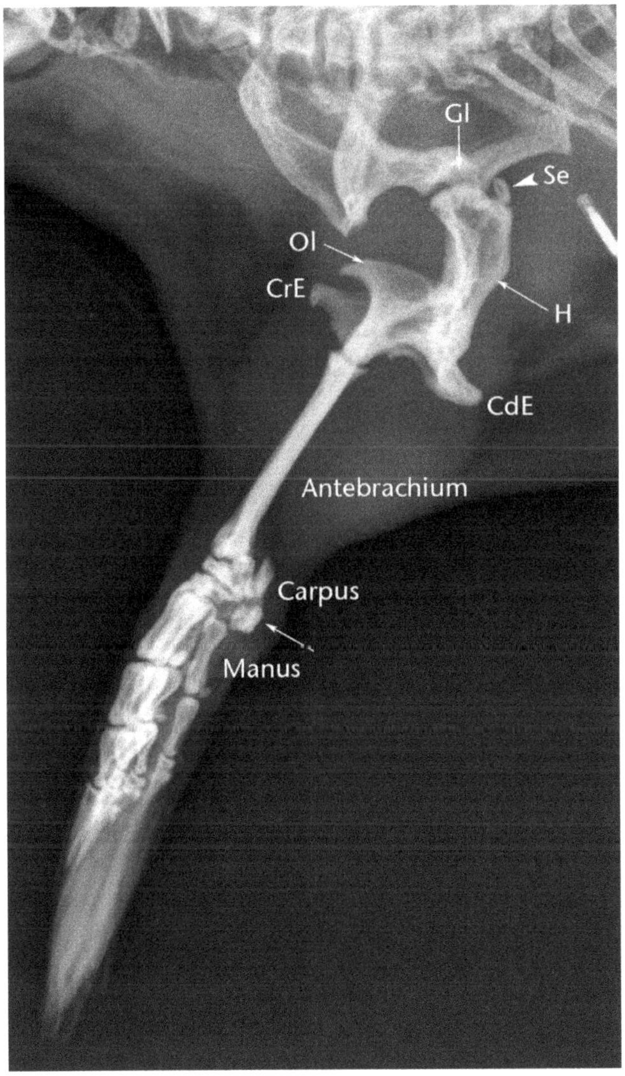

Figure 3.3: Forelimb, extended mediolateral view. There is a large sesamoid bone (Se) adjacent to the glenoid cavity (Gl) and its humeral (H) articulation. The distal end of the humerus is characterised by two flattened epicondyles: one cranially oriented (CrE), the other caudally oriented (CdE). The spade handle-shaped olecranon (Ol) projects proximal to the elbow joint. There is a caudally located carpal sesamoid bone (arrow). Note: a transponder can be seen.

The manus has five digits with webbing extending beyond the claws (Fig. 3.4).

Pelvis and hindlimb

The pelvis is small relative to the size of the long bones of the hindlimb. It has a long symphysis and pubic bones and ischia of almost equal length. A pair of large epipubic bones articulates with the cranial border of the pubic bones. Dorsal and slightly cranial to the articulations of the epipubic bones are paired, cranially projecting extensions of the pubis, the iliopectineal processes (Howes 1893) (Fig. 3.5). The acetabulum is perforated and there is no round ligament in the coxofemoral joint. The femoral epiphyses of monotremes are much larger than other mammals (Thorp and Dixon 1991). The femur has expanded extremities with prominent lateral and medial trochanters. There is a large ossified patella. The proximal end of the fibula has a broad spatulate extension that ossifies from a separate centre (Fig. 3.6).

The pes is rotated caudolaterally. In the tarsus, the talus articulates proximally with both the tibia and fibula. The calcaneus articulates more with the talus than with the fibula, and is located caudolateral to the talus (Fig. 3.6). There is a central tarsal bone that articulates proximally with the talus and distally with tarsal bones 1–3 (Szalay 1994) (Fig. 3.7). There are five digits. The metatarsal bones are short and broad, as are all the phalanges except the third phalanx, which is

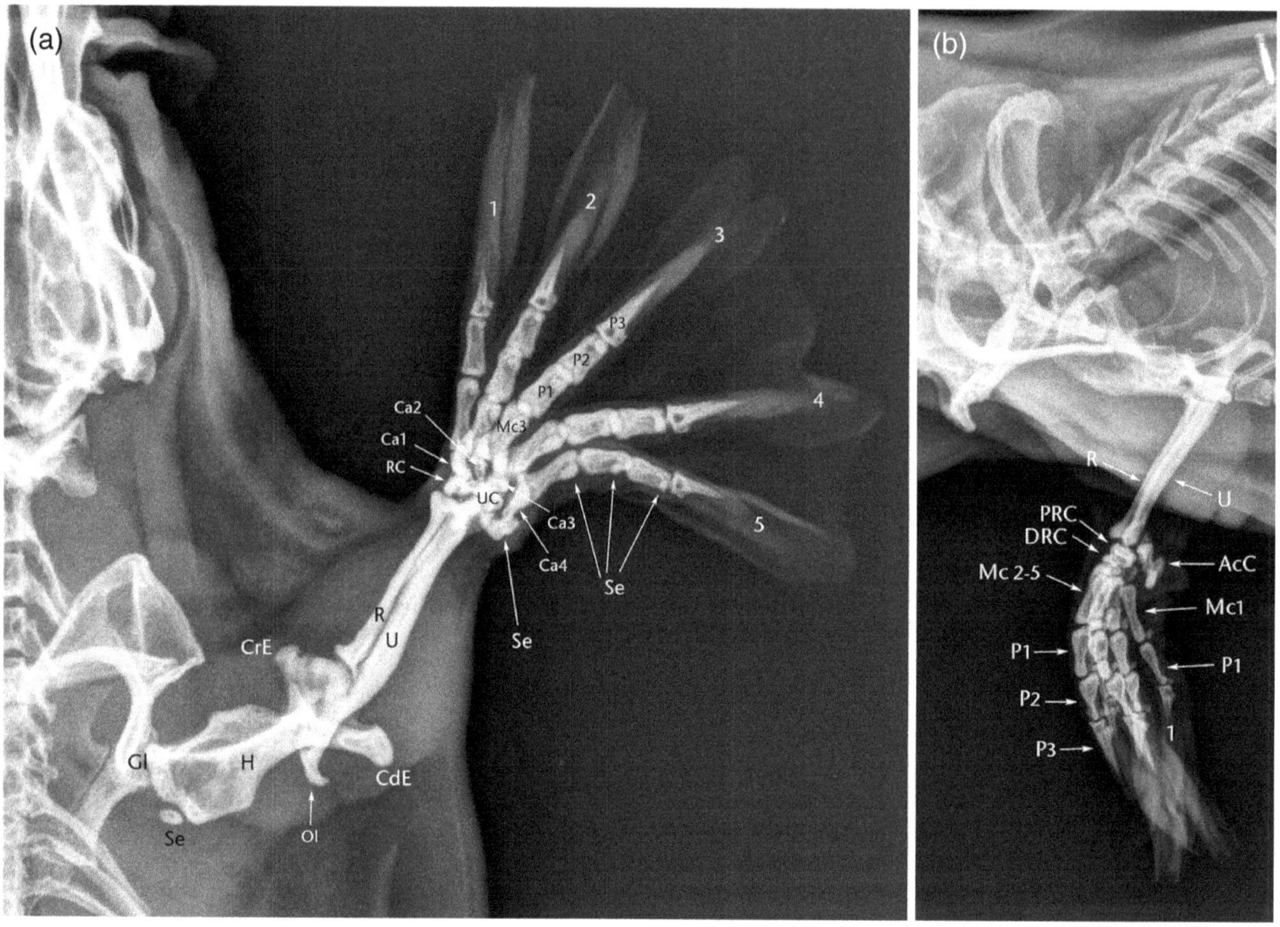

Figure 3.4: Forelimb: (a) craniocaudal and dorsopalmar view, (b) lateromedial view. The glenoid cavity (Gl), humerus (H), humeral epicondyles (CrE and CdE) and humeral sesamoid bone (Se) are labelled. The radius (R) is smaller than the ulna (U), the proximal end of which is characterised by a spade handle-shaped olecranon (Ol). The proximal row of carpal bones (PRC) contains the radial carpal bone (RC), the ulnar carpal bone (UC) and a large sigmoid shaped accessory carpal bone (AcC). There is a caudally located carpal sesamoid bone (Se). The distal row of carpal bones (DRC) contains carpal bones 1–4 (Ca1, Ca2, Ca3, Ca4), which articulate with five metacarpal bones (Mc1, Mc3, Mc 2–5), one for each digit. Of the five digits (1, 2, 3, 4, 5), four contain three phalangeal bones (P1, P2, P3) while digit 1 has only two phalangeal bones. The palmar sesamoid bones (Se) are located on the caudal surface of the antebrachiocarpal articulation, the metacarpophalangeal and the interphalangeal articulations. Note: a transponder can be seen in (b).

elongated and bears a claw (Parker and Haswell 1921; Griffiths 1978). Digit 1 has two phalanges and the others have three. The webbing of the pes does not extend beyond the claws as it does in the manus (Fig. 3.7).

Male platypuses possess a crural system consisting of a venom gland on the dorsal aspect of the thigh connected by a duct to a hollow spur projecting medially from the tarsus. The spur is made of keratin (*cornu calcaris*) and is supported by an accessory ossification, the *os calcaris*, on the medial side of the tarsus (Figs 3.6 and 3.8) (Grant and Fanning 2007). Juvenile females have a rudimentary spur sheath on the tarsus that is lost at around 8 months of age. There is no crural gland in females. The size and shape of the spur can be used to estimate the age of male platypuses (Booth and Connolly 2008).

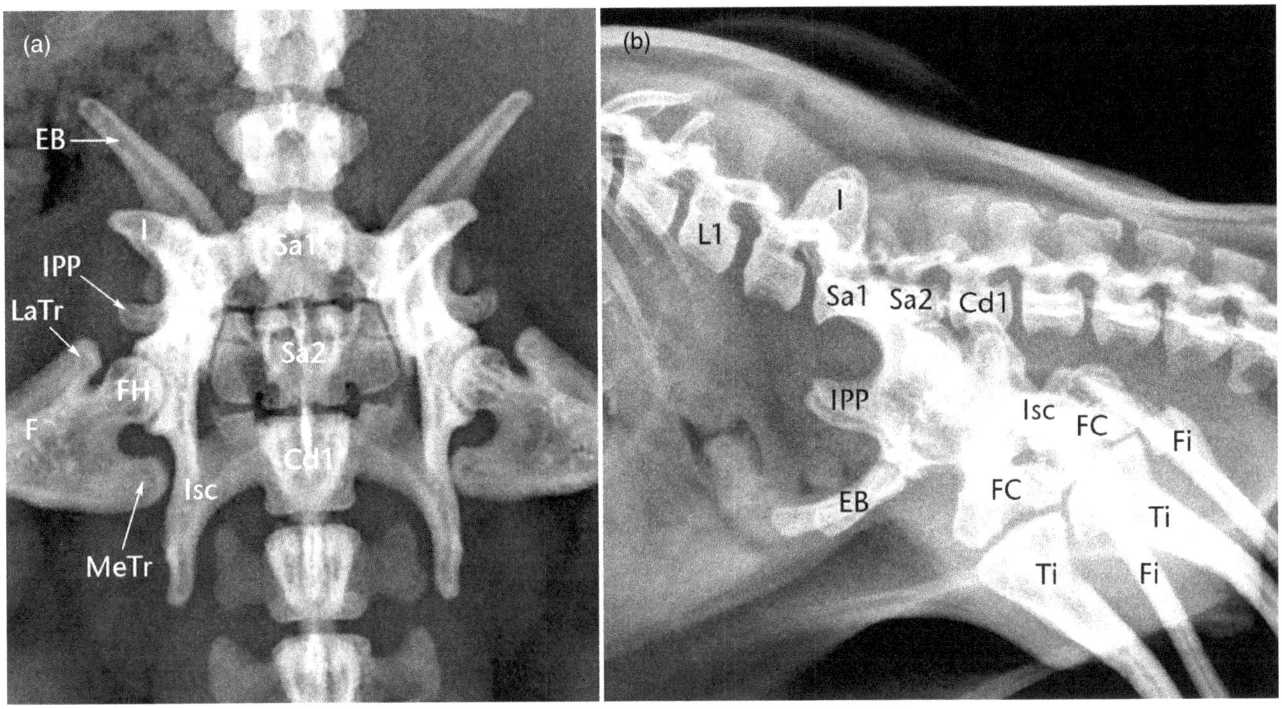

Figure 3.5: Pelvis: (a) ventrodorsal view, (b) lateral view. In the lateral view, as with the pectoral girdle, the pelvic girdle is confused by superimposition of bones that are awkwardly displayed because the limbs project laterally from the girdle. The ilium (I) is partly superimposed on the epipubic bones (EB) when seen in the ventrodorsal plane and to add to the confusion there are paired iliopectineal processes (IPP) that project craniolaterally from the pubis. The sacrum has two elements (Sa1, Sa2), which articulate with the ilia. The 1st caudal vertebra (Cd1) bears large spatulate transverse processes that are visible on the ventrodorsal image. On the lateral image the ischia (Isc) are superimposed on the femoral condyles (FC). Superimposition also complicates visualisation of the proximal ends of the tibia (Ti) and fibula (Fi) in the lateral view. The proximal femur (F) has a head (FH) and bears two trochanters: a lateral trochanter (LaTr) and a medial trochanter (MeTr). The 1st lumbar vertebra is labelled (L1).

Axial skeleton

Skull

The platypus skull has a large cranium to house the well-developed forebrain, which lacks sulci, gyri and a corpus callosum between hemispheres. A reptilian feature of the platypus skull is the presence of ectopterygoid bones. There are no tympanic bullae (Griffiths 1978). The skin of the bill is supported by elongated maxillae and premaxillae and the shorter dentaries of the mandibles (Fig. 3.9).

Adult platypuses do not have teeth. In early life, they possess 1 premolar and 2 molars in each maxilla and 2–3 molars in each mandible (Fig. 3.10). These small teeth are resorbed and replaced by keratinous grinding pads (Fig. 3.9) soon after the young platypus emerges from the nesting burrow (Green 1937).

Vertebrae

The vertebral formula of the platypus is 7 cervical, 17 thoracic, 2 lumbar, 2 sacral and 20–21 caudal vertebrae. The cervical vertebrae bear cervical ribs and each thoracic rib has ossified sternal and vertebral portions, with cartilage between them, as seen in reptiles (Griffiths 1978) (Fig. 3.11).

The thoracolumbar vertebral transverse processes are short and the ribs do not articulate with

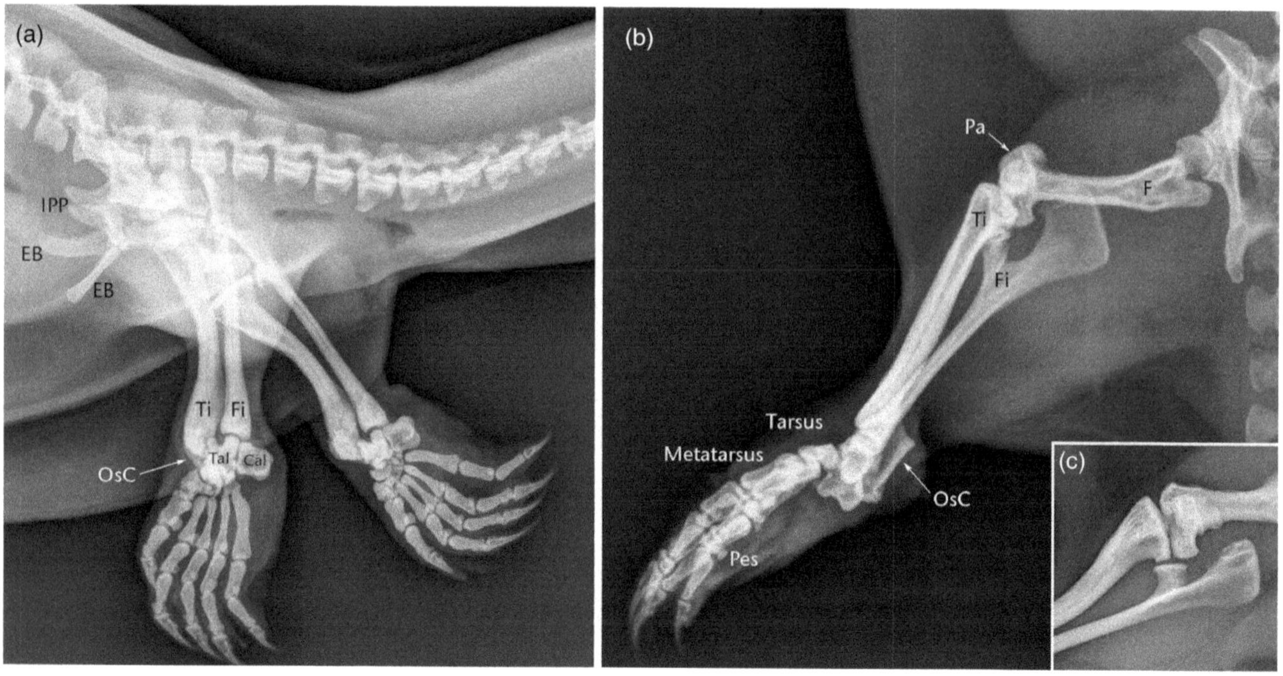

Figure 3.6: Pelvis and hindlimbs, adult male platypus (*Ornithorhynchus anatinus*): (a) lateral view, (b) mediolateral view, (c) stifle, lateral view. The epipubic bones (EB) must be distinguished from paired cranial extensions of the pubis, the iliopectineal processes (IPP). The tibia (Ti) and fibula (Fi) articulate with the talus (Tal), and the calcaneus (Cal) articulates mainly with the talus, having only a minor articulation with the fibula. The caudally located *os calcaris* (OsC) supports the spur. The femur (F) and patella (Pa) are labelled. Unusual features around the stifle are the shape of the proximal fibula, which is spatulate, and the caudally located articulation of the fibula with the femur and the tibia (c).

them. The sacrum consists of two fused sacral elements (Fig. 3.12). The caudal vertebrae have dorsal and ventral spinous processes and impressive transverse processes that diminish in size caudally (Parker and Haswell 1921) (Figs 3.12 and 3.13).

Ribs and sternum

The sternum consists of a presternum, three keeled sternebrae and a xiphisternum (Fig. 3.14). The sternal ribs are ossified and connected to the vertebral ribs by imperfectly ossified intermediate ribs (Parker and Haswell 1921) (Fig. 3.14). There are 17 ribbed vertebrae and 2 non-ribbed vertebrae in the thoracolumbar vertebral column.

Soft tissue

Head

The most prominent and remarkable soft tissue feature of the platypus head is the leathery bill and associated structures. A fold of leathery skin forms the large frontal shield at the base of the bill. The filtering lip serrations of the mandible are clearly visible on radiographs (Fig. 3.9). Insect and crustacean prey are collected into the cheek pouches, thoroughly ground using the keratinous grinding pads and sifted through the filtering lip serrations to remove much of the exoskeletons.

Thorax

Cardiovascular system

The platypus has a four-chambered heart with a closed double circulation. Radiographically, the cardiac silhouette is located between ribs 1 and 5 and appears to occupy most of the cranial thorax when viewed on a lateral radiograph (Fig. 3.14). The systemic aorta is derived from the left fourth arterial arch. Great cardiac veins open directly into the right atrium, as does a coronary vein that is not present in other mammals. A pelvic *rete mirabile* is present (Booth and Connolly 2008).

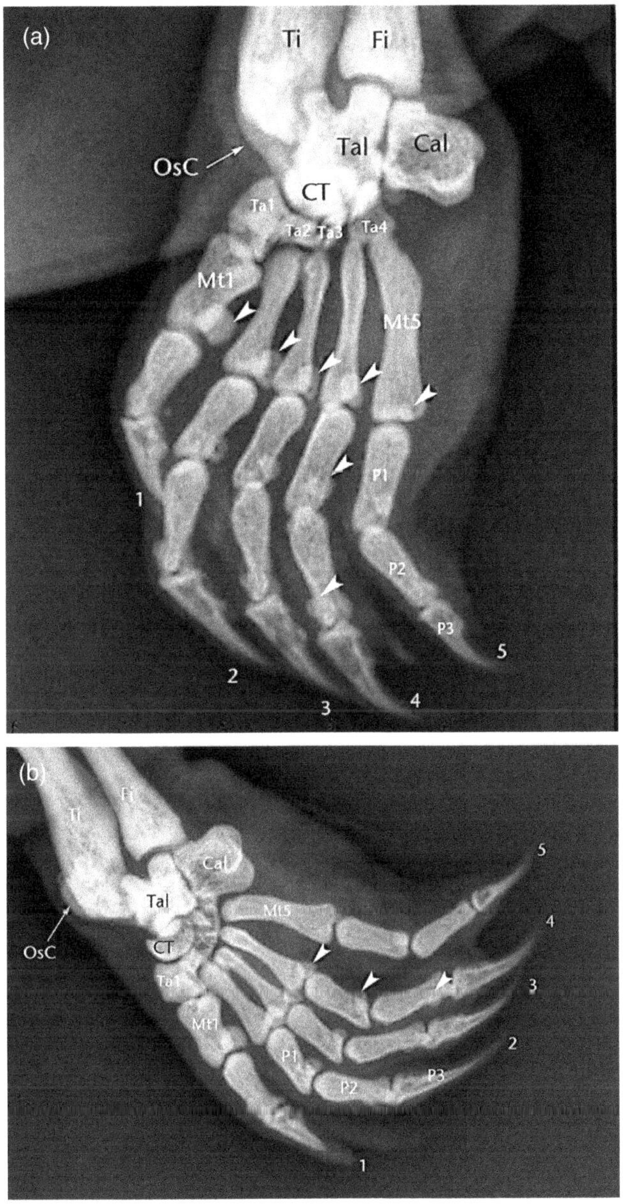

Figure 3.7: Pes, adult male platypus: (*Ornithorhynchus anatinus*): (a) dorsoplantar view, (b) plantarodorsal view. There are five digits (1, 2, 3, 4, 5), four of which have three phalanges (P1, P2, P3), while digit 1 has only two. There are five metatarsal bones (Mt1, Mt5 labelled), which are supported by four tarsal bones (Ta1, Ta2, Ta3, Ta4). The 4th tarsal bone articulates with the 4th and 5th metatarsal bones, while the other three tarsal bones in the distal row support their respective metatarsal bone. Sesamoid bones (arrowheads) are visible on the plantar surface of the metatarsophalangeal and interphalangeal articulations. The talus (Tal) is the main articulation with the tibia (Ti) and the fibula (Fi). The calcaneus (Cal) appears to have a minor articulation with the fibula and a major articulation with the talus. The central tarsal bone (CT) and caudally located *os calcaris* (OsC) are labelled.

Respiratory system

Platypus lungs are alveolar and have the typical mammalian tree-like branching system of bronchi, with one lobe on the left and two lobes on the right. On ventrodorsal images, the aerated pulmonary parenchyma can be seen between the thoracic inlet and the diaphragm, but the superimposed pectoral girdle obscures cranial thoracic pulmonary detail to some extent (Fig. 3.14).

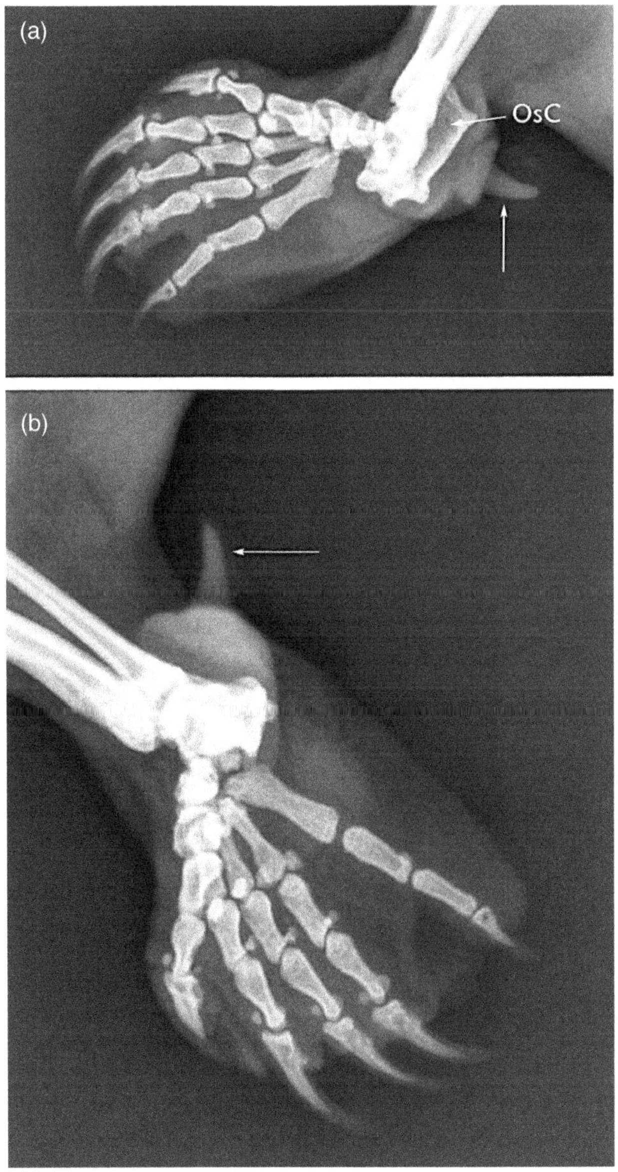

Figure 3.8: Tarsus and pes, adult male platypus: (*Ornithorhynchus anatinus*): (a) lateral view, (b) oblique lateral view. The tarsal spur (arrow) and its association with the *os calcaris* (OsC) are demonstrated.

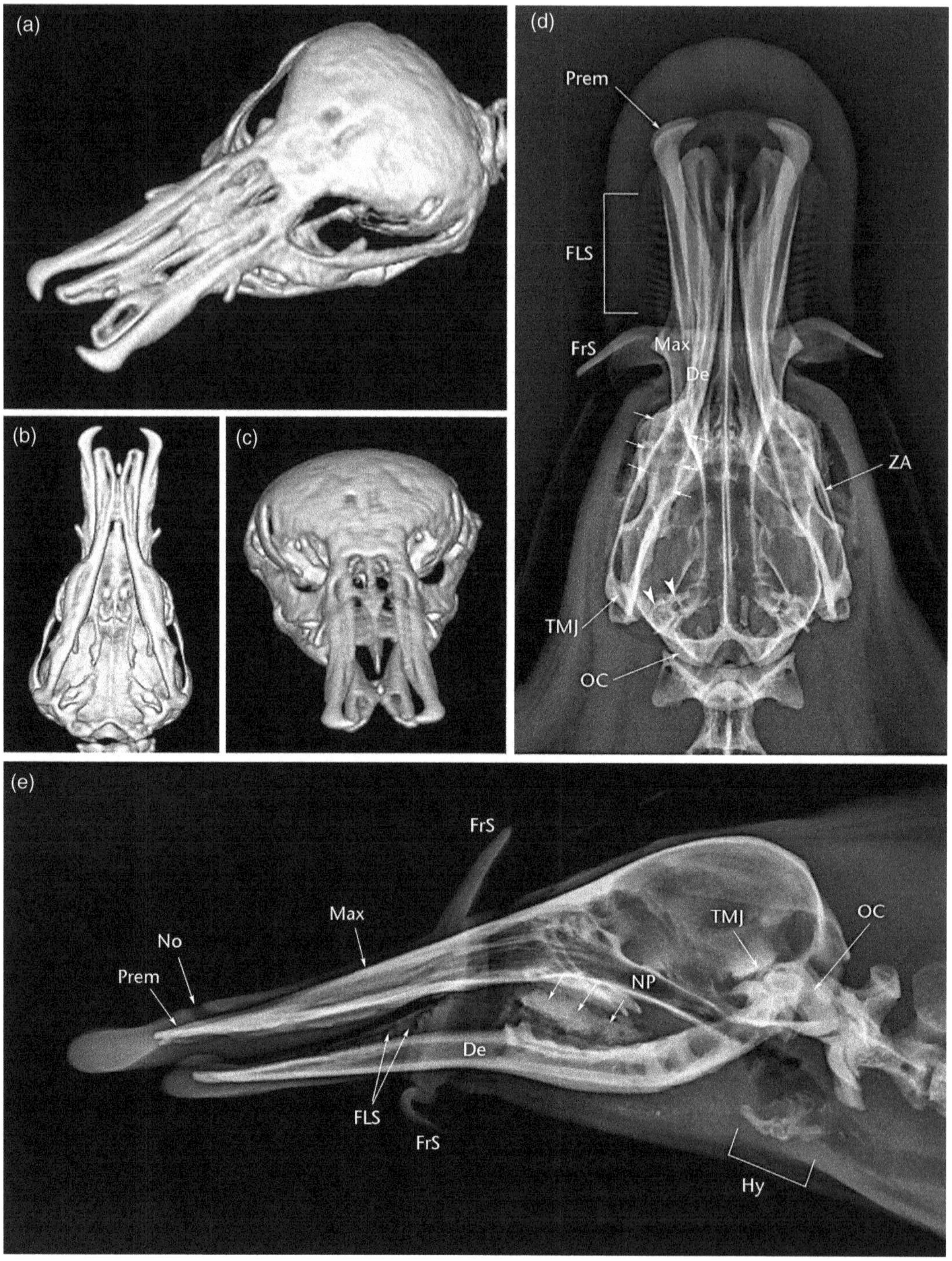

Figure 3.9: Skull: (a–c) volume-rendered three-dimensional CT reconstruction, (d) ventrodorsal radiograph, (e) lateral radiograph. The premaxilla (Prem) and maxilla (Max) are more easily recognised on the lateral view than on the ventrodorsal view because of superimposition of the dentaries (De) on the latter. The external orifice of a nostril (No) is visible dorsal to the premaxilla. The large frontal shield (FrS) is a fold of leathery skin at the base of the bill. In adults, dentition is represented by keratinous grinding pads (arrows) on both the maxilla and mandible. The filtering lip serrations (FLS) can be recognised on both projections. The zygomatic arch (ZA), temporomandibular joints (TMJ) and occipital condyles (OC) are recognisable, but the otic structures (arrowheads) less clearly so. Air in the nasopharynx (NP) and above the hyoid apparatus (Hy) can be seen on the lateral view.

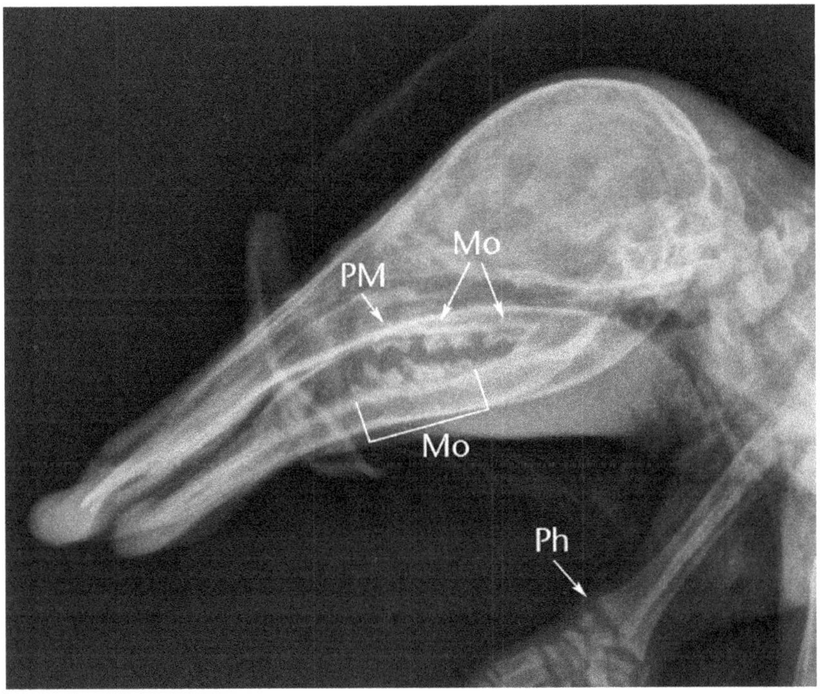

Figure 3.10: Skull, juvenile platypus (*Ornithorhynchus anatinus*): lateral view. One premolar (PM) and two molar (Mo) teeth are present in the maxilla and three molar (Mo) teeth are present in the mandible. The distal radial physis (Ph) is open, consistent with the immaturity of this platypus.

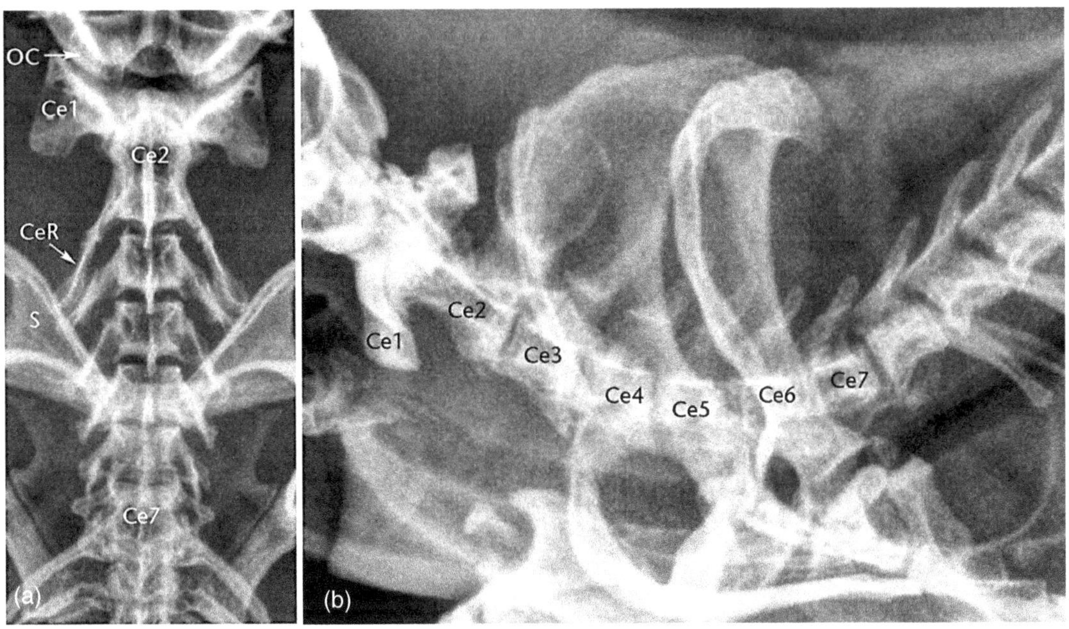

Figure 3.11: Cervical vertebral column: (a) ventrodorsal view, (b) lateral view. The occipital condyle (OC) articulates with the 1st cervical vertebra (Ce1), which has well-developed lateral wings. The bodies of the 2nd and subsequent cervical vertebrae (Ce2–Ce7) have two caudally projecting dumb-bell-shaped cervical ribs (CeR), which become progressively smaller on succeeding vertebrae. The cranially directed scapula (S) and the complex pectoral girdle are superimposed on the mid and caudal cervical vertebral column.

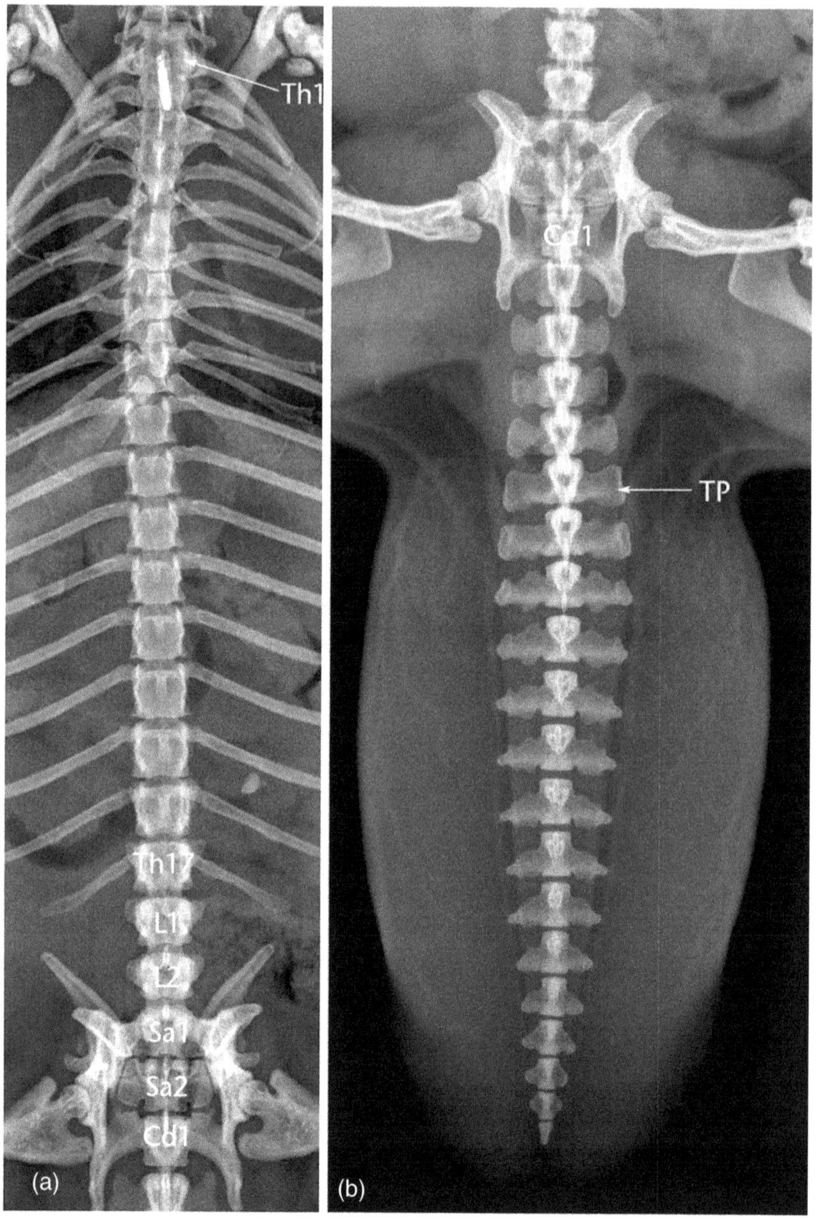

Figure 3.12: Vertebral column: (a, b) ventrodorsal views. There are 17 ribbed vertebrae which, by convention, are classed as thoracic vertebrae (Th1, Th17 labelled) and only 2 lumbar vertebrae (L1, L2). The sacral elements (Sa1, Sa2) are fused and the 1st caudal vertebra (Cd1) leads to a broad tail, with each caudal element bearing a pair of impressive transverse processes (TP). Note: a transponder can be seen in (a).

Abdomen

The platypus gastrointestinal tract is simple (Fig. 3.15). The stomach is lined with cornified stratified epithelium to further grind insect prey. It is very small and receives ground and filtered material from the cheek pouches. A small (2.5 cm) caecum is located at the ileocolic junction (Whittington 1988; Connolly *et al.* 1999). Because of the readily assimilated ingesta, the gut transit time is short. A barium study of a mature, healthy captive platypus revealed a transit time of approximately 5 hours (Booth and Connolly 2008).

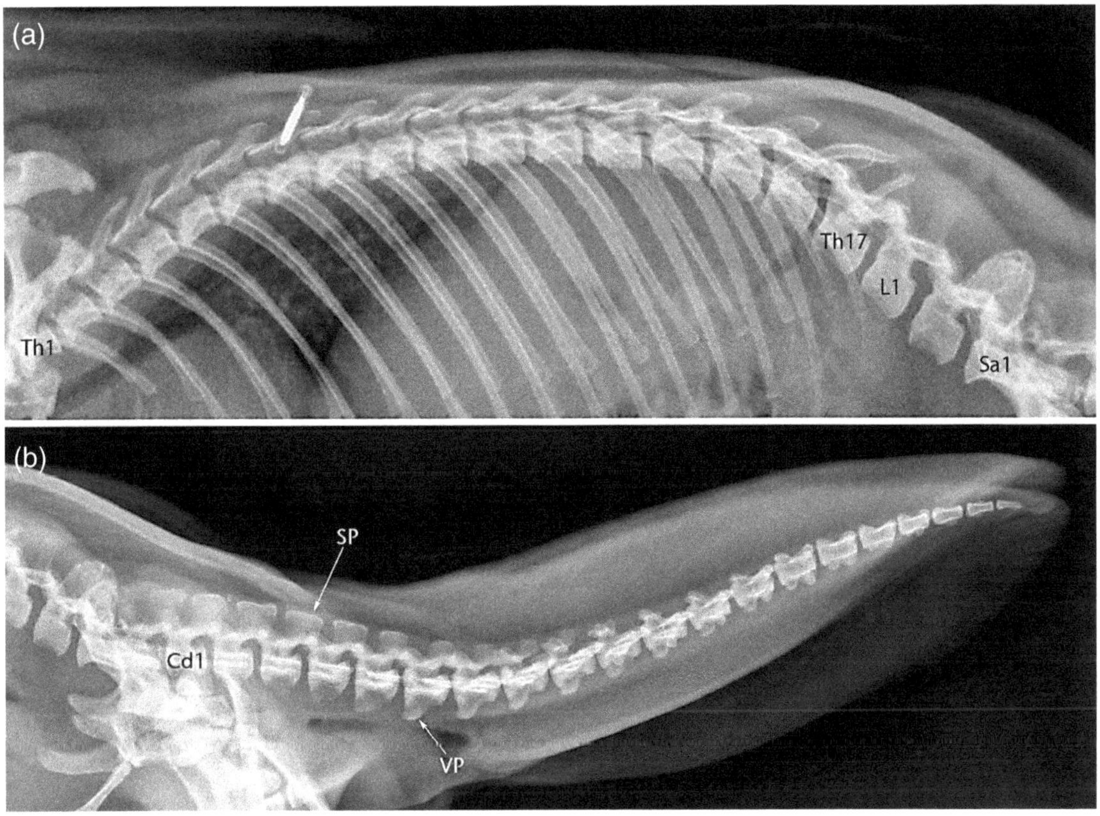

Figure 3.13: Vertebral column: (a, b) lateral views. Identified on these images are the first and last thoracic vertebrae (Th1, Th17), the first lumbar vertebra (L1) and the first sacral element (Sa1). The caudal vertebrae (Cd1 labelled) possess a dorsal spinous process (SP) and a ventral process (VP), which diminish in size caudally. Note: a transponder can be seen in (a).

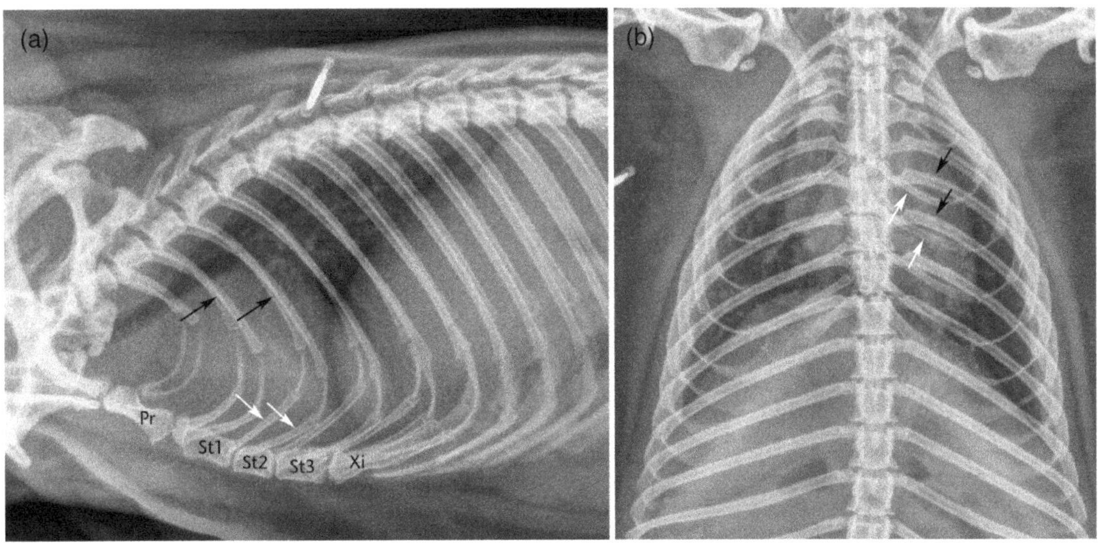

Figure 3.14: Thorax: (a) lateral view, (b) ventrodorsal view. The five sternal elements, comprising a presternum (Pr), sternal elements 1–3 (St1, St2, St3) and a xiphisternum (Xi) are identified. A comparison of sternal ribs (white arrows) and vertebral ribs (black arrows) is demonstrated in these images. Note: transponders can be seen.

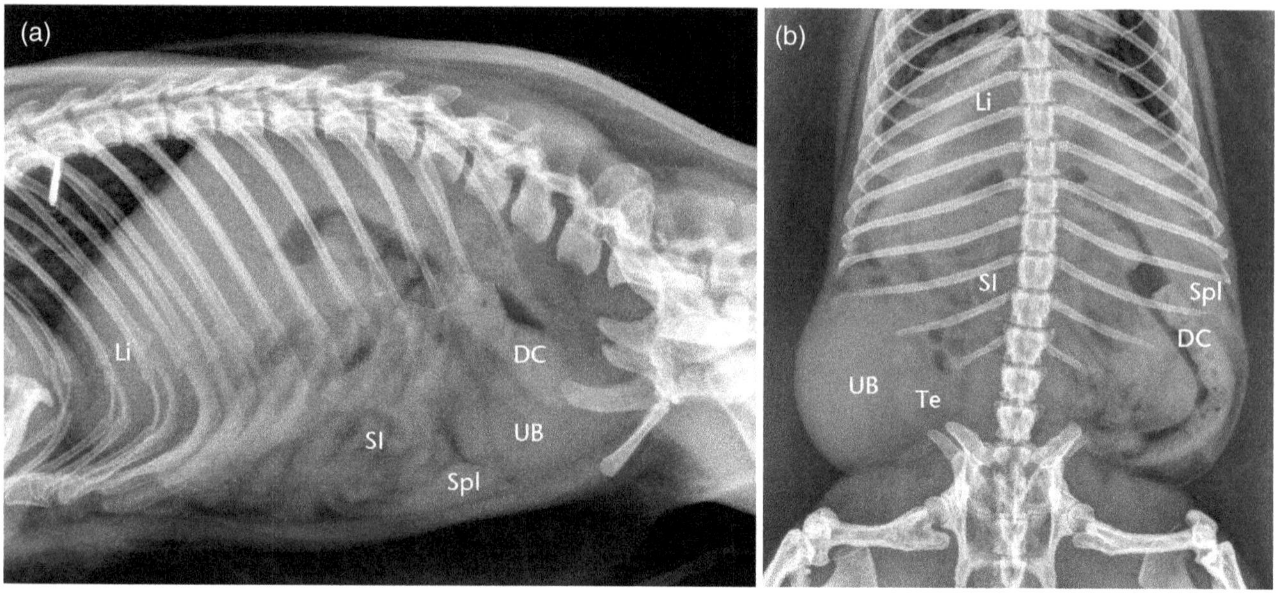

Figure 3.15: Abdomen: (a) lateral view, (b) ventrodorsal view. Definition of abdominal viscera will depend on the mixture of gas, ingesta and adipose tissue in the abdomen at the time of radiography. In these images, the following viscera can be identified: urinary bladder (UB), descending colon (DC), spleen (Spl), small intestines (SI), liver (Li), teste (Te). Note: a transponder can be seen in (a).

The platypus has well-developed, non-lobulated, bean-shaped ureotelic kidneys. Ureters drain into a common urogenital sinus opposite the neck of the bladder, which, in turn, drains into a cloaca, along with the rectum. A bladder is present as a diverticulum from the ventral surface of the urogenital sinus, but it does not receive the ureters, unlike in other mammals.

The female platypus has paired ovaries, but only the left is functional (the right ovary is rudimentary, as in birds). The ovary bears many large follicles projecting from its surface, as in reptiles. Each ovary is enclosed by the infundibulum of the poorly convoluted oviduct. Each oviduct leads to a separate uterus, which opens independently into the urogenital sinus and then into the cloaca. The platypus is oviparous, laying 1–3 leathery eggs, 14 × 17 mm in size and weighing 1.5–2 g (Griffiths 1978; Grant 1995; Booth and Connolly 2008).

The male platypus is testicond, with the abdominal testes lying caudodorsal to the kidneys (Fig. 3.15). They can be readily visualised on ultrasound examination. Seminiferous tubules in the testis drain via efferent ducts to the large epididymis and then via a short ductus deferens to the rostral end of the urogenital sinus. The penis lies in the caudal urogenital sinus, within a preputial sac ventral to the cloaca. The erect penis is approximately 7 cm long and is extruded through the cloaca and the cloacal sphincter. The shaft of the penis bears spines and each half of the bifid glans has four evertible foliate papillae. The penile urethra, which communicates with the urogenital sinus, carries only semen (not urine). There are no ampullae, coagulating glands or seminal vesicles. In addition, the prostate is disseminate, rather than discrete, but bulbourethral (Cowpers) glands are present at the base of the penile urethra (Carrick and Hughes 1978; Griffiths 1978; Booth and Connolly 2008).

References

Booth R, Connolly JH (2008) Platypus. In *Medicine of Australian Mammals.* (Eds L Vogelnest and R Woods) pp. 103–132. CSIRO Publishing, Melbourne.

Carrick FN, Hughes RL (1978) Reproduction in male monotremes. *Australian Zoologist* **20**, 211–231.

Connolly JH, Obendorf DL (1998) Distribution, captures and physical characteristics of the platypus (*Ornithorhynchus anatinus*) in Tasmania. *Australian Mammalogy* **20**, 229–235.

Connolly JH, Canfield PJ, McClure SJ, Whittington RJ (1999) Histological and immunohistological investigation of lymphoid tissue in the platypus (*Ornithorhynchus anatinus*). *Journal of Anatomy* **195**, 161–171. doi:10.1046/j.1469-7580.1999.19520161.x

Grant TR (1989) Ornithorhynchidae. In *Fauna of Australia, Volume 1B, Mammalia.* (Eds DW Walton and BJ Richardson) pp. 436–450. Australian Government Publishing Service, Canberra.

Grant TR (1995) *The Platypus: A Unique Mammal.* University of New South Wales Press Ltd, Sydney.

Grant TR, Fanning D (2007) *Platypus.* Australian Natural History Series. CSIRO Publishing, Melbourne.

Green HLH (1937) The development and morphology of the teeth of *Ornithorhynchus. Philosophical Transactions of the Royal Society of London. Series B, Biological Sciences* **228**, 367–420. doi:10.1098/rstb.1937.0015

Griffiths M (1978) *The Biology of the Monotremes.* Academic Press, New York.

Griffiths M (1989) Tachyglossidae. In *Fauna of Australia, Volume 1B, Mammalia.* (Eds DW Walton and BJ Richardson) pp. 407–435. Australian Government Publishing Service, Canberra.

Howes GB (1893) On the mammalian pelvis, with especial reference to the young of *Ornithorhynchus anatinus. Journal of Anatomy and Physiology* **27**, 542–556.

Parker JT, Haswell WA (1921) Skeleton of prototheria. In *A Text-book of Zoology, Volume II.* pp. 501–506. Macmillan and Co. Ltd, London.

Szalay FS (1994) *Evolutionary History of the Marsupials and an Analysis of Osteological Characters.* Cambridge University Press, Cambridge.

Thorp BH, Dixon JM (1991) Cartilaginous bone extremities of monotremes appear unique. *The Anatomical Record* **229**, 447–452. doi:10.1002/ar.1092290403

Weisbecker V (2011) Monotreme ossification sequences and the riddle of mammalian skeletal development. *Evolution* **65**, 1323–1335. doi:10.1111/j.1558-5646.2011.01234.x

Whittington RJ (1988) The monotremes in health and disease. In *Australian Wildlife.* Proceedings 104, pp. 727–787. Post Graduate Committee in Veterinary Science, Sydney.

4

Macropods (potoroids and macropodids)

There are 71 extant species within the super-family Macropodoidea: 21 species are endemic to New Guinea, 46 are endemic to Australia and 4 species are common to both regions (Richardson 2012). There are three extant families. The musky rat-kangaroo (*Hypsiprymnodon moschatus*) is the only living member of the family Hypsiprymnodontidae. All other species are placed in the families Potoroidae and Macropodidae. Members of the super-family Macropodoidea are often referred to as 'macropods', a term meaning 'big foot' and generally recognised as encompassing kangaroos, wallabies, potoroos, bettongs and the musky rat-kangaroo. Although not strictly correct, it is in common usage. Strictly speaking, the derivative 'macropodoids' should be used for the super-family, 'potoroids' for the family Potoroidae and 'macropodids' for the family Macropodidae (Vogelnest and Portas 2008). In this chapter and others in this book, members of the super-family Macropodoidea will be referred to as 'macropods'.

Macropods are a diverse group ranging in size from 0.5 to 90 kg bodyweight and are found throughout Australia. They occupy almost all terrestrial habitats, including tropical rainforests, woodlands, open grasslands, deserts, mountains, islands and rocky cliffs (Strahan 1995; Menkhorst 2001). Some species are arboreal, some live in caves and two species are fossorial (Tyndale-Biscoe 2005). Most are nocturnal, but the medium and larger species tend to be crepuscular. They are essentially sedentary, occupy a persistent home range and include solitary, non-social species through to gregarious species that live in well-organised societies. The degree of gregariousness generally increases with body size, openness of habitat and proportion of grasses in the diet.

The teeth (diprotodont) and feet (syndactylous) of macropods place them in the order Diprotodontia. With the exception of the musky rat-kangaroo, they are distinguished from other marsupials by foregut fermentation, hopping and embryonic diapause (although the latter does not occur in the western grey kangaroo [*Macropus fuliginosus*] or Lumholtz's tree-kangaroo [*Dendrolagus lumholtzi*]).

The musky rat-kangaroo is considered to be the closest to the ancestral stock from which all macropods evolved. Although it superficially resembles the other rat-kangaroos, it is only distantly related. Furthermore, it bounds using all its legs rather than hopping. It is the only macropod with five toes on the pes, with digit 1 being opposable. It is omnivorous and its teeth and stomach, which is simple, resemble those of possums more than those of other macropods. It does not exhibit embryonic diapause (Strahan 1995; Tyndale-Biscoe 2005).

The family Potoroidae includes a rat-kangaroo, bettongs and potoroos, which range in size from 1 to 3 kg. Rat-kangaroos hop and, in contrast to kangaroos and wallabies, can use their tails to pick up objects such as grass. An upper canine tooth is present, the premolars are very large with

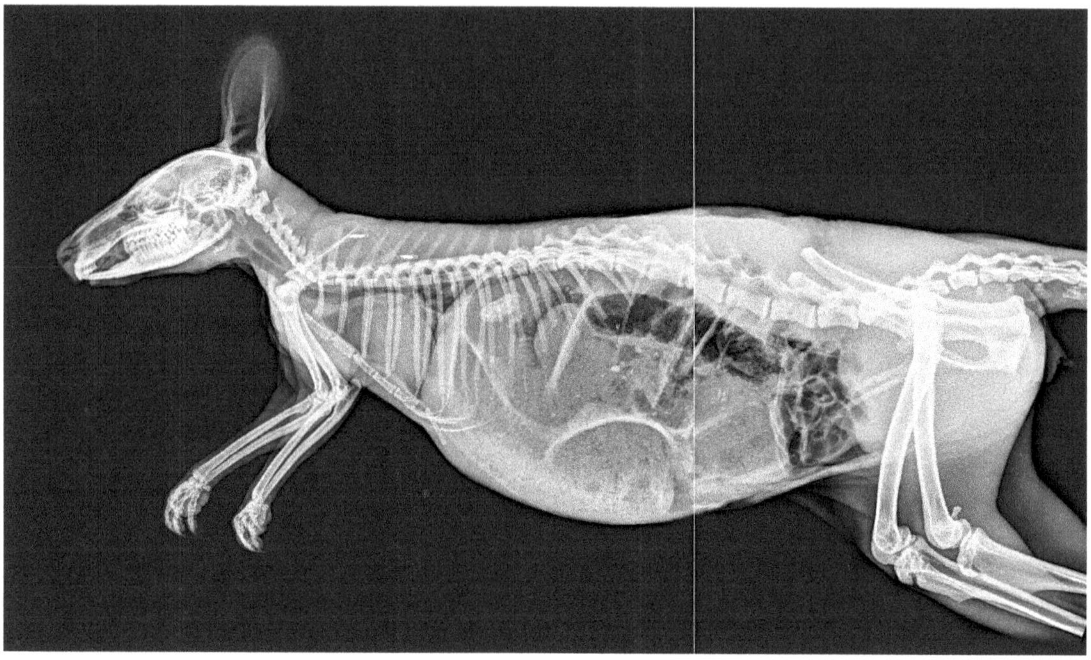

Figure 4.1: Whole body, composite lateral radiograph of a tammar wallaby (*Macropus eugenii*). The small upper body and limbs are disproportionate when compared with the abdomen and hindquarters. Note: two transponders can be seen.

prominent flutings on the sides and the four molars behind them decrease in size from rostral to caudal. The molars remain in the same position in the jaw throughout life. This is unlike wallabies and kangaroos, which exhibit molar progression. The potoroids have a sacculated forestomach and all species display postpartum oestrus and embryonic diapause (Tyndale-Biscoe 2005).

The banded hare-wallaby (*Lagostrophus fasciatus*) is the only species in the subfamily Sthenurinae. It has some of the features of the potoroids and the macropodoids (Tyndale-Biscoe 2005). One diagnostic feature is that the lower incisors bite against the upper incisors, not behind them, as in other macropodids.

The largest family is the Macropodidae, which consists of 45 species and includes the wallabies and kangaroos. Wallabies and kangaroos are distinguished by size. The six largest species are called kangaroos and the smaller species wallabies (Dawson 1995). The kangaroos are not only united as a group by their size, they are also all grazers and, unlike many of the smaller macropod species, none is endangered (Tyndale-Biscoe 2005).

There is considerable anatomical variation between the various macropod species, generally correlating with adaptations to their unique ecological niches. For example, tree-kangaroos (*Dendrolagus* spp.) have developed unique musculoskeletal adaptations to cope with a discontinuous, uneven and three-dimensional arboreal substrate (Warburton *et al.* 2011). Adaptation of the pes is characterised by relatively short and stout metatarsals and significant modification of the tarsal articular complexes, resulting in enhanced mobility and flexibility.

In this chapter, anatomical descriptions and radiographic images focus primarily on readily accessible species. When radiographically available, relevant anatomical features of other species are included. The Goodfellow's tree-kangaroo (*Dendrolagus goodfellowi*), endemic to New Guinea,

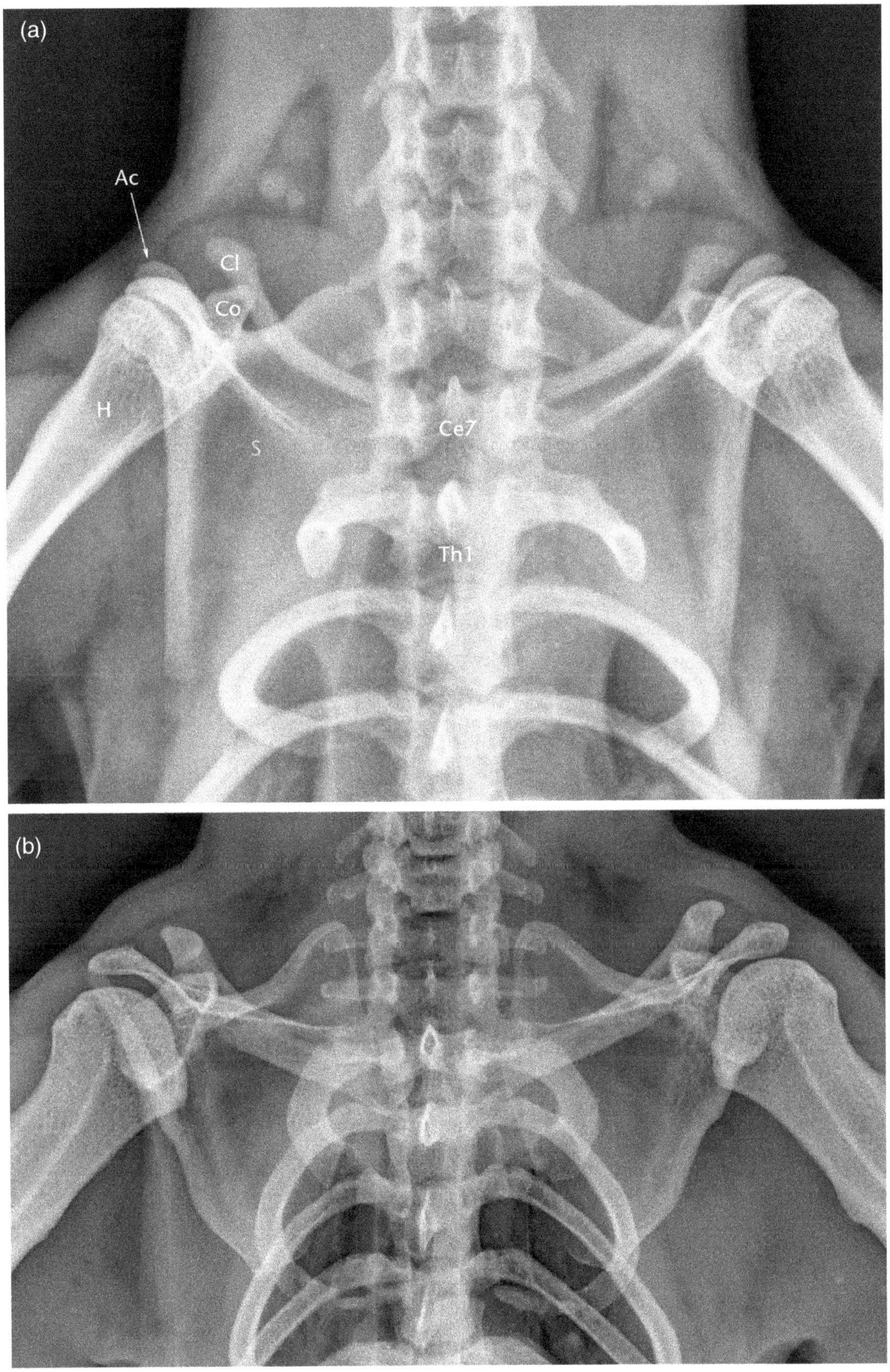

Figure 4.2: Pectoral girdle, ventrodorsal view: (a) red-necked wallaby (*Macropus rufogriseus*), (b) Goodfellow's tree-kangaroo (*Dendrolagus goodfellowi*). The clavicle (Cl) articulates with the acromion (Ac). The coracoid (Co) is directed craniomedially, arising from the edge of the glenoid cavity of the scapula (S). The humerus is visible (H). The relative location of the seventh cervical (Ce7) and the first thoracic (Th1) vertebrae is indicated. The more robust pectoral girdle structure of the tree-kangaroo can be seen in (b).

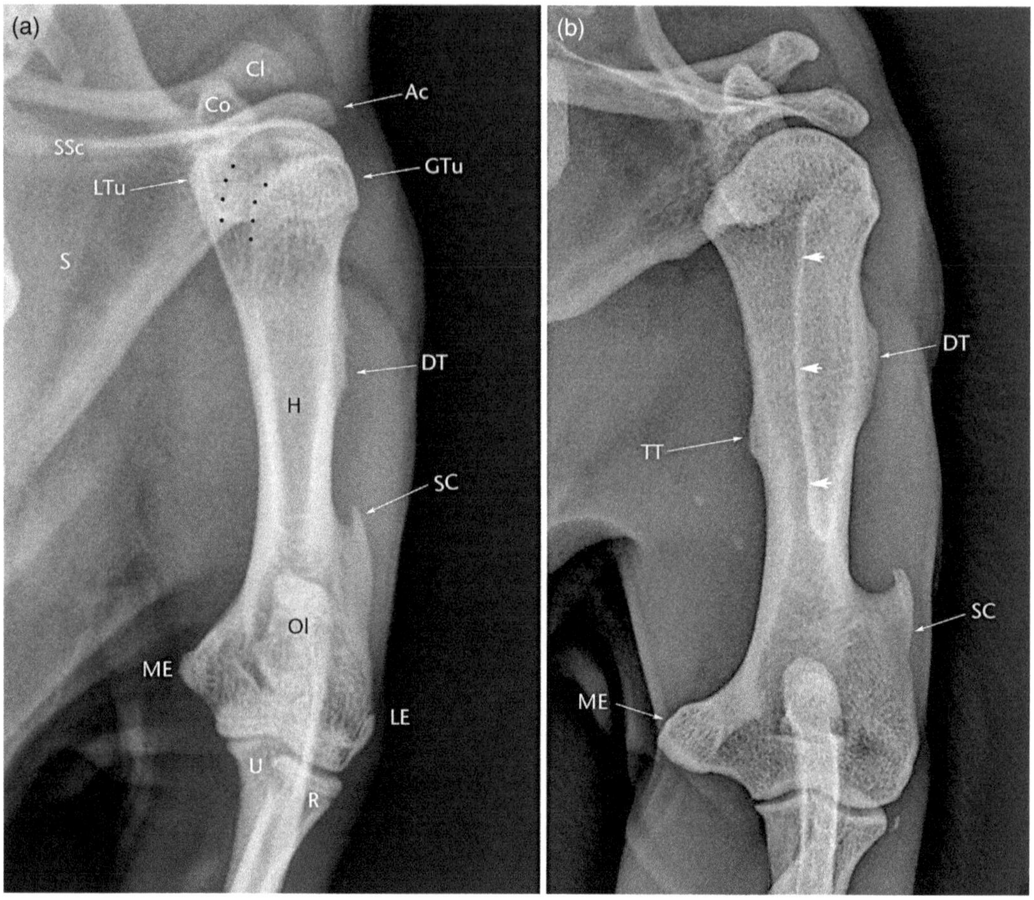

Figure 4.3: Brachium, craniocaudal view: (a) red-necked wallaby (*Macropus rufogriseus*), (b) Goodfellow's tree-kangaroo (*Dendrolagus goodfellowi*). The acromion (Ac) is located on the distal end of the spinous process (SSc) of the scapula (S). The clavicle (Cl) extends from the acromion to the sternum. The coracoid (Co) extends craniomedially from the edge of the glenoid cavity. There is a greater (GTu) and a lesser (LTu) tubercle at the proximal end of the humerus (H). The deltoid tuberosity (DT) and supinator crest (SC) are features of the diaphysis of the humerus. The humeral condyle is flanked by a medial (ME) and a lateral (LE) epicondyle. The medial epicondyle is exaggerated in the tree-kangaroo when compared with the red-necked wallaby, but appears similar to that of the red kangaroo (see Fig. 4.5). The humeral condyle articulates with the radius (R) and ulna (U). The olecranon (Ol) of the ulna is elongated. The intertubercular groove is marked by the dotted lines. The deltoid tuberosity (DT) and the teres tuberosity (TT) are prominent features of the mid-diaphysis of the tree-kangaroo, as is the pectoral ridge (arrowheads) that extends distally from the greater tubercle.

has been used to illustrate the unique anatomical features of tree-kangaroos because of the availability of an individual for radiography.

Skeletal system

Appendicular skeleton

In macropods the appendicular skeleton is highly modified for bipedal locomotion. The forelimbs are simple and small, while the hindlimbs are large and muscular (Fig. 4.1). This is particularly evident in the larger species (Richardson 2012). As growth persists well into adult life in most macropods, the epiphyses of the limb bones may be unfused even in adults (Hume *et al.* 1989).

Pectoral girdle and forelimb

Other than the tree-kangaroos, the forelimb of terrestrial macropods is disproportionately small compared with other marsupials, primarily

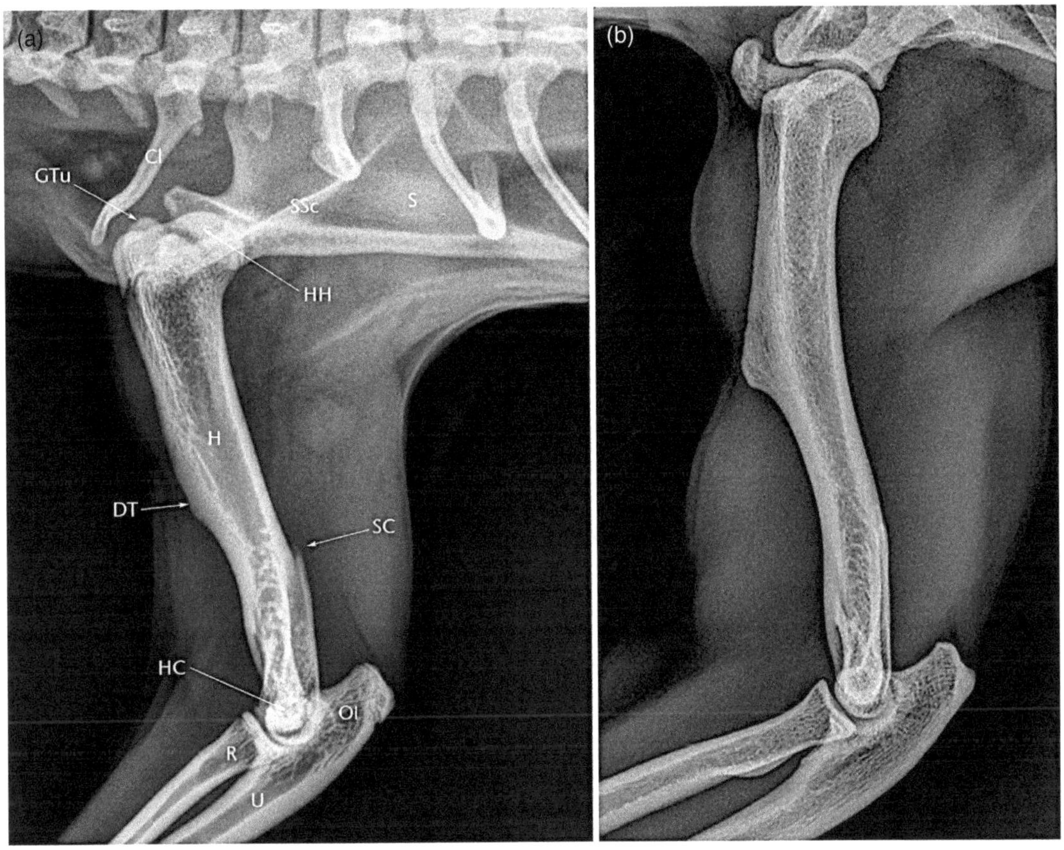

Figure 4.4: Brachium, mediolateral view: (a) red-necked wallaby (*Macropus rufogriseus*), (b) Goodfellow's tree-kangaroo (*Dendrolagus goodfellowi*). The acromion (superimposed on the proximal humerus in this view) is located on the distal end of the spinous process (SSc) of the scapula (S). The clavicle (Cl) extends from the acromioclavicular junction to the sternum. There is a greater (GTu) and a lesser tubercle (not visible in this view) adjacent to the humeral head (HH). The deltoid tuberosity (DT) and supinator crest (SC) are features of the diaphysis of the humerus (H). The humeral condyle (HC) articulates with the radius (R) and ulna (U). The olecranon (Ol) of the ulna extends proximally from the elbow joint.

because of their essentially bipedal locomotion (Hopwood 1981; Grand 1990) and the relative lack of manipulative functions (Iwaniuk *et al.* 2000; Weisbecker and Archer 2008). Tree-kangaroos have relatively more equal ratios of forelimb to hindlimb length and muscle weight than terrestrial macropods (Grand 1990). They have greater manual dexterity than terrestrial macropods, with a high degree of freedom of movement of the upper forelimb (Warburton *et al.* 2011).

The scapula is simple and roughly trapezoidal in terrestrial macropods, whereas in tree-kangaroos it is more triangular (Fig. 4.2). The vertebral border of the scapula is relatively wide compared with its length. Terrestrial macropods have a more circular glenoid cavity compared with tree-kangaroos, in which it is more oval in outline. The acromion is elongated and protuberant for articulation with the clavicle. The clavicles are relatively short, robust and strongly curved in large kangaroo species, but slighter and straighter in small species (Richardson 2012). The clavicle is robust in tree-kangaroos, with a distinct dorsoventral curvature, an expanded and bulbous sternal articulation and modestly expanded acromial extremity (Fig. 4.2). Medially, the clavicle is dorsoventrally compressed (Warburton *et al.* 2011).

The proximal humerus comprises the head and two low and broad tuberosities. The intertubercular groove is shallow and relatively wide. The shaft

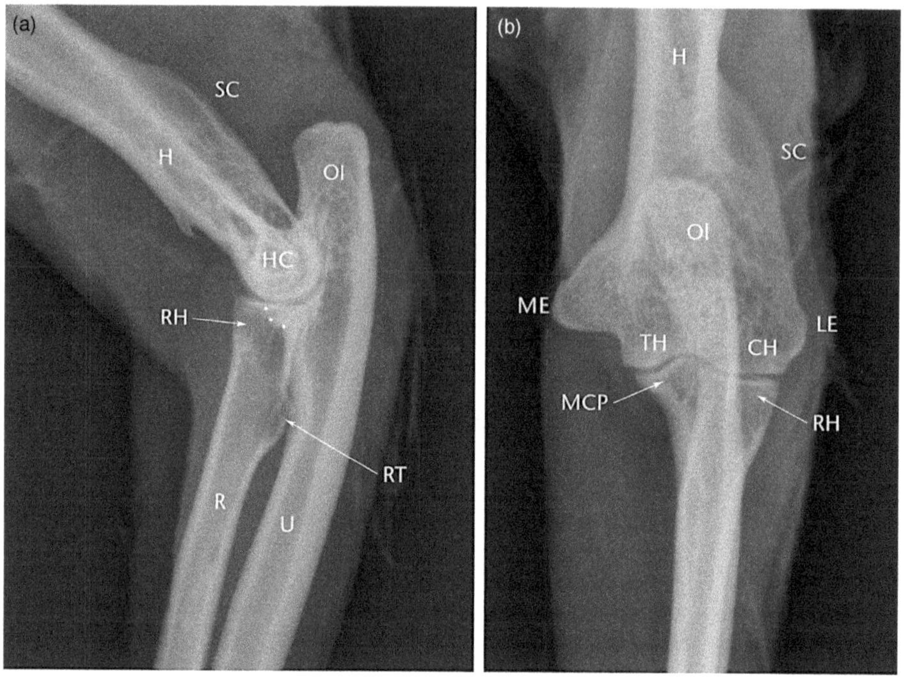

Figure 4.5: Elbow, red kangaroo (*Macropus rufus*): (a) mediolateral view, (b) craniocaudal view. The humeral condyle (HC) and the trochlear notch of the ulna form a stable hinge joint. The radial head (RH) articulates with the laterally located *capitulum humeri* (CH) while the ulna and its medial coronoid process (MCP) articulates with the *trochlea humeri* (TH) as well as the trochlear notch of the humerus. The humeral condyle is flanked by a medial (ME) and a lateral (LE) epicondyle. The ulna extends proximally as the olecranon (Ol). The medial coronoid process can be visualised on lateral images of the elbow (dotted line). The supinator crest (SC) of the humerus (H) is visible on the caudolateral aspect of the distal humerus. A radial tuberosity (RT) extends caudally from the proximal radius (R) in the direction of the ulna (U).

is robust with a well-developed pectoral ridge, extending from the greater tubercles along the medial edge of the humerus in tree-kangaroos. The distal humerus is broad mediolaterally and the medial epicondyle is large. The supinator ridge is long and extends in the form of a hooked process above the groove of the radial nerve. There is an epicondylar foramen penetrating the distal end of the humerus. The deltoid tuberosity of the humerus is strongly developed (Owen 1839; Hume *et al.* 1989; Warburton *et al.* 2011) (Figs 4.3, 4.4 and 4.5).

The radius and ulna are elongated and slender in terrestrial macropods, but more robust in tree-kangaroos (Richardson 2012) (Fig. 4.6). They are separate and of similar thickness and both articulate with the humerus.

There is no intermediate carpal bone (Parker and Haswell 1921). The manus has five digits that are not opposable. The first and fifth digits are approximately two-thirds of the dimensions of the third digit, while the second and fourth digits are intermediate in size (Richardson 2012) (Fig. 4.7). The metacarpal bones of tree-kangaroos differ from those of terrestrial kangaroos by their enlarged heads. The phalanges are more elongated than those of terrestrial kangaroos and the articular facets reflect a strongly flexed posture (Warburton *et al.* 2011). In tree-kangaroos, the distal phalanges have long, laterally compressed and sharply curved claws. In contrast, the claws of terrestrial kangaroos are shorter, broader and less curved. The nails of potoroids are particularly long and robust because they are used for digging.

Pelvis and hindlimb

Powerful hindlimbs with a long, narrow pes and powerful fourth digit are characteristic of most macropod species. Rock-wallabies and tree-

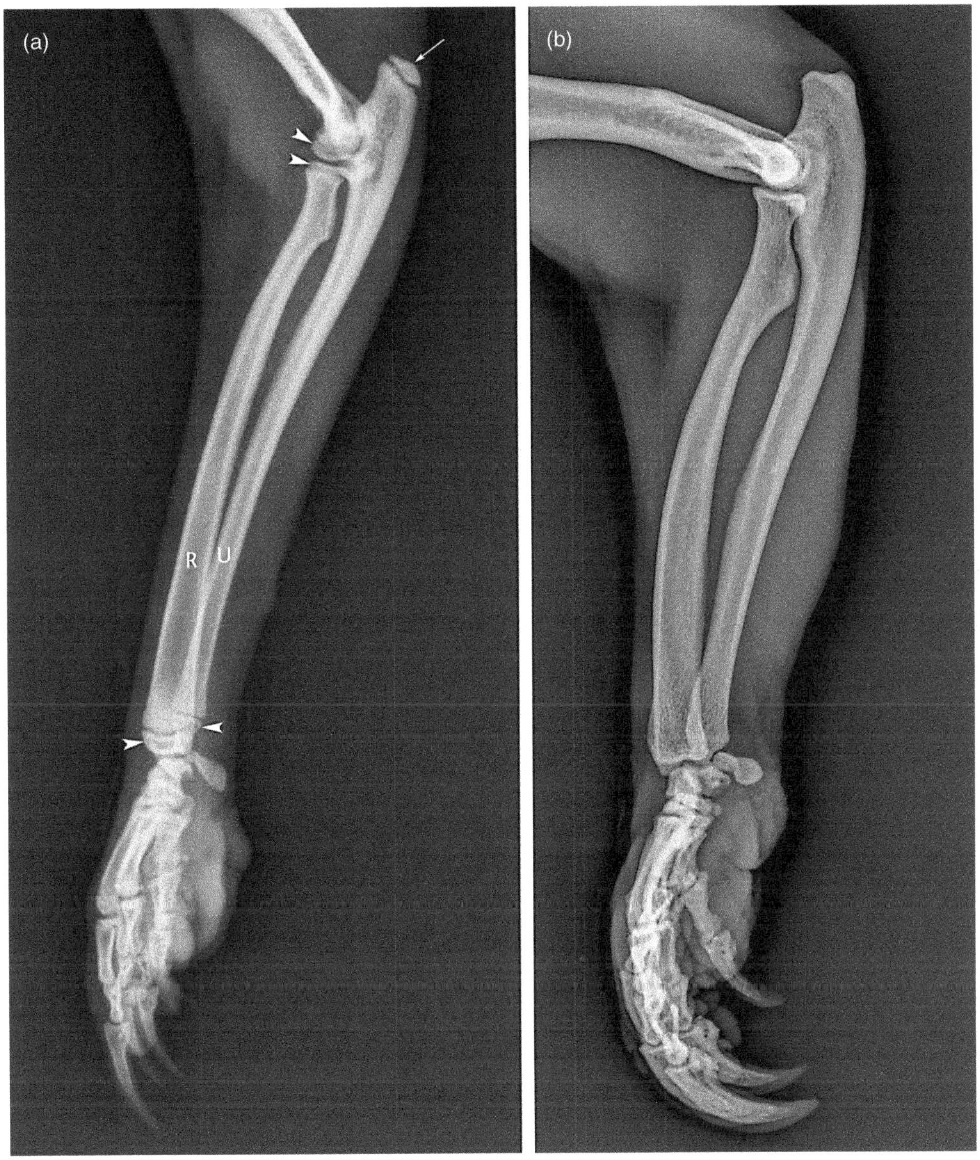

Figure 4.6: Antebrachium and manus, mediolateral view: (a) swamp wallaby (*Wallabia bicolor*), (b) Goodfellow's tree-kangaroo (*Dendrolagus goodfellowi*). As growth persists well into adult life in most macropods, the epiphyses of the limb bones may be unfused even in adults. Unfused epiphyses (arrowheads) are present at the ends of the radius (R), ulna (U) and humerus and an apophysis is present on the proximal end of the olecranon (arrow) in (a). In (b) note the curvilinear shape of the manus and the slightly bowed radius in the tree-kangaroo. These shape variations are regarded as adaptive changes to arboreal living.

kangaroos are an exception, having a pes that is shorter and broader. The hindlimbs and pelvic girdle are much longer, larger and more heavily muscled than the forelimbs and pectoral girdle. The epipubic bones, which extend from the cranial aspect of the pubis, are proportionately small and slender in macropods (Fig. 4.8).

The femur, tibia, fibula and bones of the pes are long and relatively slender in the large macropods (Hume *et al.* 1989). The tibia and fibula are in close contact and the fibula is reduced distally to a thin splint; in tree-kangaroos the tibia and fibula are relatively short and separated, allowing a high degree of rotation (see Fig. 4.11). As is the case in all

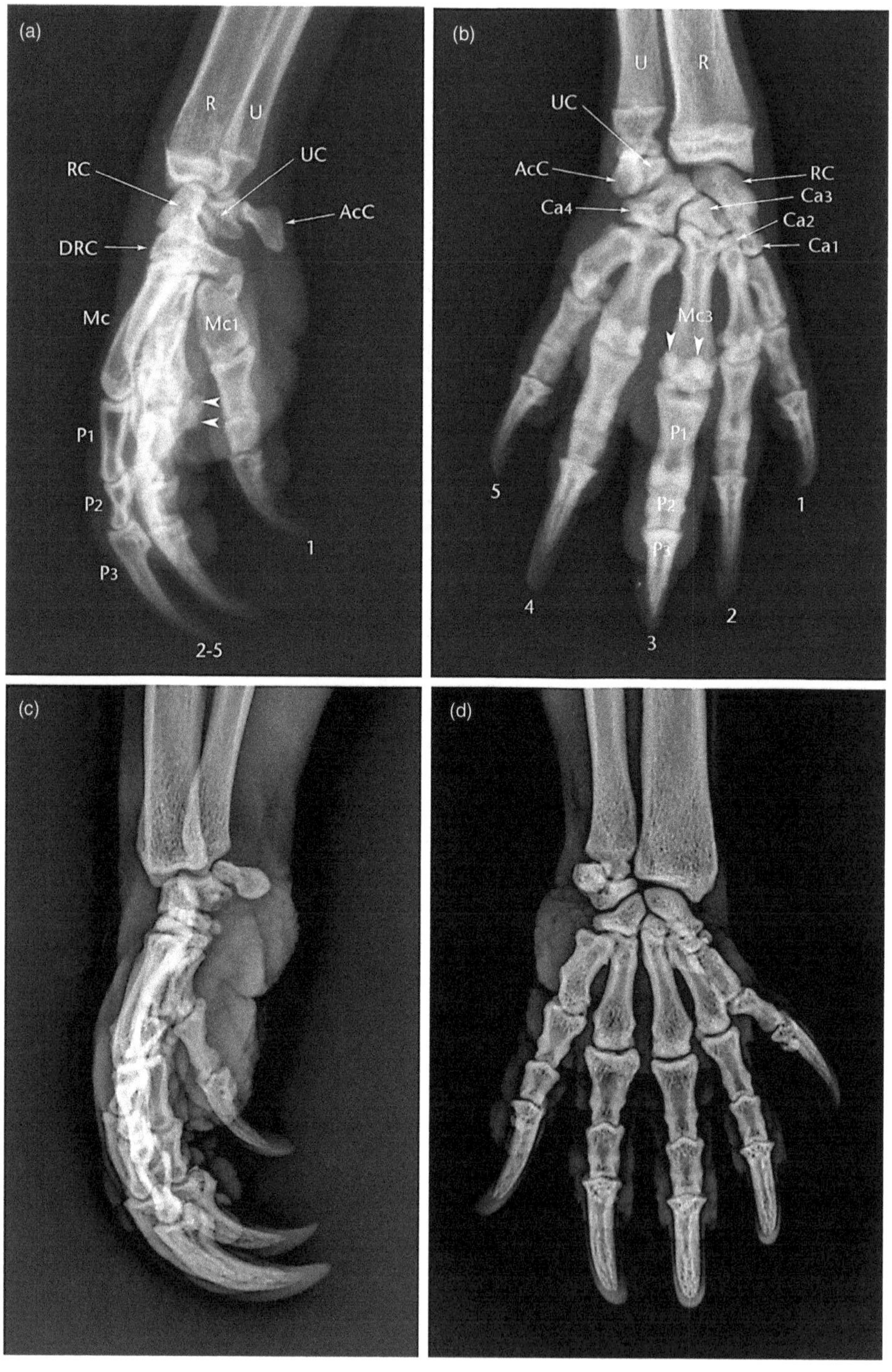

Figure 4.7: Manus: (a) swamp wallaby (*Wallabia bicolor*), mediolateral view, (b) red-necked wallaby (*Macropus rufogriseus*), dorsopalmar view, (c) Goodfellow's tree-kangaroo (*Dendrolagus goodfellowi*), mediolateral view, (d) Goodfellow's tree-kangaroo, dorsopalmar view. The antebrachiocarpal articulation is between the radius (R) and ulna (U) and the ulnar (UC), the 4th (Ca4) and the radial (RC) carpal bones. The carpometacarpal articulation is between the distal row of carpal bones (DRC), being the 1st (Ca1), 2nd (Ca2), 3rd (Ca3) and 4th (Ca4) carpal bones and the heads of the five metacarpal bones (Mc, Mc1, Mc3 labelled). Paired sesamoid bones are present on the palmar surface of the distal end of each metacarpal (arrowheads). Each of the four main digits (2, 3, 4, 5) consists of three phalanges (P1, P2, P3). Digit 1 (1) has two phalanges. An accessory carpal bone (AcC) is present on the lateral side of the caudal surface of the carpus.

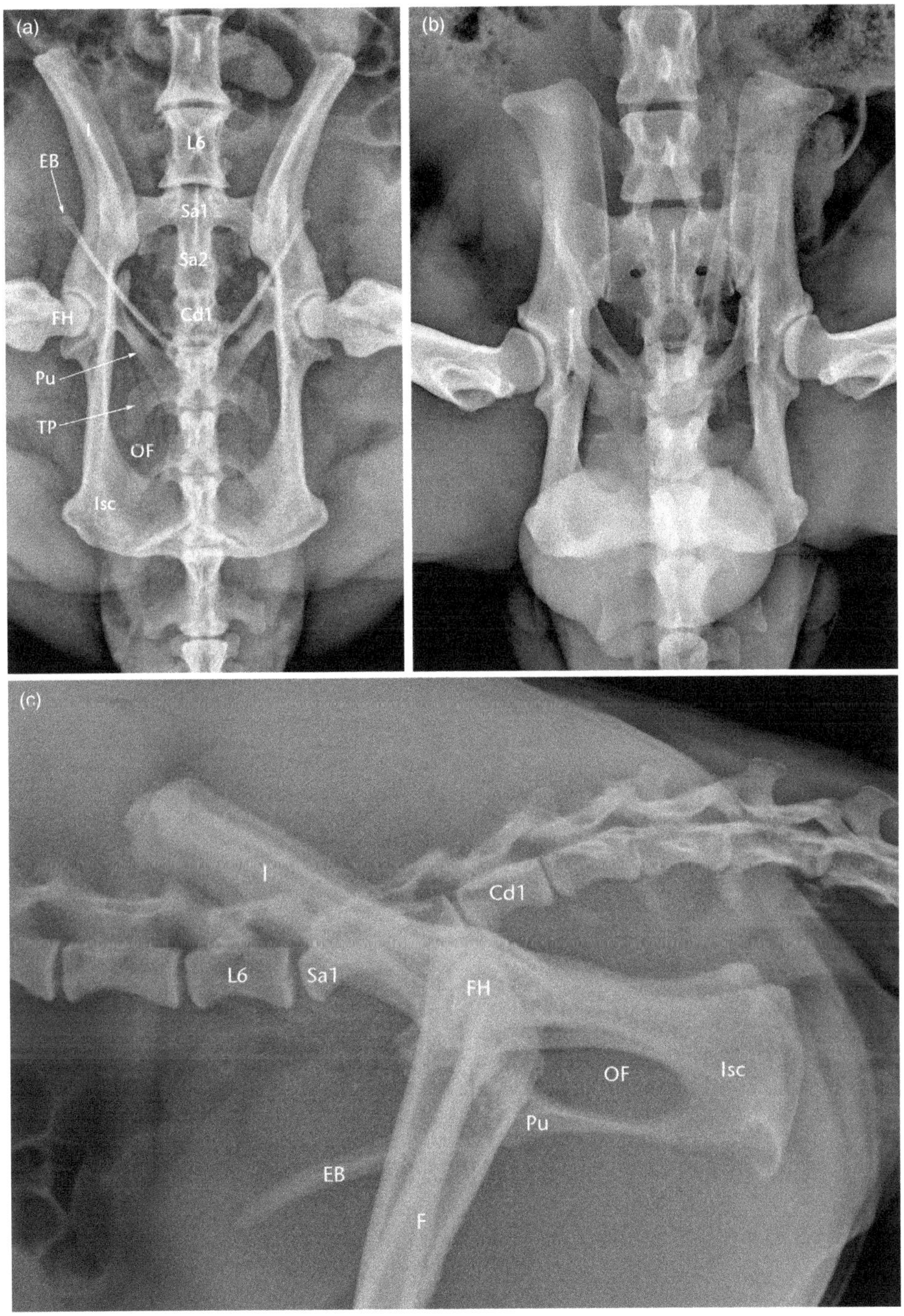

Figure 4.8: Pelvis: (a) tammar wallaby (*Macropus eugenii*), (b) Goodfellow's tree-kangaroo (*Dendrolagus goodfellowi*), ventrodorsal view, (c) red-necked wallaby (*Macropus rufogriseus*), lateral view. In most macropods the ilium (I) is elongated. In the tree-kangaroo the ilium is broad and stocky. The ilium articulates with the two sacral elements (Sa1, Sa2). The pubis (Pu) articulates with the paired epipubic bones (EB). There are large obturator foraminae (OF). The ischium (Isc) and ilium (I) are the main contributors to the acetabulum, which houses the femoral head (FH). The lumbosacral articulation is between the sixth (L6) lumbar and first sacral (Sa1) elements. The first caudal vertebra (Cd1) is the first of six that have well developed transverse processes (TP). The femur (F) is labelled. (b) Note that the right acetabulum (left side of the image) of the coxofemoral joint of the tree-kangaroo is shallower than the left and there are osteophytes on its cranial rim. These changes are consistent with degenerative joint disease.

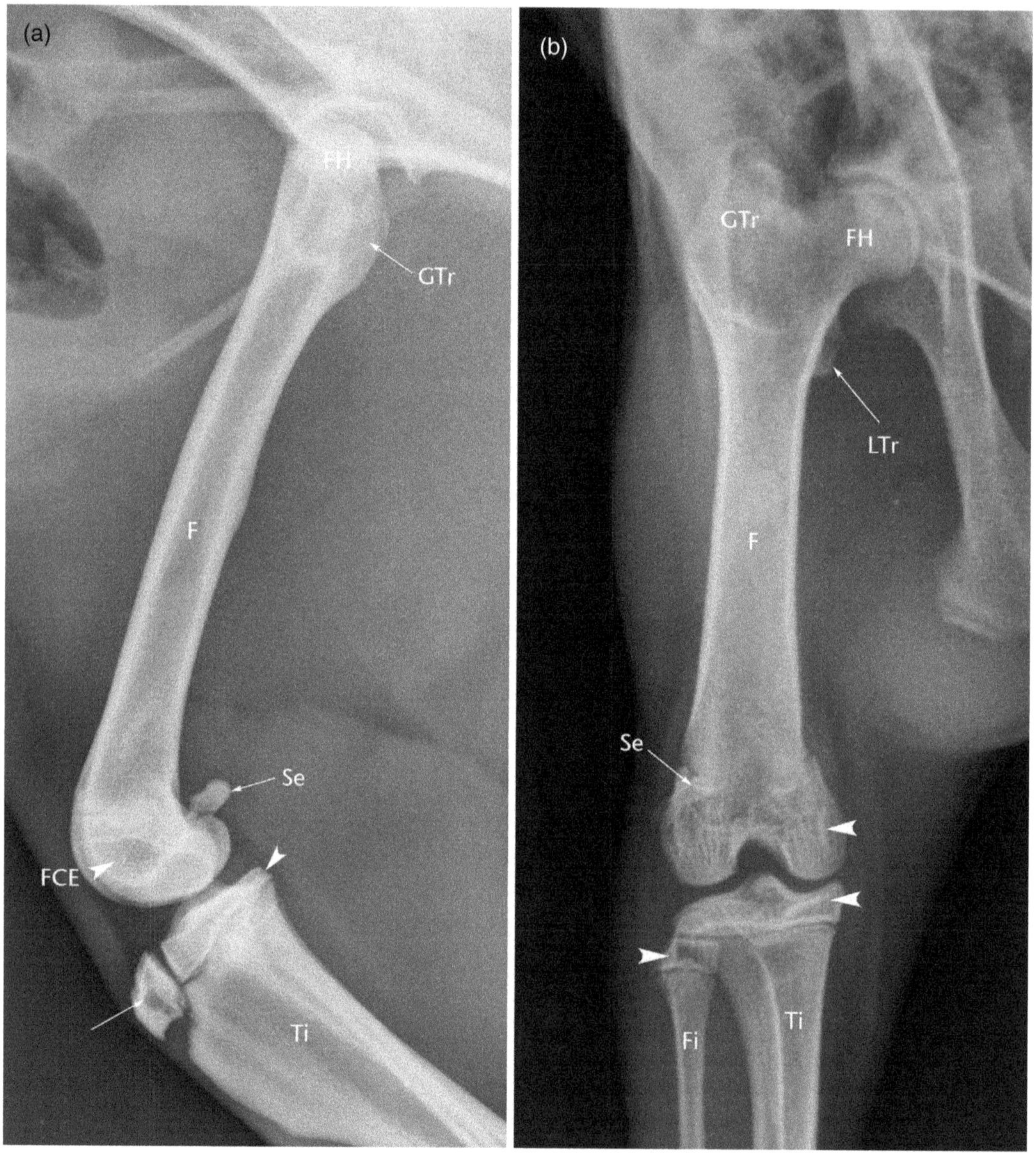

Figure 4.9: Femur and stifle: (a) swamp wallaby (*Wallabia bicolor*), mediolateral view. (b) quokka (*Setonix breviceps*), caudocranial view. The femoral head (FH) articulates with the acetabulum. There is a well developed greater trochanter (GTr) and a small lesser trochanter (LTr) at the proximal end of the femur (F). The distal end of the femur forms a condyle that incorporates the femoral condylar epiphysis (FCE). It consists of a lateral and a medial condyle. There is a solitary, caudally located sesamoid bone (Se) on the lateral aspect of the femoral condyle. There is no patella. The ends of the femur, tibia (Ti) and fibula (Fi) adjacent to the stifle form the epiphyses (arrowheads). The tibial crest apophysis (arrow) initially appears as a separate centre of ossification on the cranial surface of the proximal tibia but eventually fuses with the proximal tibial epiphysis.

marsupials other than bandicoots and the bilby, macropods do not have an ossified patella. Instead, there is a patelloid, which is a fibrocartilaginous induration of the quadriceps femoris extensor tendon (Reese *et al.* 2001; Richardson 2012) (Figs 4.9–4.11).

The load-bearing calcaneus is elongated and strongly developed. The talus makes contact with the central tarsal bone distally (Figs 4.12 and 4.13). The articulations of the talus and tibia and of the talus and calcaneus are such that movement is confined to the craniocaudal plane (the pes can move

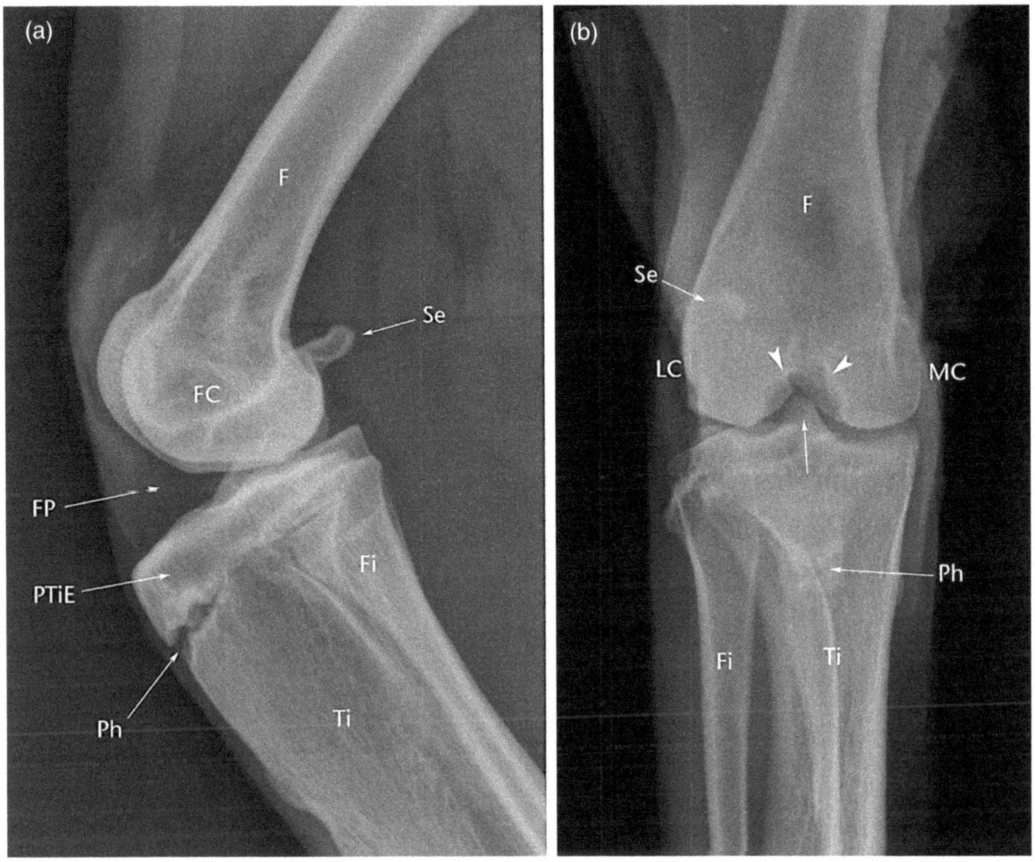

Figure 4.10: Stifle, red kangaroo (*Macropus rufus*): (a) mediolateral view, (b) caudocranial view. The rounded femoral condyles (FC) articulate with the flat proximal tibial articular surface. The physis (Ph) between the proximal tibial epiphysis (PTiE) and the body of the tibia (Ti) remains open well into adult life in macropods. A fat pad (FP) is visible between the femoral condyle and the proximal tibial epiphysis on the cranial aspect of the stifle. A well-developed intercondyloid eminence (arrow) is housed by the intercondyloid fossa (arrowheads) of the femur (F), which is located between its medial (MC) and lateral (LC) condyles. The fibula (Fi) is visible on the caudolateral aspect of the tibia. The single caudolateral sesamoid (Se) is visible in both views.

back and forth only; it cannot twist sideways). This is achieved partly by an expansion of the lateral process on the calcaneus to increase contact with the fourth tarsal bone (Hume *et al.* 1989). In tree-kangaroos, the tibial articulation with the talus is smooth, allowing for some lateral movement (Fig. 4.13c). The calcaneus has a smooth, broad articulation with the talus proximally and the fourth tarsal bone distally. These adaptations allow adduction of the pes (Richardson 2012). In macropods, the fourth and fifth metatarsal bones appear to articulate exclusively with the fourth tarsal bone (Fig. 4.13).

Other than in the musky rat-kangaroo, the first digit of the pes is absent and the second and third digits are syndactylous and bound together for most of their length by soft tissue. They are very small and take no part in support or propulsion, but are used for grooming. Digit 4 is long and robust in large macropods, while digit 5 is shorter and more slender (Figs 4.12 and 4.13). The bones of the pes of rock-wallabies and tree-kangaroos are shorter and more robust (Vogelnest and Portas 2008). In tree-kangaroos digits 4 and 5 are almost equally developed (Fig. 4.13c).

Axial skeleton

Skull

The skull of macropods is slender and elongated in most species. The skulls of potoroids are short, broad and shallow in most species, although in some species of *Potorous* the skull is elongated and narrow (Seebeck and Rose 1989). The rostral and ventral parts of the zygomatic arch extend ventrally to form an inverted conical process, the jugal process, the tip of which is at the level of the caudal maxillary molars (Owen 1839) (Fig. 4.14). The jugular process of the occipital bone is elongated. The tympanic bulla is augmented by a continuation of air-cells (pneumatisation) into the caudal component of the temporal bone and zygomatic arch (Owen 1839) (Figs 4.15–4.17). The mandibular condyle forms a broad and slightly convex surface, allowing free rotation of the mandible in every direction (Owen 1839).

Palatine vacuities (fenestrations of the palatal portion of the palatine bone) are present. The lateral surface of the ramus of the mandible is marked by the deep masseteric fossa into which the masseter muscle inserts. Rostrally, it becomes the masseteric canal, at the base of which the masseteric foramen communicates with the pterygoid fossa on the medial aspect of the mandible (Fig. 4.14). This is a distinctive feature of macropods and most prominent in the macropodoids (Hume *et al.* 1989; Tomo *et al.* 2007). It is not seen in other herbivorous marsupials (Sanson 1989). Caudomedially, the mandible may be concave for attachment of the medial pterygoid muscle. At the ventral margin is a characteristic, medially directed

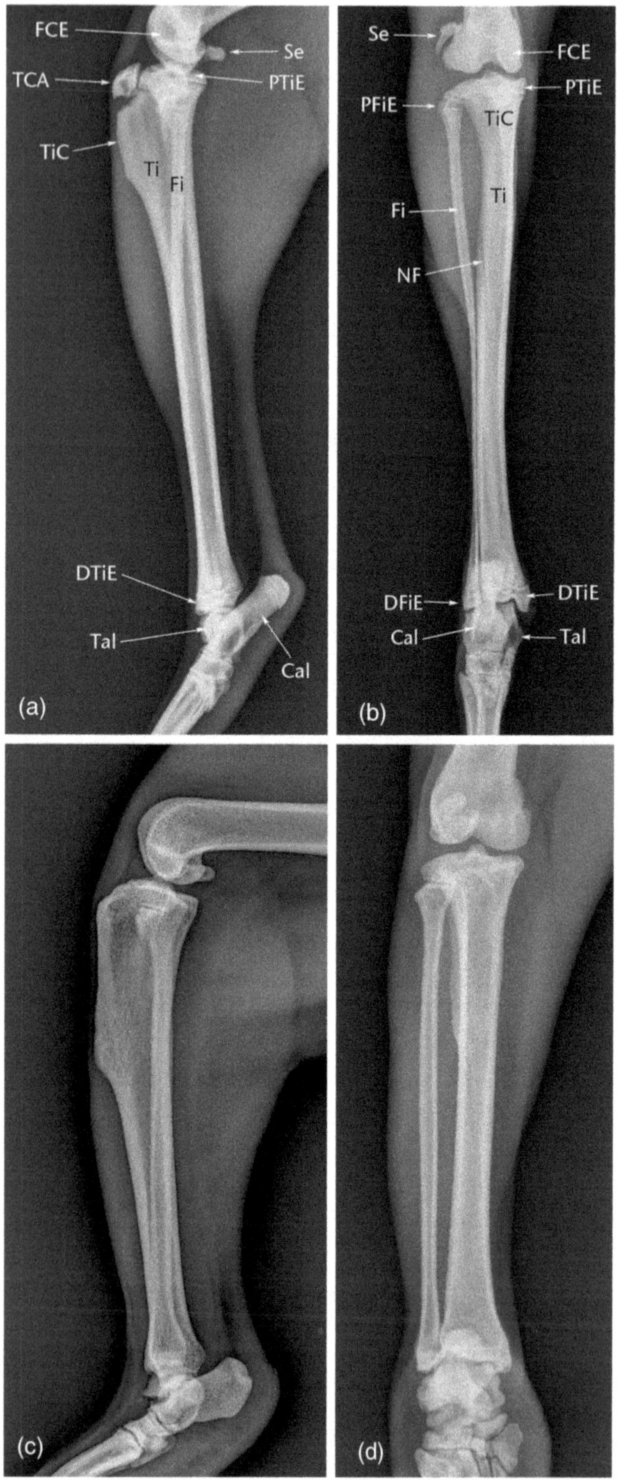

Figure 4.11: Crus: (a) sub-adult swamp wallaby (*Wallabia bicolor*), mediolateral view, (b) sub-adult swamp wallaby caudocranial view, (c) Goodfellow's tree-kangaroo (*Dendrolagus goodfellowi*), mediolateral view, (d) Goodfellow's tree-kangaroo, caudocranial view. Unfused epiphyses are visible at the ends of the long bones in (a) and (b). They are the femoral condylar epiphysis (FCE), the proximal tibial (PTiE) and the proximal fibular (PFiE) epiphyses. In addition, the tibial crest apophysis (TCA) is located at the proximal end of the tibial crest (TiC). A nutrient foramen (NF) is visible crossing the lateral cortex of the tibia (Ti) adjacent to the fibula (Fi). The tarsocrural articulation involves the distal tibial (DTiE) and the distal fibular (DFiE) epiphyses, which articulate with the calcaneus (Cal) and the talus (Tal). The distal femoral sesamoid (Se) is visible. (c, d) Features of the tree- kangaroo crus are illustrated. The fibula is thicker and the talus and calcaneus appears stout and more robust than in the wallaby.

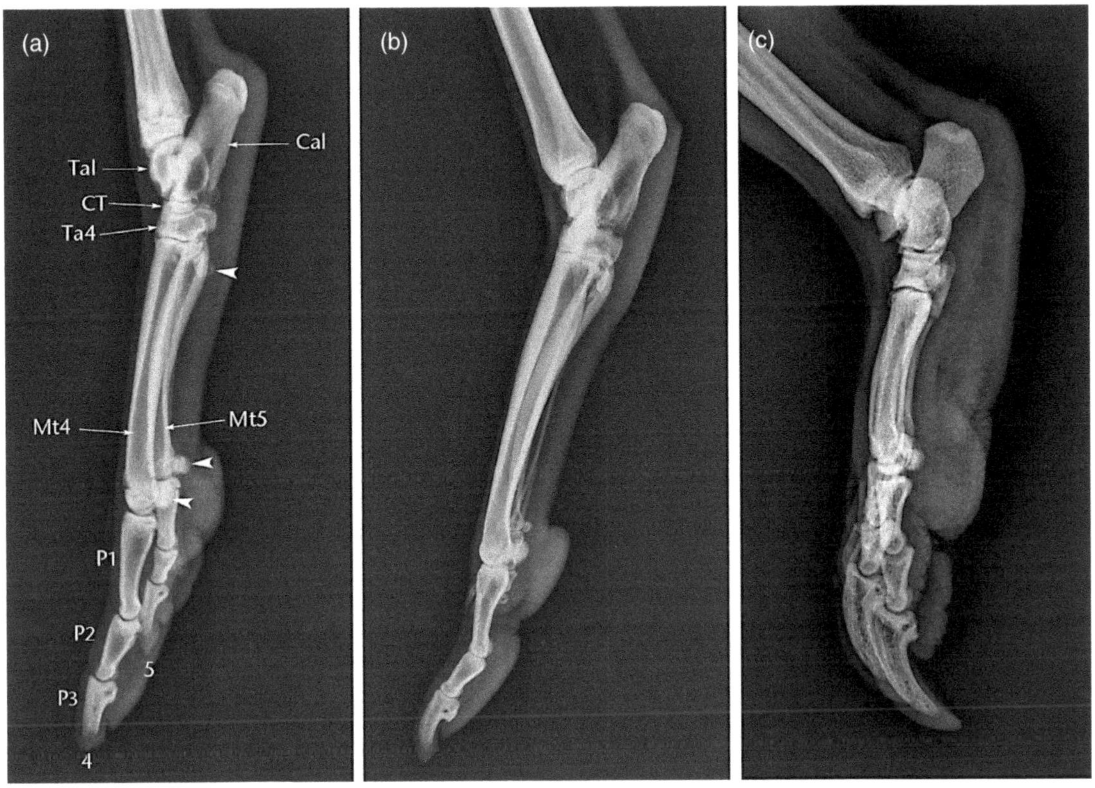

Figure 4.12: Pes, mediolateral view: (a) swamp wallaby (*Wallabia bicolor*), (b) red kangaroo (*Macropus rufus*), (c) Goodfellow's tree-kangaroo (*Dendrolagus goodfellowi*). The calcaneus (Cal), talus (Tal), central tarsal (CT), 4th tarsal bone (Ta4), 4th metatarsal (Mt4), 5th metatarsal (Mt5), sesamoid bone (arrowhead), phalanges 1–3 (P1, P2, P3), digits 4 and 5 (4, 5) are labelled in (a). Note that the calcaneus is thicker and a relatively more substantial bone in the tree-kangaroo than in the wallaby and kangaroo. The metatarsals and first two phalanges are also relatively short while P3 is relatively large and curved.

inflected angle forming a shelf, which is characteristic of most marsupials (Dawson *et al.* 1989) (Fig. 4.17e). In most macropods the mandibular symphysis is not fused and is flexible, which allows the lower incisors to spread apart and shear against the upper incisors. In the banded hare-wallaby and the potoroids the symphysis is more rigid and may be fused. Macropods have a wide range of dental adaptations reflecting their different diets. Dentition and dental radiology are covered in Chapter 11.

Vertebrae

The vertebral formula of macropods is 7 cervical, 13 thoracic, 6 lumbar, 2 sacral and 21–25 caudal vertebrae (Owen 1839; Chen *et al.* 2005; Richardson 2012) (Figs 4.18–4.20). The vertebrae are opisthocoelous (the vertebral body is convex cranially and concave caudally) (Richardson 2012). The ventral arch of the atlas (first cervical vertebra) is incompletely ossified. The atlanto-occipital articulation is very flexible, allowing the muzzle to remain horizontal whether the neck is horizontal or vertical. This flexibility may be associated with some weakness at this articulation and may account for the high incidence of traumatic cervical fractures. As in most mammals, the axis (second cervical vertebra) is formed by 6–7 separate ossification centres, with the cranial buttress and dens being formed by 2–3 separate ossification centres. In juvenile macropods, a physis between the cranial and caudal buttress is clearly visible on radiographs and may be

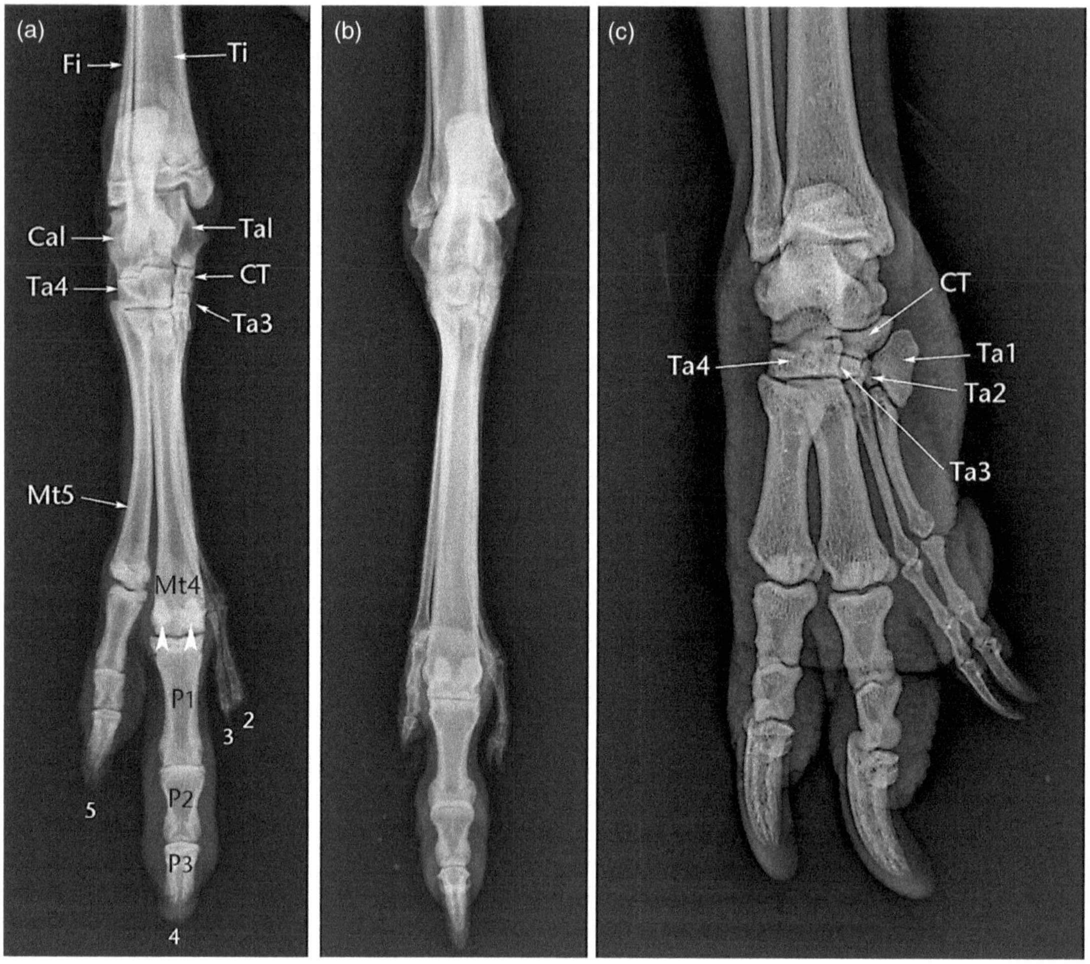

Figure 4.13: Pes, dorsoplantar view: (a) swamp wallaby (*Wallabia bicolor*), (b) red kangaroo (*Macropus rufus*), (c) Goodfellow's tree-kangaroo (*Dendrolagus goodfellowi*). The talus (Tal) and calcaneus (Cal) articulate with the crus proximally and the central (CT) and 4th tarsal bone (Ta4) distally. The 3rd (Ta3) and 4th tarsal bones articulate with the proximal ends of the 4th (Mt4) and 5th (Mt5) metatarsal bones, which are the important structural elements of the pes. The 3rd tarsal bone also articulates with the 2nd (2) and 3rd (3) digits, which are underdeveloped and syndactylous. In the tree-kangaroo, the 2nd tarsal bone (Ta2) is visible and the 1st tarsal bone (Ta1) is a large eccentrically located bone that moulds to the edge of the central tarsal proximally and the 2nd metatarsal distally. There are three phalanges (P1, P2, P3) in each digit and paired plantar metatarsophalangeal sesamoids (arrowheads) are visible on the main weight-bearing digits (4 and 5). The tibia (Ti) and Fibula (Fi) are labelled. In the tree-kangaroo, the metatarsals and phalanges of digits 4 and 5 are shorter and stockier than in the wallaby and kangaroo and both digits are well developed.

confused with a fracture. This does, however, appear to be a common site for fractures, which can be distinguished from the normal physis by the presence of displacement in either the lateral or ventrodorsal projections or both (see Chapter 12). In kangaroos, the vertebral column has evolved to accommodate distribution of the forces acting on it during hopping. There is an increase in size of the intervertebral discs from those of the thoracic vertebrae to those at the lumbosacral junction. For example, in a medium-sized western grey kangaroo, the intervertebral disc depth at the thoracolumbar junction is approximately 4 mm, but at the lumbosacral junction it approximately 17 mm. The transverse processes of the thoracic vertebrae are well developed and the ribs articulate with them, as well as with the bodies. In kangaroos, there are large dorsolateral extensions of the transverse

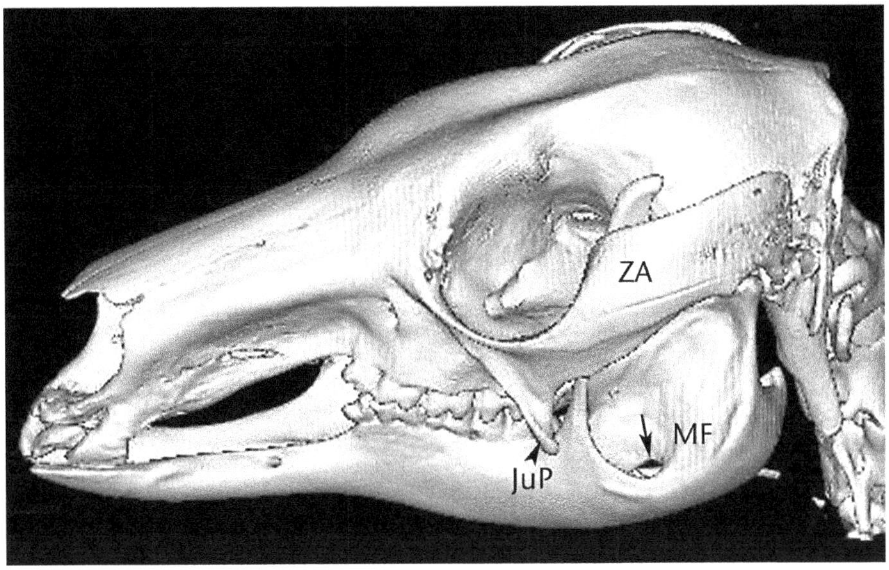

Figure 4.14: Skull, volume-rendered three-dimensional CT reconstruction, eastern grey kangaroo (*Macropus giganteus*). At the level of the orbit, a sharply defined jugal process (JuP) extends ventrally. The zygomatic arch (ZA) is thin and flat caudal to the orbit. The lateral surface of the mandible is marked by the masseteric fossa (MF) into which the masseter muscle inserts. At the base of the fossa, the masseteric foramen (arrow) communicates with the pterygoid fossa on the medial aspect of the mandible.

processes of the lumbar vertebrae. These provide extensive attachments for the hypaxial and epaxial musculature. Similarly, the first six caudal vertebrae have well-developed processes that allow for the complex muscular and tendinous arrangements of the tail (Richardson 2012). The sacrum consists of two ankylosed vertebrae (Figs 4.8 and 4.20). In the larger species of macropods, the first caudal vertebra has an extensive articulation with the sacrum, which allows the tail to swivel during pentapedal locomotion. In contrast, smaller macropods that do not have a bounding gait, such as the potoroids and the quokka (*Setonix brachyurus*), have shorter, less muscular hindquarters, lumbar and caudal vertebrae without enlarged mammillary or transverse processes and simpler intervertebral joints. Haemal arches (chevron bones) lie ventrally and transversally across sequential intervertebral discs for most of the length of the tail (Richardson 2012) (Fig. 4.19).

Ribs and sternum

There are 13 pairs of ribs, with the first being the shortest and broadest (Owen 1839). In kangaroos, the ribs are slender and rounded, except at their sternal extremities, which are flattened for the attachment of the cartilages. The seven cranial pairs of ribs articulate directly with the sternum (Owen 1839). The sternum consists of five or six segments, to which the clavicles and ribs have cartilaginous attachments (Fig. 4.21).

Soft tissue

Head and neck

Macropods possess both a cervical and thoracic thymus. The cervical is the larger and more dominant one at all stages of development (Yadav 1973). The cervical thymus is a paired structure found subcutaneously on the ventral neck. Its shape and extent vary with species. In healthy pouch young, they are firm, bulging structures with the caudal border level with, or just caudal to, the clavicles. The medial border is close to the midline but separate from the opposite lobe. The lateral border does not extend beyond the ventral surface of the neck and the cranial limit is close to the angle of the

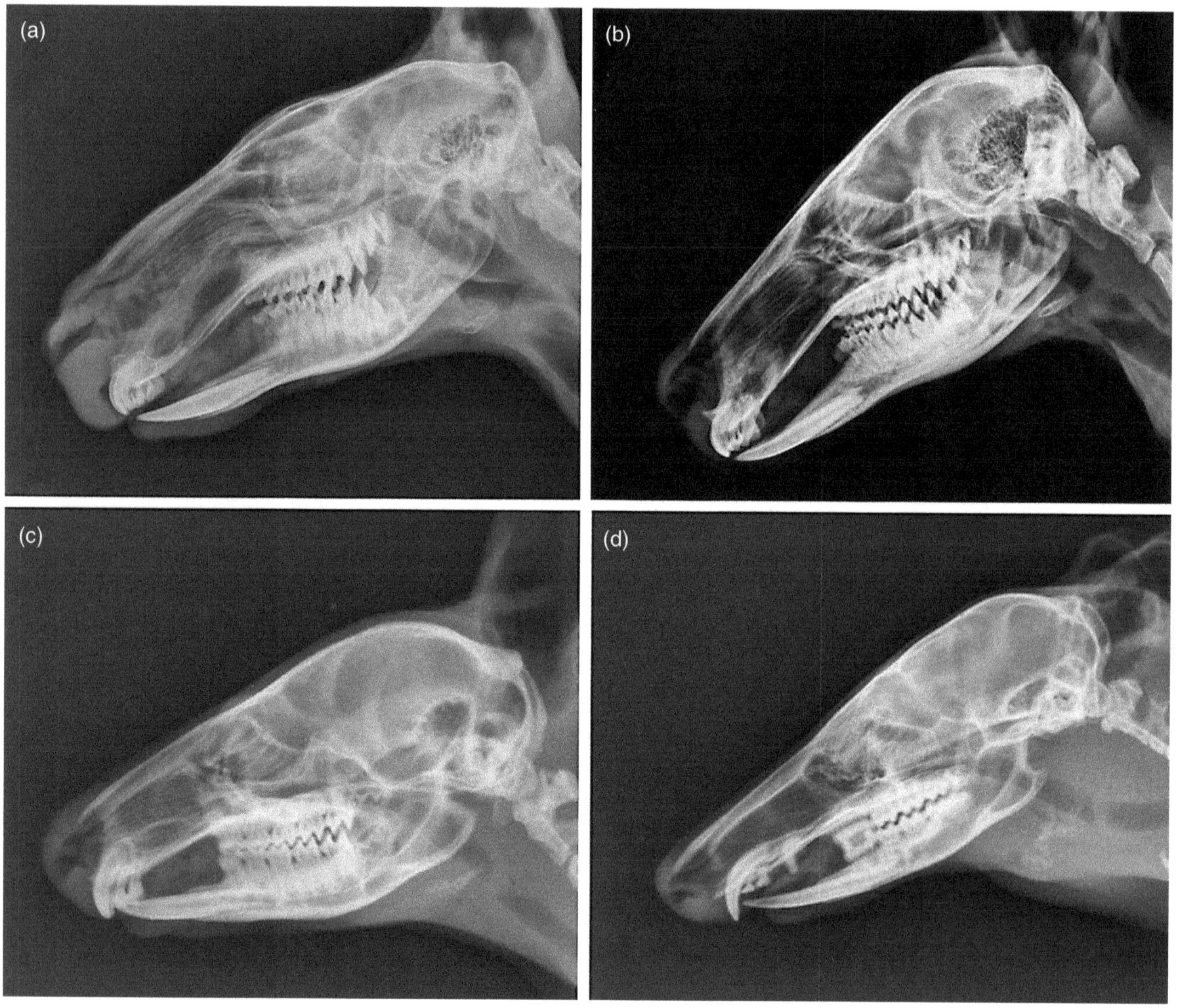

Figure 4.15: Skull, lateral view: (a) red kangaroo (*Macropus rufus*), (b) red-necked wallaby (*Macropus rufogriseus*), (c) quokka (*Setonix brachyurus*), (d) long-nosed potoroo (*Potorous tridactylus*).

mandible. In most species, the left and right lobes are pear-shaped structures with the narrow part cranially. In unfurred or early furred pouch young, the cervical thymic lobes can be visualised, but they regress fully by sexual maturity (Speare 1988; Tyndale-Biscoe 2005). They may be visible radiographically as soft tissue masses in the neck.

Thorax

The crura of the diaphragm originate from the first and second lumbar vertebrae. The right crus is the larger of the two and constitutes over half of the lumbar portion of the diaphragm. The left crus does not extend quite as far along the vertebrae as the right either caudally or cranially (Boardman 1941). In kangaroos, the thoracic portion of the trachea lies just to the right of the midline (Boardman 1941) (Fig. 4.21). The mediastinum is complete in the wallaroo (*M. robustus*), separating the left and right pleural cavities (Boardman 1941). Although not specifically described, this is likely to be the case in other macropod species.

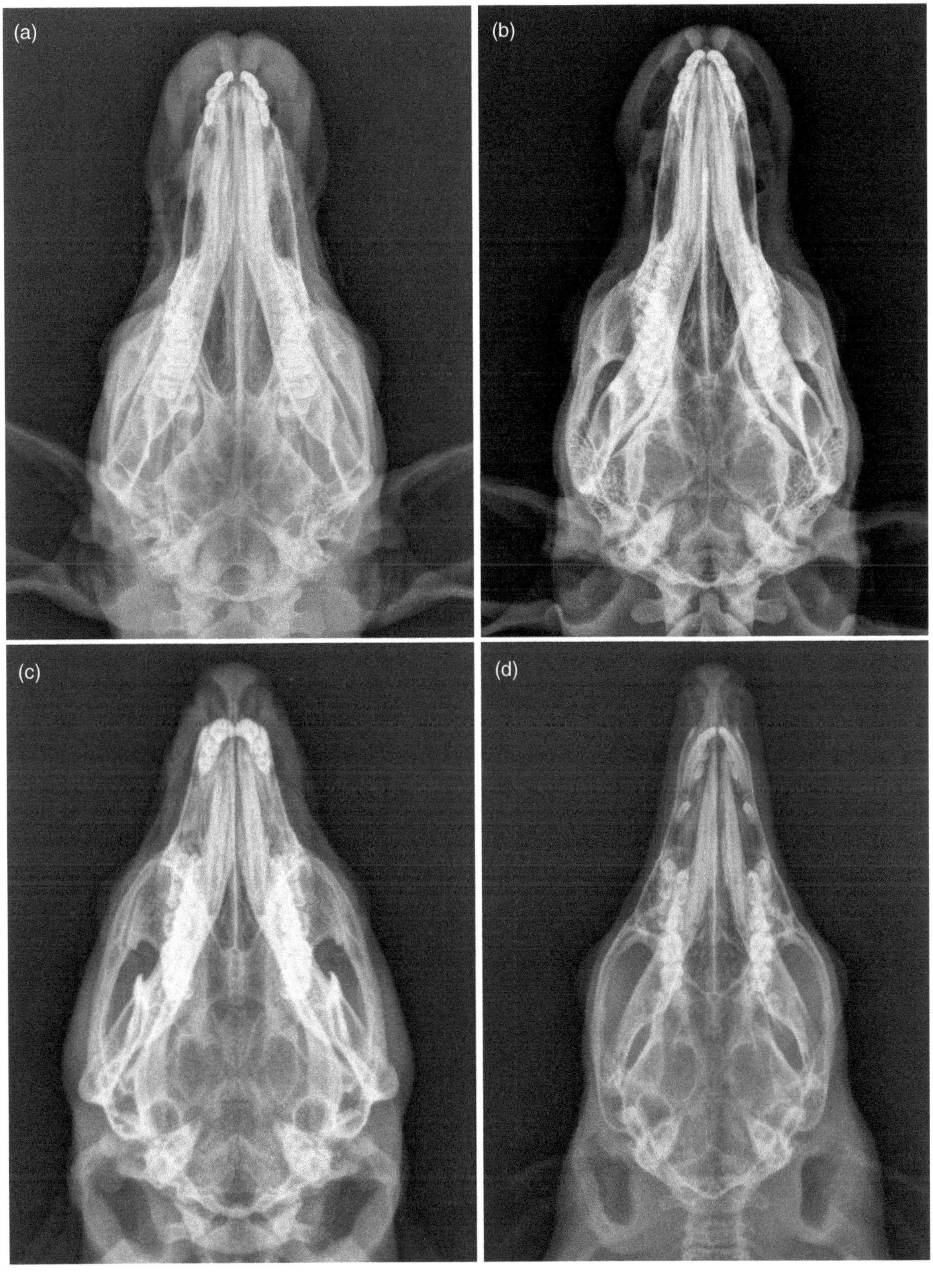

Figure 4.16: Skull, ventrodorsal view: (a) red kangaroo (*Macropus rufus*), (b) red-necked wallaby (*Macropus rufogriseus*), (c) quokka (*Setonix brachyurus*), (d) long-nosed potoroo (*Potorous tridactylus*).

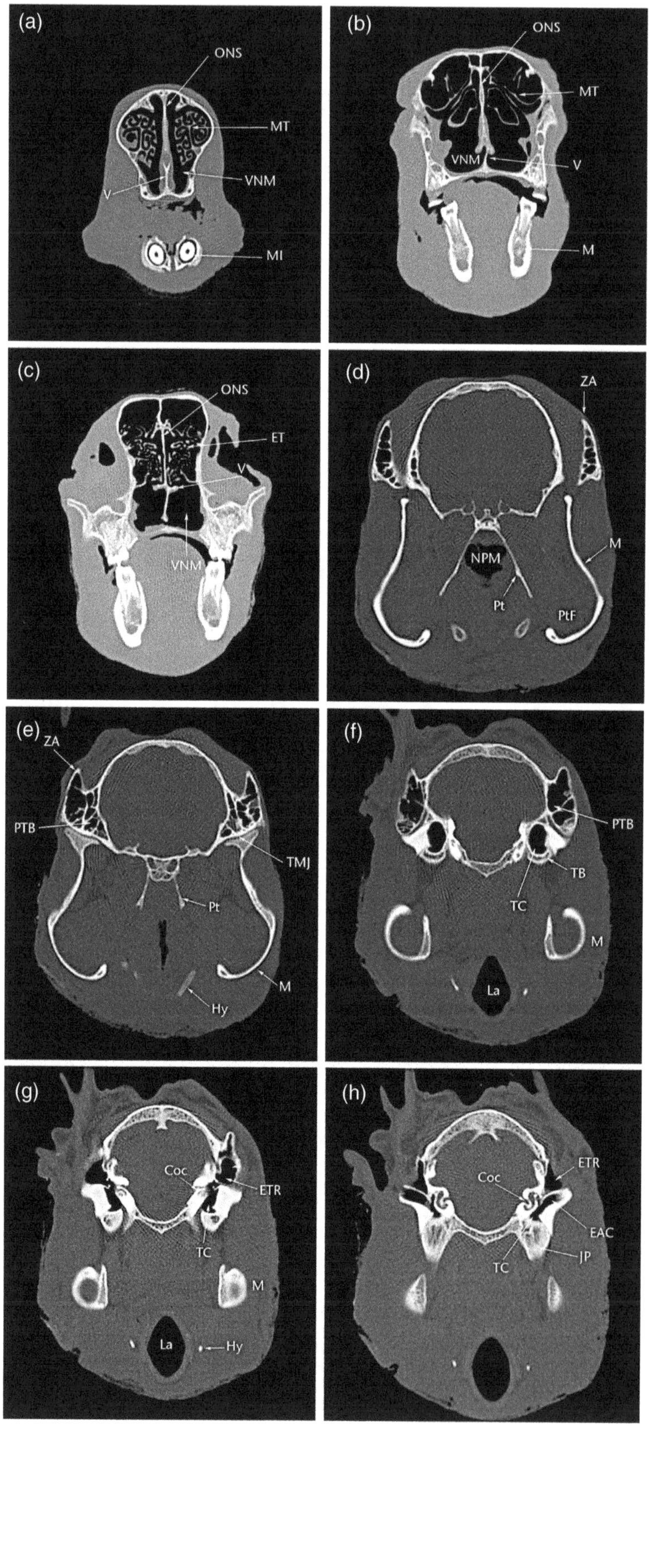

Figure 4.17: *(Continued).*

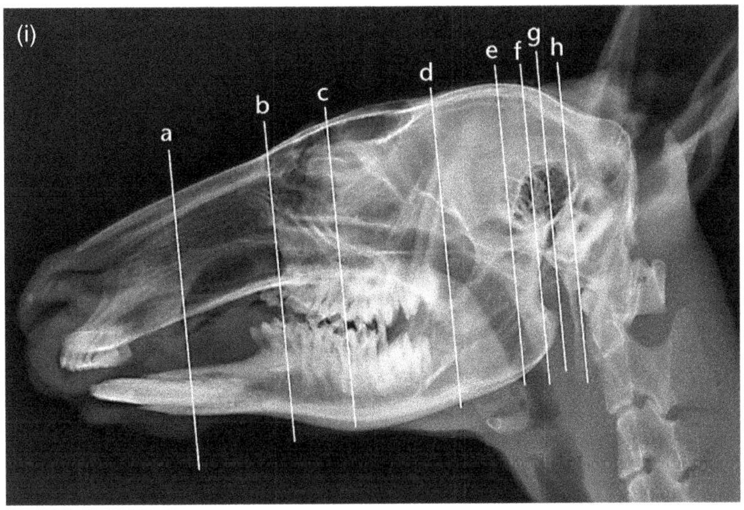

Figure 4.17: Skull, eastern grey kangaroo (*Macropus giganteus*): (a–h) transverse reconstructed CT images, (i) skull, lateral radiograph, demonstrating the image planes acquired in (a–h). The ossified nasal septum (ONS) divides the nasal cavity and extends ventrally to the vomer (V). The scrolls of maxillary turbinates (MT) and ethmoid turbinates (ET) are visible. Ventrally, the mandible (M) and mandibular incisor (MI) are visible. The ventral nasal meatus (VNM) is visible ventral to the maxillary and ethmoid turbinates and further caudally the nasopharyngeal meatus (NPM) is visible. The zygomatic (ZA) process of the temporal bone is pneumatised and communicates with the pneumaticised temporal bone (PTB) to become the epitympanic recess (ETR), which communicates with the tympanic cavity (TC). The temporomandibular joint (TMJ) can be seen ventral to the pneumaticised temporal bone. The pterygoid (Pt) bone extends ventrally and is directed towards the pterygoid fossa (PtF), which serves as the insertion of the medial pterygoid muscle. The cochlea (Coc) is visible medial to the tympanic cavity. The jugular process (JP) of the occipital bone extends ventrally from the tympanic bulla (TB). The external auricular canal (EAC) is lateral to the tympanic cavity. Parts of the hyoid apparatus (Hy) and the laryngeal lumen (La) are visible ventrally.

Cardiovascular system

The macropod heart is 30% larger than that of eutherians of similar bodyweight (Dawson and Needham 1981), almost as wide as it is long and pointed at the apex (Runciman *et al.* 1992) (Fig. 4.21). Other unusual features of the cardiovascular system of macropods compared with that of eutherians include that the heart is separated from the diaphragm by lung tissue; the cranial *vena cava* is paired (persistent left cranial *vena cava*); the right atrium has two auricular appendages surrounding the base of the aorta (seen in all *Macropus* species other than the wallaroo) (Boardman 1941; Runciman *et al.* 1992); there is a more crescentric right ventricular lumen; there is a common origin of the cranial and caudal mesenteric arteries and there are separate aortic origins of the internal and external iliac arteries (Gannon *et al.* 1989). The pericardium is suspended from the dorsal aspect of the sternum (by the sternopericardiac ligament), as seen in all marsupials other than the koala (Gannon *et al.* 1989).

Respiratory system

In the common wallaroo (*M. robustus*), the right lung is nearly double the size of the left. The left lung is elongated, laterally compressed because of the depth of the cardiac impression and not divided into lobes. The cranial third of the right lung is thin and laterally flattened while the rest is thick, especially dorsally. It is triangular in shape with a well defined cardiac notch (Boardman 1941). Generally, macropods have few deep fissures dividing the lungs into lobes. However, notches of variable depth are common. The degree of lobation of the right lung varies with species. In the common

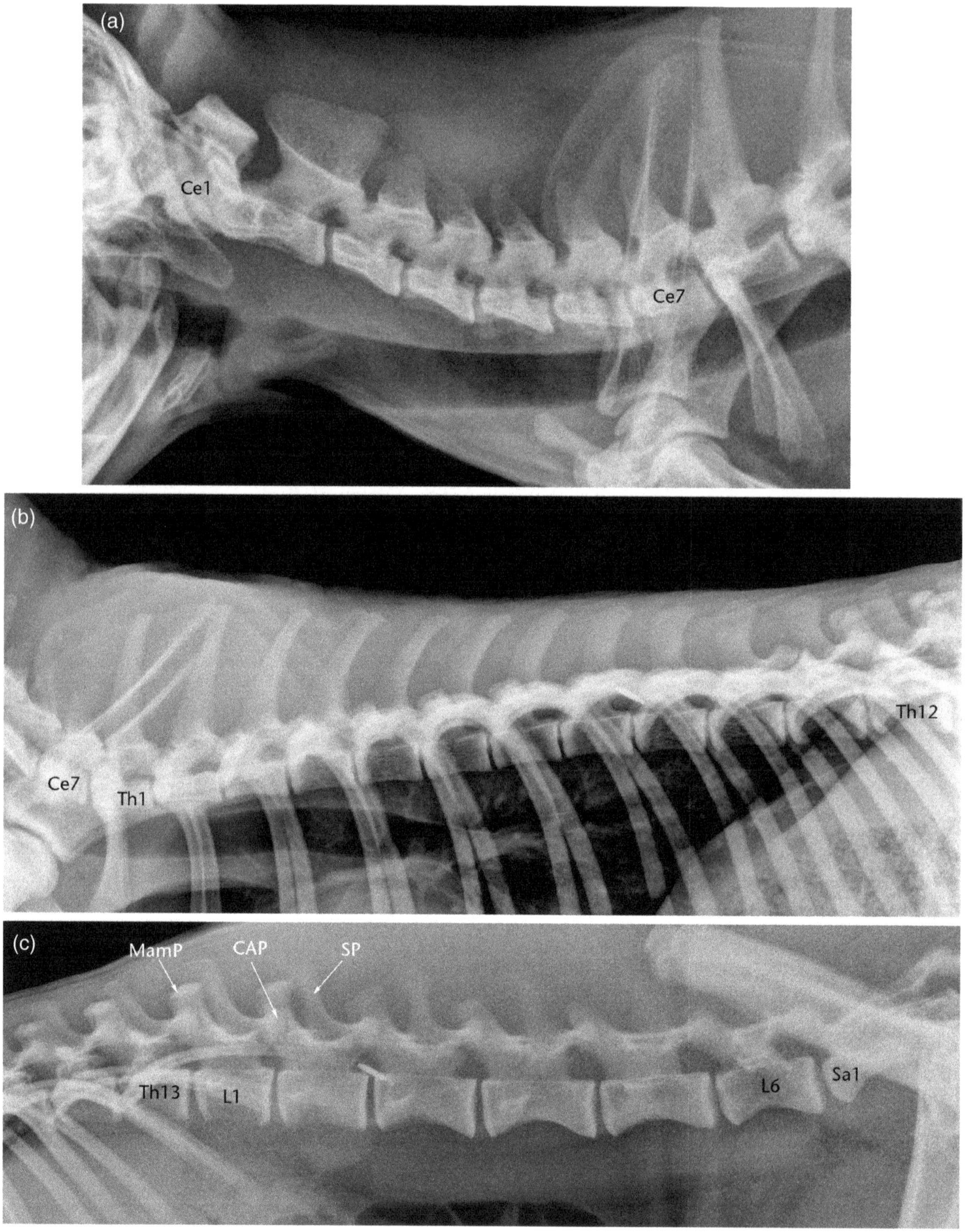

Figure 4.18: Vertebral column, red-necked wallaby (*Macropus rufogriseus*): (a–c) lateral views. There are 7 cervical (Ce1, Ce7 labelled), 13 thoracic (Th1, Th12, Th13 labelled) and 6 lumbar (L1, L6 labelled) vertebrae and 2 sacral (Sa1 labelled) elements. In the thoracolumbar region, an enlarged mammillary process (MamP) articulates with the caudal articular process (CAP) of the preceding vertebra. With the exception of Ce1, each cervical, thoracic and lumbar vertebra has a dorsal spinous process (SP).

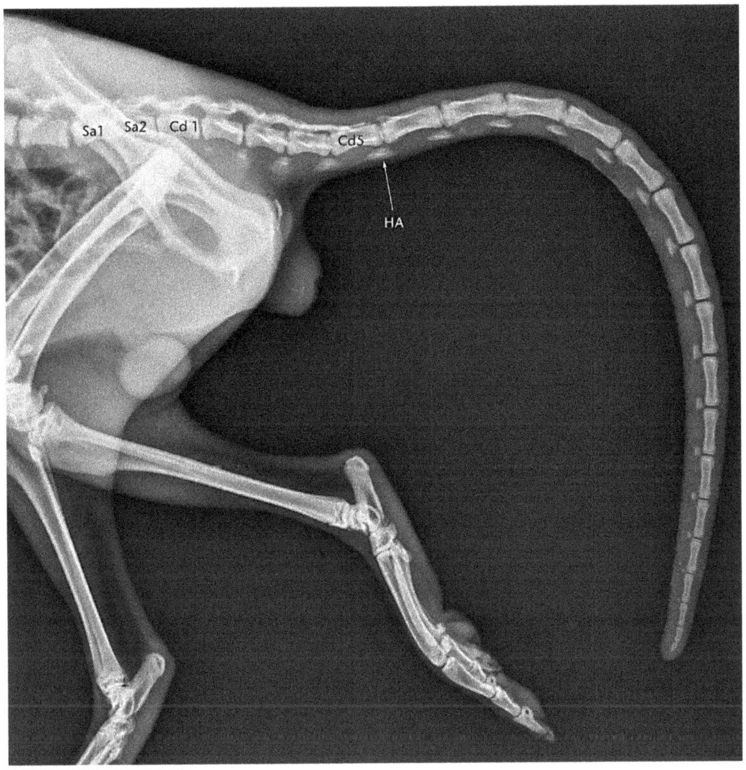

Figure 4.19: Sacrum and caudal vertebrae, quokka (*Setonix brachyurus*), lateral view. The normal dorsally located articular arrangement between vertebrae ceases around the 5th caudal vertebra (Cd5). Haemal arches (HA) are present ventral to the intervertebral spaces. The 2 sacral elements (Sa1, Sa2) and 1st caudal vertebra (Cd1) are labelled.

wallaroo, the right lung is bilobate (Boardman 1941). The right lung of the potoroo (species not specified) is described as having two to three deep fissures and an accessory lobe. This may indicate that it has four lobes, with one of them perhaps further subdivided (Owen 1868). The right lung of the tree-kangaroo (species not specified), eastern grey kangaroo (*M. giganteus*) and red-necked wallaby (*M. rufogriseus*), is described as being trilobate, with a deep median sulcus incompletely dividing the lung into cranial and caudal parts and an accessory lobe (Sonntag 1921). The right lung of the pretty-faced wallaby (*M. parryi*) has one or two notches, possibly indicating the absence of distinct divisions between lobes based on external appearance (Owen 1868). As in many marsupials, the left lung of macropods is considered unilobate, having an absence of distinct lobation, although fissures or clefts may be present (Owen 1852; Owen 1868; Sonntag 1921). An eparterial or apical bronchus is present on the right side.

Abdomen

Macropods are foregut fermenters. The stomach of most species is large and sacculated, and radiographically is the dominant feature in the cranial two-thirds of the abdomen (Fig. 4.22). It contains a mixture of digesta and gas, obscuring most other structures within the abdomen. The caecum and proximal colon are relatively small. Individual segments of small intestine can be identified by their size and shape but, generally, they are displaced and distorted by the large stomach. Individual viscera such as the reproductive organs, kidneys, liver, spleen and bladder are poorly distinguished because of the absence of body fat and superimposition of the gas and ingesta that dominate the appearance of the abdominal contents. If a pouch

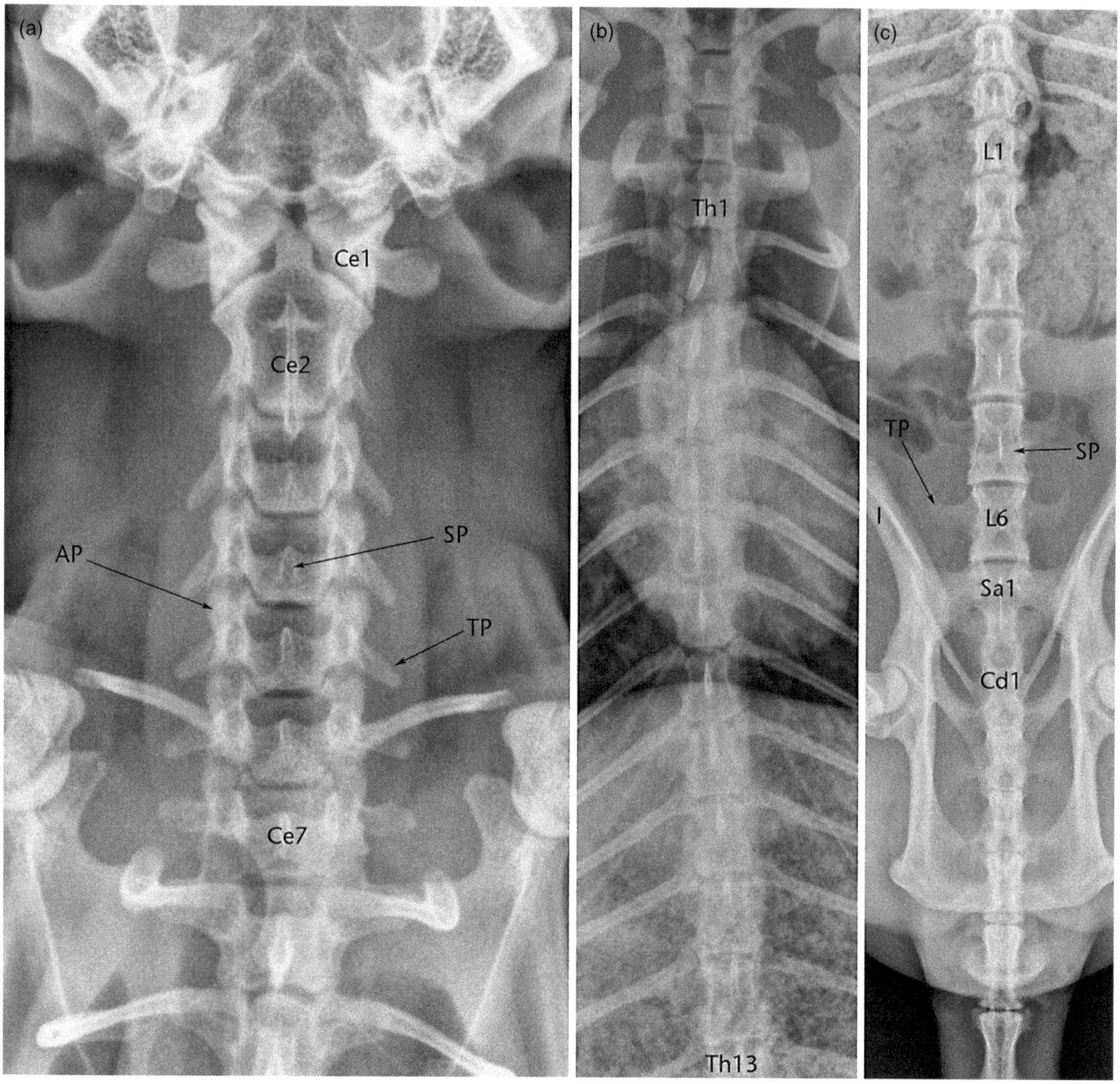

Figure 4.20: Vertebral column, red-necked wallaby (*Macropus rufogriseus*): (a–c) ventrodorsal views. The 7 cervical vertebrae (Ce1, Ce2, Ce7 labelled) have dorsolaterally located articular processes (AP), transverse processes (TP) and, with the exception of Ce1, spinous processes (SP). The 13 thoracic vertebrae (Th1, Th13 labelled) have centrally located articular arrangements between the 1st and 10th thoracic vertebrae, becoming dorsolaterally located from 11th thoracic vertebra caudally. Each thoracic vertebra articulates with a pair of ribs. The 6 lumbar vertebrae (L1, L6 labelled) have dorsolateral articular processes, a spinous process and transverse processes. The paired sacral vertebrae (Sa1 labelled) articulate with the ilia (I). The first caudal vertebra is labelled (Cd1).

young is present, it may be seen radiographically and if being suckled, one of the two mammary glands may be large and radiographically visible.

The anatomy of the macropod male reproductive tract is similar to that of other marsupials. Paired testes and epididymides are located

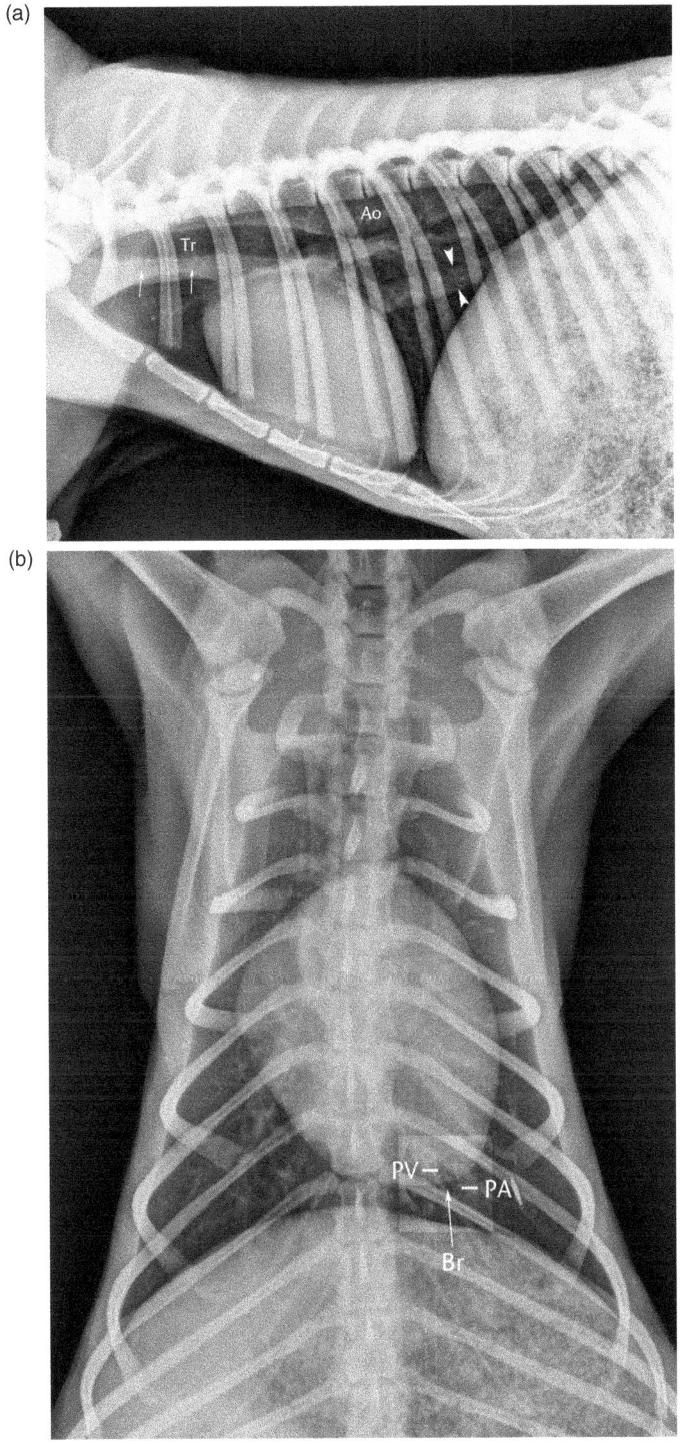

Figure 4.21: Thorax, red-necked wallaby (*Macropus rufogriseus*): (a) lateral view, (b) ventrodorsal view. The heart is centrally located in the thorax. Individual lung lobation is not radiographically apparent. The cranial mediastinum (arrows) contains the cranial *vena cavae*, the trachea (Tr), the oesophagus and the great arteries, but generally the vascular structures cannot be distinguished individually. Caudal to the heart, the descending aorta (Ao) and the caudal *vena cava* (arrowheads) can be distinguished. The pulmonary arteries (PA) are located dorsal and lateral to their companion bronchus (Br), while the pulmonary veins (PV) are located ventral and medial to their companion bronchus.

(a)

(b)

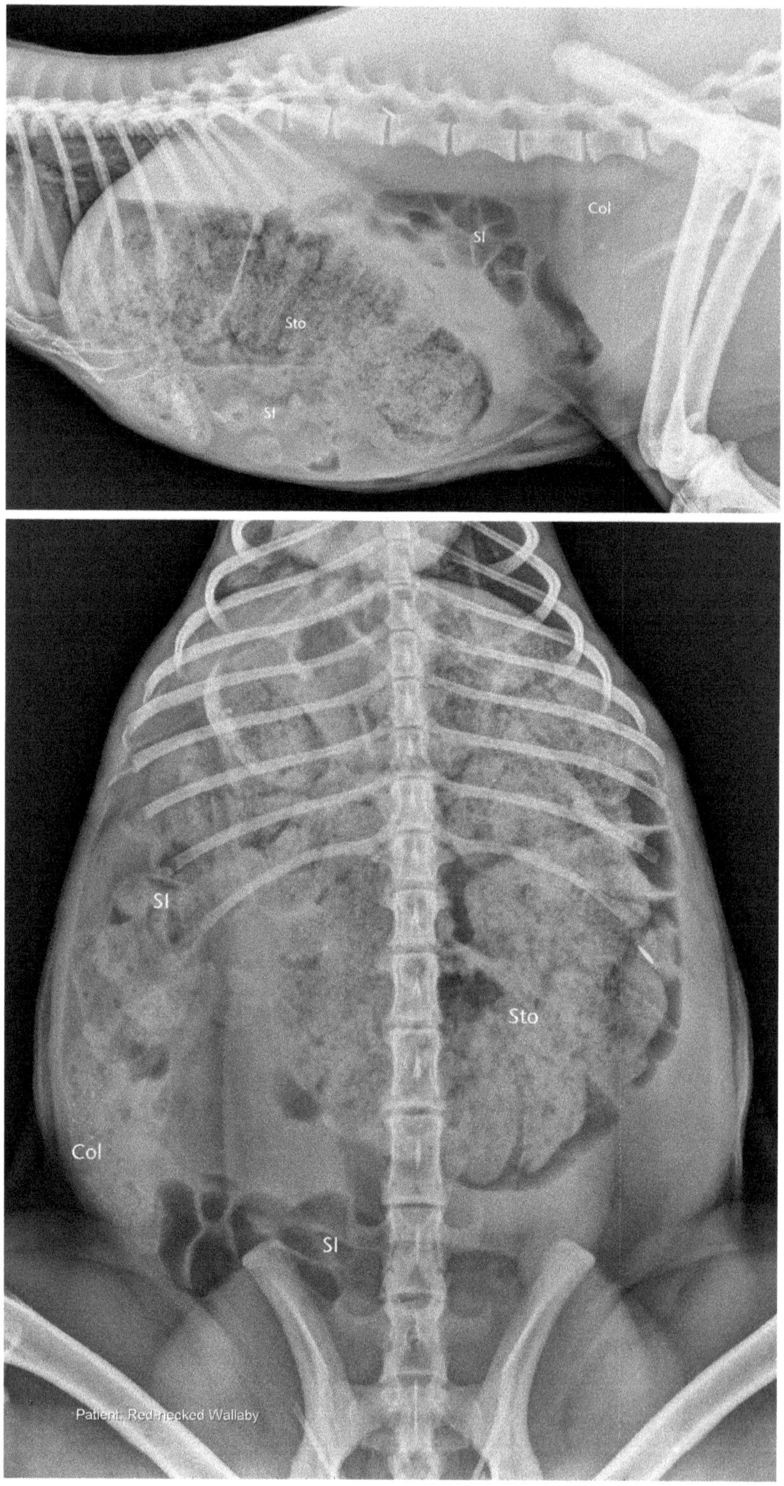

Figure 4.22: Abdomen, red-necked wallaby (*Macropus rufogriseus*): (a) lateral view, (b) ventrodorsal view. The stomach and its contents (Sto) dominate the image of abdominal viscera, occupying two-thirds of the abdomen. The stomach wall contains haustra, causing a sacculated appearance. Segments of small intestine (SI) and the colon (Col) can be recognised in regions not occupied by the stomach. Macropods have little abdominal adipose tissue, which means that other abdominal viscera are poorly defined. Note: a transponder can be seen.

externally in a pendulous, pedunculated, prepenile scrotum. A strong, active cremaster muscle encloses the spermatic cord and retracts the scrotal contents tightly against the pelvic body wall in cold weather or when stressed. When relaxed, in hot weather or under anaesthesia, the scrotum hangs very low. The ductus deferentia join the prostatic urethra close to the neck of the bladder. Seminal vesicles and ampullae are lacking in marsupials. In macropods, three pairs of bulbourethral glands and a relatively large, carrot-shaped, disseminate prostate are the only accessory sex glands (Dawson *et al.* 1989; Hume *et al.* 1989).

The anatomy of the macropod female reproductive tract is similar to that of other marsupials. The completely separate uteri (no marsupials have a single chambered uterus) open by separate cervices into the vaginal cul-de-sac, which is divided by a median septum (Dawson *et al.* 1989). Two separate lateral vaginae arise from the vaginal cul-de-sac. The caudal ends of the lateral vaginae open into the urogenital sinus, which also receives the urethra. Caudally, the two lateral vaginae may unite for a short distance before joining the urogenital sinus. The two lateral vaginae are separated by a mass of fibrous connective tissue called the urogenital strand. Prior to each parturition, a pseudovaginal canal forms within the urogenital strand, connecting the vaginal cul-de-sac cranially with the urogenital sinus caudally to permit the passage of the fetus (Tyndale-Biscoe and Renfree 1987; Dawson *et al.* 1989). The urogenital sinus together with the rectum opens into the common vestibule or cloaca.

References

Boardman W (1941) On the anatomy and functional adaptation of the thorax and pectoral girdle in the wallaroo (*Macropus robustus*). *Proceedings of the Linnean Society of New South Wales* **66**, 349–387.

Chen X, Milne N, O'Higgins P (2005) Morphological variation of the thoracolumbar vertebrae in *Macropodidae* and its functional relevance. *Journal of Morphology* **266**, 167–181. doi:10.1002/jmor.10370

Dawson TJ (1995) *Kangaroos: The Biology of the Largest Marsupials.* University of New South Wales Press: Sydney.

Dawson TJ, Needham AD (1981) Cardiovascular characteristics of two resting marsupials: An insight into the cardio-respiratory allometry of marsupials. *Journal of Comparative Physiology* **145**, 95–100. doi:10.1007/BF00782599

Dawson TJ, Finch E, Freedman L, Hume ID, Renfree MB, Temple-Smith PD (1989) Morphology and physiology of the metatheria. In *Fauna of Australia, Volume 1B, Mammalia.* (Eds DW Walton and BJ Richardson) pp. 451–504. Australian Government Publishing Service, Canberra.

Gannon BJ, Baudinette RV, Runciman SIC, Frappell PB (1989) Cardiorespiratory structure and function in macropodid marsupials. In *Kangaroos, Wallabies and rat-kangaroos, Volume 1.* (Eds G Grig, P Jarman and I Hume) pp. 407–418. Surrey Beatty and Sons, Chipping Norton.

Grand TI (1990) Body composition and the evolution of the Macropodidae (Potorous, Dendrolagus and Macropus). *Anatomy and Embryology* **182**, 85–92. doi:10.1007/BF00187530

Hopwood PR (1981) Carcass muscle weight distribution and yield: A comparison between grey kangaroos, *Macropus giganteus*, and red kangaroos, *M. rufus. Australian Wildlife Research* **8**, 263–268. doi:10.1071/WR9810263

Hume ID, Jarman PJ, Renfree MB, Temple-Smith PD (1989) Macropodidae. In *Fauna of Australia, Volume 1B, Mammalia.* (Eds DW Walton and BJ Richardson) pp. 679–715. Australian Government Publishing Service, Canberra.

Iwaniuk AN, Nelson JE, Whishaw IQ (2000) The relationships between brain regions and forelimb dexterity in marsupials (Marsupialia): A comparative test of the principle of proper mass. *Australian Journal of Zoology* **48**, 99–110. doi:10.1071/ZO99043

Menkhorst PW (2001) *A Field Guide to the Mammals of Australia.* Oxford University Press, Melbourne.

Owen R (1839) On the osteology of Marsupialia. *The Transactions of the Zoological Society of London* **2**, 379–408. doi:10.1111/j.1469-7998.1839.tb00030.x

Owen R (1852) Notes on the anatomy of the tree-kangaroo (*Dendrolagus inustus*). *Proceedings of the Zoological Society of London* pp. 103–107, London.

Owen R (1868) *On the Anatomy of Vertebrates, Volume 3, Mammals.* pp. 513–603. Longmans, Green and Co., London.

Parker JT, Haswell WA (1921) Skeleton of metatheria. In *A text-book of Zoology, Volume II.* pp. 501–506. Macmillan and Co. Ltd, London.

Reese S, Pfuderer UR, Bragulla H, Loeffler K, Budras KD (2001) Topography, structure and function of the patella and the patelloid in marsupials. *Anatomia, Histologia, Embryologia* **30**, 289–294. doi:10.1046/j.1439-0264.2001.00334.x

Richardson K (2012) *Australia's Amazing Kangaroos: Their Conservation, Unique Biology and Coexistence with Humans.* CSIRO Publishing, Melbourne.

Runciman SIC, Baudinette RV, Gannon BJ (1992) The anatomy of the adult marsupial heart: an historical review. *Australian Journal of Zoology* **40**, 21–34. doi:10.1071/ZO9920021

Sanson GD (1989) Morphological adaptations of teeth to diets and feeding in the Macropodoidea. In *Kangaroos, Wallabies and Rat-kangaroos.* (Eds GC Grigg, P Garman and ID Hume) pp. 151–168. Surrey Beatty, Sydney.

Seebeck JH, Rose RW (1989) Potoroidae. In *Fauna of Australia, Volume 1B, Mammalia.* (Eds DW Walton and BJ Richardson) pp. 716–739. Australian Government Publishing Service, Canberra.

Sonntag CF (1921) Contributions to the visceral anatomy and myology of the marsupialia. *Proceedings of the Zoological Society of London* pp. 851–882, London.

Speare R (1988) Clinical assessment, diseases and management of the orphaned macropod PY. In *Australian Wildlife.* Proceedings 104, pp. 211–296. Post Graduate Committee in Veterinary Science, Sydney.

Strahan R (1995) *The Australian Museum complete book of Australian mammals.* Reed New Holland, Sydney.

Tomo S, Tomo I, Townsend GC, Hirata K (2007) Masticatory muscles of the great-gray kangaroo (*Macropus giganteus*). *The Anatomical Record* **290**, 382–388. doi:10.1002/ar.20508

Tyndale-Biscoe CH (2005) *Life of marsupials.* CSIRO Publishing, Melbourne.

Tyndale-Biscoe CH, Renfree MB (1987) *Reproductive physiology of marsupials.* Cambridge University Press, Cambridge.

Vogelnest L, Portas T (2008) Macropods. In *Medicine of Australian Mammals.* (Eds L Vogelnest and R Woods) pp. 133–225. CSIRO Publishing, Melbourne.

Warburton NM, Harvey KJ, Prideaux GJ, O'Shea JE (2011) Functional morphology of the forelimb of living and extinct tree-kangaroos (Marsupialia: Macropodidae). *Journal of Morphology* **272**, 1230–1244. doi:10.1002/jmor.10979

Weisbecker V, Archer M (2008) Parallel evolution of hand anatomy in kangaroos and vombatiform marsupials: functional and evolutionary implications. *Palaeontology* **51**, 321–338. doi:10.1111/j.1475-4983.2007.00750.x

Yadav M (1973) The presence of the cervical and thoracic thymus lobes in marsupials. *Australian Journal of Zoology* **21**, 285–301. doi:10.1071/ZO9730285

5

Koala

Larry Vogelnest, Graeme Allan and Susan Hemsley

The koala (*Phascolarctos cinereus*) is the only member of the family Phascolarctidae. Koalas are found in eastern Australian sclerophyll forests and woodlands, have a widespread distribution and are abundant in some areas. The distribution, however, is largely fragmented, with populations separated by unsuitable habitat or cleared land (Martin and Handasyde 1999). Koalas in the southern part of their range are larger and have longer, darker fur than their northern counterparts. Bodyweight ranges and averages are: Queensland females 4.1–7.3 kg (av. 5.1 kg), males 4.2–9.1 kg (av. 6.5 kg); Victorian females 7.0–11.0 kg (av. 8.5 kg), males 9.5–14.9 kg (av. 12.0 kg) (Martin and Handasyde 1995). Free-ranging koalas commonly live for 13–18 years, with males having a shorter life expectancy than females. Longevity of captive koalas is greater, with some animals living to over 20 years (Blanshard and Bodley 2008).

Koalas are largely nocturnal, with some activity during the day. They are one of the largest and most specialised of the arboreal folivores, feeding principally on the leaves of certain species of *Eucalyptus* (Blanshard and Bodley 2008). The koala shares some characteristics with wombats, distinguishing them from other Australian diprotodont marsupials: the pouch opens caudally; cheek pouches are present; the tail is vestigial; the first and second premolars are absent; the palatine vacuities are restricted to the palatine bones rather than extending beyond their boundaries; the mandibular symphysis is fused; the ventral arch of the axis vertebra consists of cartilage rather than bone; the transverse process of the seventh cervical vertebra is perforate; the supratragus of the ear is small; there is a distinctive gastric gland; and the pericardium adheres to the diaphragm, rather than being connected by membranous sheets (Lee and Carrick 1989; Grand and Barboza 2001).

The koala is primarily arboreal, having skeletal adaptations for arboreal locomotion, sitting and sleeping: orthograde trunk; head flexed forward; shoulders directed craniad and laterad; elbows internally rotated; palmar and plantar surfaces of the manus and pes turned towards the midline; an axis of closure between digits 2 and 3; and long, strong, sharp claws on all digits except the broad digit 1 of the pes (Grand and Barboza 2001).

Skeletal system

Appendicular skeleton

Pectoral girdle and forelimb

Paired clavicles articulate with the acromion of the scapula abaxially and the first sternal element ventrally and axially. The first sternal element also articulates with the first pair of ribs (Fig. 5.1). The first pair of ribs is thicker and more radio-opaque than succeeding pairs and together with the first sternal element, the clavicles and the scapulae are joined to provide a semi-rigid pectoral girdle.

In the resting position, the dorsal border of the koala's scapula lies parallel to the spines of the upper thoracic vertebrae. The glenoid cavity is directed laterally and cranially, so the scapular spine lies at almost 90° to the vertebral border. The

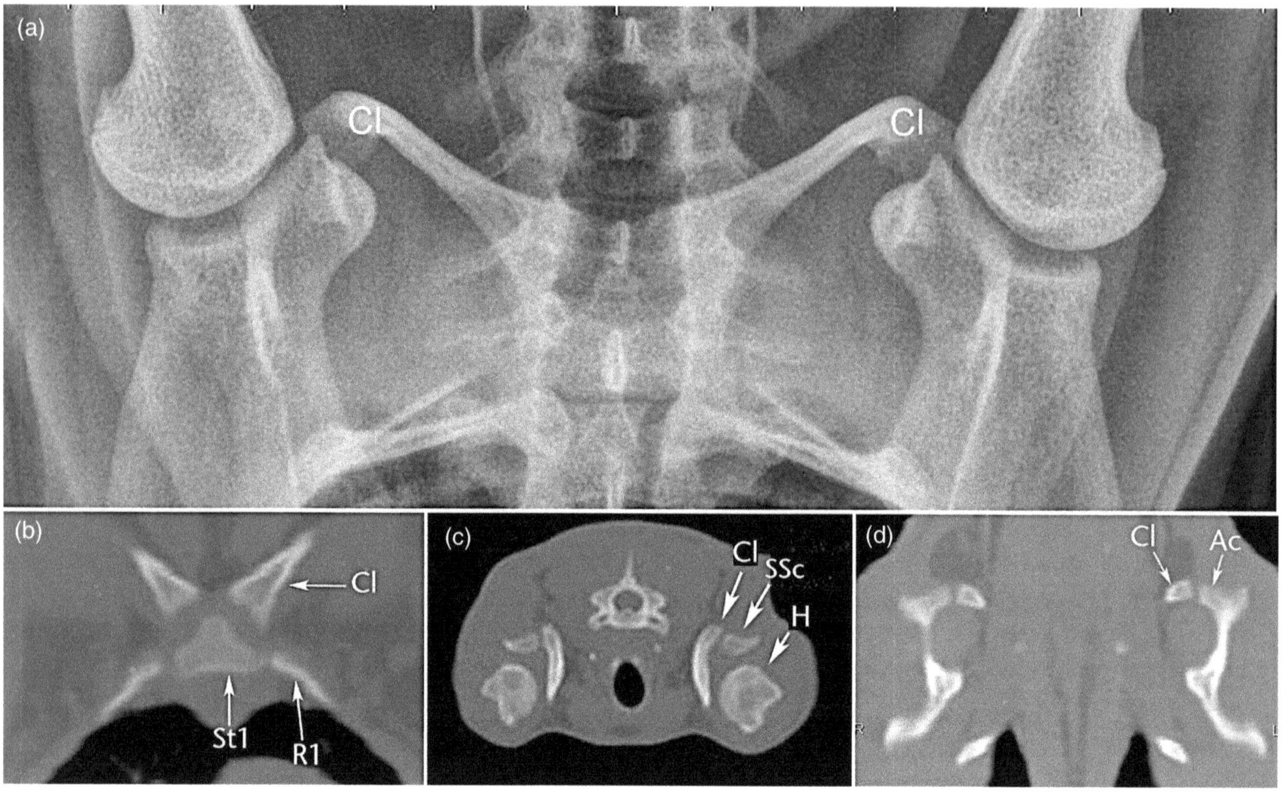

Figure 5.1: Pectoral girdle: (a) ventrodorsal view. The clavicles (Cl) articulate axially and ventrally with the 1st sternal element and abaxially with the acromion of the scapula. (b) CT reconstruction in the dorsal plane to demonstrate the axial and ventral articulation of the clavicle (Cl) with the 1st sternal element (St1), cranial to the 1st rib (R1). (c, d) CT reconstructions in the transverse plane to demonstrate the abaxial articulation of the clavicle (Cl) with the acromion (Ac) of the cranially angulated spinous process (SSc) of the scapula. Note the adjacent head of the humerus (H).

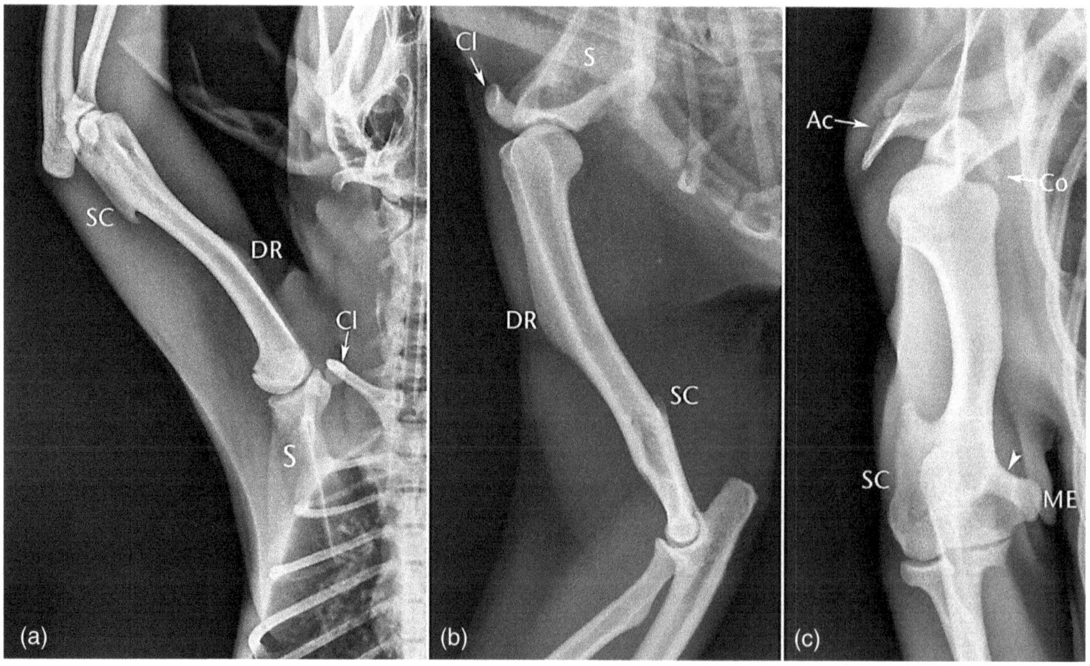

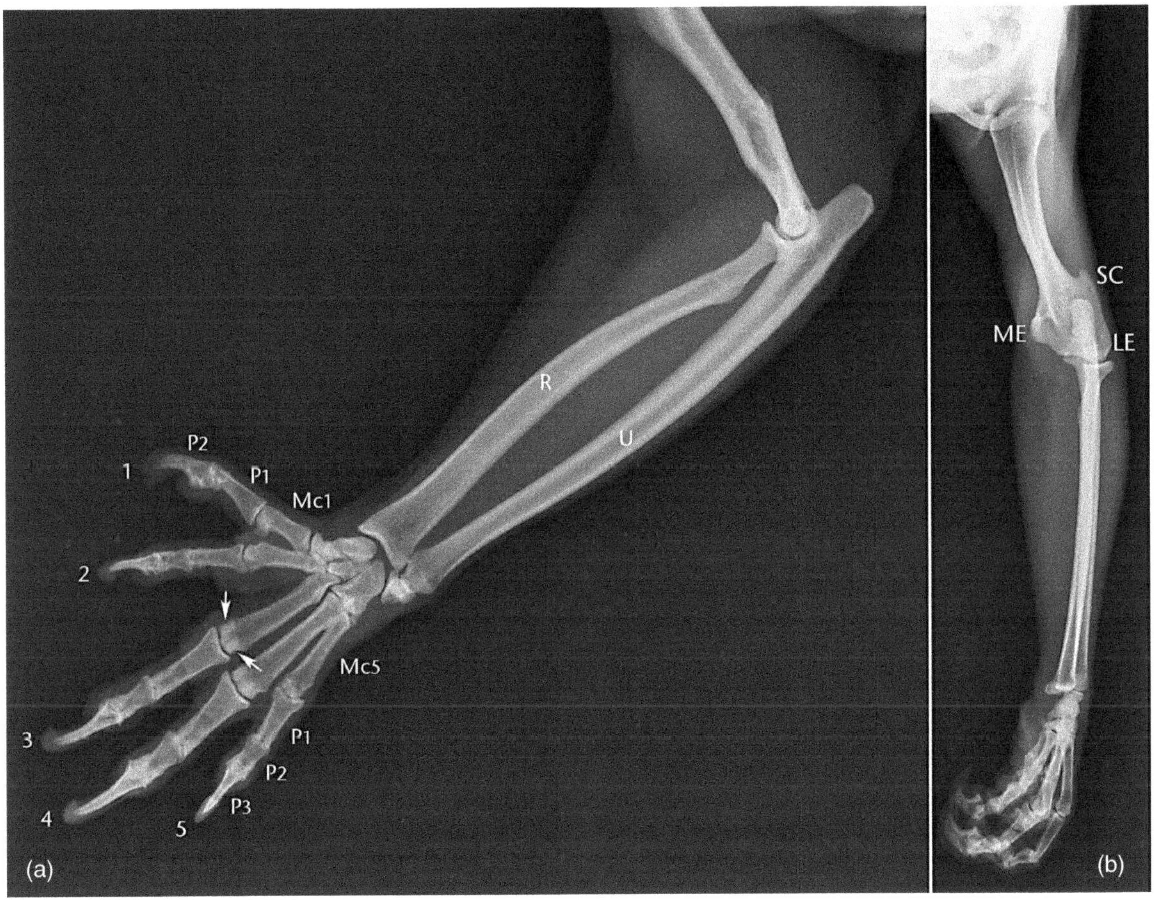

Figure 5.3: Antebrachium and manus: (a) lateral view of elbow and antebrachium and craniocaudal view of manus, (b) craniocaudal view of elbow and antebrachium and oblique lateral view of manus. Note the lateral (LE) and medial epicondyles (ME) and supinator crest (SC) of the humerus. Koalas have a bowed radius (R) and ulna (U), which enhances supination, and an internally rotated manus. The 1st two digits (1, 2) of the manus oppose the other three (3, 4, 5). Paired palmar metacarpophalangeal sesamoids (arrows) define the distal end of the metacarpal bones (Mc1, Mc5 labelled). There are three phalangeal bones in digits 2–5 (P1, P2, P3) and two in digit 1. The terminal phalangeal bone forms a claw for each digit.

supraspinous fossa is small and the axillary border unremarkable (Grand and Barboza 2001).

The humerus of the koala is long and slender; the greater and lesser tuberosities and the deltoid tuberosity are prominent, but not extreme. The deltoid ridge marks the proximal third of the humerus (Fig. 5.2). The epicondylar ridges are moderate. As measured by humeral torsion, from the glenohumeral to the humero-radio-ulnar joints, the koala elbow rotates internally. The proximal articular surfaces of the radius and ulna are antiflexed (straightened), so the antebrachium may achieve full extension. The olecranon is abbreviated in comparison with other marsupials, such as the wombat. Relative to body size, the koala's radius and ulna are long. The deep bow of the radius facilitates supination (Fig. 5.3) (Owen 1839; Lee and Carrick 1989; Grand and Barboza 2001).

The manus is relatively large. Digits 1 and 2 oppose digits 3, 4 and 5, effecting a forcipate grip

Figure 5.2: Brachium: (a) oblique lateral view, (b) lateromedial view, (c) craniocaudal view. Note the articulation of the clavicle (Cl) with the acromion (Ac) of the scapula (S). The humerus has a prominent supinator crest (SC) distally. The deltoid ridge (DR) is visible on the cranial surface of the proximal half of the humerus. There is a large medial humeral epicondyle (ME). The coracoid process (Co) of the scapula and a supracondylar foramen (arrowhead) are visible on the craniocaudal view.

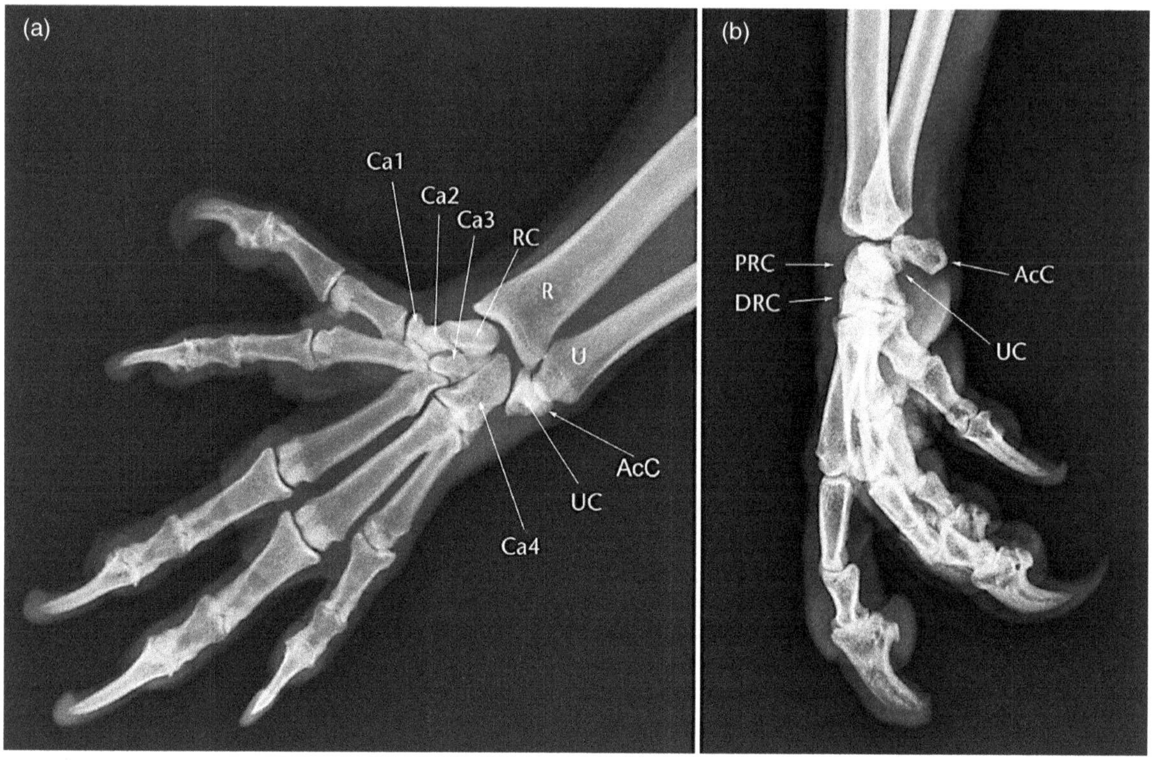

Figure 5.4: Manus: (a) dorsopalmar view, (b) mediolateral view. The radius (R) and ulna (U) articulate with the proximal row of carpal bones (PRC), namely the radial carpal (RC), 4th carpal (Ca4) and ulnar carpal bones (UC). The distal row of carpal bones (DRC), consists of the 1st (Ca1), 2nd (Ca2), 3rd (Ca3) and 4th carpal bones (Ca4). The accessory carpal bone (AcC) is superimposed on the ulnar carpal bone in (a), but is seen clearly in (b).

(Figs 5.3 and 5.4). The metacarpals and phalanges of digits 1 and 2 are rotated on their long axes in opposition to those of digits 3–5 (Grand and Barboza 2001). All digits have sharp claws (Cork 1987; Lee and Martin 1988).

Shoulder dysplasia has been reported in captive koalas in the United States (Pye 2009), but is rarely reported in wild or captive koalas in Australia. To fully evaluate the shoulder, three standard radiographic views should be obtained: lateral (left and right) with the forelimb extended at a 90° angle to the spine (Fig. 5.5a); ventrodorsal with the forelimbs extended cranially and the humerus at a 90° angle to the spine and the radius and ulna at a 90° angle to the humerus (Fig. 5.5b); and ventrodorsal with the forelimbs extended caudally as parallel as possible to the spine (Fig. 5.5c). A subjective system was developed to grade shoulder radiographs based on the degree of malformation of the supraglenoid and infraglenoid tubercles and the coracoid process, shallowing or loss of the glenoid cavity, flattening or loss of the humeral head, malformation of the greater and lesser tubercles, loss of the intertubercular groove and degree of humeral diaphyseal abnormalities. Shoulders are graded as normal or as mildly, moderately or severely dysplastic (Pye 2009) (see Chapter 12).

Pelvis and hindlimb

The koala pelvis is unremarkable other than the moderate outward flaring of the ilia (Fig. 5.6; see also Fig. 5.24). Flat, broad and slightly curved epipubic bones extend cranially from the cranial aspect of the pubis (Figs 5.6 and 5.7; see also Fig. 5.24). There is no depression for the round ligament on the femoral head (Owen 1839). The koala's hip joints tend to be flexed, the knees externally rotated and the pes turned towards the mid-line in inversion. As in all marsupials other than bandicoots and the bilby, there is not a bony patella (Fig.

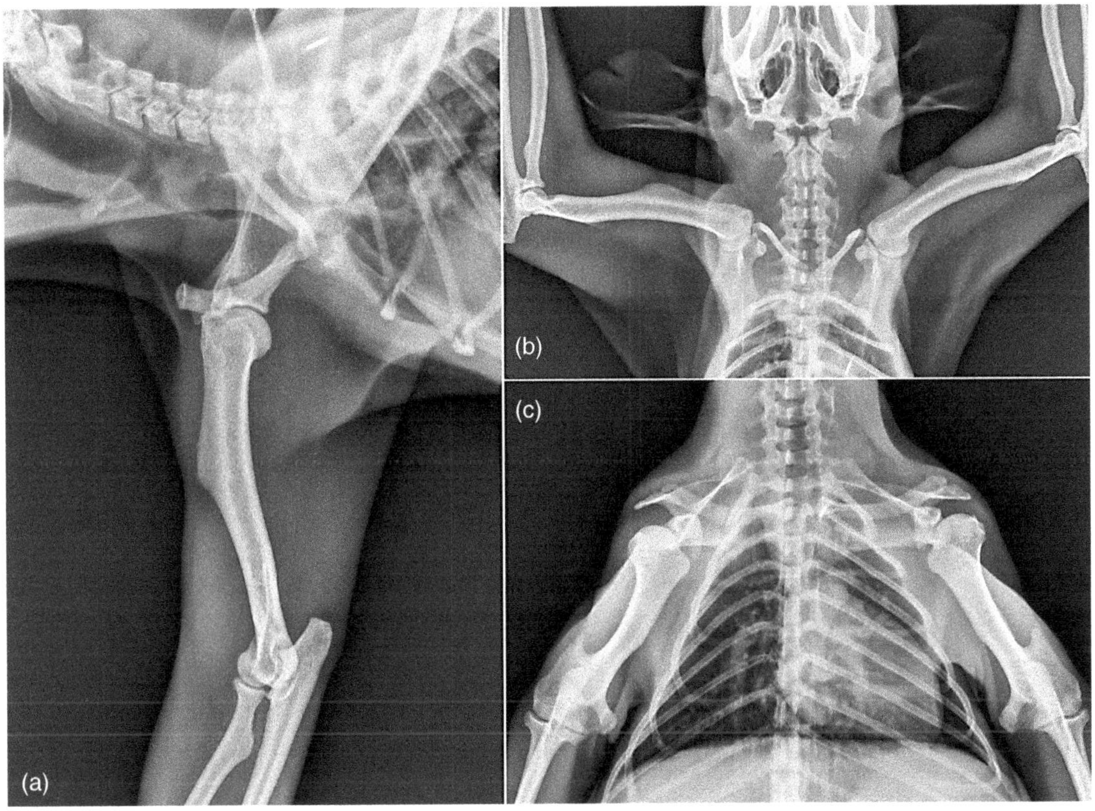

Figure 5.5: Standard radiographic views to evaluate the shoulder (Pye 2009): (a) lateral view with the forelimb extended at a 90° angle to the spine, (b) ventrodorsal view with the forelimbs extended cranially with the humerus at a 90° angle to the spine and the radius and ulna at a 90° angle to the humerus, (c) ventrodorsal view with the forelimbs extended caudally as parallel as possible to the spine. Note: a transponder can be seen in (a) and (b).

5.8). Instead, there is a patelloid, which is a fibrocartilaginous induration of the quadriceps femoris extensor tendon (Reese *et al.* 2001).

The functional axis of the pes lies between digit 1 and remaining four digits. Digits 2 and 3 are fused by skin (syndactylous) (Figs 5.8 and 5.9) (Grand and Barboza 2001). Digit 1 is opposed to the other digits. All digits of the pes, other than digit 1, have sharp claws (Cork 1987; Lee and Martin 1988).

Hip dysplasia has been reported in koalas in the United States (Pye *et al.* 2008), but rarely reported in wild or captive koalas in Australia. To evaluate the hip joints fully, three standard radiographic views should be used: ventrodorsal with the hindlimbs extended and rotated medially so that the lateral fabellae are positioned under the lateral condyles of the femurs (Fig. 5.10a); ventrodorsal in a frog-leg position obtained with the femurs at a 90° angle to the spine and the tibias at a 90° angle to the femurs (Fig. 5.10b); and right lateral with the hindlimbs extended and overlaying each other (Fig. 5.10c). Care should be taken to avoid pelvic rotation around the longitudinal axis. Slight rotation can affect the appearance of the acetabulum in the ventrodorsal view, because normal koalas have relatively shallow acetabula. A scoring system is used to grade the radiographic appearance of the hip joint. Points ranging from 0 = unaffected to 5 = severely affected are assigned to each of the following areas: acetabulum, femoral head, femoral neck and femur, giving a total out of 20 for each hip and a total out of 40 for each koala (see Chapter 12).

Axial skeleton

Skull

The koala has a box-like skull. It is oblong in shape, parallel-sided with the zygomatic arches running straight caudally from their broadest points at the

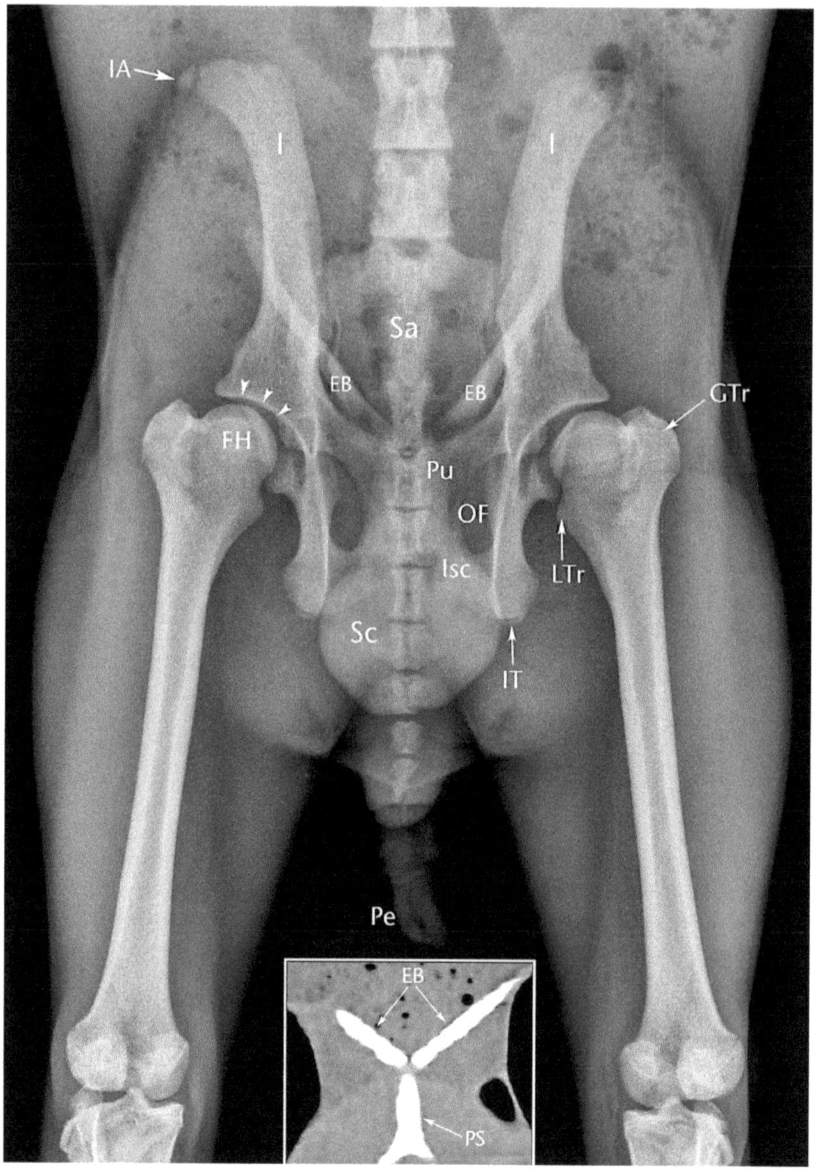

Figure 5.6: Pelvis and femurs, male koala (*Phascolarctos cinereus*), extended ventrodorsal view. Each cup-shaped acetabulum (arrowheads) is directed caudolaterally and accepts the femoral heads (FH). The ilia (I) are elongated. An iliac apophysis (IA) is present on the iliac crest. The epipubic bones (EB) diverge from their articulation with the pubis (Pu). Three sacral elements (Sa) are visible. The scrotum (Sc) is superimposed on the ischia (Isc) and caudal vertebrae, and the bifid penis (Pe) is visible. The obturator foramen (OF), ischial tuberosity (IT) and the greater (GTr) and lesser (LTr) trochanters of the femur are labelled. Inset: reconstructed CT in the dorsal plane to demonstrate the relationship of the epipubic bones (EB) with the pubic symphysis (PS).

orbits, rather than curving outwards. The zygomatic arches are long relative to the length of the skull (Fig. 5.11). The jugular process of the occipital bone is elongated relative to most other marsupials, but not as long as in macropods (Fig. 5.11; see also 5.20). There is virtually no inflection of the angle of the mandible (Owen 1839). Koala orbits (as in all marsupials) lack a bony roof and, hence, are open dorsally; the orbital floor is principally muscular in composition (Kempster and Hirst 2002).

Koalas have diprotodont dentition, with a dental formula of I 3/1, C 1/0, PM 1/1, M 4/4 (total 30) (Eberhard 1972). The first maxillary incisors are well developed, while the other upper pairs are

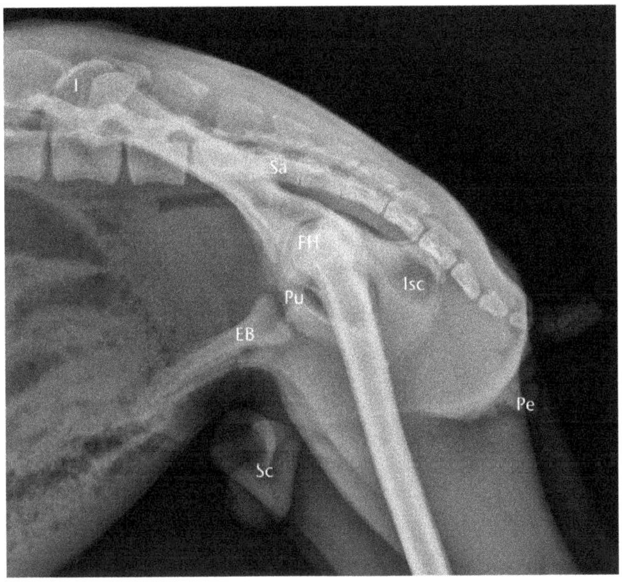

Figure 5.7: Pelvis and femurs, male koala (*Phascolarctos cinereus*), lateral view. Note the articulation of the epipubic bones (EB) with the pubis (Pu) and the relative locations of the scrotum (Sc) and penis (Pe). In this example, there are three sacral elements (Sa) and nine caudal vertebrae. The elongated ilia (I) are superimposed on the caudal lumbar vertebrae and the sacrum. The femoral heads (FH) are superimposed, as are the ilia and ischia (Isc).

smaller and quite slender. Canines are present only in the maxilla and are small, peg-like structures. There is a single pair of premolars (premolar 4) bearing a relatively sharp longitudinal ridge in each arcade (Lee and Martin 1988). Each of the molars has four pyramidal cusps, which interdigitate with those on the opposing arcade (Lanyon and Sanson 1986). Throughout a koala's life, there is continual wear of the occlusal surfaces of the cheek teeth (premolars and molars). The longitudinal cutting ridge of each premolar and the pyramidal cusps of each molar are gradually worn flat. Wear of the rostral cheek teeth is more advanced than that of the more caudal teeth (Eberhard 1972; Martin 1981).

The nasal anatomy of the koala has some distinctive features compared with that of other mammals (Negus 1958). Most notably, the nasal cavity is relatively simple, with a lack of complex scrolling of the dorsal and ventral nasal conchae, but there is complex and extensive bony cavitation (Hemsley *et al.* 2013).

The nasal vestibule arches dorsocaudally and is largely occupied by the alar fold. The ventral wall of the nasal cavity is formed by the incisive bone rostrally, where the palate is thick and houses the long, curved roots of the incisor teeth, the maxilla centrally and the palatine bone, which has large left and right vacuities, caudally (Fig. 5.12).

The nasal septum is cartilaginous in the rostral third and osseous in the caudal two-thirds of the nasal cavity. It generally follows a straight course, but in some koalas there is a degree of curvature away from the midline. Dorsal and ventral conchae divide the majority of the length of the nasal cavity into dorsal, middle and ventral meatuses, which communicate via a common meatus (Fig. 5.13). Ethmoidal conchae separated by narrow ethmoidal meatuses are located caudally (Fig. 5.12).

The dorsal concha consists of a simple longitudinal ridge projecting ventrally from the nasal bone and terminating at the rostromedial border of the frontal sinus. The ventral concha consists, for the most part, of a much larger, curved scroll that occupies much of the rostral nasal cavity (Fig. 5.12). Rostrally, it takes the form of a fine, wide bony ridge with a laterally curved free margin. Caudal to the roots of the incisor teeth, the floor of the nasal cavity medial to the ventral concha descends abruptly (Fig. 5.12). At this point, the conchal attachment to the nasal wall is in a ventrolateral position and the conchal plate forms a scroll (Fig. 5.13). Caudal to this, the conchal attachment divides into dorsolateral and ventrolateral components, between which is a depression (the rostral maxillary recess). At approximately the level of molar 1, the ventral portion of the concha re-establishes contact with the ventrolateral nasal wall, thereby forming a paranasal sinus (the ventral conchal sinus). Caudally, this sinus also incorporates the scrolled dorsal portion of the concha. At the level of molar 2, the horizontal plates of the vomer extend laterally from the nasal septum to fuse with the nasal wall. The conchal sinus terminates just caudal to this point.

The vomer divides the caudal nasal cavity into a dorsal component housing the ethmoidal conchae and an open, domed ventral component, the

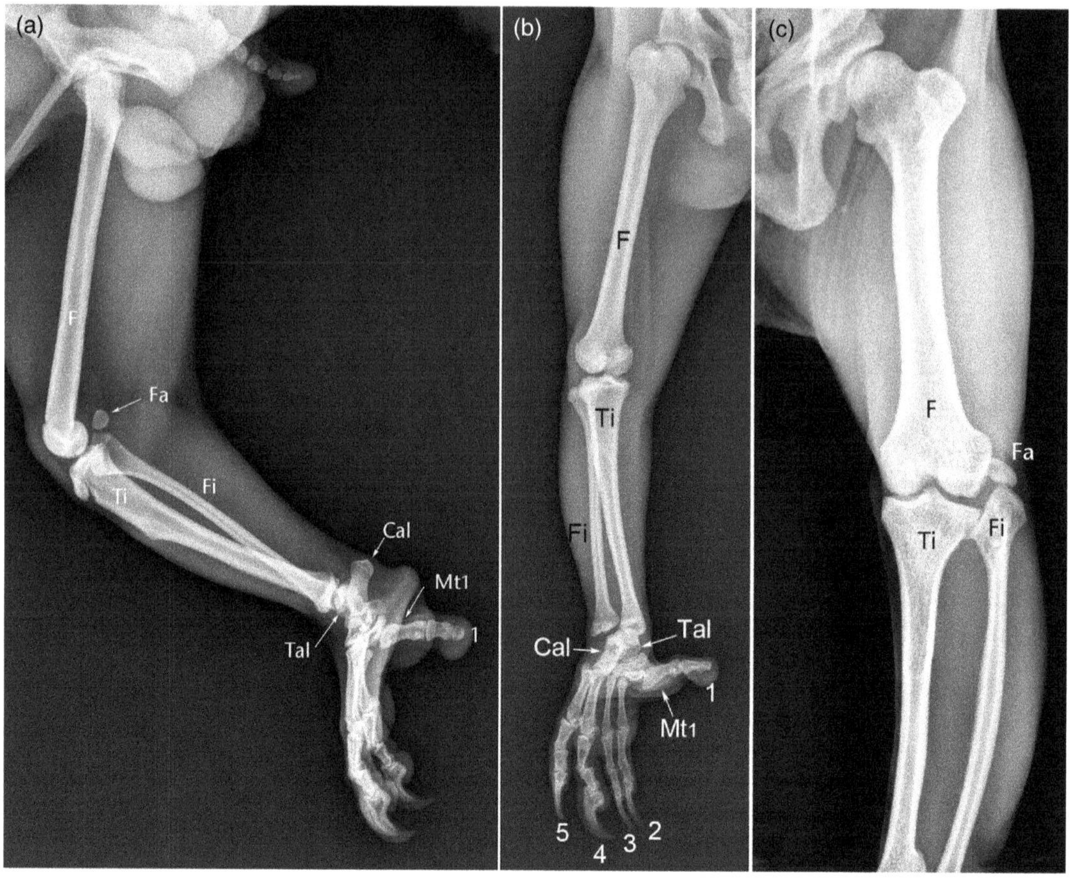

Figure 5.8: Hindlimb: (a) mediolateral view, (b) oblique craniocaudal view, (c) craniocaudal view. Note the absence of an ossified patella and the presence of a single (laterally located) fabella (Fa) caudal to the stifle. The long bones are the femur (F), tibia (Ti) and fibula (Fi). The talus (Tal) and calcaneus (Cal) contribute to the tarsocrural articulation. Digits 2 (2) and 3 (3) are syndactylous. The 1st metatarsal bone (Mt1) is the largest bone in digit 1 (1). Digits 4 and 5 (4, 5) are labelled.

nasopharyngeal meatus (Figs 5.12 and 5.14). The ethmoidal conchae consist of several short scrolls ventral to parts of both the frontal and ventral conchal sinuses (Fig. 5.12). The scrolls extend rostrally from the cribriform plate of the ethmoid bone, which has foramina visible clearly in transverse CT images (Fig. 5.15). The caudal nasal cavity is a common site for cryptococcal granuloma formation. It is a difficult region to access, but well suited to investigation by cross-sectional imaging (Malik *et al.* 1997; Bercier *et al.* 2012; Wynne *et al.* 2012).

The narrow dorsal and middle nasal meatuses are contiguous with the region of the nasal cavity occupied by the ethmoidal conchae. The ventral nasal meatus commences as a narrow passage, enlarges after it descends sharply caudal to the palatine plate of the incisive bone (Fig. 5.13) and then reduces considerably in height in the mid-third of the nasal cavity where the ventral conchal sinus is at its maximum size (Fig. 5.12). The ventral nasal meatus continues caudally as the nasopharyngeal meatus, which communicates with the nasopharynx via the choana (Fig. 5.12).

There are four pairs of bilaterally symmetrical paranasal sinuses. The ventral conchal, frontal and caudal maxillary sinuses are large, while the sphenoidal sinus is much smaller (Fig. 5.12). The ventral conchal sinus occupies the mid-third of the nasal cavity proper, reducing the height of the ventral nasal meatus and narrowing the caudal components of the dorsal, middle and common meatuses (Fig. 5.14). The roof of the ventral conchal sinus is separated from the nasal and frontal bones by a narrow space.

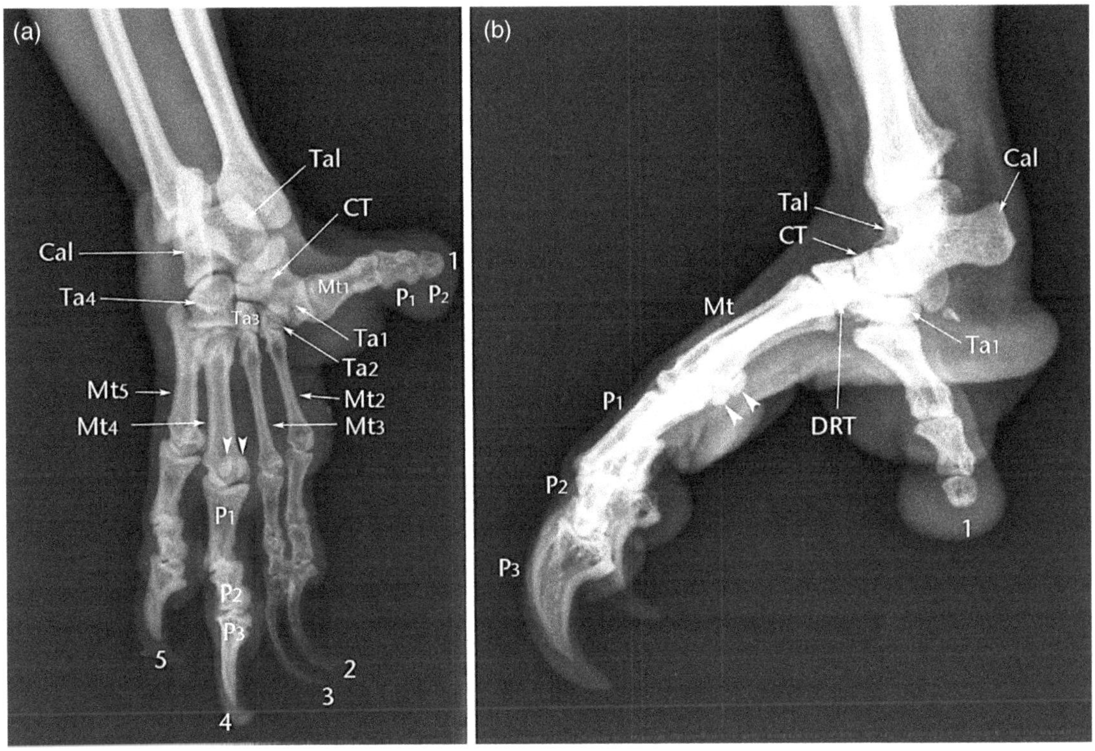

Figure 5.9: Tarsus and pes: (a) dorsoplantar view, (b) lateral view. The talus (Tal) and calcaneus (Cal) form the proximal row of tarsal bones. The central tarsal bone (CT) articulates distally with the 1st, 2nd and 3rd tarsal bones (Ta1, Ta2, Ta3). The 2nd, 3rd and 4th (Ta4) tarsal bones articulate distally with metatarsal bones 2–5 (Mt2, Mt3, Mt4, Mt5). Digit 1 (1) has two phalanges (P1, P2) and the 1st metatarsal bone (Mt1). The remaining digits (2–5) have three phalanges (P1, P2, P3). There are paired metatarsophalangeal sesamoids (arrowheads) on each digit. On the lateral view, the distal row of tarsal bones is indicated (DRT).

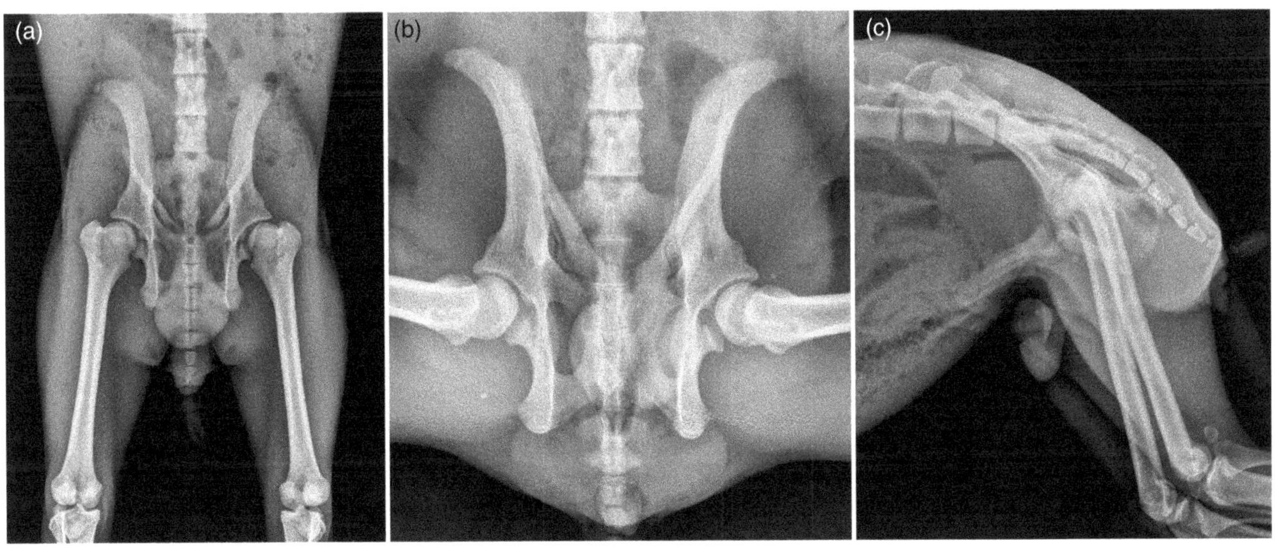

Figure 5.10: Standard radiographic views to evaluate the hip (Pye *et al.* 2008): (a) ventrodorsal view with the hindlimbs extended and rotated medially so that the lateral fabellae are positioned under the lateral condyles of the femurs, (b) ventrodorsal view in a frog-leg position obtained with the femurs at a 90° angle to the spine and the tibias at a 90° angle to the femurs, (c) right lateral view with the hindlimbs extended and overlaying each other.

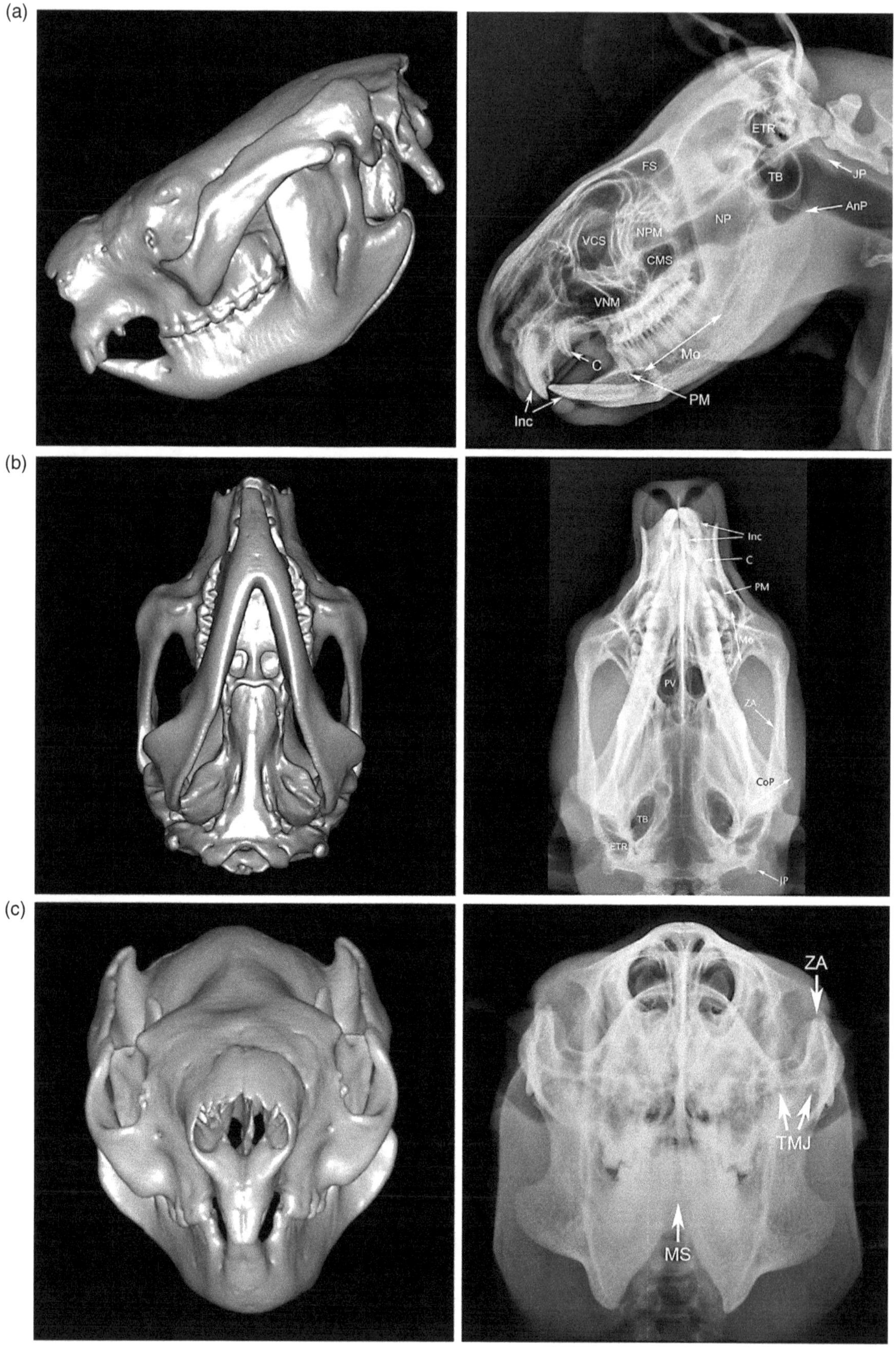
(a)
ETR
FS
TB
JP
AnP
NP
VCS
NPM
CMS
VNM
Mo
C
PM
Inc
(b)
Inc
C
PM
Mo
PV
ZA
CoP
TB
ETR
JP
(c)
ZA
TMJ
MS

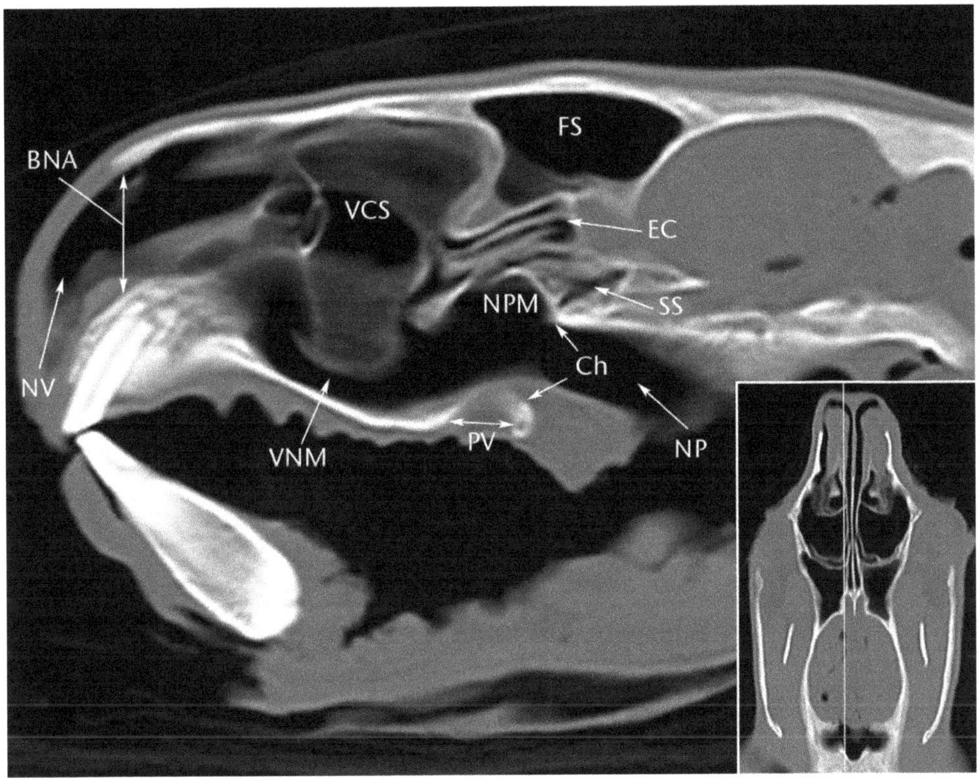

Figure 5.12: Head: sagittal CT image at the level indicated on the dorsal image taken at approximately mid-height of the bony nasal aperture (inset). Bony nasal aperture (BNA); dorsal and ventral margins of the right choana (Ch); ethmoidal conchae (EC) separated by narrow ethmoidal meatuses; frontal sinus (FS); nasopharynx (NP); nasopharyngeal meatus (NPM); nasal vestibule (NV); palatine vacuity (PV); sphenoidal sinus (SS); ventral conchal sinus (VCS); ventral nasal meatus (VNM). (Image reproduced with permission from Hemsley *et al.* 2013, Computed tomographic anatomy of the nasal cavity, paranasal sinuses and tympanic cavity of the koala. *Australian Veterinary Journal* **91**, 353–365, Wiley, Chichester, UK.)

The frontal sinus is undivided and extends from the caudal limit of the orbit to the midpoint of the zygomatic arch. Dorsomedial to the ventral conchal sinus it is only a narrow space, but is more expanded caudally (Fig. 5.12). Rostromedially, there is a narrow communication between the frontal sinus and the ethmoidal meatuses.

The caudal maxillary sinus is ventral to the caudal component of the ventral conchal sinus and dorsal to molars 3 and 4, the roots of which indent

Figure 5.11: Skull. (a) Lateral view of volume-rendered three-dimensional CT reconstruction and comparative lateral radiograph. The skull possesses large air-filled intranasal spaces, particularly the ventral conchal sinus (VCS), nasopharyngeal meatus (NPM), frontal sinus (FS), nasopharynx (NP), ventral nasal meatus (VNM) and caudal maxillary sinus (CMS). Also note the tympanic bulla (TB), epitympanic recess (ETR), jugular process of the occipital bone (JP) and angular process (AnP) of the mandible. Dentition: incisors (Inc); canine (C); premolar (PM); molars (Mo). (b) Ventrodorsal view of volume-rendered three-dimensional CT reconstruction and comparative ventrodorsal radiograph. The air-filled TB and ETR are visible on this view. Note the arrangement of the dentition (Inc., incisors; C, canine; PM, premolar; Mo, molars) and the bulbous condyloid process of the mandible (CoP). The jugular process (JP) is viewed end-on. The palatine vacuities (PV) are accentuated by the underlying intranasal air spaces. Zygomatic arch (ZA). (c) Rostrocaudal view of volume-rendered three-dimensional CT reconstruction and comparative rostrocaudal radiograph. The skull is box-like in this view. The mandibles appear disproportionately large and the temporomandibular joint (TMJ) can be seen as a lucent ellipse. Note the zygomatic arch (ZA) and the fused mandibular symphysis (MS).

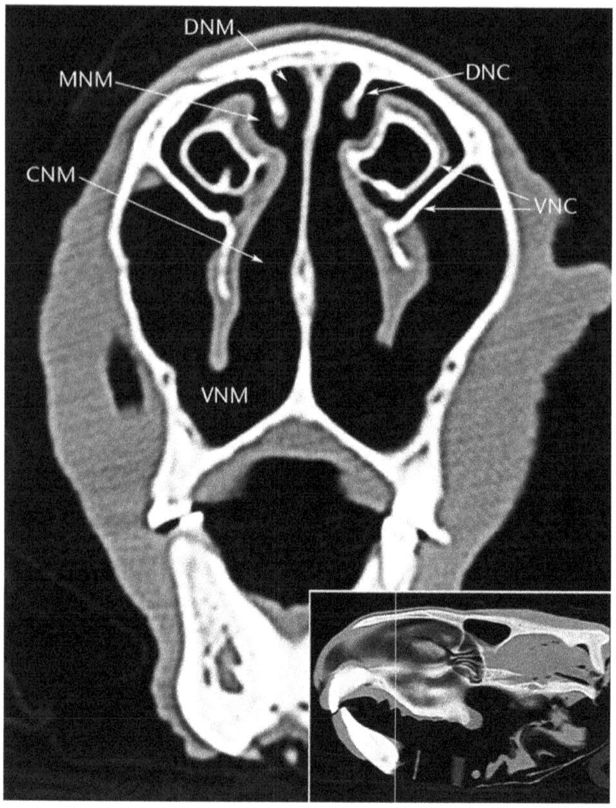

Figure 5.13: Head: transverse CT image at the level indicated on the midline sagittal image (inset). Common nasal meatus (CNM); dorsal nasal concha (DNC); dorsal nasal meatus (DNM); middle nasal meatus (MNM); ventral nasal concha (VNC); ventral nasal meatus (VNM). (Image reproduced with permission from Hemsley *et al.* 2013, Computed tomographic anatomy of the nasal cavity, paranasal sinuses and tympanic cavity of the koala. *Australian Veterinary Journal* **91**, 353–365, Wiley, Chichester, UK.)

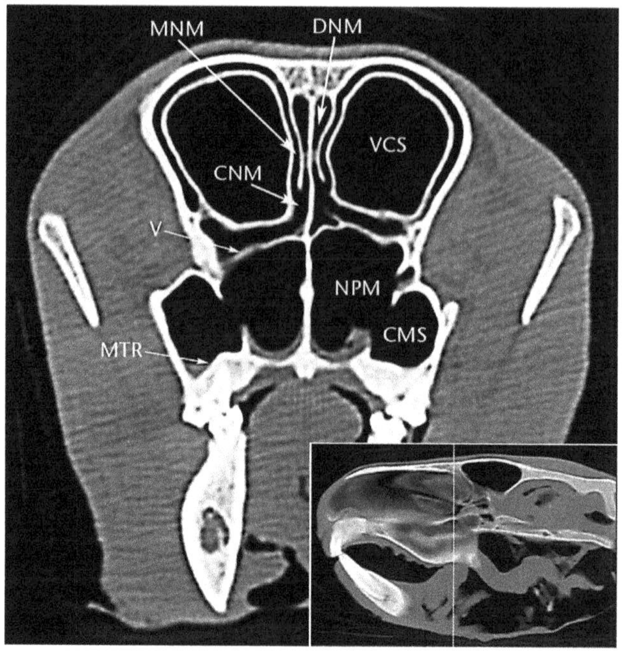

Figure 5.14: Head: transverse CT image at the level indicated on the midline sagittal image (inset). Note the wide communication between the caudal maxillary sinus (CMS) and the nasopharyngeal meatus (NPM). In the CMS, a molar tooth root (MTR) causes indentation of the sinus floor; vomer (V); ventral conchal sinus (VCS); nasopharyngeal meatus (NPM) are labelled. The dorsal (DNM), middle (MNM) and common (CNM) nasal meatuses are defined. (Image reproduced with permission from Hemsley *et al.* 2013, Computed tomographic anatomy of the nasal cavity, paranasal sinuses and tympanic cavity of the koala. *Australian Veterinary Journal* **91**, 353–365, Wiley, Chichester, UK.)

the sinus floor (Fig. 5.14). A large opening communicates medially with the nasopharyngeal meatus (Fig. 5.14).

The small, narrow sphenoidal sinus is located adjacent to the median plane (Fig. 5.12). Dorsorostrally, it has a narrow communication with the ethmoidal meatuses that is appreciable in transverse CT images.

The laterally compressed tympanic bullae, which are formed by the basisphenoid bone, are located medial to the temporomandibular joints and the retroarticular processes of the temporal bone. The epitympanic portion of the tympanic cavity is expansive and consists of medial and lateral compartments separated by a thin bony plate (Fig. 5.16). The medial compartment is located within the temporal bone, predominantly rostral to the petrous part, which houses the inner ear. Ventrally, it communicates widely with the central portion of the tympanic cavity (Figs 5.16 and 5.17). The lateral compartment occupies the base of the zygomatic process of the temporal bone, dorsal to the retroarticular process and external acoustic meatus (Figs 5.16 and 5.17). A round, soft tissue-covered opening is present in the bony wall between the lateral compartment and the external acoustic meatus (Fig. 5.17) and a narrow opening

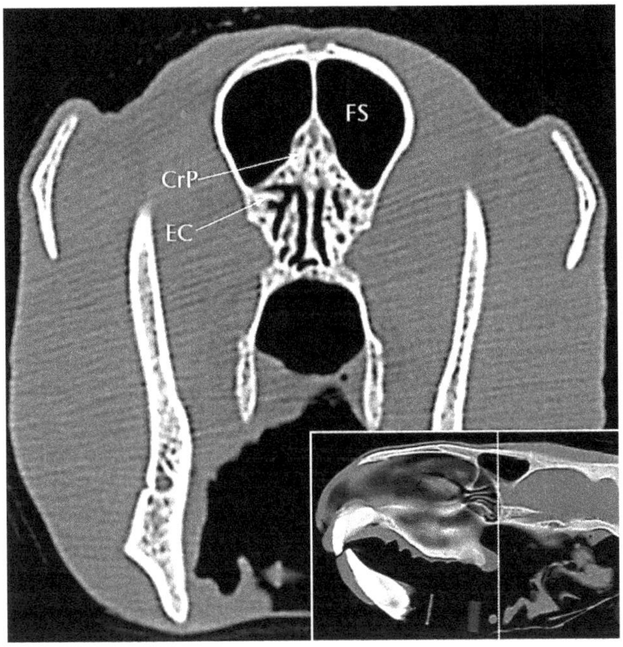

Figure 5.15: Head: transverse CT image at the level indicated on the midline sagittal image (inset). Cribriform plate (CrP); ethmoidal concha (EC); frontal sinus (FS). (Image reproduced with permission from Hemsley *et al.* 2013, Computed tomographic anatomy of the nasal cavity, paranasal sinuses and tympanic cavity of the koala. *Australian Veterinary Journal* **91**, 353–365, Wiley, Chichester, UK.)

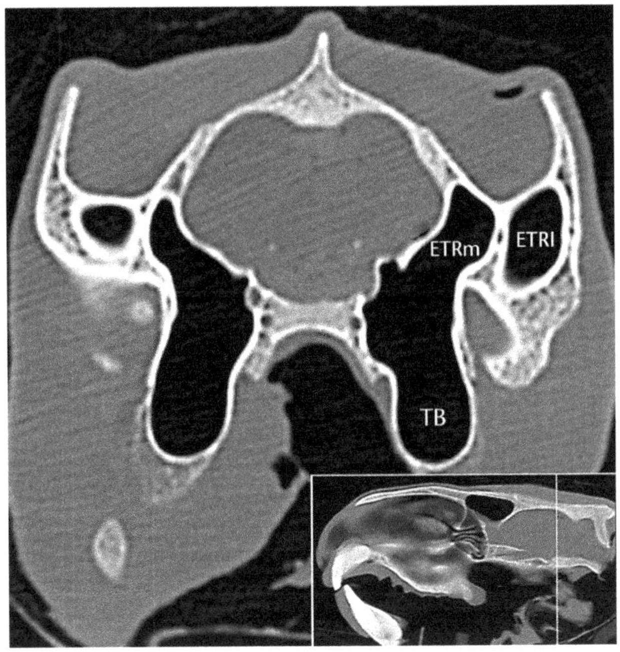

Figure 5.16: Head: transverse CT image, at the level indicated on the midline sagittal image (inset), demonstrating the extensive nature of the epitympanic recess, its separation from the cranial cavity by a thin bony plate and wide communication with the tympanic bulla via the central portion of the tympanic cavity. Epitympanic recess (lateral compartment) (ETRl); epitympanic recess (medial compartment) (ETRm); tympanic bulla (TB). (Image reproduced with permission from Hemsley *et al.* 2013, Computed tomographic anatomy of the nasal cavity, paranasal sinuses and tympanic cavity of the koala. *Australian Veterinary Journal* **91**, 353–365, Wiley, Chichester, UK.)

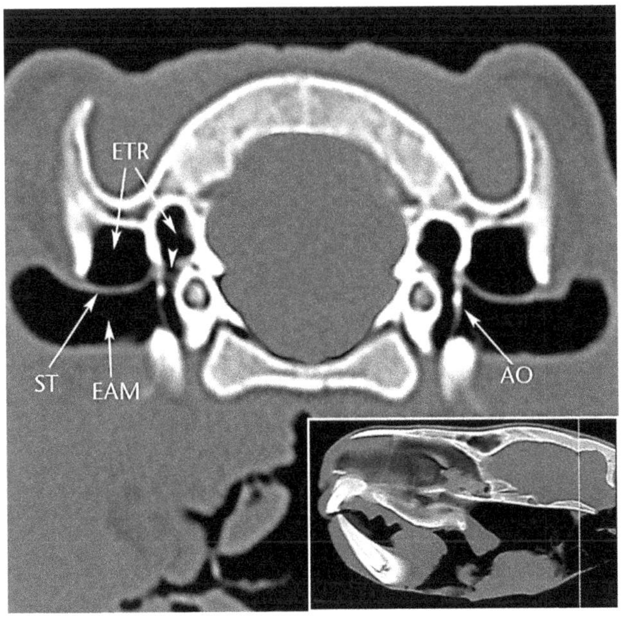

Figure 5.17: Head: transverse CT image, at the level indicated on the midline sagittal image (inset), demonstrating the relationships between the external acoustic meatus and components of the tympanic cavity. Auditory ossicle (AO); external acoustic meatus (EAM); epitympanic recess (ETR); soft tissue (ST) between the lateral compartment of the ETR and the EAM; communication between the medial compartment of epitympanic portion of the tympanic cavity and central portion of the tympanic cavity (arrowhead). (Image reproduced with permission from Hemsley *et al.* 2013, Computed tomographic anatomy of the nasal cavity, paranasal sinuses and tympanic cavity of the koala. *Australian Veterinary Journal* **91**, 353–365, Wiley, Chichester, UK.)

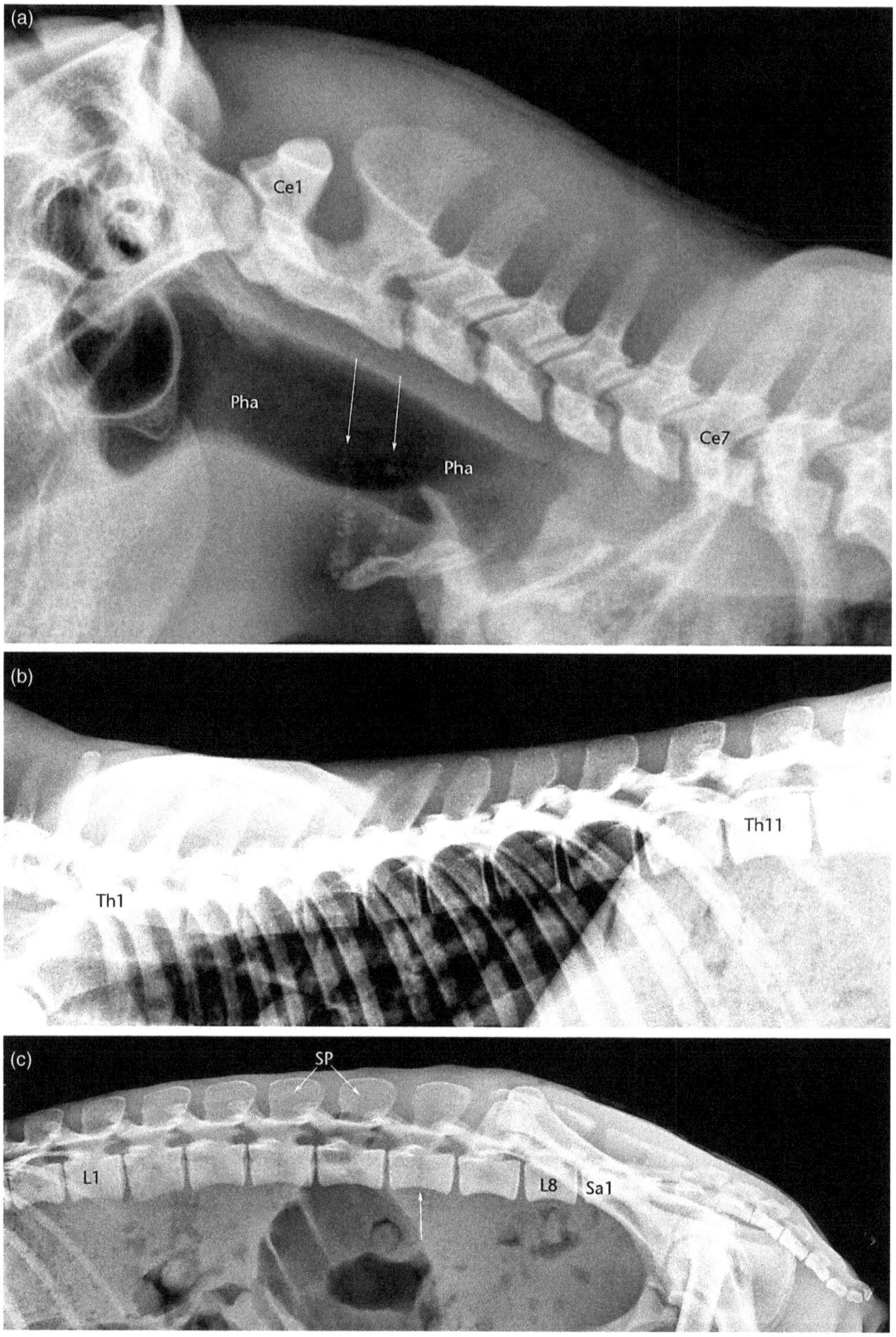

Figure 5.18: (a) Cervical spine, lateral view. There are 7 cervical vertebrae (Ce1, Ce7 labelled). Note the elongated pharynx (Pha) and the two rows of poorly mineralised hyoid bones (arrows). (b) Thoracic spine, lateral view. There are 11 thoracic vertebrae in most koalas (as in this example; Th1 and Th11 labelled), although occasionally there are 12. (c) Lumbar spine, lateral view. There are 8 lumbar vertebrae (L1 and L8 labelled). Note the spatulate-shaped dorsal spinous processes (SP) of the lumbar vertebrae and sacral elements (Sa1 labelled). A vascular foramen perforates the centre of each vertebral body (arrow). The 2nd lumbar vertebral body is slightly trapezoid, compared with the oblong shape of the bodies of the remaining lumbar vertebrae. Note: a transponder can be seen in (b).

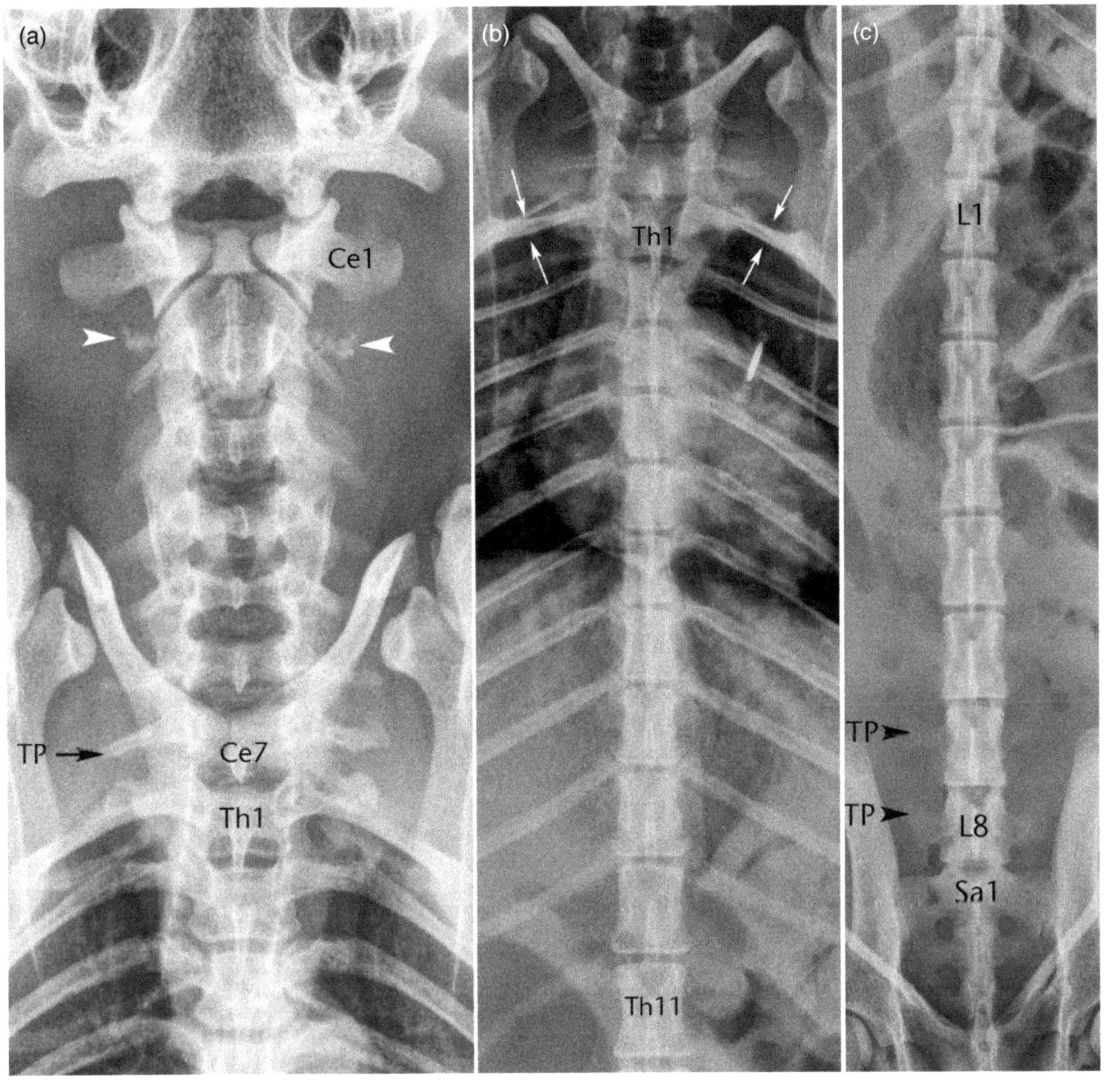

Figure 5.19: Vertebral column: (a) cervical, (b) thoracic, (c) lumbar, ventrodorsal views. Transverse processes (TP) are present on the cervical and lumbar vertebrae. The 1st pair of ribs (white arrows) is thicker and more opaque than succeeding pairs of ribs. Also note the poorly mineralised hyoid bones (white arrowheads) in (a). The 1st and 7th cervical (Ce1, Ce7), thoracic (Th1, Th11) and lumbar (L1, L8) vertebrae and 1st sacral (Sa1) element are labelled. Note: a transponder can be seen in (b).

in the compartment's medial wall communicates with the central component of the tympanic cavity.

The pharyngeal opening of the auditory tube is medial to the hamulus of the pterygoid bone, near the caudal limit of the soft palate. The tube courses adjacent to the pterygoid bone and medial to the dorsal part of the tympanic bulla. At the caudal aspect of the bulla, it passes through a bony foramen adjacent to the ventral margin of the tympanic membrane. This foramen and the bony components associated with the temporal bone and tympanic bulla are more readily identifiable on CT scans than the more rostral components.

Vertebrae

The vertebral column is made up of 7 cervical, 11 or 12 thoracic and 8 lumbar vertebrae, 3 sacral elements and 7–9 caudal vertebrae (the fewest of any marsupial) (Figs 5.6, 5.7, 5.18, 5.19). The first cervical vertebra is incomplete ventrally and the ventral arch of the second cervical vertebra consists of cartilage rather than bone.

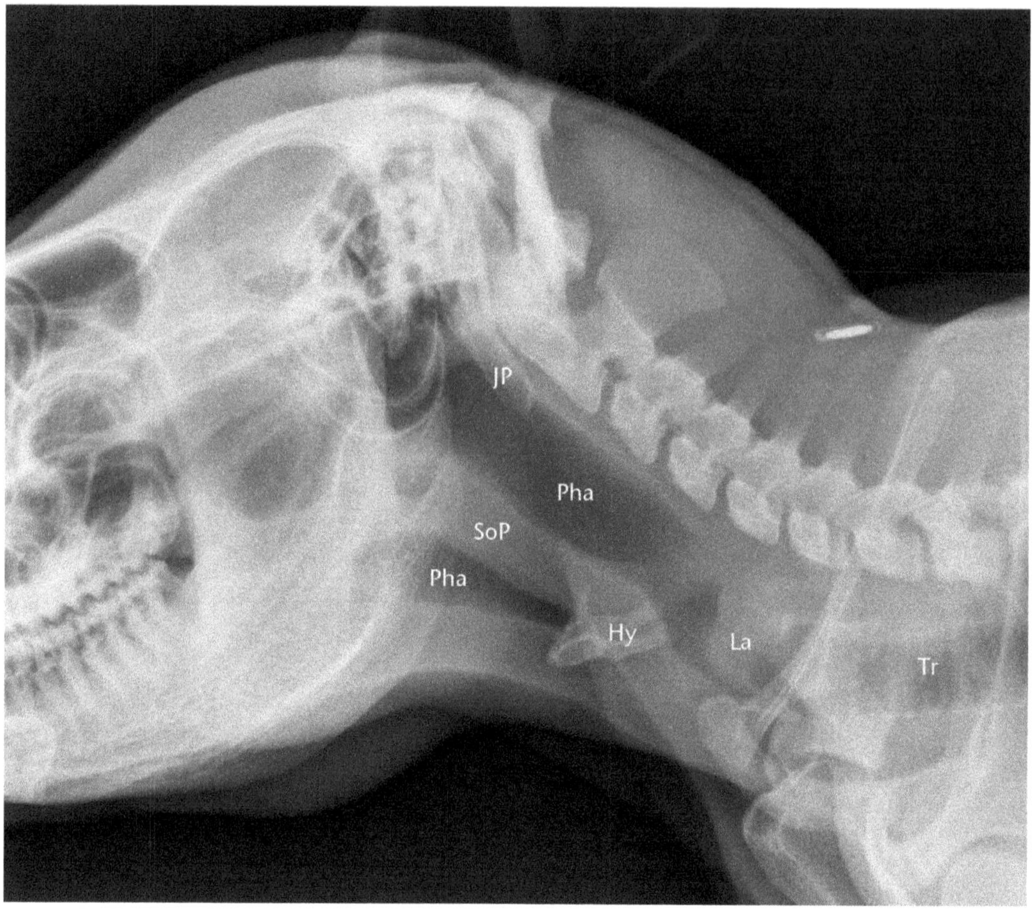

Figure 5.20: Neck: lateral view. The tubular pharynx (Pha), bisected by the elongated soft palate (SoP), defines the cranial end of the caudally descended larynx (La) and caudally displaced trachea (Tr). The hyoid bones (Hy) have descended with the larynx. Note the elongated jugular process (JP). Note: a transponder can be seen.

Ribs and sternum

The koala is exceptional in having only 11 pairs of ribs (occasionally 12), the smallest number for any Australian marsupial (Jones 1923). The first pair of ribs is thicker and more radio-opaque than succeeding pairs of ribs (Fig. 5.19). There are five sternebrae (four plus the xiphisternum). The clavicles articulate ventrally with the cranial end of the first sternal element and dorsally with the spinous process of the scapula (Fig. 5.1).

Soft tissue

Head, neck and thorax

Typically, the thyroid glands, which lack an isthmus, lie between the fourth and ninth tracheal rings.

Pharynx and larynx

The koala has a permanently descended larynx positioned at the level of the third and fourth cervical vertebrae and a well-developed sternothyroid muscle anchored very deep in the thoracic cavity, on the fourth sternebra (Charlton *et al.* 2011).

The pharynx is tubular and extends from the root of the tongue to the cricoid cartilage. For almost its entire extent, it is divided into an upper and a lower channel by the remarkably elongated soft palate (Fig. 5.20), which extends caudally for a distance of 90 mm and reaches as far as the thyroid cartilage (Young 1881).

Radiographically, the basihyoid and thyrohyoid bones appear well defined. The more rostral hyoid bones are represented by two rows of short, poorly mineralised bones (Fig. 5.20). There is a long gap

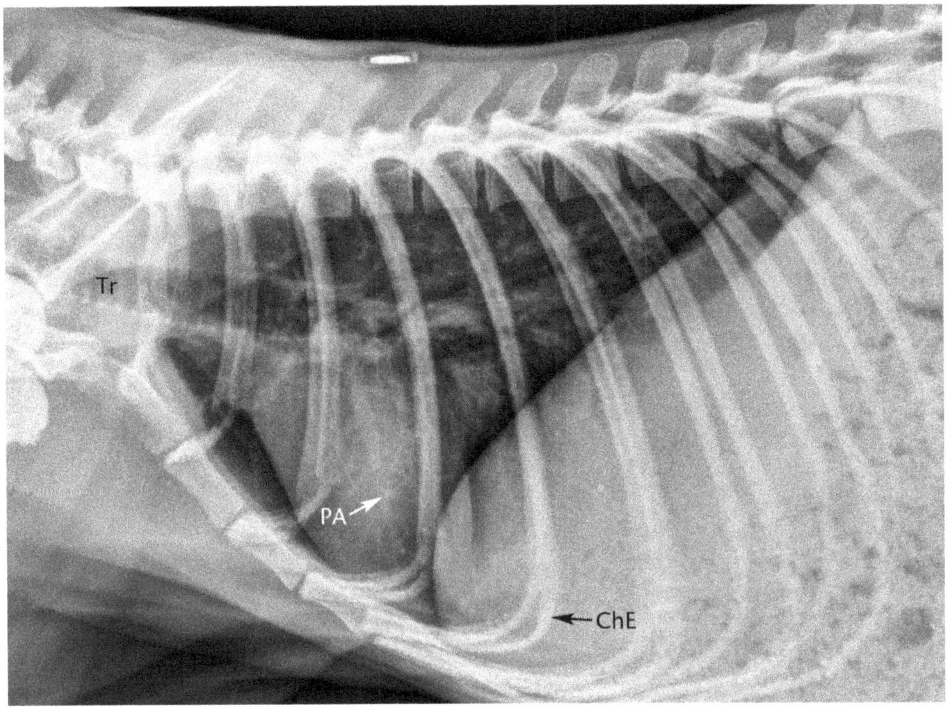

Figure 5.21: Thorax: lateral view. The thoracic cavity appears relatively small compared with the overall size of the torso. The tracheal (Tr) rings and chondral elements (ChE) of the ribs are mineralised. A pulmonary artery (PA) can be distinguished. Note: a transponder can be seen.

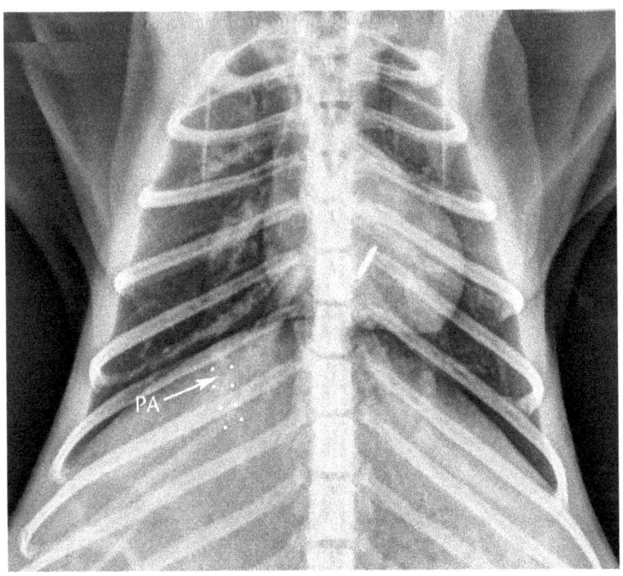

Figure 5.22: Thorax: ventrodorsal view. Lung lobation is not visible. A pulmonary artery (PA, dotted lines) can be defined in this view. Note: a transponder can be seen.

between them and the base of the skull. It appears that these rudimentary bones may represent the stylohyoid, epihyoid and ceratohyoid bones.

Thorax

Koalas have an intact mediastinum, so some pathological conditions can be confined to one side of the thoracic cavity (Blanshard and Bodley 2008). Both superficial cervical and thoracic lobes of the thymus may be present, although usually the thoracic lobes are absent or very small (the reverse of the usual situation in eutherians and some other marsupials, where only the thoracic component is present) (Yadav 1973).

Cardiovascular system

The heart is approximately 50 mm long and rounded at the apex, which is positioned level with the caudal border of the fourth left costal cartilage

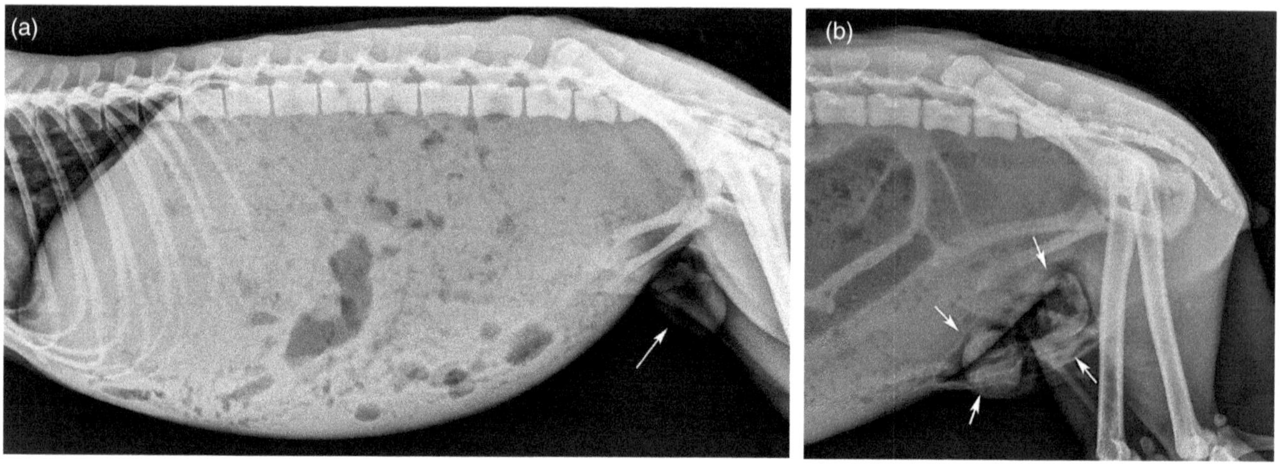

Figure 5.23: Abdomen: (a) male koala (*Phascolarctos cinereus*), lateral view, (b) caudal abdomen, female koala, lateral view. Intraluminal contents (gas and digesta) dominate the radiographic image of the abdomen. Absence of abdominal adipose tissue produces a 'ground-glass' opacity that blends the viscera, rendering distinction of individual abdominal organs impossible. Testes (arrow) are evident ventral to the epipubic bones in the male (a). An immature joey (arrows) is in the pouch of the female (b). Note that the amount of intestinal intraluminal gas in (b) is excessive.

(Fig. 5.21). As in most marsupials, right auricular extensions encompass the aorta and there is no trace of a fossa ovalis. The right atrioventricular valve appears to have one very large, non-septal cusp. There is no obvious phrenicopericardiac ligament, so the pericardium appears directly adherent to the diaphragm (Sonntag 1921; 1922). The caudal *vena cava* is located to the right of the thoracic caudal aorta, rather than the most common arrangement of being ventral to it.

Respiratory system

The capacity of the thorax is very small in comparison with that of the abdomen (Fig. 5.22). The lungs are simple in form, the right lung having three lobes and the left, two (Martin 1836; Forbes 1881). The caudal lobes of each side are approximately equal in size, but are half the size of the two cranial lobes. Accessory lobes are not present.

Abdomen

Koalas are monogastric hindgut fermenters. The stomach is simple and has a large conglomerate gastric gland situated on its lesser curvature. The stomach contents are relatively dry, densely packed, finely chewed leaf material. Approximately 250 mm of small intestine precede the caecum. The mixture of digesta and gas in the gastrointestinal tract presents a confusing image of the abdominal viscera. The viscera cannot be distinguished easily because absence of abdominal adipose tissue results in positive silhouetting of adjacent viscera, camouflaging their serosal surfaces (Figs 5.23 and 5.24).

Koalas possess the largest caecum (up to 2 m long) relative to body size for any known mammal (MacKenzie 1918). The caecum and proximal colon occupy the entire ventral portion of the abdomen. Cylindrical masses of digesta are divided into pellets of increasing dryness and firmness in the distal colon and may be seen on radiographs.

All liver lobe edges (except on a few of the small processes in the centre of the visceral surface) are normally very sharp. The gall bladder is a finger-like sac, approximately 10–15 mm in diameter and 50–60 mm long. The spleen is narrow and strap-like, with a slightly broader and thicker cranial end, and is normally no more than 10 mm thick. The kidneys are simple (not lobulated). When empty, the urinary bladder is contracted, rather thick-walled and lies within the pelvic cavity (Blanshard and Bodley 2008).

The female reproductive tract is typical of marsupials, with paired uteri and cervices and two (nulliparous koalas) or three (multiparous koalas) vaginae. As in other marsupials, male koalas have a prominent scrotum that is cranial to the opening of the common vestibule (cloaca) and penis (Figs 5.6 and 5.7). The scrotum and testes are clearly visible on radiographs of the pelvis and may obscure pelvic structures (Figs 5.6 and 5.7). The unlobed prostate encircles the urethra immediately caudal to the urinary bladder and there are three pairs of bulbourethral glands (Young 1879; Temple-Smith and Taggart 1990).

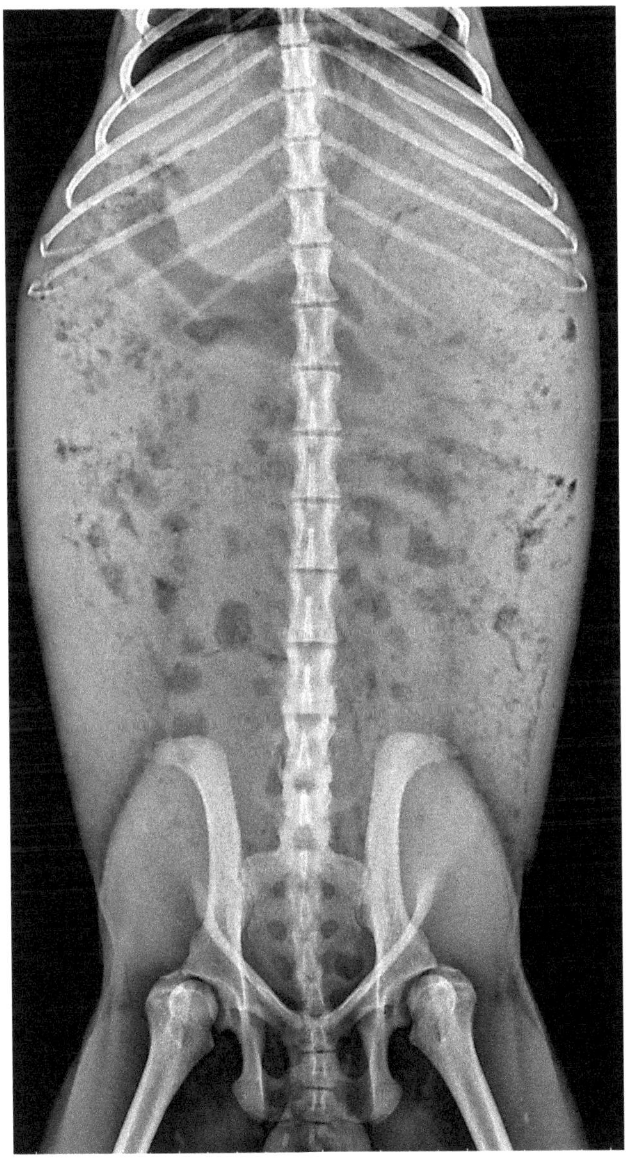

Figure 5.24: Abdomen: ventrodorsal view: Intraluminal contents (gas and digesta) dominate the radiographic image of the abdomen. Absence of abdominal adipose tissue produces a 'ground-glass' opacity that blends the viscera, rendering distinction of individual abdominal organs impossible.

The pouch contains two teats and associated mammary glands. It is not a radiographic feature unless it contains a joey with a mineralised skeleton (Fig. 5.23b). The enlarged mammary gland of a lactating female may be visible radiographically as a mass with soft-tissue density.

Unless enlarged, lymph nodes are not generally discernible on radiographs. The superficial inguinal, superficial axillary, rostral mandibular, mandibular and facial lymph nodes are palpable in healthy koalas (Hanger and Heath 1991; Blanshard 1994). The mandibular nodes are quite small and difficult to discern when not enlarged. The superficial inguinal nodes are like flat lima beans, approximately 10–20 mm in length, and lie in the groin near the origin of the epipubic bones. Several small axillary nodes form a tapered, cord-like chain 15–20 mm long, deep in the axilla. The rostral mandibular lymph node is approximately 5–8 mm long, positioned caudal and slightly ventral to the commissure of the lips, overlying the mandible at the rostral border of the masseter muscles. Facial nodes may sometimes be present over the maxilla, 10–20 mm below the eye. Popliteal and subiliac lymph nodes are absent. The ileocaecal junction is flanked by two circular caecocolic lymphoid patches, and there is a cluster of small mesenteric lymph nodes (Hanger and Heath 1991; Hemsley *et al.* 1996).

References

Bercier M, Wynne J, Klause S, Stadler CK, Gorow A, Pye GW (2012) Nasal mass removal in the koala (*Phascolarctos cinereus*) *Journal of Zoo and Wildlife Medicine* **43**, 898–908. doi:10.1638/2011-0266R.1

Blanshard WH (1994) Medicine and husbandry of koalas. In *Wildlife*. Proceedings 233, pp. 547–623.

Post Graduate Committee in Veterinary Science, Sydney.

Blanshard WH, Bodley KB (2008) Koalas. In *Medicine of Australian Mammals.* (Eds L Vogelnest and R Woods) pp. 227–327. CSIRO Publishing, Melbourne.

Charlton BD, Ellis WAH, McKinnon AJ, Cowin GJ, Brumm J (2011) Cues to body size in the formant spacing of male koala (*Phascolarctos cinereus*) bellows: honesty in an exaggerated trait. *The Journal of Experimental Biology* **214**, 3414–3422. doi:10.1242/jeb.061358

Cork S (1987) Form and function in the koala. In *Koala: Australia's Endearing Marsupial.* (Ed L Cronin) pp. 31–55. Reed Books, Sydney.

Eberhard H (1972) Ecology of the koala, *Phascolarctos cinereus* (Goldfuss), on Flinders Chase, Kangaroo Island. PhD thesis. University of Adelaide, Australia.

Forbes WA (1881) On some points in the anatomy of the koala (*Phascolarctos cinereus*). *Proceedings of the Zoological Society of London* **1881**, 180–195.

Grand TI, Barboza PS (2001) Anatomy and development of the koala, *Phascolarctos cinereus*: an evolutionary perspective on the superfamily Vombatoidea. *Anatomy and Embryology* **203**, 211–223. doi:10.1007/s004290000153

Hanger JJ, Heath TJ (1991) Topography of the major superficial lymph nodes and their efferent lymph pathways in the koala (*Phascolarctos cinereus*). *Journal of Anatomy* **177**, 67–73.

Hemsley S, Canfield PJ, Husband AJ (1996) The distribution of organised lymphoid tissue in the alimentary tracts of koalas (*Phascolarctos cinereus*) and possums (*Trichosurus vulpecula* and *Pseudocheirus peregrinus*). *Journal of Anatomy* **188**, 269–278.

Hemsley S, Palmer H, Canfield RB, Stewart MEB, Krockenberger MB, Malik R (2013) Computed tomographic anatomy of the nasal cavity, paranasal sinuses and tympanic cavity of the koala. *Australian Veterinary Journal* **91**, 353–365. doi:10.1111/avj.12098

Jones FW (1923) *The Mammals of South Australia.* Part I. The Monotremes and the Carnivorous Marsupials (The Ornithodelphia and didactylous Didelphia). pp. 1–131. Government Printer, Adelaide.

Kempster RC, Hirst LW (2002) Bony orbital anatomy of the koala (*Phascolarctos cinereus*). *The Anatomical Record* **267**, 288–291. doi:10.1002/ar.10117

Lanyon JM, Sanson GD (1986) Koala (*Phascolarctos cinereus*) dentition and nutrition. I. Morphology and occlusion of cheek teeth. *Journal of Zoology* **209**, 155–168. doi:10.1111/j.1469-7998.1986.tb03572.x

Lee AK, Carrick FN (1989) Phascolarctidae. In *Fauna of Australia, Volume 1B, Mammalia.* (Eds DW Walton and BJ Richardson) pp. 740–754. Australian Government Publishing Service, Canberra.

Lee A, Martin R (1988) *The Koala: A Natural History.* NSW University Press, Sydney.

MacKenzie WC (1918) *The Gastro-intestinal Tract in Monotremes and Marsupials.* Critchley Parker, Australia.

Malik R, Martin P, Wigney DI, Church DB, Bradley W, Allan GS, Bellenger CR, Lamb WA, Barrs VR, Foster S, Hemsley S, Canfield PJ, Love DN (1997) Nasopharyngeal cryptococcosis. *Australian Veterinary Journal* **75**, 483–488. doi:10.1111/j.1751-0813.1997.tb14377.x

Martin W (1836) Notes on the anatomy of Koala, *Phascolarctos fuscus*, Desm. *Proceedings of the Zoological Society of London* **1836**, 109–113.

Martin RW (1981) Age-specific fertility in three populations of the koala *Phascolarctos cinereus* Goldfuss, in Victoria. *Australian Wildlife Research* **8**, 275–283. doi:10.1071/WR9810275

Martin RW, Handasyde KA (1995) Koala *Phascolarctos cinereus* (Goldfuss, 1817). In *The Mammals of Australia.* (Ed R Strachan) pp. 196–198. Australian Museum / Reed Books, Sydney.

Martin RW, Handasyde KA (1999) *The Koala: Natural History, Conservation and Management.* UNSW Press, Sydney.

Negus V (1958) *The Comparative Anatomy and Physiology of the Nose and Paranasal Sinuses.* E and S Livingstone, Edinburgh.

Owen R (1839) On the osteology of Marsupialia. *The Transactions of the Zoological Society of London*

2(5), 379–408. doi:10.1111/j.1469-7998.1839.tb00030.x

Pye GW (2009) Shoulder dysplasia in koalas (*Phascolarctos cinereus*) at San Diego Zoo. *Journal of Zoo and Wildlife Medicine* **40**, 453–457. doi:10.1638/2008-0107.1

Pye GW, Hamlin-Andrus C, Moll J (2008) Hip dysplasia in koalas (*Phascolarctos cinereus*) at the San Diego Zoo. *Journal of Zoo and Wildlife Medicine* **39**, 61–68. doi:10.1638/2007-0106.1

Reese S, Pfuderer UR, Bragulla H, Loeffler K, Budras KD (2001) Topography, structure and function of the patella and the patelloid in marsupials. *Anatomia, Histologia, Embryologia* **30**, 289–294. doi:10.1046/j.1439-0264.2001.00334.x

Sonntag CF (1921) The comparative anatomy of the koala (*Phascolarctos cinereus*) and vulpine phalanger (*Trichosurus vulpecula*). *Proceedings of the Zoological Society of London* **91**, 547–577. doi:10.1111/j.1096-3642.1921.tb03280.x

Sonntag CF (1922) On the myology and classification of the wombat, koala and the phalangers. *Proceedings of the Zoological Society of London* **92**, 863–896. doi:10.1111/j.1469-7998.1922.tb07085.x

Temple-Smith PD, Taggart DA (1990) On the male generative organs of the koala (*Phascolarctos cinereus*): an update. In *Biology of the Koala*. (Eds AK Lee, KA Handasyde and GD Sanson) pp. 33–54. Surrey Beatty and Sons, Chipping Norton, NSW.

Wynne J, Klause S, Stadler CK, Pye GW, Meyer W, Sykes JE (2012) Preshipment testing success: resolution of a nasal sinus granuloma in a captive koala (*Phascolarctos cinereus*) caused by *Cryptococcus gatti*. *Journal of Zoo and Wildlife Medicine* **43**, 939–942. doi:10.1638/2012-0045R.1

Yadav M (1973) The presence of the cervical and thoracic thymus lobes in marsupials. *Australian Journal of Zoology* **21**, 285–301. doi:10.1071/ZO9730285

Young AH (1879) On the male generative organs of the Koala (*Phascolarctos cinereus*). *Journal of Anatomy and Physiology* **13**, 305–317.

Young AH (1881) Anatomy of the Koala (*Phascolarctos cinereus*). *Journal of Anatomy and Physiology* **15**, 466–474.

6

Wombats

There are three species of wombat: the common wombat (*Vombatus ursinus*), the southern hairy-nosed wombat (*Lasiorhinus latifrons*) and the northern hairy-nosed wombat (*Lasiorhinus krefftii*). The three species are broadly similar in appearance. The common wombat has bare nasal skin and a coarse, short hair coat. The ears are small and rounded. In contrast, hairy-nosed wombats have longer, silky fur, soft, furred muzzles and longer, pointier ears. Northern hairy-nosed wombats are similar in appearance to the southern species, but with a larger, square muzzle (Strahan 1995). Adult wombats of all three species weigh approximately 20–35 kg. Males are larger than females except in the northern hairy-nosed species, the females of which are larger in every dimension except neck girth (Johnson and Crossman 1991).

Wombats are fossorial grazers and many of their physical characteristics reflect this. They have large, broad, dorsoventrally flattened heads, thick skulls, solid jaws and a short neck. Both pectoral and pelvic girdles are strong. They have stocky, powerful bodies with short, stout, strong muscular limbs and strong, clawed feet, ideally suited for digging (Fig. 6.1). The sacrum is large and broad. They are expert burrowers, using their forelimbs for digging and their incisors for cutting and removing obstacles. The hindlimbs move soil away from the excavation site (Wells 1989). Wombats walk on the soles of their feet (plantigrade) with a characteristic rolling gait. They walk, trot or bound and are capable of moving at high speed (40 km/h) for short distances when necessary (Wells 1989). Wombats of all three species have a characteristic 'sacral plate' on the rump that is almost as rigid as bone. The plate is used to help compact burrow walls and may also provide protection against predators.

In this chapter, anatomical descriptions and radiographic images focus primarily on the common and southern hairy-nosed wombats. No specimens of northern hairy-nosed wombats were available for examination.

Skeletal system

Appendicular skeleton

Pectoral girdle and forelimb

As in most marsupials, clavicles are present. They are long (up to 8 cm), robust and have a double curvature. They are expanded at the sternal extremity, where the articular surface has a deep notch. They are compressed distally towards the acromion and the articular surface with the acromion is extended and narrow. The acromion is located at the distal end of the slender, curved spinous process of the scapula (Figs 6.2 and 6.3). The coracoid process of the scapula, which protrudes from the craniomedial margin of the glenoid cavity, is well developed (Owen 1838) (Fig. 6.2). The scapula is rectangular and thickened. The glenoid cavity is directed slightly cranially and medially and the scapular spine lies at 60° to the vertebral border. The supraspinous fossa is large (Grand and Barboza 2001).

The humerus is compact and robust and the attachments for the rotator cuff and deltoid muscles (greater and lesser tubercles and the deltoid tuberosity) are strongly developed (Figs 6.3 and

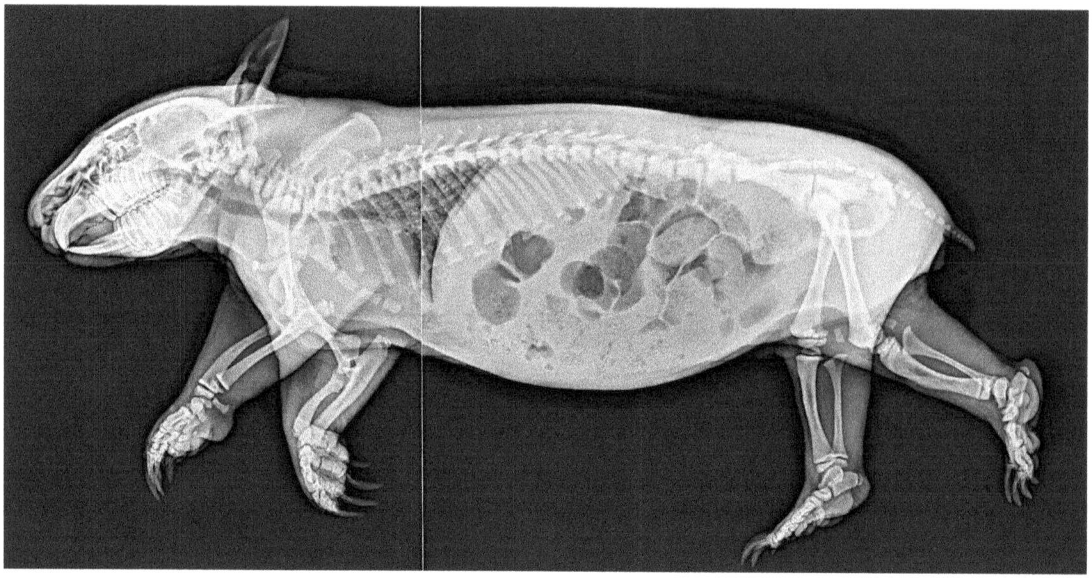

Figure 6.1: Whole body, composite lateral radiograph of an immature common wombat (*Vombatus ursinus*).

6.4). The deltoid tuberosity extends from the greater tubercle to the mid-diaphysis of the humerus (Fig. 6.4). The epicondylar origins of the extensor and flexor muscles are large (Grand and Barboza 2001). The supinator crest (Fig. 6.4) is also strongly developed and laterally it extends in the form of a hooked process above the groove of the radial nerve (Owen 1839).

Relative to body size, the wombat's radius and ulna are short. The olecranon of the ulna is elongated (Fig. 6.5). The radius is compact. The pronounced olecranon, the prominent deltoid tuberosity of the humerus, the strong scapular spine and acromion and the robust clavicle are indicative of strong forelimb musculature (Wells 1989).

The manus is broad and bears five digits with strong, flattened claws (Wells 1989). The distal radius articulates with a broad radial carpal bone, which represents combined radial and intermediate carpal bones. The distal ulna articulates with a depression formed by the ulnar and accessory carpal bones. The second (distal) row of carpal bones consists of four bones. The first carpal bone supports digit 1 and has a small sesamoid bone articulating with its radial carpal surface. The second carpal bone articulates with digit 2 and is wedged between the radial carpal bone and the first and third carpal bones. The fourth carpal bone is the largest of the second row of carpal bones and articulates distally with the fourth and fifth metacarpal bones, medially with the third carpal bone and proximally with the ulnar carpal bone. The five metacarpal bones are thick and short. Digit 1 has two phalanges while the others (2–5) have three (Owen 1839) (Fig. 6.6).

Pelvis and hindlimb

The pelvis is strong and broad, with a marked outward flaring of the ilia in the southern hairy-nosed wombat and to a lesser extent in the common wombat. The tuberosities of the ischia incline laterally and have a distinct and strong process that extends laterally and curves cranially (Fig. 6.7). The epipubic bones articulate with the pubis at two points (Owen 1839). The acetabula are deep and their articulation with the femoral head is congruent.

All three wombat species have a characteristic 'sacral plate', which is almost as rigid as bone, on their rump, dorsal to the pelvis. This 'plate' comprises thick, dense skin (Fig. 6.8).

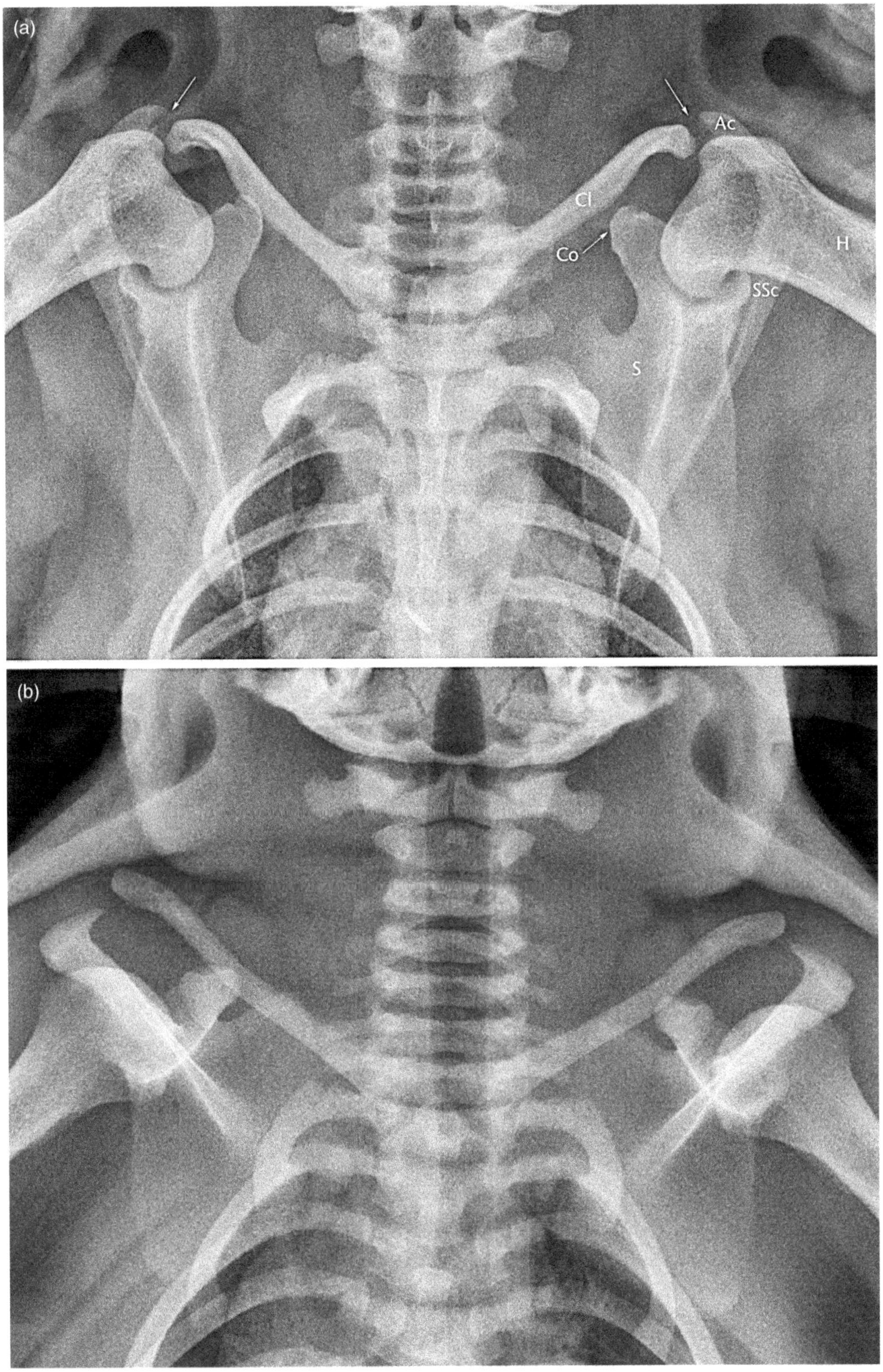

Figure 6.2: Pectoral girdle, ventrodorsal view: (a) southern hairy-nosed wombat (*Lasiorhinus latifrons*), (b) immature common wombat (*Vombatus ursinus*). The arrows indicate the junction of the clavicle (Cl) and acromion (Ac), superimposed on the humerus (H) at the distal end of the spinous process (SSc) of the scapula (S). The craniomedial edge of the glenoid cavity is marked by the coracoid process (Co) of the scapula. Note: a transponder can be seen in (a).

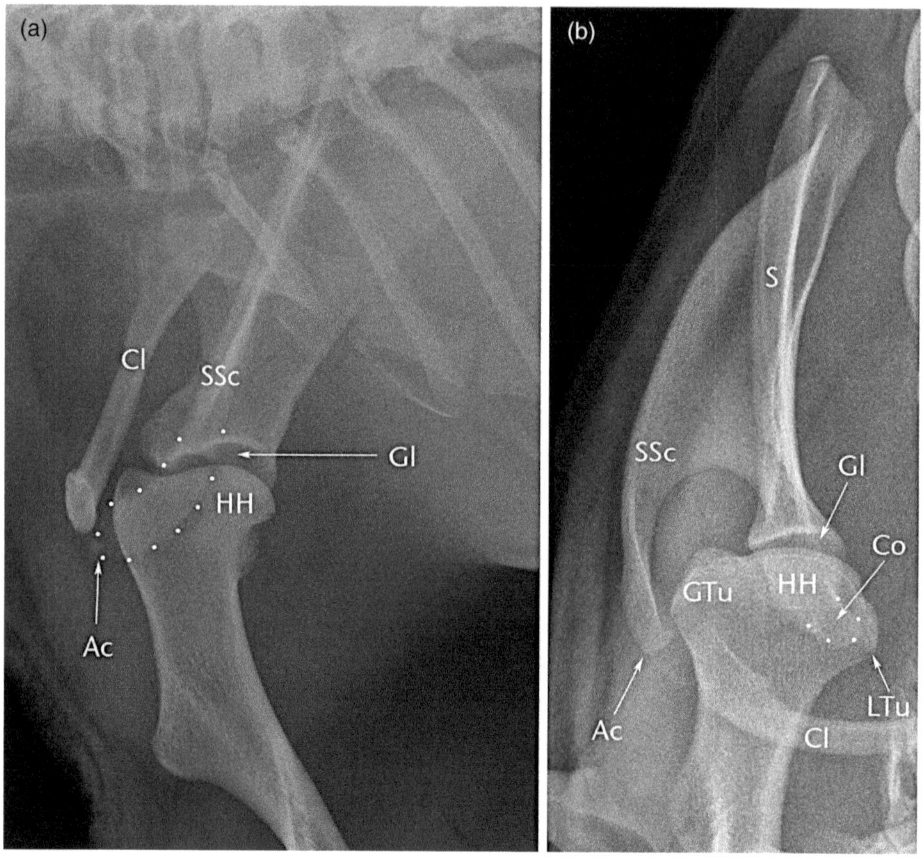

Figure 6.3: Shoulder, common wombat (*Vombatus ursinus*): (a) mediolateral view, (b) craniocaudal view. The elongated, curved spinous process (SSc, dotted line) of the scapula (S) extends distal to the shoulder joint where the acromion (Ac) articulates with the clavicle (Cl). The concave glenoid cavity (Gl) articulates with the humeral head (HH). The coracoid process (Co, dotted line) is superimposed on the humeral head and lesser tubercle (LTu). The greater tubercle (GTu) is large.

The femur has a round head, with the round ligament occupying a large, deep acetabular fossa. The greater and lesser trochanters are features of the proximal end of the femur (Fig. 6.9). At the distal end, the lateral and medial condyles articulate with the tibia. As is the case in all marsupials other than bandicoots and the bilby, wombats do not have an ossified patella. It is represented only by a patelloid, which is a fibrocartilaginous induration of the quadriceps femoris extensor tendon (Reese *et al.* 2001) (Figs 6.9 and 6.10).

The fibula expands to be nearly equivalent in size to the tibia at the distal extremity and takes a large share in the formation of the tarsocrural joint. The articular surface of the fibula is slightly convex, while that of the tibia is slightly concave. The proximal extremity of the fibula is enlarged, but compressed, and has two tuberosities laterally. A large sesamoid bone articulates with the proximal end of the fibula (Owen 1839) (Fig. 6.10).

The talus articulates with the tibia, fibula, calcaneus and the central tarsal bone. The articular surface of the tibia is concavo-convex and articulates mainly with the talus. The fibula articulates with the calcaneus. There is a central tarsal bone and four tarsal bones in the distal row of the tarsus. These articulate with the five metatarsal bones, with the fourth tarsal bone sharing an articulation with the fourth and fifth metatarsal bones. The pes is long, narrow and slightly rotated, which results in a pigeon-toed stance. There are five digits. Digit 1 is reduced to a small nub and is clawless. Digits 2

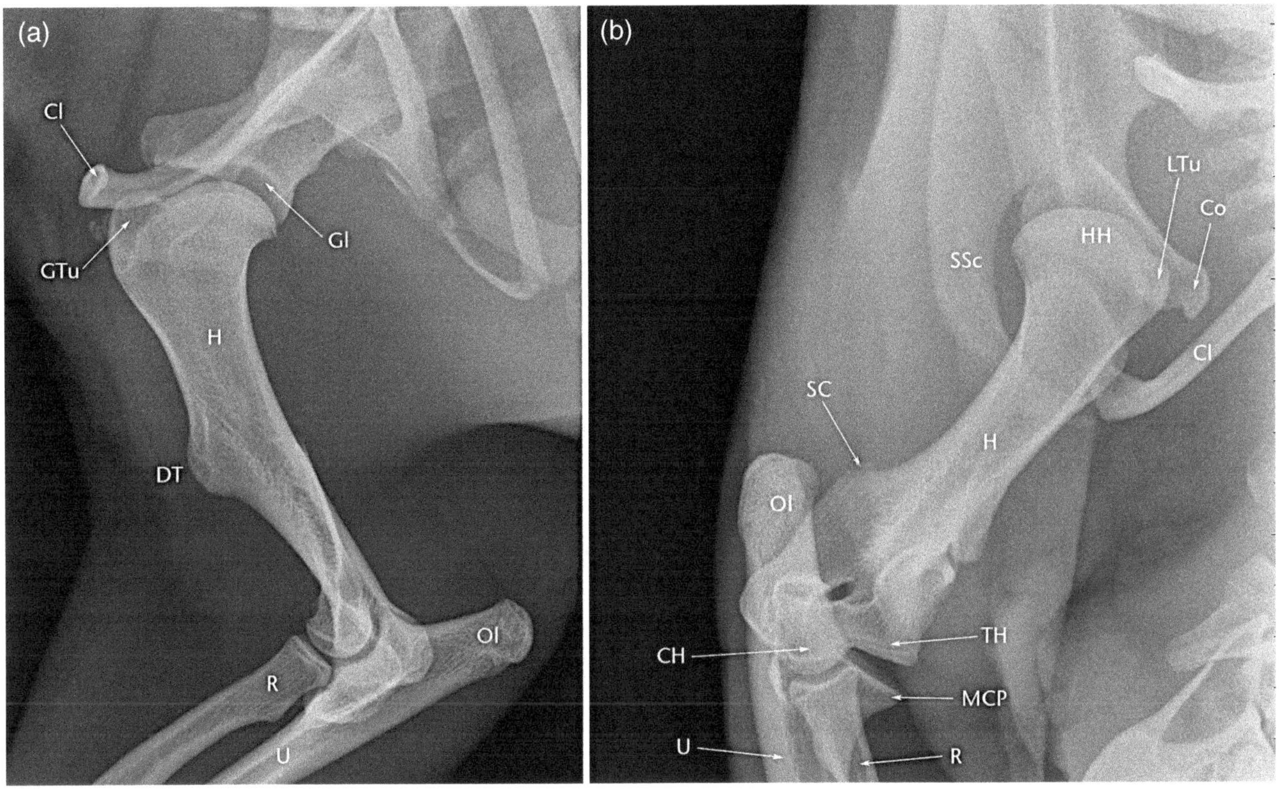

Figure 6.4: Brachium, common wombat (*Vombatus ursinus*): (a) mediolateral, (b) oblique craniocaudal view. The proximal humerus (H) features a head (HH), a greater tubercle (GTu) and a lesser (LTu) tubercle. The HH articulates with the glenoid cavity (Gl) of the scapula. The coracoid process (Co) extends craniomedially from the margin of the glenoid cavity. The spinous process (SSc) of the scapula articulates with the clavicle (Cl) through its distal process, the acromion (obscured in these views). The body of the humerus (H) features a prominent deltoid tuberosity (DT) cranially and a supinator crest (SC) laterally. The elbow articulations are between the trochlea humeri (TH) and the medial coronoid process (MCP) of the ulna (U) and the capitulum humeri (CH) and the radius (R). The olecranon (Ol) is elongated and extends proximal to the elbow joint.

and 3 are syndactylous and have long, strong claws. Digit 4 is slightly shorter and digit 5 is much shorter than digit 3. Both digits also have well-developed claws (Wells 1989). Paired plantar metatarsophalangeal sesamoid bones are present in all digits and sesamoid bones may also be located on the plantar surface of the proximal interphalangeal articulation (Fig. 6.11).

Axial skeleton

Skull

The skull is large, broad and dorsoventrally flattened (Fig. 6.12). There are significant differences between the skulls of the two genera of wombats. Some examples include a more prominent zygomatic process in the hairy-nosed wombats, the palate being narrow between the first molars of the common wombat, but wide in hairy-nosed wombats, and the nasal bones of hairy-nosed wombats being broader than those of the common wombat (Scott *et al.* 1988; Wells 1989; Triggs 2009) (Figs 6.13 and 6.14).

The zygomatic arch is complete and the orbit, as in all marsupials, is open dorsally. Palatine vacuities are present in the caudal palate (Parker and Haswell 1921) (Fig. 6.12). The auditory bulla is not formed by the tympanic, or any other element of the temporal bone, but by the expansion of the base of the greater wing of the sphenoid bone (Owen 1839). The mandibular symphysis is fused. The

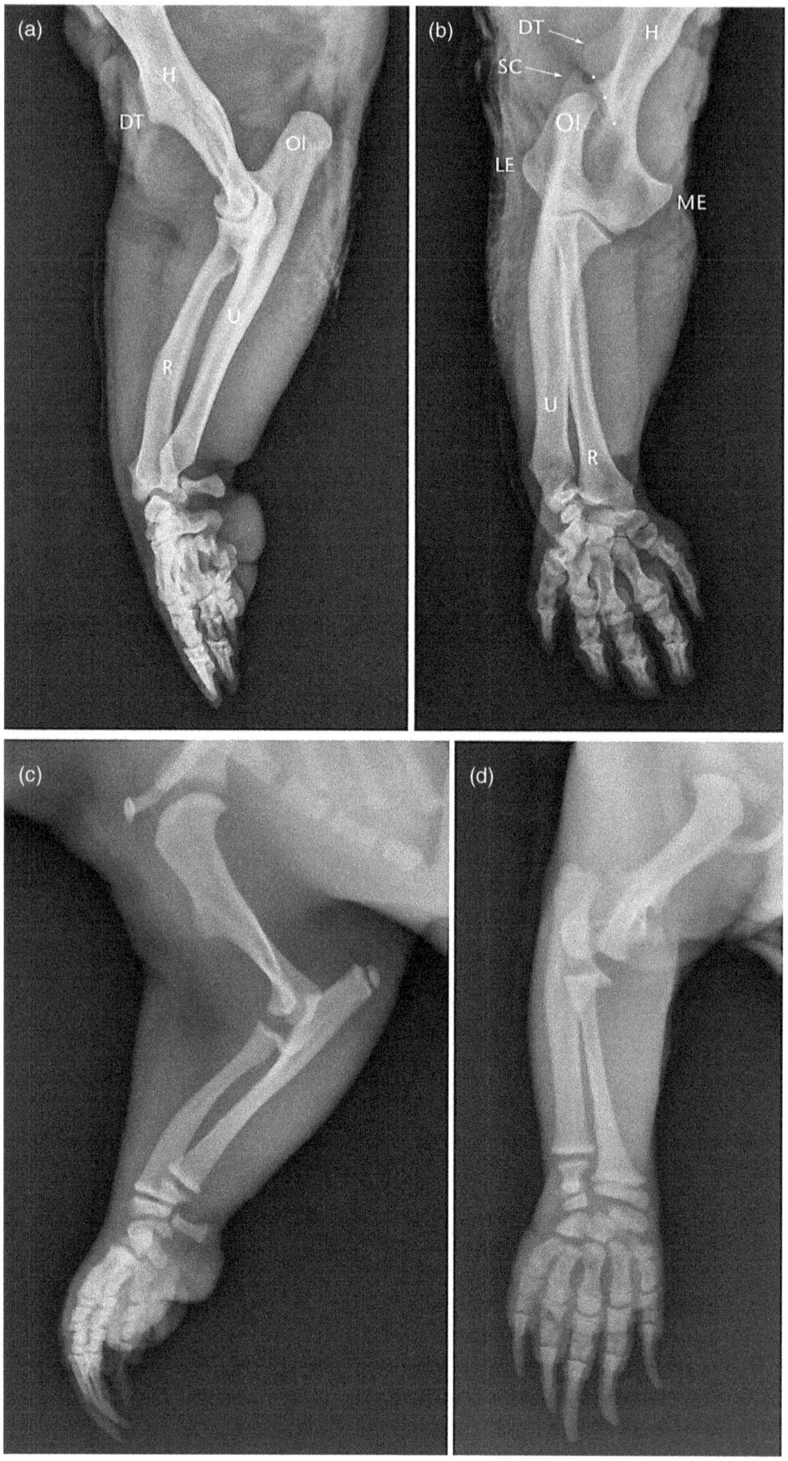

Figure 6.5: Brachium and antebrachium, adult southern hairy-nosed wombat (*Lasiorhinus latifrons*): (a) mediolateral view, (b) craniocaudal view and dorsopalmar view. Immature common wombat (*Vombatus ursinus*): (c) mediolateral view, (d) craniocaudal and dorsopalmar view. The deltoid tuberosity (DT) of the humerus (H) projects craniolaterally and extends distomedially as the deltoid ridge (dotted line). The supinator crest (SC) extends proximally from the lateral epicondyle (LE). There is a large medial epicondyle (ME). The olecranon (Ol) of the ulna (U) is elongated. The ulna (U) and the radius (R) form the antebrachium. Note the southern hairy-nosed wombat images were obtained from disarticulated limbs of a cadaver.

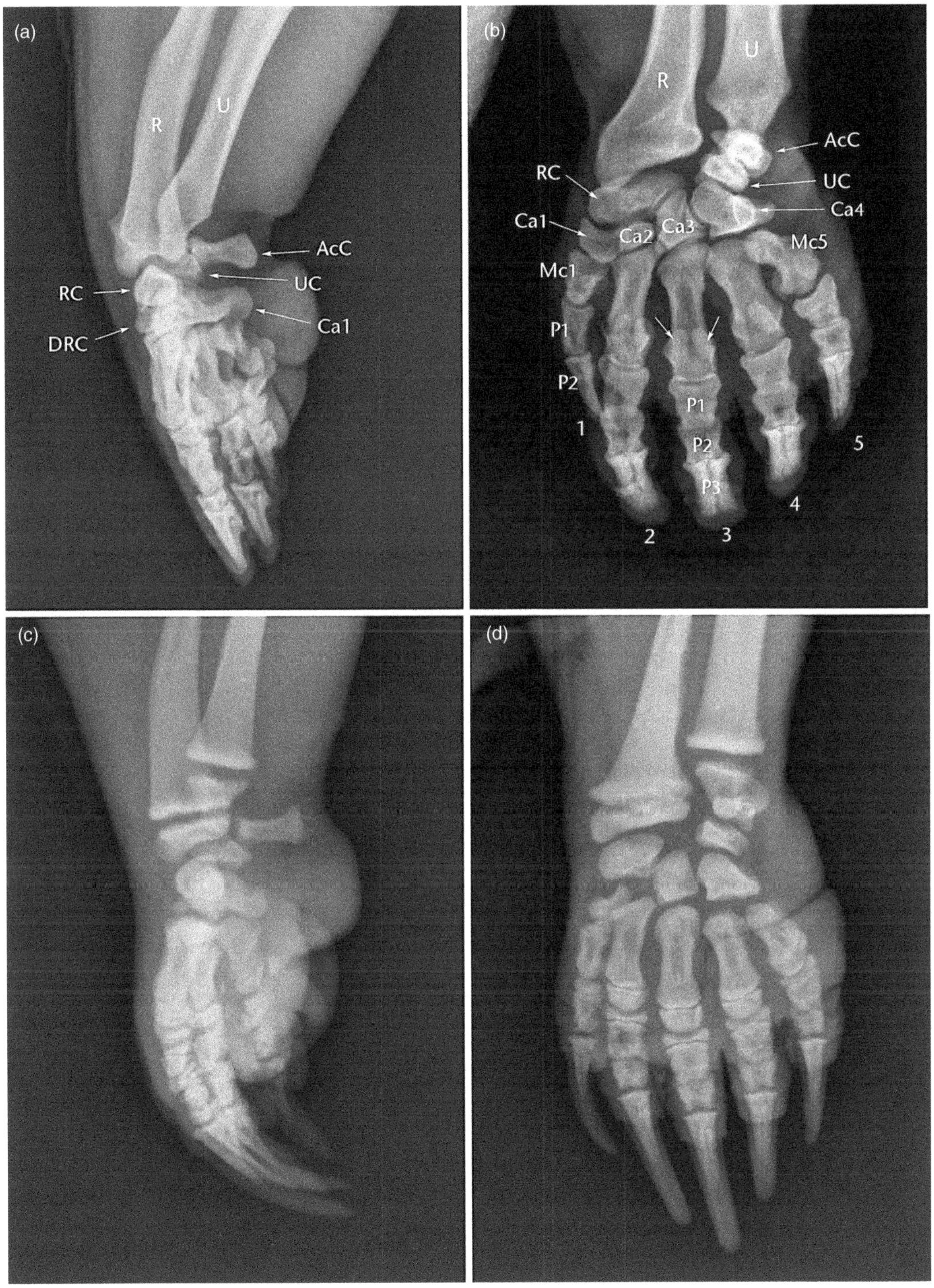

Figure 6.6: Carpus and manus: (a) southern hairy-nosed wombat (*Lasiorhinus latifrons*), mediolateral view, (b) dorsopalmar view; (c) immature common wombat (*Vombatus ursinus*), mediolateral view, (d) dorsopalmar view. The radius (R) and ulna (U) provide the antebrachial articulation with the proximal row of carpal bones, which comprises the accessory carpal (AcC), the ulnar carpal (UC) and the radial carpal (RC) bones. The distal row of carpal bones (DRC) consists of carpal bones 1–4 (Ca1, Ca2, Ca3, Ca4), which articulate with digits 1–5 (1, 2, 3, 4, 5). The metacarpal bones (Mc1, Mc5 labelled) articulate with the 1st phalangeal bone (P1) of each digit. Paired sesamoid bones are present on the palmar surface of the distal end of each metacarpal bone (arrows). There are two phalanges (P1, P2) in digit 1 and three (P1, P2, P3) in digits 2–5.

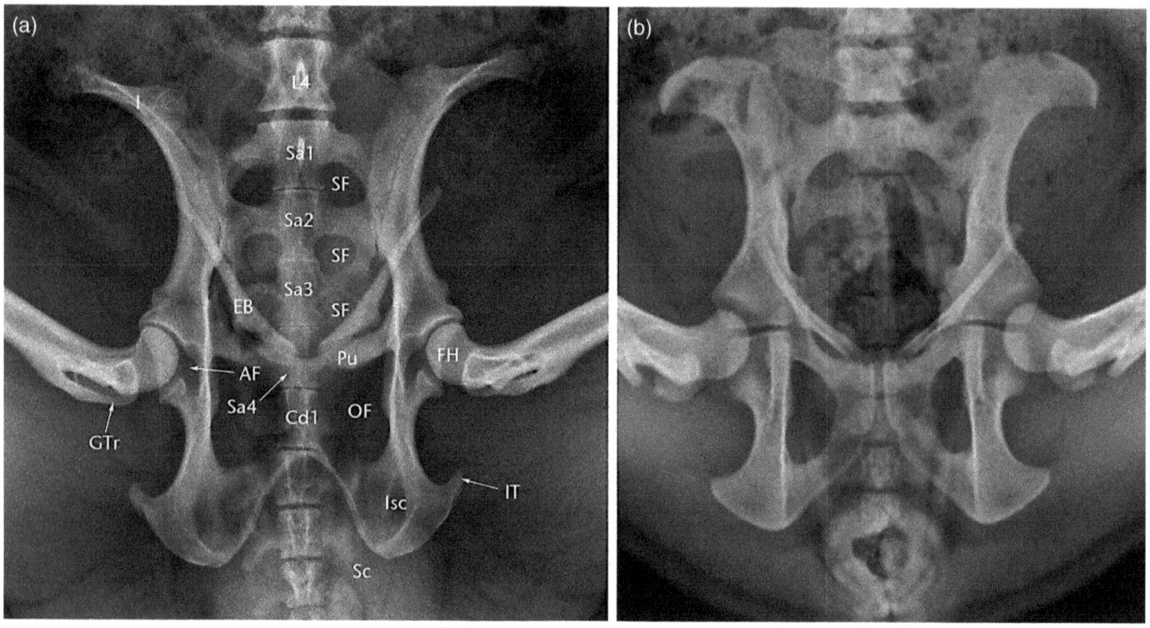

Figure 6.7: Pelvis, ventrodorsal view: (a) male southern hairy-nosed wombat (*Lasiorhinus latifrons*), (b) immature female common wombat (*Vombatus ursinus*). The four fused sacral elements (Sa1, Sa2, Sa3, Sa4) are partially separated by sacral foraminae (SF), while their fused lateral processes contribute to each sacroiliac articulation. The ilium (I) is markedly curved in the southern hairy-nosed wombat and the ischium (Isc) has a well-defined, hooked ischial tuberosity (IT). The epipubic bones (EB) articulate with the pubis (Pu). An obturator foramen (OF) separates the ischium and pubis. There is a deep acetabular fossa (AF) medial to the coxofemoral joint. The femoral head (FH) articulates congruently with the acetabulum. A well-defined greater trochanter (GTr) is visible. The scrotum (Sc) of the southern hairy-nosed wombat is superimposed on the caudal vertebrae. Key vertebral elements are labelled: 4th lumbar vertebra (L4), sacral elements 1–4 (Sa1, Sa2, Sa3, Sa4), 1st caudal vertebra (Cd1).

angle of the mandible is obtuse and there are prominent, laterally flattened angular processes (Figs 6.12 and 6.13; see also Fig. 6.17).

Wombats are unique among marsupials in that all teeth are open-rooted and continuously growing (elodont), with the core of soft dentine and margin of hard enamel ensuring maintenance of an efficient dental mill. They are diprotodont (one pair of lower incisors), but are also unique by being the only diprotodont marsupial with only one pair of upper incisors. Dentition and dental radiology are covered in Chapter 11. Features of wombat dentition are illustrated in Figs 6.13–6.16.

Vertebrae

Hairy-nosed wombats have 7 cervical, 13 thoracic, 6 lumbar, 4 sacral and 15–18 caudal vertebrae. The common wombat differs by having 15 thoracic, 4 lumbar and 10–12 caudal vertebrae. The tail is vestigial (Wells 1989). Part of the body of the atlas remains permanently cartilaginous (Owen 1839). The cervical vertebrae appear tall, short and wide, with long, thin, transverse processes (Fig. 6.17). The thoracic vertebrae have tall, caudally directed dorsal spinous processes (Fig. 6.18). In common wombats, large, cranially directed mammillary processes become visible around T12 and persist to L1 (Fig. 6.18). The lumbar vertebrae have well-developed transverse processes (Fig. 6.19).

There are four sacral elements, the bodies of which are fused. Their transverse processes are long. They are separated centrally but fused abaxially. There is an ovoid sacral foramen where they are separate. The lateral ends of the transverse processes of the first two sacral elements fuse to form the sacral part of the sacroiliac articulation

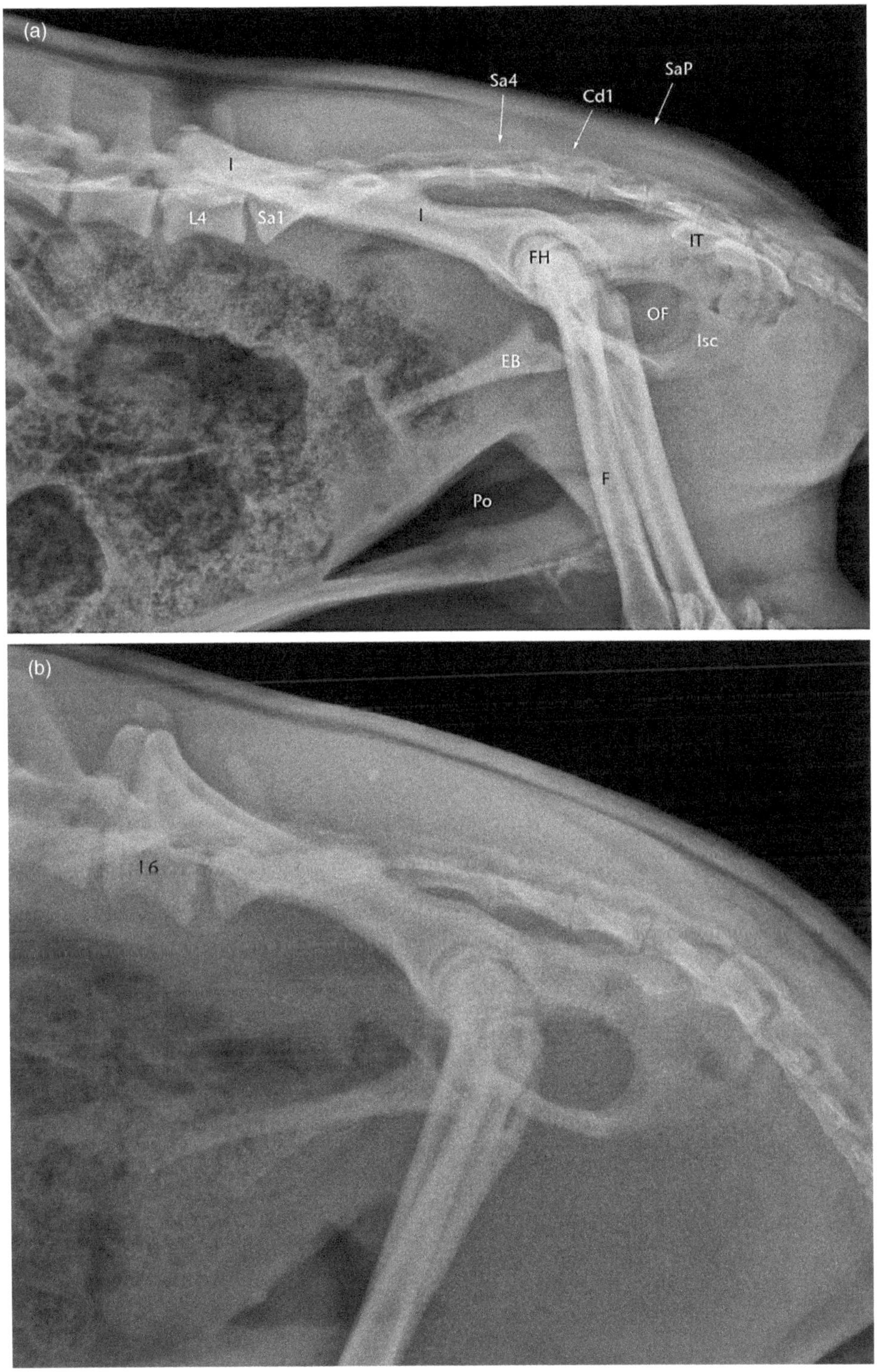

Figure 6.8: Pelvis, lateral view: (a) female common wombat (*Vombatus ursinus*), (b) female southern hairy-nosed wombat (*Lasiorhinus latifrons*), (c) immature common wombat.

(Continued)

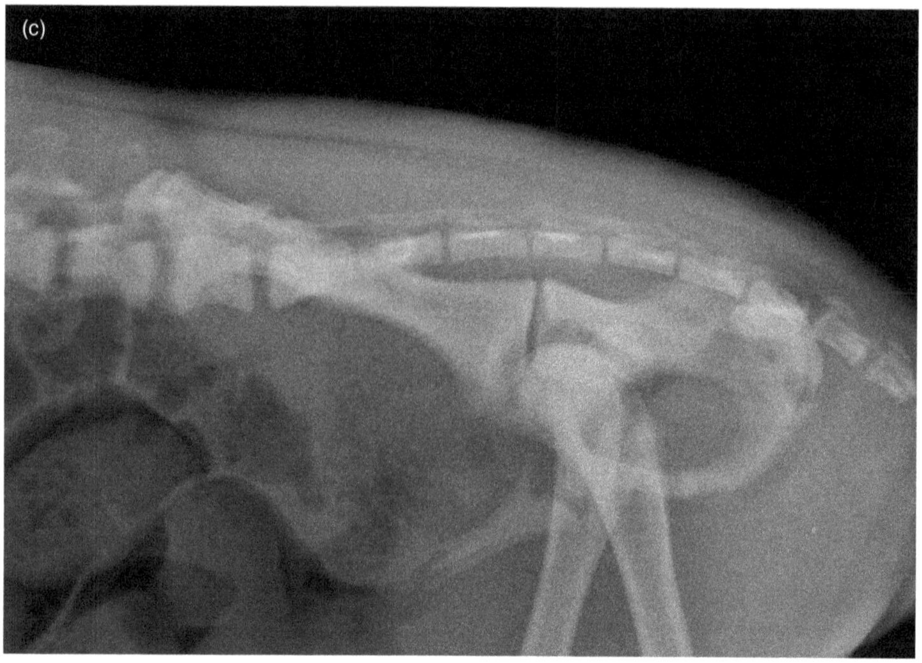

Figure 6.8: (*Continued*). The four fused sacral elements (Sa1 and Sa4 labelled) articulate cranially with the 4th lumbar vertebra (L4) in the common wombat and 6th (L6) in the southern hairy-nosed wombat and caudally with the 1st caudal vertebra (Cd1). The ilium (I) is elongated and the ischium (Isc) has a hooked ischial tuberosity (IT). The epipubic bones (EB) articulate with the pubis. An obturator foramen (OF) separates the ischium and pubis. The femoral head (FH) articulates congruently with the acetabulum. The sacral plate (SaP) and air in the pouch (Po) are visible.

(Owen 1839; Grand and Barboza 2001) (Figs 6.7 and 6.19).

There are usually approximately 12 caudal vertebrae. The first six or seven caudal vertebrae have transverse processes. Those further caudally exist as bodies only. The spinal canal is complete in all but the last three. There are no haemal arches in the caudal vertebral column (Fig. 6.8) (Owen 1839).

Ribs and sternum

Hairy-nosed wombats have 13 pairs of ribs and common wombats have 15. Only six pairs articulate with the sternum in common wombats. There are six sternebrae (Owen 1839) (Fig. 6.20).

Soft tissue

Head, neck and thorax

Pharynx and larynx

The hyoid apparatus and larynx conform to the standard arrangement seen in other mammals. They are located in the upper cervical region, with the larynx supported by paired sets of hyoid bones that are attached to the base of the skull (Fig. 6.13). This differs from the closely related koala, in which the larynx is located in the caudal cervical region and the hyoid apparatus is not a well-defined radiographic feature.

Thorax

The thoracic cavity is relatively small, which is probably a result of the excessively large abdominal cavity.

Cardiovascular system

As a consequence of the small thoracic cavity, the cardiac silhouette appears large and visually occupies almost 30% of the thoracic volume. On lateral images, the heart is located between the first and fifth ribs. Its caudal border contacts the cranially displaced diaphragm (Fig. 6.20). The pericardium

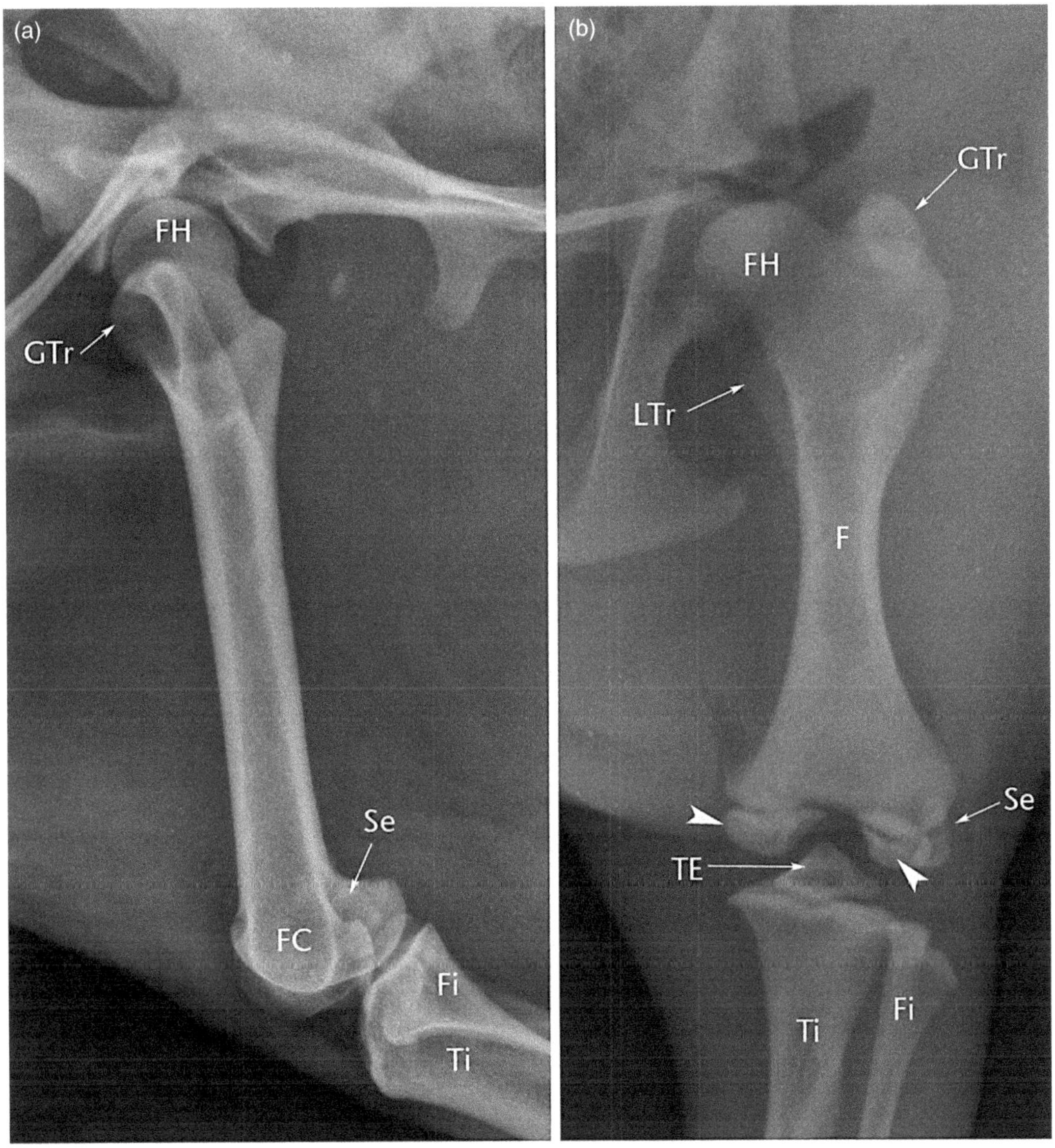

Figure 6.9: Femur and stifle: (a) southern hairy-nosed wombat (*Lasiorhinus latifrons*), mediolateral view, (b) immature common wombat (*Vombatus ursinus*), craniocaudal view. The proximal end of the femur (F) has a well-defined femoral head (FH), a greater trochanter (GTr) and a lesser (LTr) trochanter. (b) The femoral condylar epiphyses (arrowheads) and the proximal tibial epiphysis (TE) are visible. There is a single lateral sesamoid (Se) caudal to the lateral condyle. The tibia (Ti) and fibula (Fi) are visible. Note the absence of a patella cranial to the femoral condyle (FC).

adheres to the diaphragm (Grand and Barboza 2001).

Respiratory system

The arboreal arrangement of pulmonary arteries, veins and bronchi is typical of most mammalian species. The tracheal cartilages do not appear mineralised and the air-filled lungs are featureless, other than the visible vascular and bronchial structures. The left lung has a small cardiac impression cranially and is not divided into lobes. The right lung has several fissures. There is a fissure

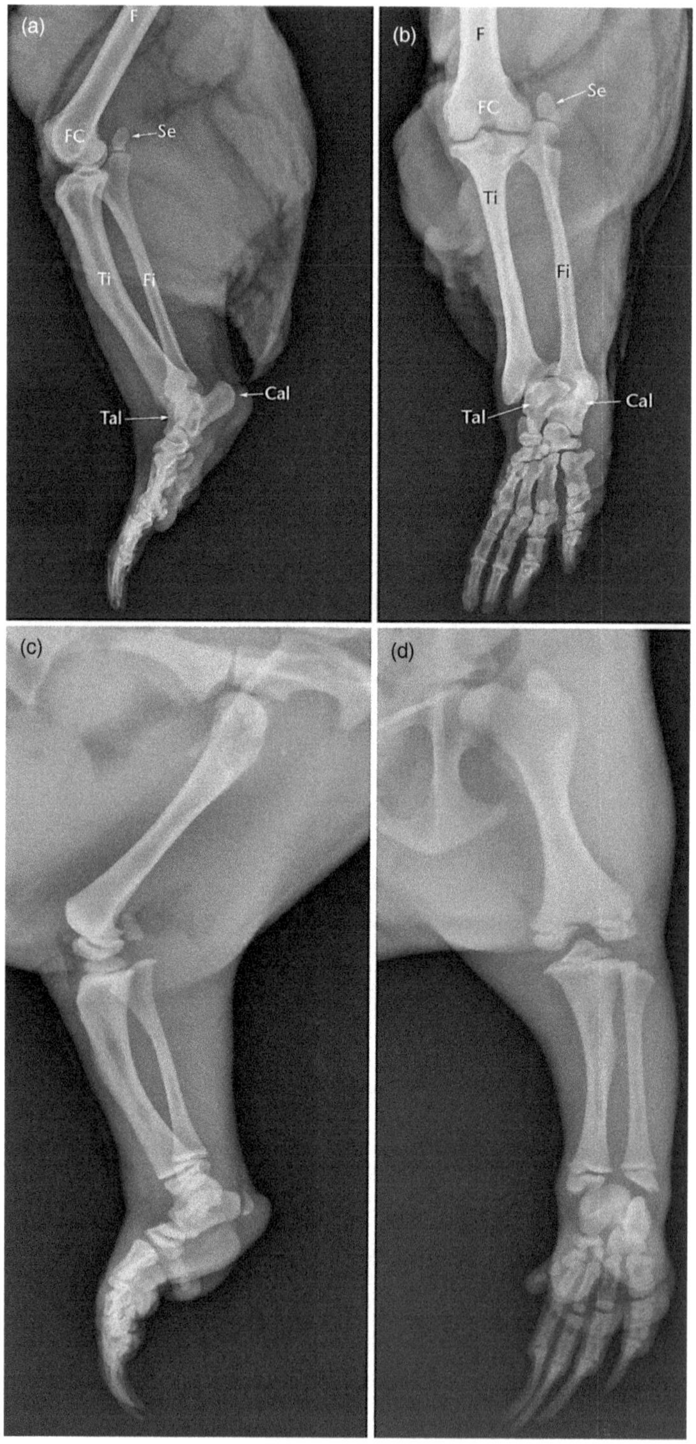

Figure 6.10: Crus and tarsus: (a) southern hairy-nosed wombat (*Lasiorhinus latifrons*), mediolateral view, (b) southern hairy-nosed wombat, craniocaudal and dorsoplantar view, (c) hindlimb, immature common wombat (*Vombatus ursinus*), mediolateral view, (d) immature common wombat, caudocranial and plantarodorsal view. The distal femur (F) articulates through the femoral condyle (FC) mainly with the tibia (Ti). The caudolaterally located fibula (Fi) has a proximal tuberosity that is associated with the single sesamoid (Se) bone of the stifle. The tibia and fibula articulate distally with the talus (Tal) and calcaneus (Cal). Note the southern hairy-nosed wombat images were obtained from disarticulated limbs of a cadaver.

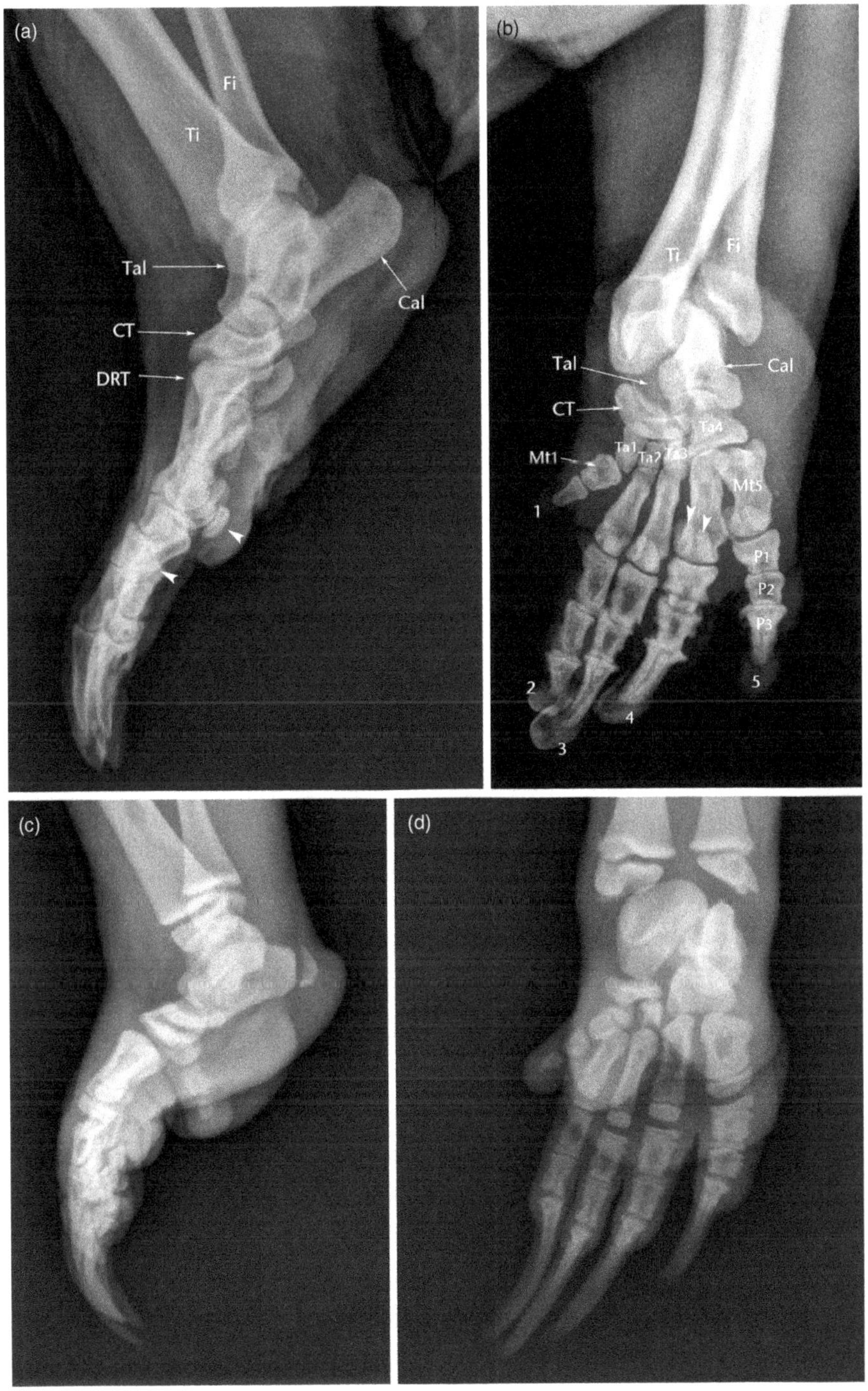

Figure 6.11: Tarsus and pes: (a) southern hairy-nosed wombat (*Lasiorhinus latifrons*), mediolateral view, (b) dorsoplantar view; (c) immature common wombat (*Vombatus ursinus*), mediolateral view, (d) plantarodorsal view. The articulation of the tibia (Ti) and fibula (Fi) with the talus (Tal) is most clearly illustrated on the mediolateral image. The articulation of the fibula and calcaneus (Cal) is more clearly visualised on the dorsoplantar view. The relationship between the central tarsal bone (CT) and distal row of tarsal bones (DRT) is illustrated best in the dorsoplantar view. The distal row of tarsal bones (Ta1, Ta2, Ta3, Ta4) articulate distally with the five metatarsal bones, two of which are labelled (Mt1, Mt5). Note the presence of paired sesamoid bones (arrowheads) on the plantar surface of the metatarsophalangeal and some interphalangeal articulations. There are five digits (1, 2, 3, 4, 5). Digit 1 has two phalanges and the remaining digits have three (P1, P2, P3).

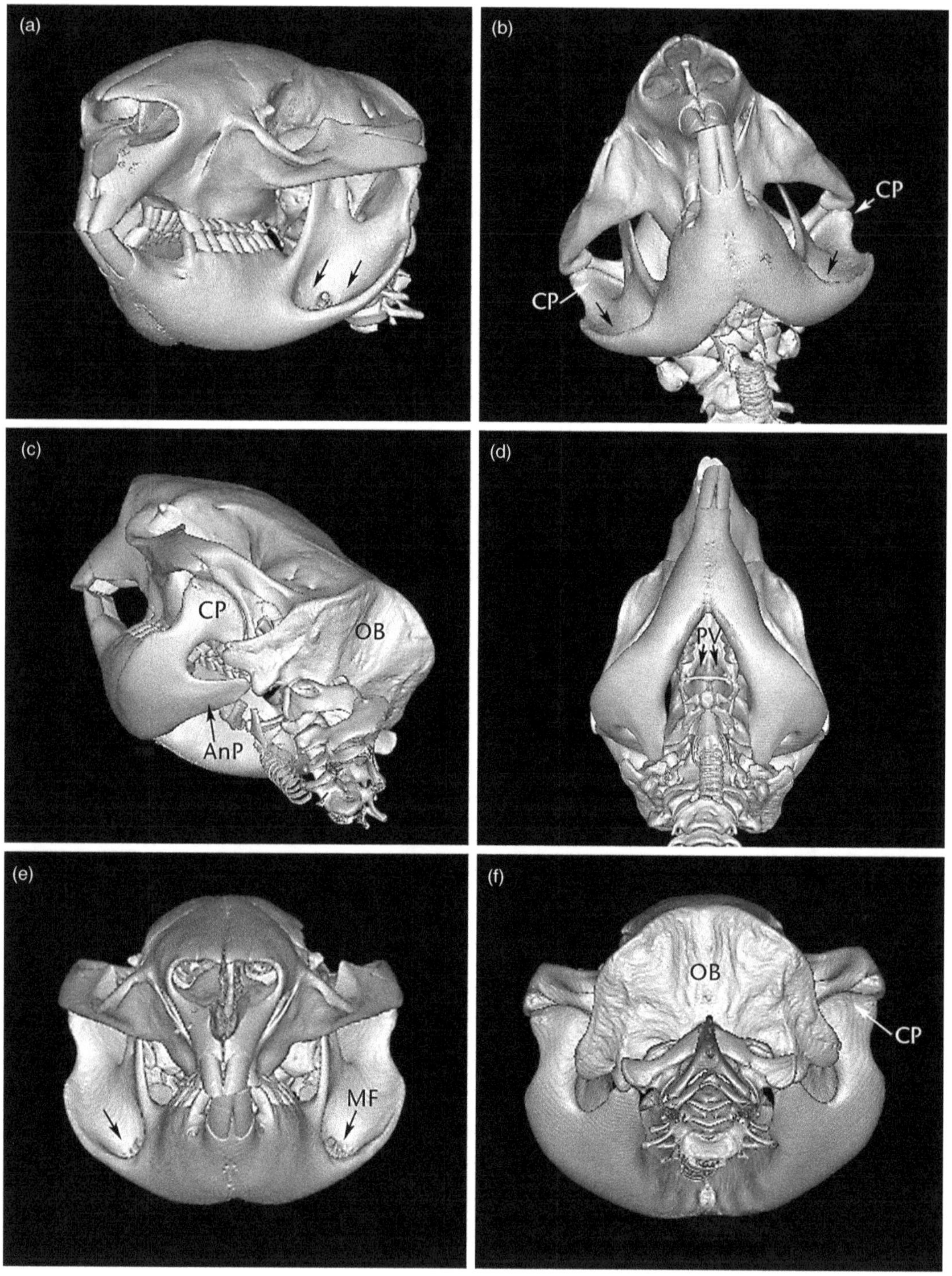

Figure 6.12: Skull, common wombat (*Vombatus ursinus*): (a–f) volume-rendered three-dimensional CT reconstructions demonstrate several unique features of the wombat skull: the broad, flattened shape; sturdy mandibles with a shovel-shaped condyloid process (CP) that has a broad articular surface; the buttress-shaped caudal surface of the occipital bone (OB); the flattened angular processes (AnP); deep recess in the lateral surfaces of the mandibular rami (masseteric fossa, MF) that accept the muscles of mastication (black arrows); wide nostrils without visible turbinate structures; one pair of upper incisors; and palatine vacuities (PV). Note the slight lateral deviation of the rostral mandible and malocclusion in this individual.

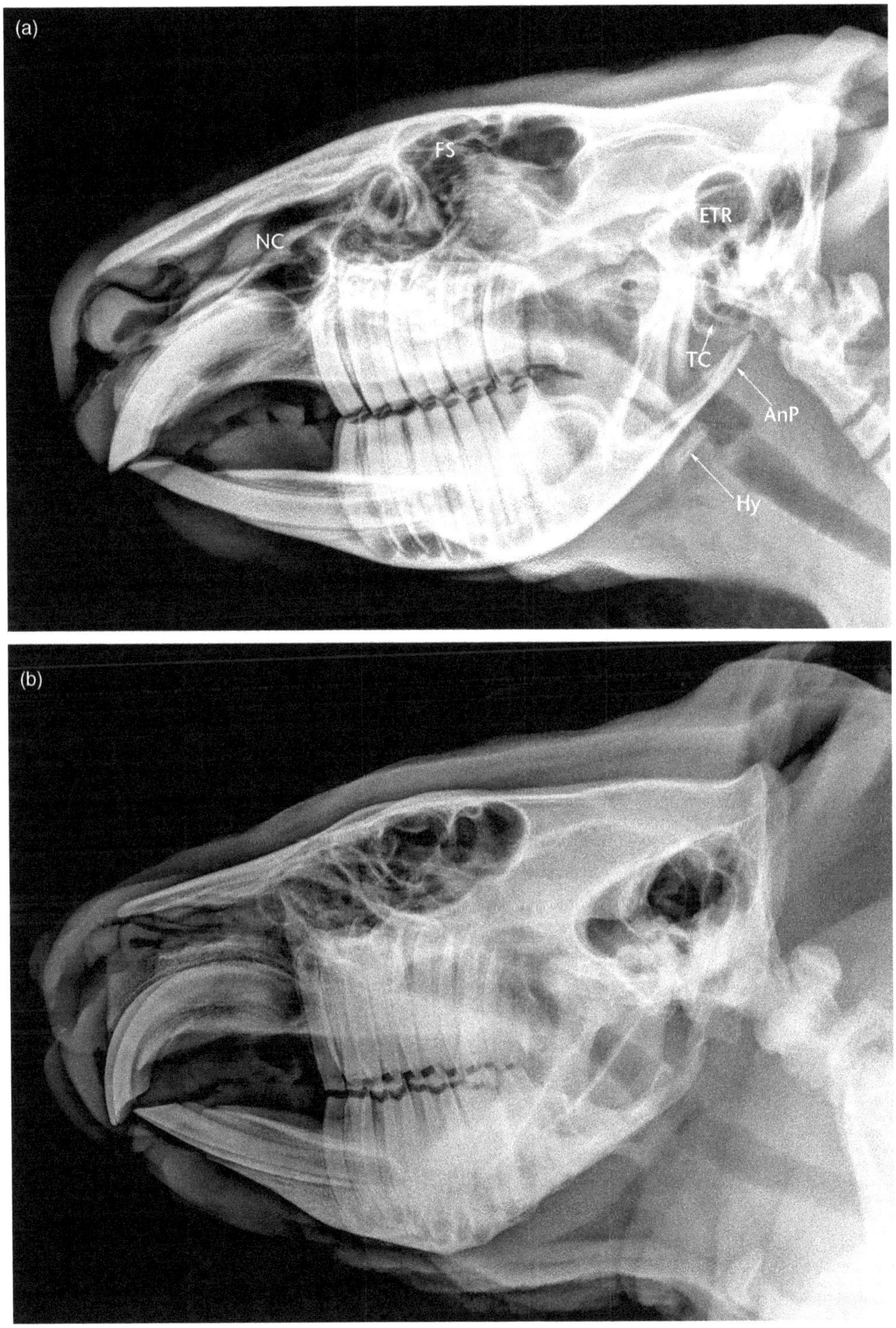

Figure 6.13: Skull, lateral view: (a) common wombat (*Vombatus ursinus*), (b) southern hairy-nosed wombat (*Lasiorhinus latifrons*). The paired maxillary and mandibular incisors dominate the rostral end of the skull while the open-rooted premolars and molars dominate the middle of the skull. The nasal cavity (NC) and the frontal sinuses (FS) contain mainly air, as there is a paucity of maxillary turbinates. Caudally, air is present in the tympanic cavities (TC) and epitympanic recesses (ETR). A prominent angular process (AnP) projects caudally from each mandible. The hyoid apparatus (Hy) is partially obscured by the mandible.

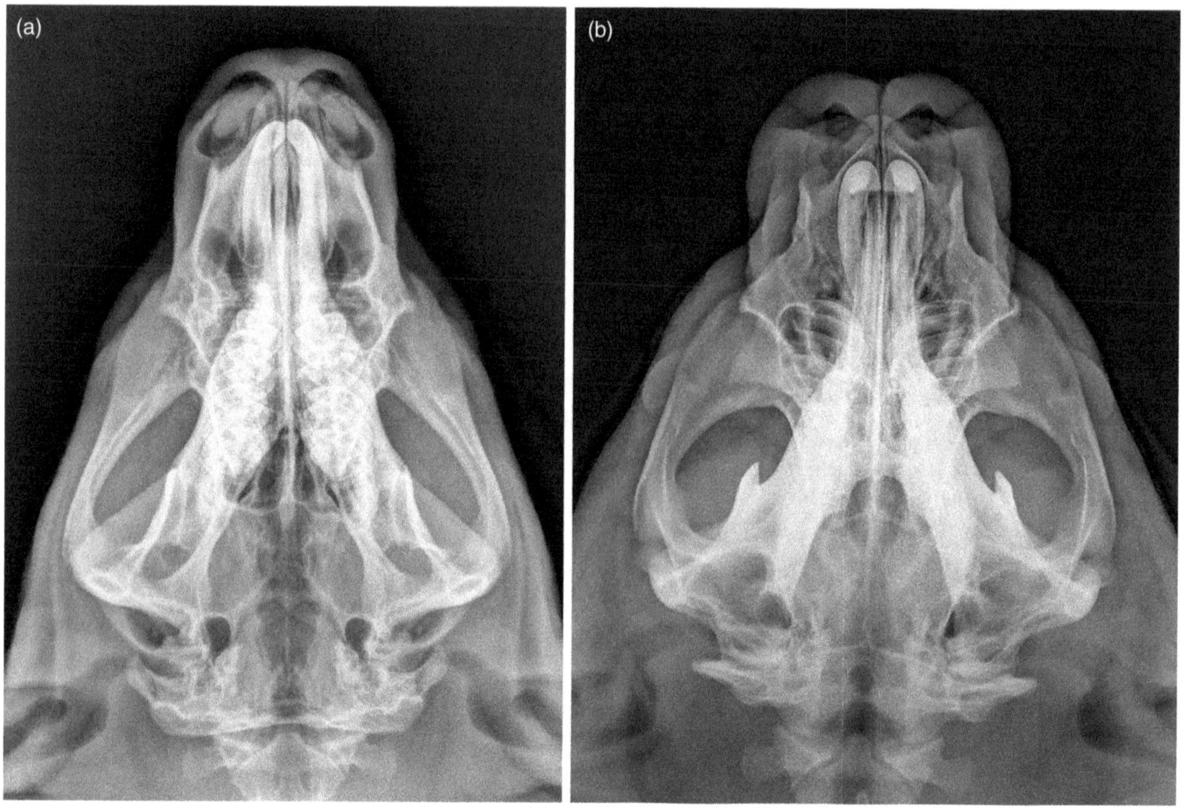

Figure 6.14: Skull, ventrodorsal view: (a) common wombat (*Vombatus ursinus*), (b) southern hairy-nosed wombat (*Lasiorhinus latifrons*). The skull of the common wombat is narrower than the skull of the hairy-nosed wombat.

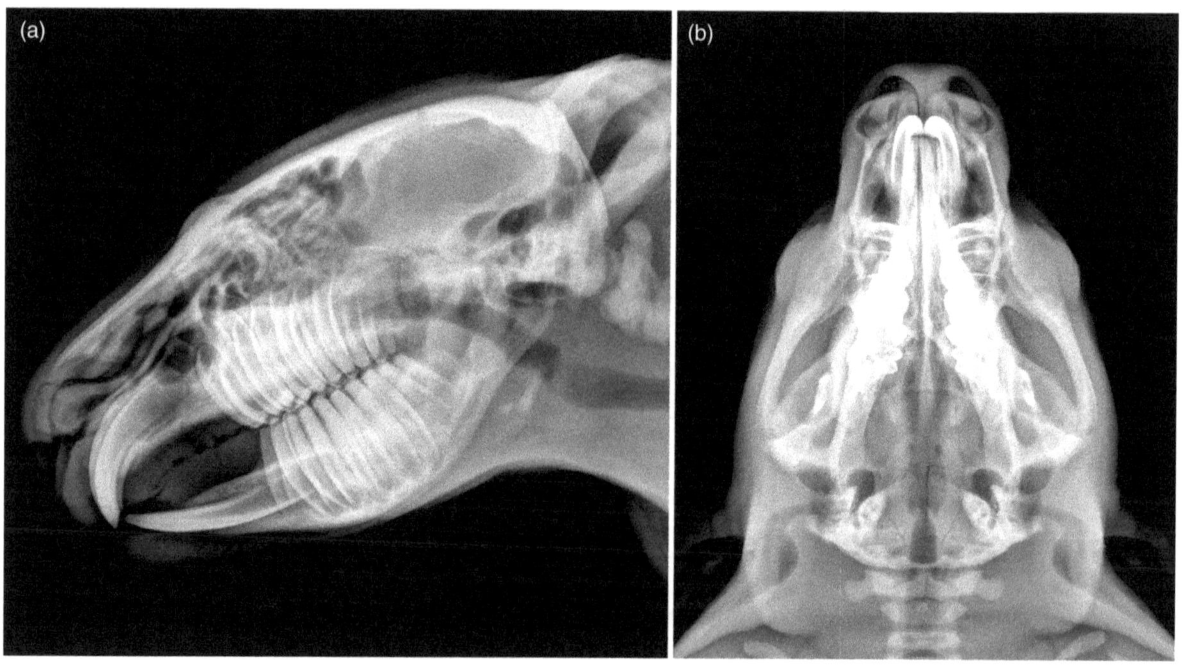

Figure 6.15: Skull, immature common wombat (*Vombatus ursinus*): (a) lateral view, (b) ventrodorsal view.

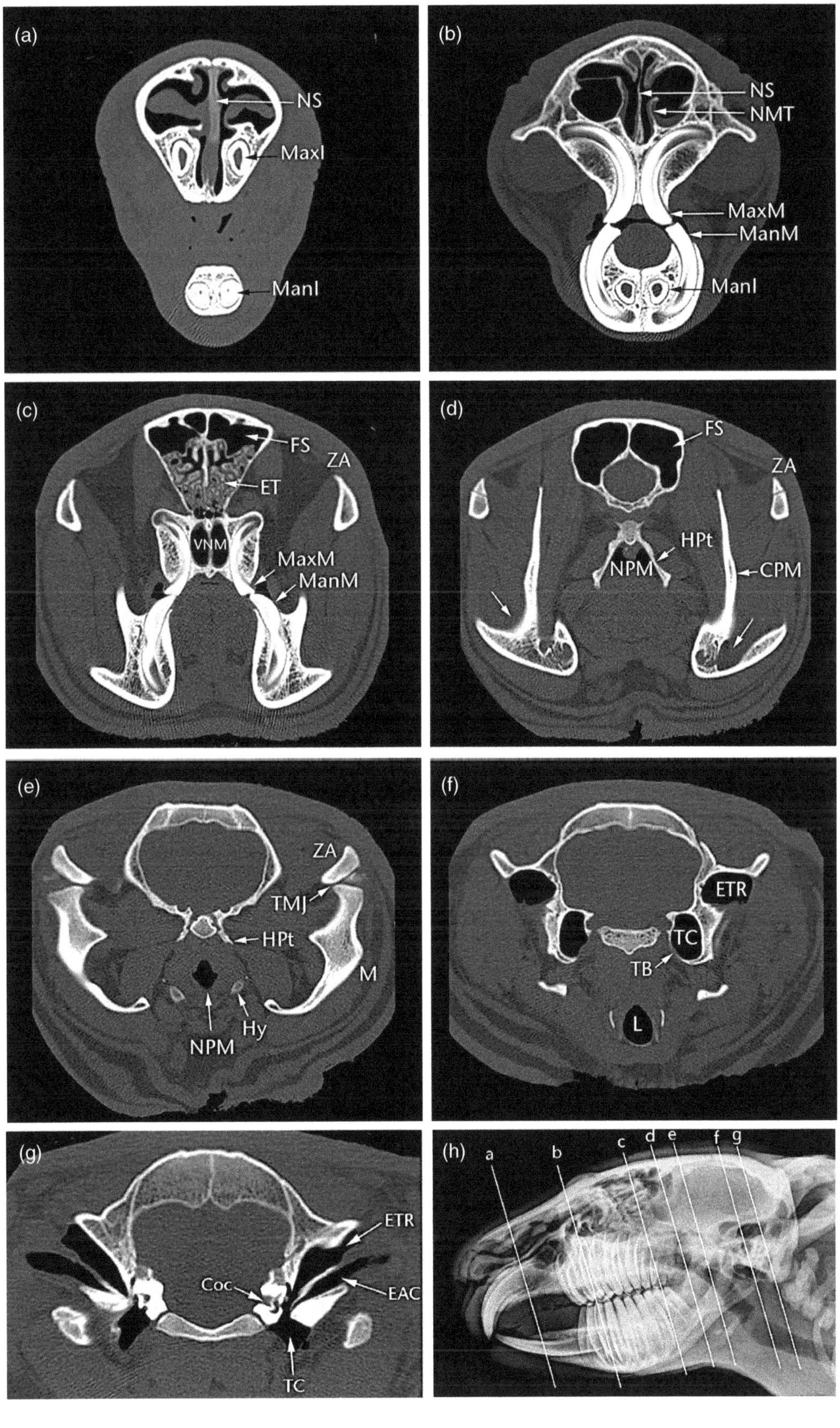
(a)
NS
MaxI
ManI
(b)
NS
NMT
MaxM
ManM
ManI
(c)
FS
ZA
ET
VNM
MaxM
ManM
(d)
FS
ZA
HPt
NPM
CPM
(e)
ZA
TMJ
HPt
M
Hy
NPM
(f)
ETR
TC
TB
L
(g)
ETR
EAC
Coc
TC
(h)
a
b
c
d
e
f
g

Figure 6.16

Figure 6.16: Skull, common wombat (*Vombatus ursinus*): (a–f) transverse reconstructed CT images, (g) transverse reconstructed CT image, (h) lateral radiograph to demonstrate the image planes acquired in (a–g). The mandibular (ManI) and maxillary (MaxI) incisors are prominent features of the rostral skull. There is a well-defined nasal septum (NS) dividing the nasal cavity. Nasomaxillary turbinates (NMT) are sparse, so the nasal cavity is capacious. The mandibular (ManM) and maxillary (MaxM) premolars and molars are curved and when apposed take on a sigmoid shape. The frontal sinus (FS) is bordered ventrally by complex ethmoidal turbinates (ET). Caudally, the nasal cavity connects to the nasopharyngeal meatus (NPM), which is located in part between the hamuli of the pterygoid bone (HPt), through the ventral nasal meatus (VNM). The coronoid process of the mandible (CPM) extends dorsally between the zygomatic arch (ZA) and the skull. The deep lateral recesses in the mandibles that accept the masticatory muscles are readily visible on these cross-sectional images (d, white arrows). The tympanic bulla (TB) contains the tympanic cavity (TC). Medially, the cochlear (Coc) is visible. The TC communicates laterally with the epitympanic recess (ETR) and the external auricular canal (EAC). The temperomandibular joint (TMJ), larynx (L) and parts of the hyoid apparatus (Hy) are visible.

extending from the ventral border at the cardiac notch towards the hilus, separating the relatively small cranial lobe from the body of the lung. A broad, well-defined middle lobe extends ventrally, just caudal to the apical portion of the lung (Owen 1868). Based on a single dissection of a common wombat, one of the authors (L.V.) found an accessory lobe arising from the medial aspect of the dorsal region caudal to the hilus. Fissures and lobes are not discernible radiographically.

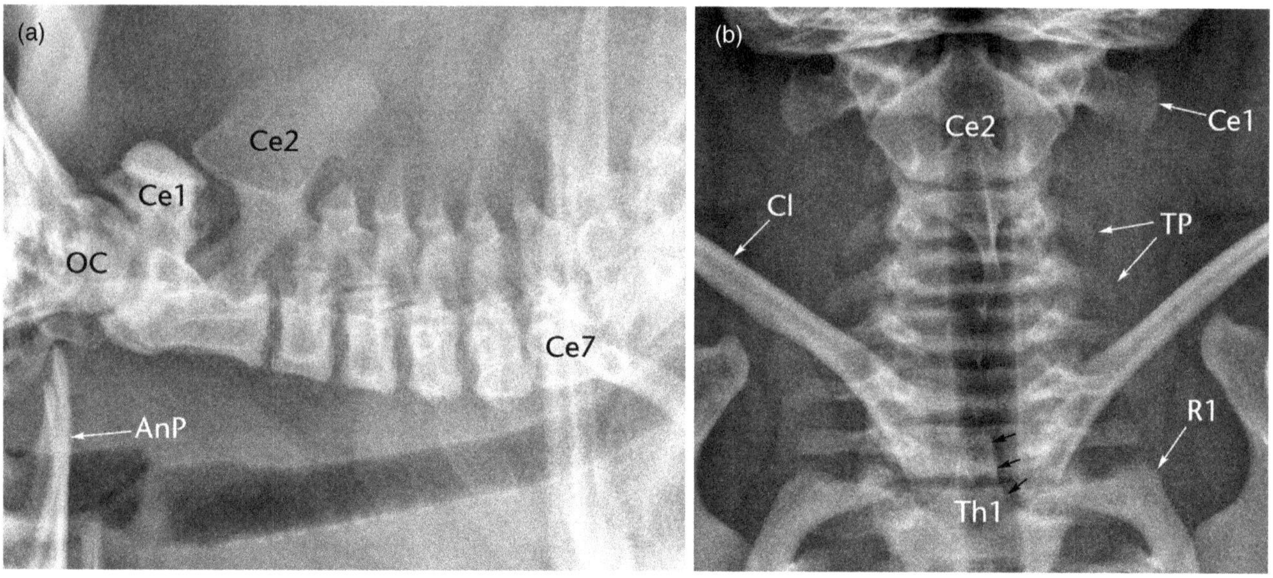

Figure 6.17: Cervical vertebral column, common wombat (*Vombatus ursinus*): (a) lateral view, (b) ventrodorsal view. There are 7 cervical vertebrae (Ce1, Ce2, Ce7 labelled). They are tall and short. The occipital condyles (OC) are well developed, while the atlas appears narrow and less substantial. The pointed angular processes (AnP) of the mandible intrude into image (a). The cervical vertebrae have transverse processes (TP) that become progressively longer caudally. The clavicle (Cl) and 1st rib (R1) are labelled. The location of the 1st thoracic vertebra is identified (Th1). It is superimposed on the 1st sternal element. Where the concave articular surface of the 1st sternal element articulates with the clavicle is visible (black arrows).

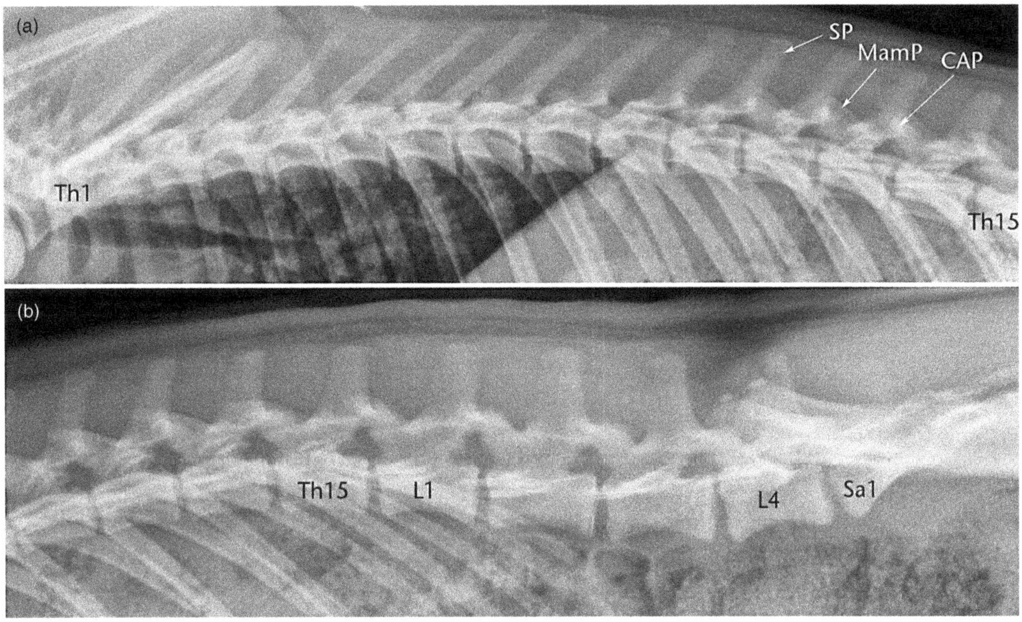

Figure 6.18: Thoracic and lumbar vertebral column, common wombat (*Vombatus ursinus*): (a, b) lateral views. There are 15 ribbed thoracic vertebrae (Th1, Th15 labelled) and 4 lumbar vertebrae in the common wombat (L1, L4 labelled). The southern hairy nosed wombat has fewer ribs and 6 lumbar vertebrae. The sacrum (Sa1) is largely obscured by the pelvis. Each vertebral element has a well-defined dorsal spinous process (SP). In the thoracolumbar region, cranially directed mammillary processes (MamP) are superimposed on the caudal articular process (CAP) of the preceding vertebra.

Abdomen

Wombats are hindgut fermenters. There is a simple (monogastric) stomach with a cardio-gastric gland, a short small intestine, a very long, sacculated colon and a rudimentary caecum. Radiographically, the mixture of digesta and gas in the gastrointestinal tract presents a confusing image of the abdominal viscera. The viscera cannot be easily distinguished (Figs 6.21 and 6.22).

Female reproductive anatomy is similar to that of other marsupials. The ovaries occupy an ovarian bursa. There are paired lateral uteri, each of which opens into a vaginal cul-de-sac through its own cervix. The paired lateral vaginae both open separately into the urogenital sinus at the level of the urethral opening and are separated along their length by a soft tissue median septum. The urogenital sinus enters the common vestibule or cloaca ventral to the termination of the rectum (Tyndale-Biscoe and Renfree 1987).

There are two mammary glands located over the caudal ventral abdominal wall within the pouch (Wells 1989). The wombat pouch is rear-opening and has a well-developed sphincter muscle.

The male has a pre-penile, non-pendulous scrotum, a relatively large, carrot-shaped prostate and three pairs of bulbourethral glands. There are no seminal vesicles or ampullae (MacCallum 2004). The non-erect penis has a sigmoid flexure and occupies a preputial sac adjacent to the opening of the common vestibule.

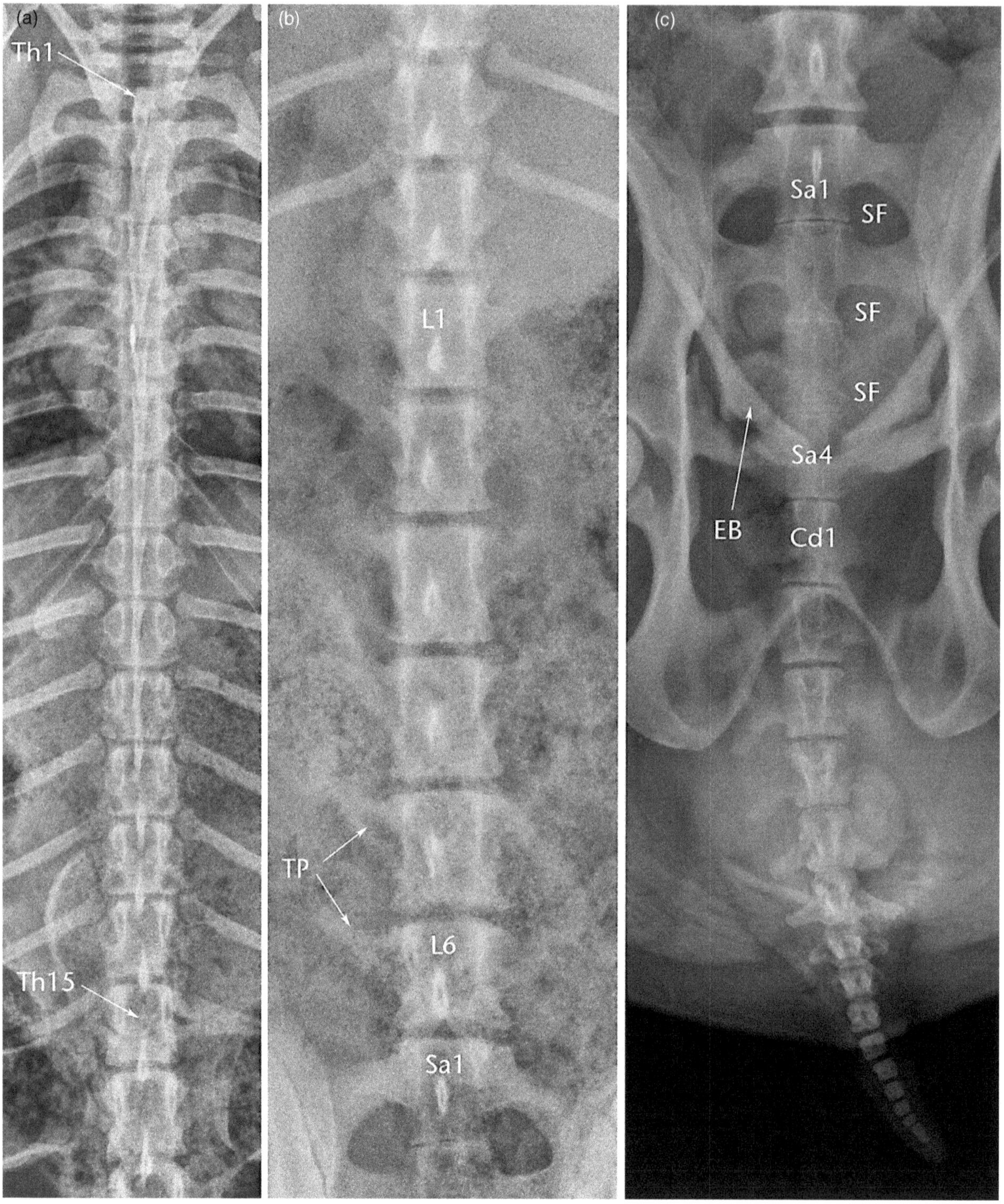

Figure 6.19: Vertebral column, ventrodorsal views: (a) common wombat (*Vombatus ursinus*), (b, c) southern hairy-nosed wombat (*Lasiorhinus latifrons*). (a) The 15 thoracic vertebrae of the common wombat are shown (Th1, Th15 labelled). (b) The 6 lumbar vertebrae of the southern hairy-nosed wombat are shown (L1, L6 labelled). Each lumbar vertebra has a pair of transverse processes (TP). The complex sacrum, comprising 4 sacral elements (Sa1, Sa4 labelled) and 3 sacral foraminae (SF) precede the caudal vertebrae (Cd1). The epipubic bones (EB) are superimposed on the 4th sacral element. Note: a transponder can be seen in (a).

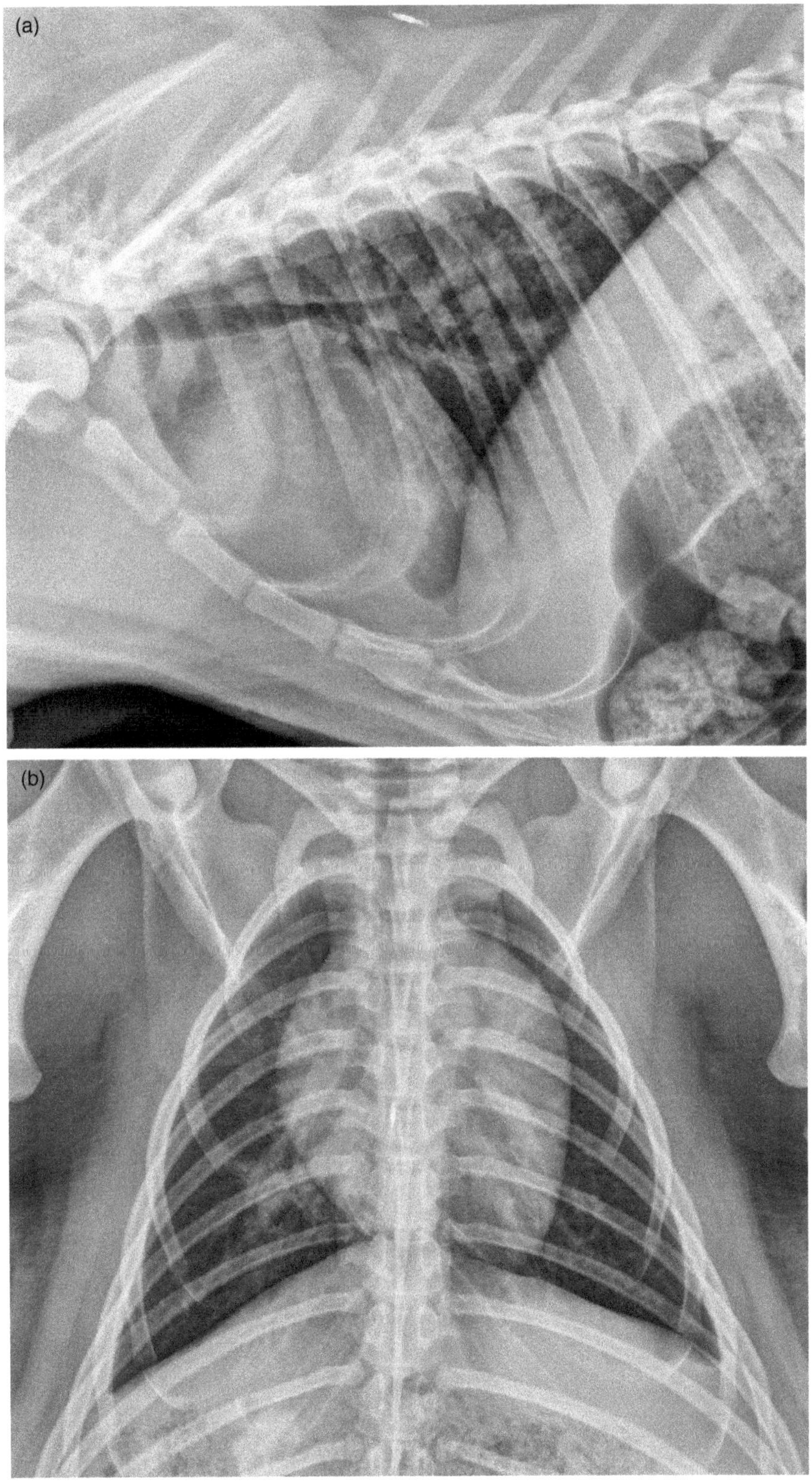

Figure 6.20: Thorax, southern hairy-nosed wombat (*Lasiorhinus latifrons*): (a) lateral view, (b) ventrodorsal view; (c) immature common wombat (*Vombatus ursinus*), lateral view. Note: a transponder can be seen in (a) and (b).

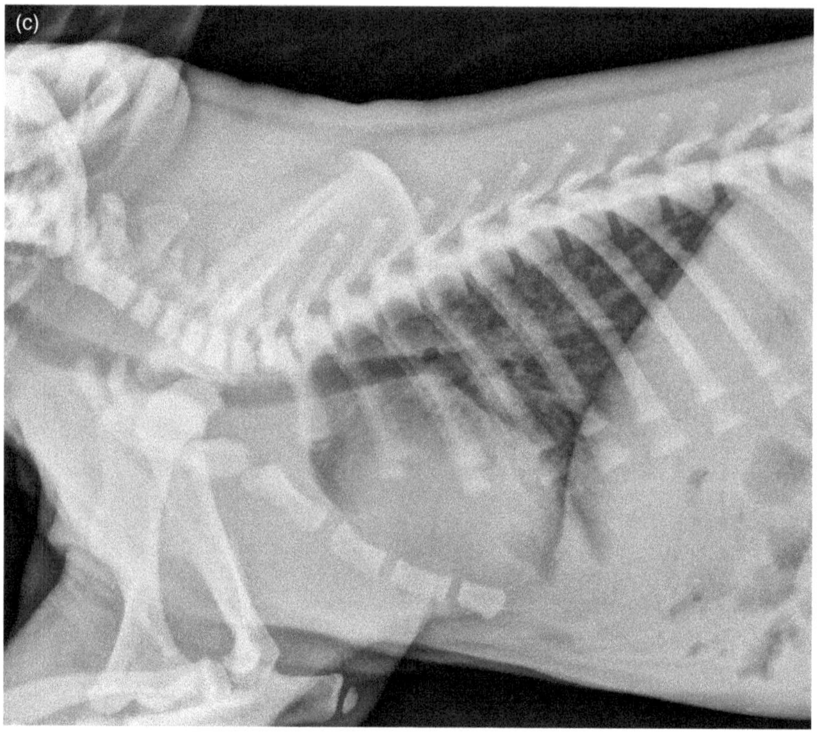

Figure 6.20: (*Continued*).

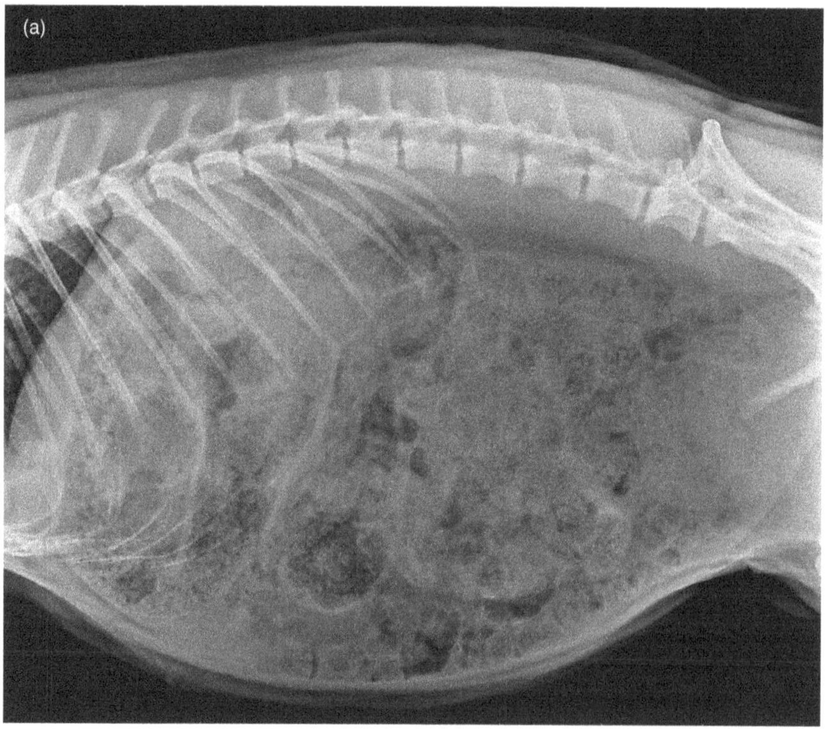

Figure 6.21: Abdomen, lateral view: (a) southern hairy-nosed wombat (*Lasiorhinus latifrons*), (b) adult common wombat (*Vombatus ursinus*), (c) immature common wombat. Intraluminal contents (gas and digesta) dominate the radiographic image of the abdomen and obscure the viscera.

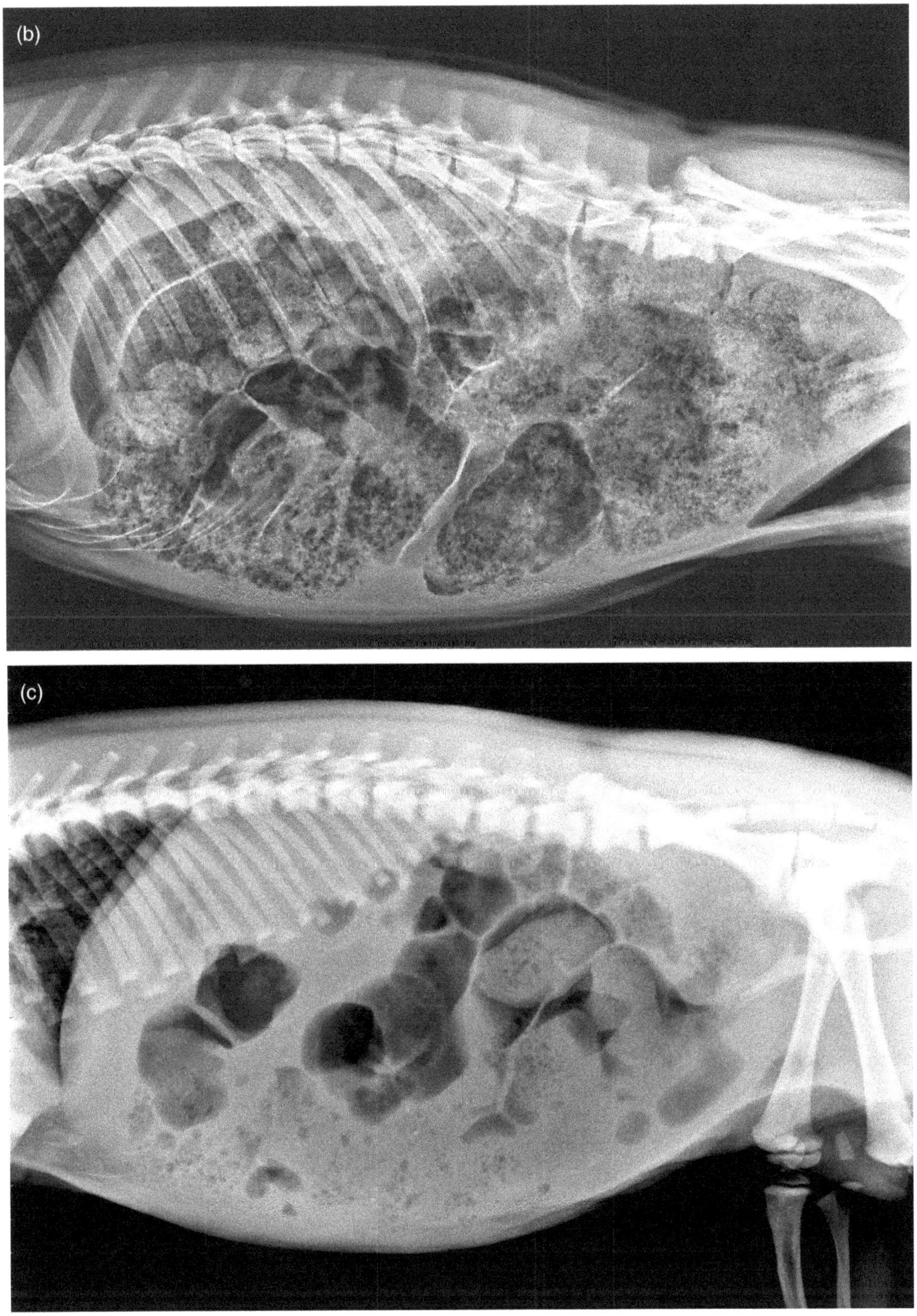

Figure 6.21: *(Continued).*

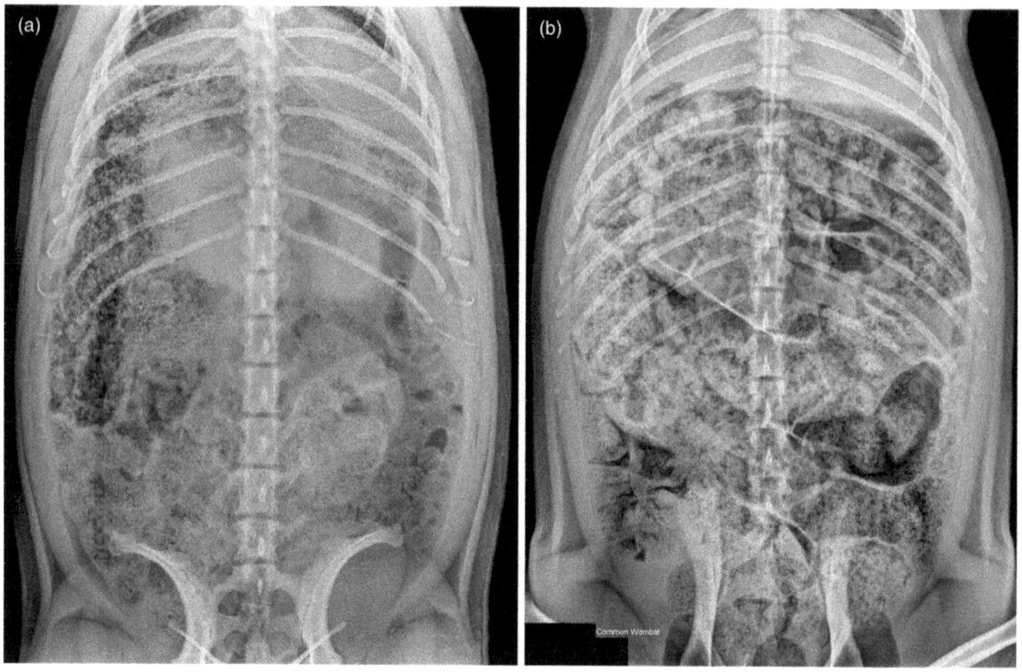

Figure 6.22: Abdomen, ventrodorsal view: (a) southern hairy-nosed wombat (*Lasiorhinus latifrons*), (b) common wombat (*Vombatus ursinus*). Intraluminal contents (gas and digesta) dominate the radiographic image of the abdomen and obscure the viscera.

References

Grand TI, Barboza PS (2001) Anatomy and development of the koala, *Phascolarctos cinereus*: an evolutionary perspective on the superfamily Vombatoidea. *Anatomy and Embryology* **203**, 211–223. doi:10.1007/s004290000153

Johnson C, Crossman D (1991) Sexual dimorphism in the northern hairy-nosed wombat, *Lasiorhinus krefftii* (Marsupialia: Vombatidae). *Australian Mammalogy* **14**, 145–146.

MacCallum CA (2004) The reproductive biology of the common wombat (*Vombatus ursinus*): studies toward the development of an artificial insemination protocol. MPhil thesis. The University of Queensland, Australia.

Owen R (1839) On the osteology of Marsupialia. *The Transactions of the Zoological Society of London* **2**(5), 379–408. doi:10.1111/j.1469-7998.1839.tb00030.x

Owen R (1868) *On the Anatomy of Vertebrates, Volume 3, Mammals.* pp. 513–603. Longmans, Green and Co., London.

Parker JT, Haswell WA (1921) Skeleton of metatheria. In *A Text-book of Zoology, Volume II.* pp. 501–506. Macmillan and Co. Ltd, London.

Reese S, Pfuderer UR, Bragulla H, Loeffler K, Budras KD (2001) Topography, structure and function of the patella and the patelloid in marsupials. *Anatomia, Histologia, Embryologia* **30**, 289–294. doi:10.1046/j.1439-0264.2001.00334.x

Scott GG, Richardson KC, Groves CP (1988) Osteological differences of the skulls of *Lasiorhinus latifrons* Owen, 1845 and *Vombatus ursinus* Shaw, 1800 (Marsupialia: Vombatidae). *Australian Journal of Zoology* **36**, 599–609. doi:10.1071/ZO988 0599

Strahan R (1995) *The Australian Museum Complete Book of Australian Mammals.* Reed New Holland, Sydney.

Triggs B (2009) *Wombats.* CSIRO Publishing, Melbourne.

Tyndale-Biscoe H, Renfree MR (1987) *Reproductive Physiology of Marsupials.* Cambridge University Press, Cambridge.

Wells RT (1989) Vombatidae. In *Fauna of Australia, Volume 1B, Mammalia.* (Eds DW Walton and BJ Richardson) pp. 755–768. Australian Government Publishing Service, Canberra.

7

Dasyurids

The order Dasyuromorphia represents the carnivorous marsupials and is divided into three families: Thylacinidae, with the extinct thylacine (*Thylacinus cynocephalus*); Myrmecobiidae, with the numbat (*Myrmecobius fasciatus*); and Dasyuridae, with the remaining 66 species (commonly collectively referred to as 'dasyurids'). The Dasyuridae includes the Tasmanian devil (*Sarcophilus harrisii*), quolls, antechinus, dunnarts, kowari (*Dasyuroides byrnei*) and phascogales (Krajewski and Westerman 2003). They are found throughout Australia and New Guinea, occupying environments ranging from high alpine regions (agile antechinus [*Antechinus agilis*], spot-tailed quoll [*Dasyurus maculatus*]) to coastal forests (quolls, Tasmanian devils, antechinus, phascogales) and deserts (crest-tailed mulgara [*Dasycercus cristicauda*], little red kaluta [*Dasykaluta rosamondae*], kowari).

All members of the Dasyuridae have four pairs of upper and three pairs of approximately equal sized lower incisors. The numerous incisors (polyprotodont) immediately distinguishes the dasyurids from herbivorous marsupials such as wallabies, kangaroos, wombats and koala and from the possums and gliders, all of which have only one pair of lower incisors (diprotodont). In addition, dasyurids have at least four separately developed digits on the pes, with no fusion between the second and third digits (syndactyly), a characteristic distinguishing them from the polyprotodont bandicoots and bilby, which are syndactylous (Strahan 1995). The numbat has five pairs of structurally simple upper and lower molar teeth, which distinguishes them from the Dasyuridae, which have four pairs of upper and lower molars with well developed cusps (Morton *et al.* 1989).

Dasyurids range in size from the narrow-nosed planigale (*Planigale tenuirostris*) (4–9 g) to the Tasmanian devil (up to 10 kg) (Fig. 7.1). They are predominantly nocturnal but many species also forage during the day. Many species are arboreal, others are both arboreal and terrestrial and some are completely terrestrial (Holz 2008).

Dasyurids are not long lived. The longevity of Tasmanian devils is up to 6 years in the wild and 8 years in captivity and of quolls is up to 3 years in the wild and 6 years in captivity. The smaller dasyurids generally live for no more than 5 years in captivity, usually up to 2 years in the wild and occasionally 3 years. Males of some species, such as *Antechinus* spp., all die in the wild after their first year.

As with eutherian carnivores, apart from size there is little variation in body shape and anatomy of the dasyurids (Fig. 7.1). The most extreme variation in body shape is found in the kultarr (*Antechinomys laniger*). This species has elongated hindlimbs and a long, tufted tail. One noticeable morphological feature that is highly developed among dasyurids is caudal fat storage. Fat tails occur in desert-dwelling species of *Dasycercus*, *Pseudantechinus* and *Sminthopsis*. This allows increased fat storage while allowing unhindered manoeuvrability during hunting (Morton 1980; Morton *et al.* 1989).

In this chapter, anatomical descriptions and radiographic images focus primarily on readily accessible species. However, when radiographically available, relevant anatomical features of other species are included.

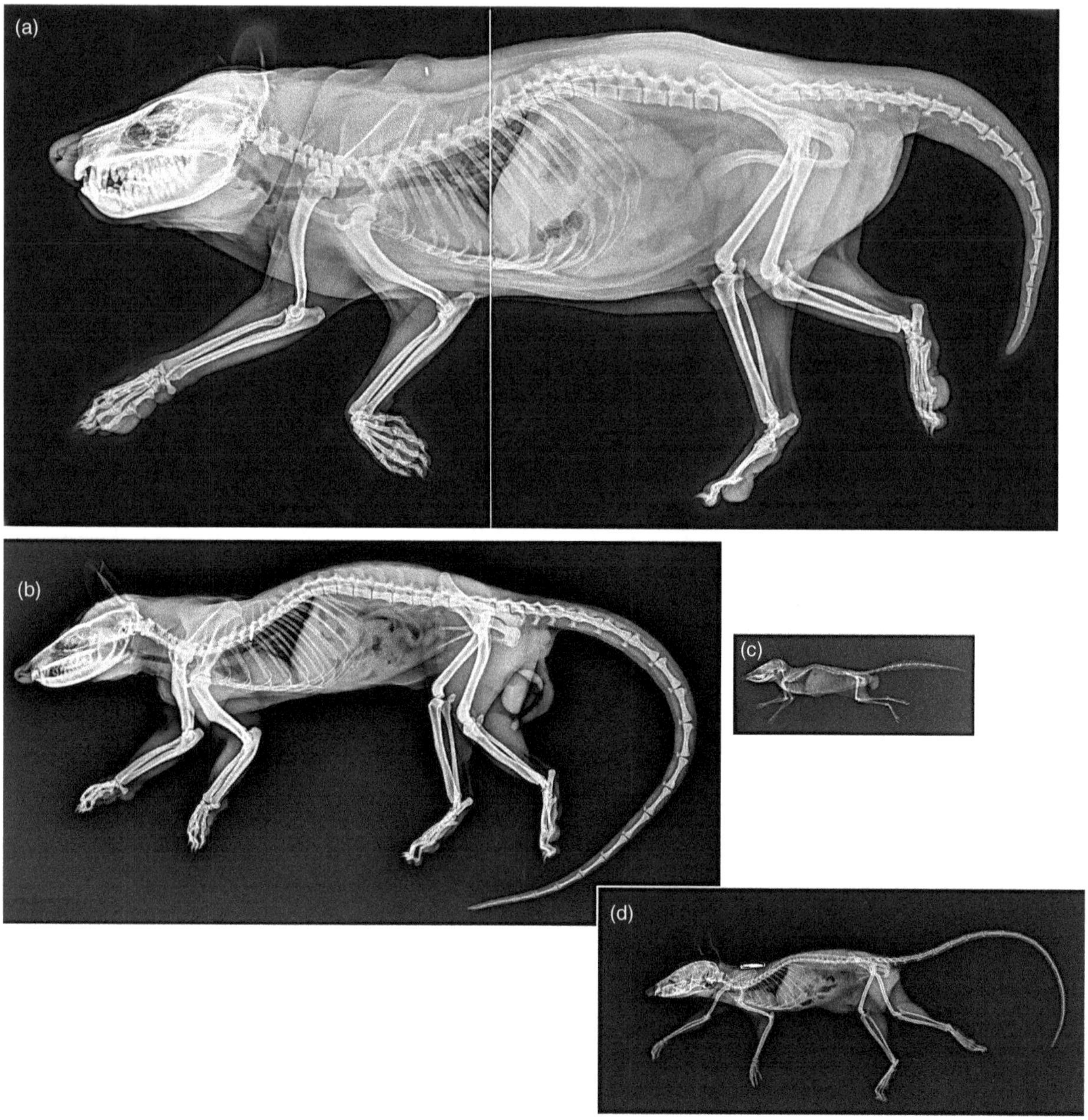

Figure 7.1: Whole body, lateral radiographs: (a) Tasmanian devil (*Sarcophilus harrisii*), (b) northern quoll (*Dasyurus hallucatus*), (c) fat-tailed dunnart (*Sminthopsis crassicaudata*), d) red-tailed phascogale (*Phascogale calura*). The images are proportionately scaled, based on relative body size and illustrate the wide range of body sizes in the Dasyuridae. Note: a transponder can be seen in (a), (b) and (d).

Skeletal system

Appendicular skeleton

Pectoral girdle and forelimb

As in all marsupials other than bandicoots and the bilby, a clavicle is present. The clavicle of the crest-tailed mulgara is slender and somewhat expanded at its sternal and acromial articular extremities. It is relatively straight at its mediolateral axis, convex in a ventral direction towards the midline and concave laterally at its acromial end (Jones 1949). Radiographically, the clavicles of the Tasmanian devil and spot-tailed quoll also appear slender and straight with slight curvature and expansion at the

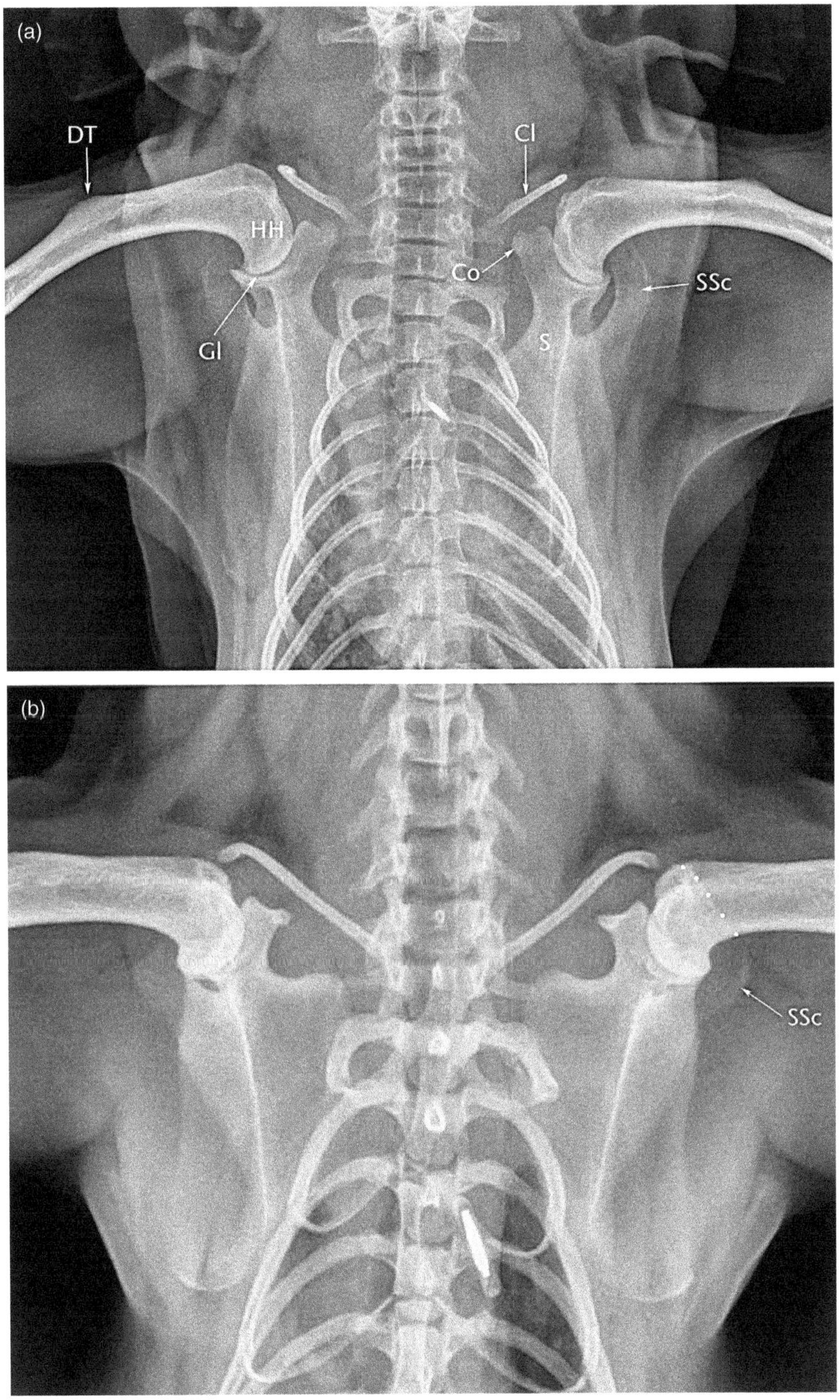

Figure 7.2: Pectoral girdle, dorsoventral view: (a) Tasmanian devil (*Sarcophilus harrisii*), (b) spot-tailed quoll (*Dasyurus maculatus*). The clavicle (Cl) is slender in the Tasmanian devil and in this species appears to have an indirect articulation with the sternum and the acromion. The spinous process (SSc) of the scapula (S) is slender and curved with its distal end directed towards the clavicle, forming the acromioclavicular articulation in the spot-tailed quoll (dotted line). The coracoid (Co) is directed craniomedially from the margin of the glenoid cavity (Gl). The humeral head (HH) articulates with the glenoid cavity. The deltoid tuberosity (DT) of the humerus is not well developed in the Tasmanian devil. Note: transponders can be seen.

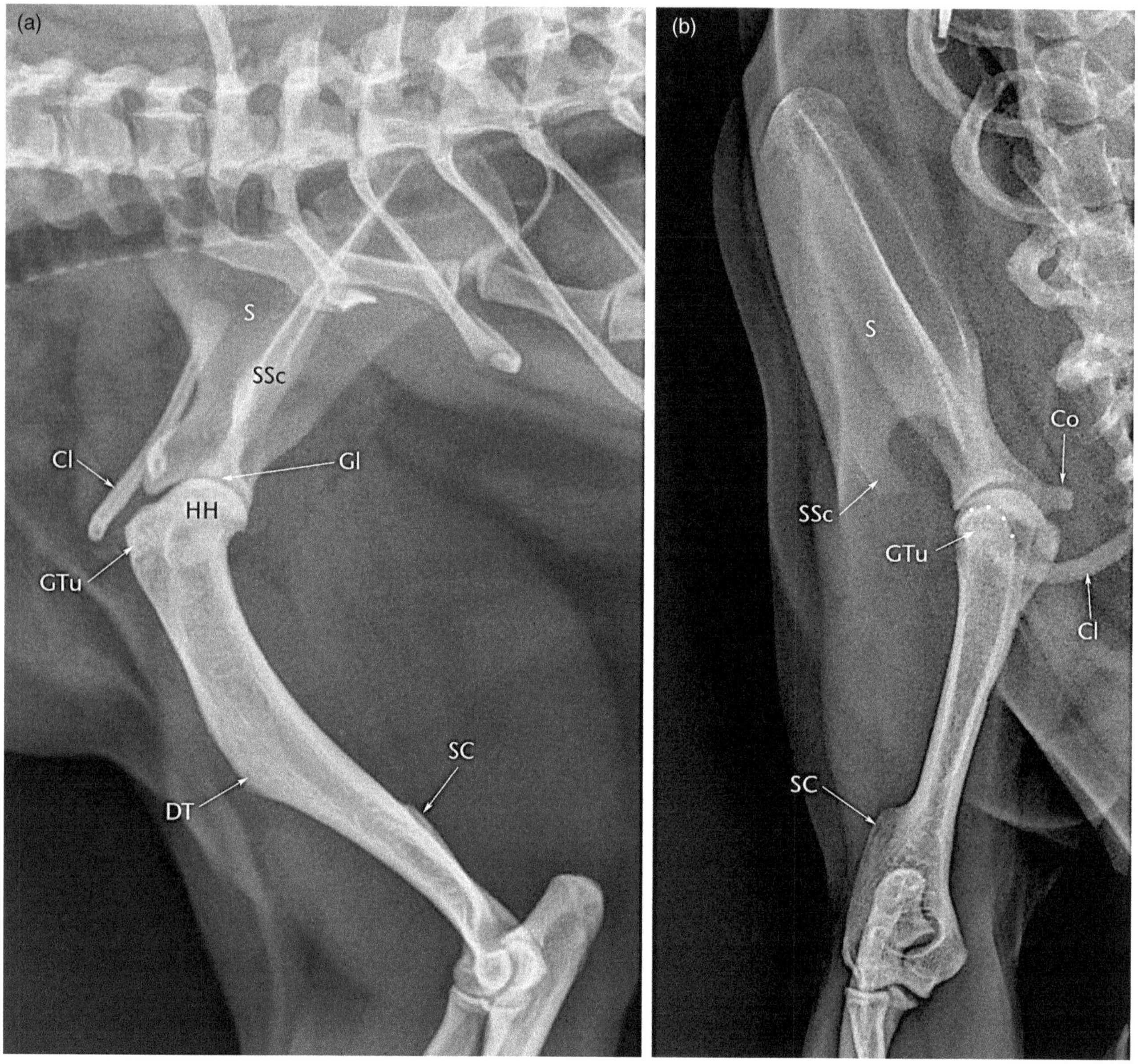

Figure 7.3: Shoulder and brachium, spot-tailed quoll (*Dasyurus maculatus*): (a) mediolateral view, (b) craniocaudal view. The spinous process (SSc) of the scapula (S) curves distally towards the clavicle (Cl), but cannot be traced in its entirety because of its relative radiolucency. The glenoid cavity (Gl) and humeral head (HH) create a congruent articulation for the shoulder joint. The greater tubercle (GTu) of the humerus is well developed; its proximal edge is marked with a dotted line in (b). The coracoid (Co) is visible on the medial margin of the glenoid. The deltoid tuberosity (DT) is visible in the mediolateral view and the supinator crest (SC) of the humerus can be seen on the craniocaudal view of the brachium.

sternal and acromial articular extremities (Fig. 7.2). The scapula is unremarkable and of similar structure in all dasyurids. It is similar to that of eutherian carnivores. The supraspinous fossa is much larger than the infraspinous fossa. The spine and acromion are curved and form a scroll-like covering of the infraspinous fossa and the axillary border of the bone is deflected in a dorsal direction, converting the infraspinous fossa into an incomplete osseous tube. The acromion is well developed and at its extremity articulates with the clavicle. The coracoid is well developed and projects

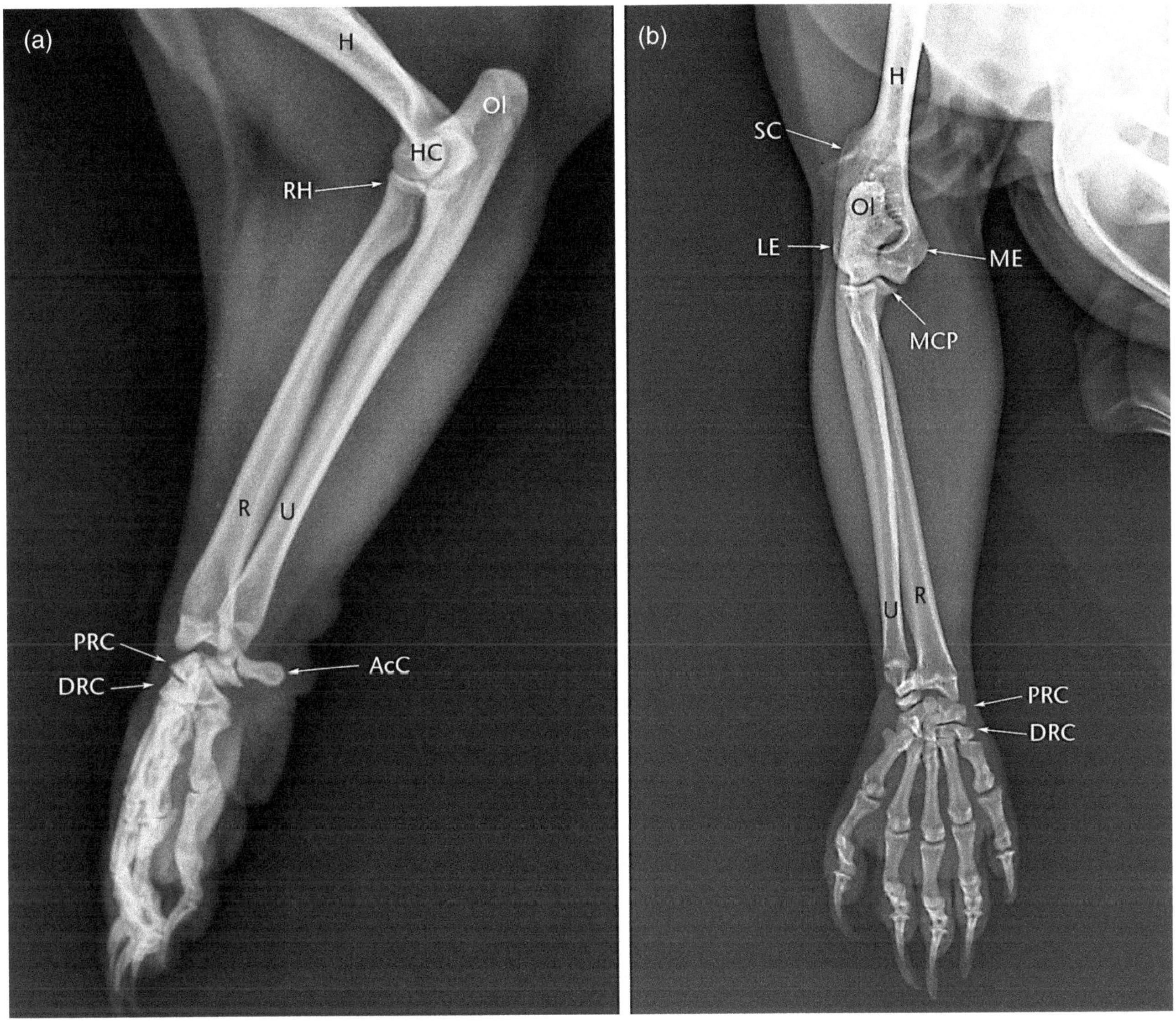

Figure 7.4: Antebrachium and manus, spot-tailed quoll (*Dasyurus maculatus*): (a) mediolateral view, (b) craniocaudal view. The radius (R) and ulna (U) are long and slender. The olecranon (Ol) projects proximal to the elbow joint. The supinator crest (SC) is prominent. The humerus (H) contributes to the elbow through the humeral condyle (HC), which has lateral (LE) and medial (ME) epicondyles. The antebrachial contribution to the elbow is through the radial head (RH) laterally and the medial coronoid process (MCP) and proximal ulna medially. Distally, the antebrachium articulates with the proximal row of carpal (PRC) bones to provide the antebrachiocarpal articulation. The distal row of carpal (DRC) bones articulates proximally with the proximal row of carpal bones and distally with the manus. The accessory carpal bone (AcC) is slender and projects caudally from the proximal row of carpal bones.

craniomedially from the margin of the glenoid cavity (Jones 1949) (Figs 7.2 and 7.3).

The humerus, radius and ulna are unremarkable and structurally similar to those of eutherian carnivores. The humerus is short and resembles that of canids in that the medial condyle is imperforate, but differs in the more marked development of ridges for muscular insertion (Owen 1839). The lateral supinator crest is prominent (Figs 7.3 and 7.4) and the deltoid ridge is well developed in many species. The radius and ulna are slender bones (Fig. 7.4) and the styloid

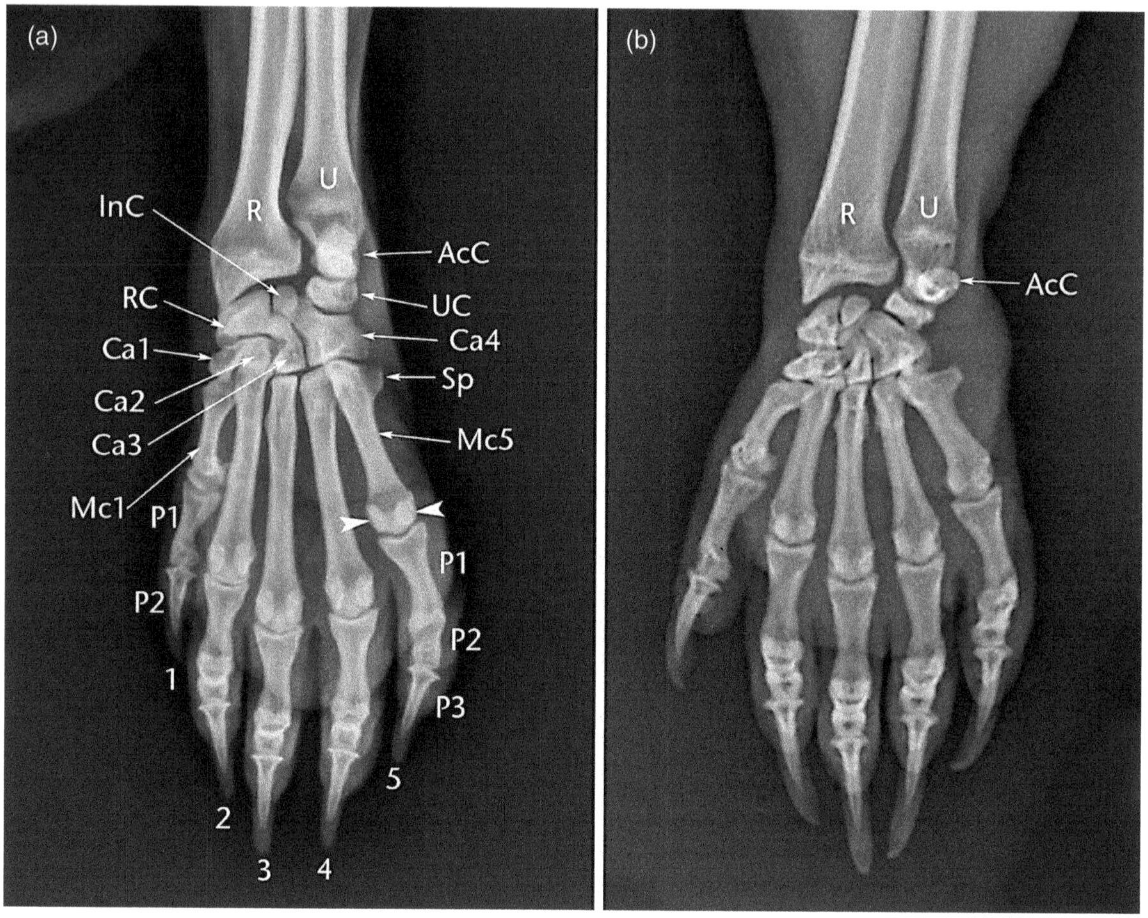

Figure 7.5: Manus, dorsopalmar view: (a) Tasmanian devil (*Sarcophilus harrisii*), (b) spot-tailed quoll (*Dasyurus maculatus*). The radius (R) and ulna (U) articulate with the carpus. The proximal row of carpal bones contains a radial (RC), an ulnar (UC), an intermediate (InC) and an accessory (AcC) carpal bone. In the distal row, the 1st, 2nd and 3rd carpal bones (Ca1, Ca2, Ca3) articulate distally with their respective metacarpals (Mc1, Mc5 labelled) while the 4th carpal bone (Ca4) articulates with the 4th and 5th metacarpals. There are five digits (1, 2, 3, 4, 5). Digit 1 has two phalanges and digits 2–5 each have three (P1, P2, P3). A pair of metacarpophalangeal sesamoid bones (arrowheads) is located on the palmar surface of each metacarpophalangeal articulation. There is an osseous spur (Sp) on the lateral surface of the proximal end of Mc5.

processes of the two bones are essentially at the same level.

The manus has five clawed digits (Holz 2008). The carpus has a large radial carpal bone and a small intermediate carpal bone articulating with the radius and an ulnar carpal bone articulating with the ulna. The intermediate carpal bone articulates with the radial carpal bone, the ulnar carpal bone and the fourth carpal bone. The accessory carpal bone is well developed. The third metacarpal bone is the longest and most slender of the metacarpal bones (Jones 1949) (Fig. 7.5).

Pelvis and hindlimb

The pelvis is unremarkable and similar in form to that of other marsupials. The ilia are elongated and articulate with the transverse processes of the first sacral vertebra. The acetabula have wide ventral notches and the round ligament is well developed (Jones 1949). Two epipubic bones (rudimentary in the thylacine) extend from the cranial aspect of the pubis (Figs 7.6 and 7.7).

The bones of the hindlimb are unremarkable and structurally similar to those of eutherian carnivores. The greater trochanter of the femur is well

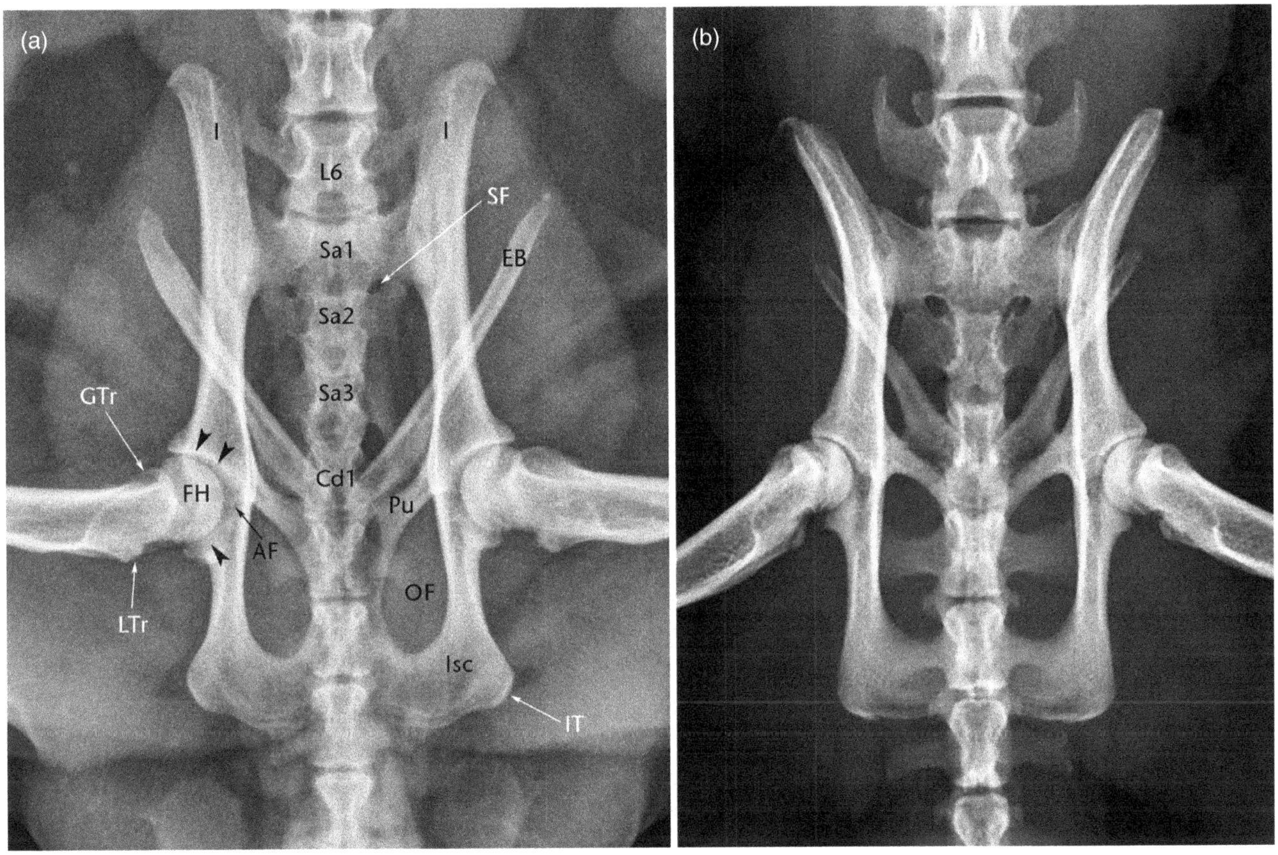

Figure 7.6: Pelvis, ventrodorsal view: (a) Tasmanian devil (*Sarcophilus harrisii*), (b) spot-tailed quoll (*Dasyurus maculatus*). The lumbosacral junction is between the 6th lumbar vertebra (L6) and the sacrum. There are three sacral elements (Sa1, Sa2, Sa3). Sa1 and Sa2 have fused transverse processes, partially separated by a sacral foramen (SF). The ilium (I) articulates with the 1st sacral element. The paired epipubic bones (EB) are larger in the Tasmanian devil than in the spot-tailed quoll and articulate with the pubis (Pu). The pubis and ischium (Isc) are separated by the obturator foramen (OF). The ischial tuberosity (IT) is not well developed. The cranial and caudal articular surfaces of the acetabulum (arrowheads) are perforated by the acetabular fossa (AF). The proximal femur has a spherical femoral head (FH) that has a congruent articulation with the acetabulum, a greater trochanter (GTr) and a lesser trochanter (LTr). The 1st caudal vertebra (Cd1) articulates proximally with the 3rd sacral element.

developed. As is the case in all marsupials other than bandicoots and the bilby, dasyurids do not have an ossified patella and it is represented only by a patelloid, which is a fibrocartilaginous induration of the quadriceps femoris extensor tendon (Jones 1949; Reese *et al.* 2001) (Figs 7.8 and 7.9).

The tibia and fibula are approximately the same length. They articulate with each other proximally and distally, but elsewhere are separated by a space approximately as wide as the diameter of the tibia. The proximal end of the fibula articulates with an oblong sesamoid bone that is associated with the caudolateral aspect of the lateral femoral condyle. The distal tibia and fibula articulate with their respective tarsal bones (Fig. 7.9). In the crest-tailed mulgara, the tibia and fibula articulate only with the talus, which is a short, compact bone (Jones 1949). In *Dasyurus* spp., the first tarsal bone is well developed and supports the first metatarsal bone. The fourth tarsal bone supports the fourth and fifth metatarsal bones.

In the *Dasyurus* spp., the first digit has two phalanges. It is relatively short and slender, distally reaching the level of the metacarpophalangeal

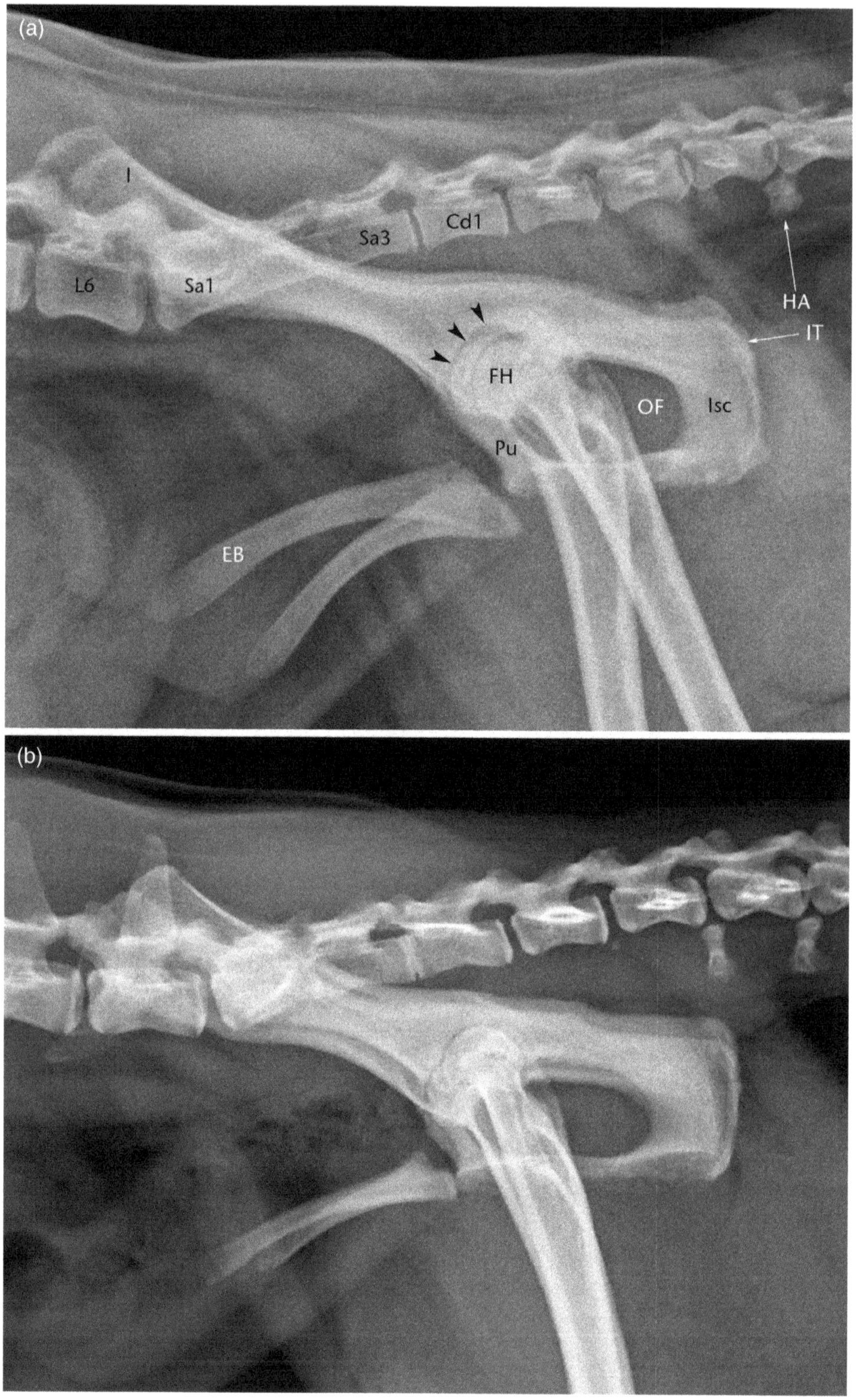

Figure 7.7: Pelvis, lateral view: (a) Tasmanian devil (*Sarcophilus harrisii*), (b) spot-tailed quoll (*Dasyurus maculatus*). The sacrum (Sa1, Sa3 labelled) is more easily seen in this view, as are its articulations cranially with the 6th lumbar vertebra (L6) and caudally with the 1st caudal vertebra (Cd1). The ilium (I) of the Tasmanian devil is elongated. The pubis (Pu) and its articulation with the epipubic bones (EB) are clear. The ischia (Isc) and their tuberosities (IT) and the obturator foraminae (OF) are superimposed in this view. Haemal arches (HA) are present ventral to the articulations of the caudal vertebrae.

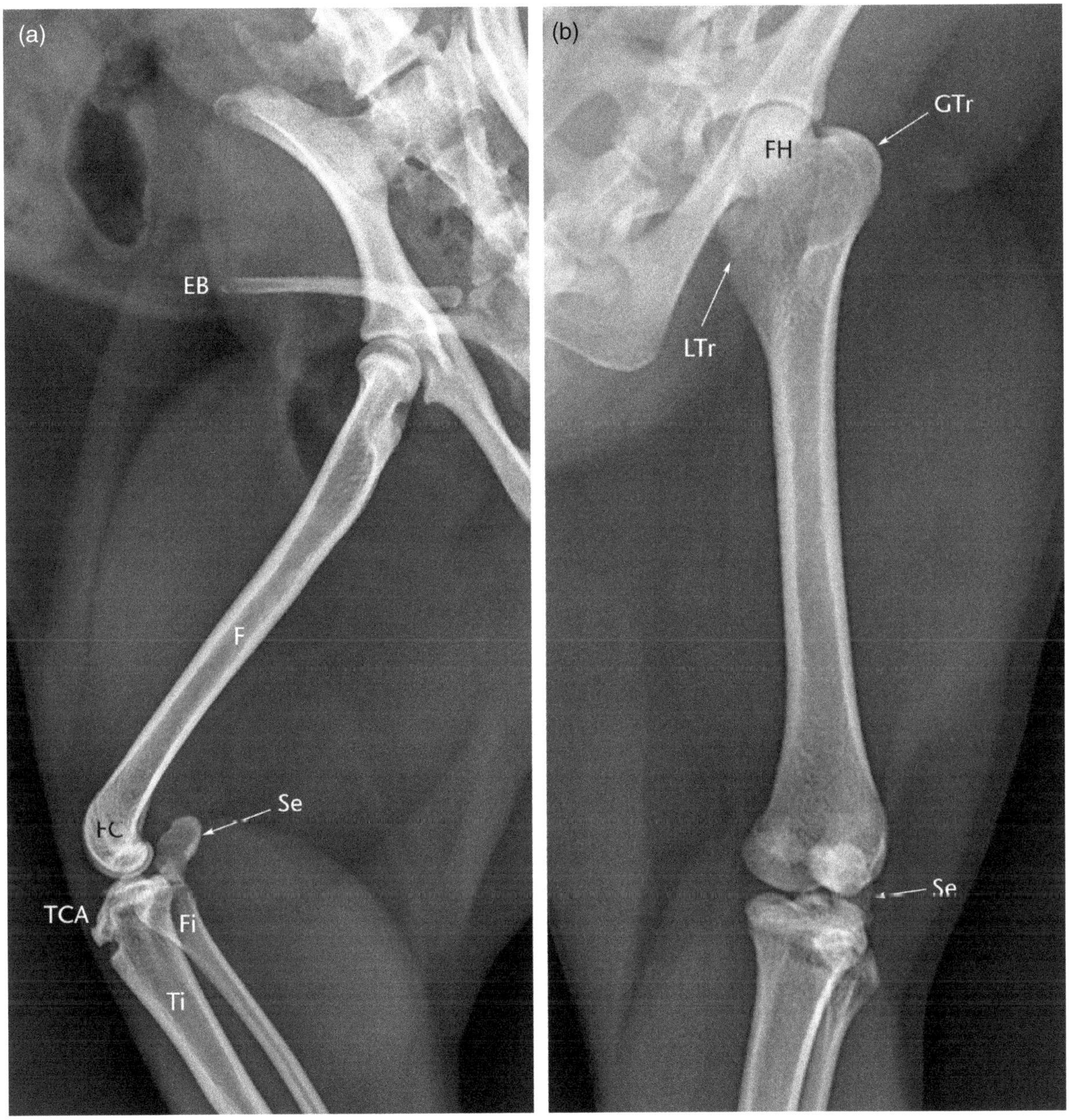

Figure 7.8: Femur and stifle, spot-tailed quoll (*Dasyurus maculatus*): (a) mediolateral view, (b) oblique caudocranial view. The proximal femur (F) has well developed greater (GTr) and lesser (LTr) trochanters and a femoral head (FH). The distal femur articulates with the tibia (Ti) and fibula (Fi). A large sesamoid bone (Se) is adjacent to the proximal end of the fibula, located on the lateral aspect of the femoral condyle (FC). In the lateral image, an incompletely fused tibial crest apophysis (TCA) can be seen. An epipubic bone (EB) is visible crossing the ilium.

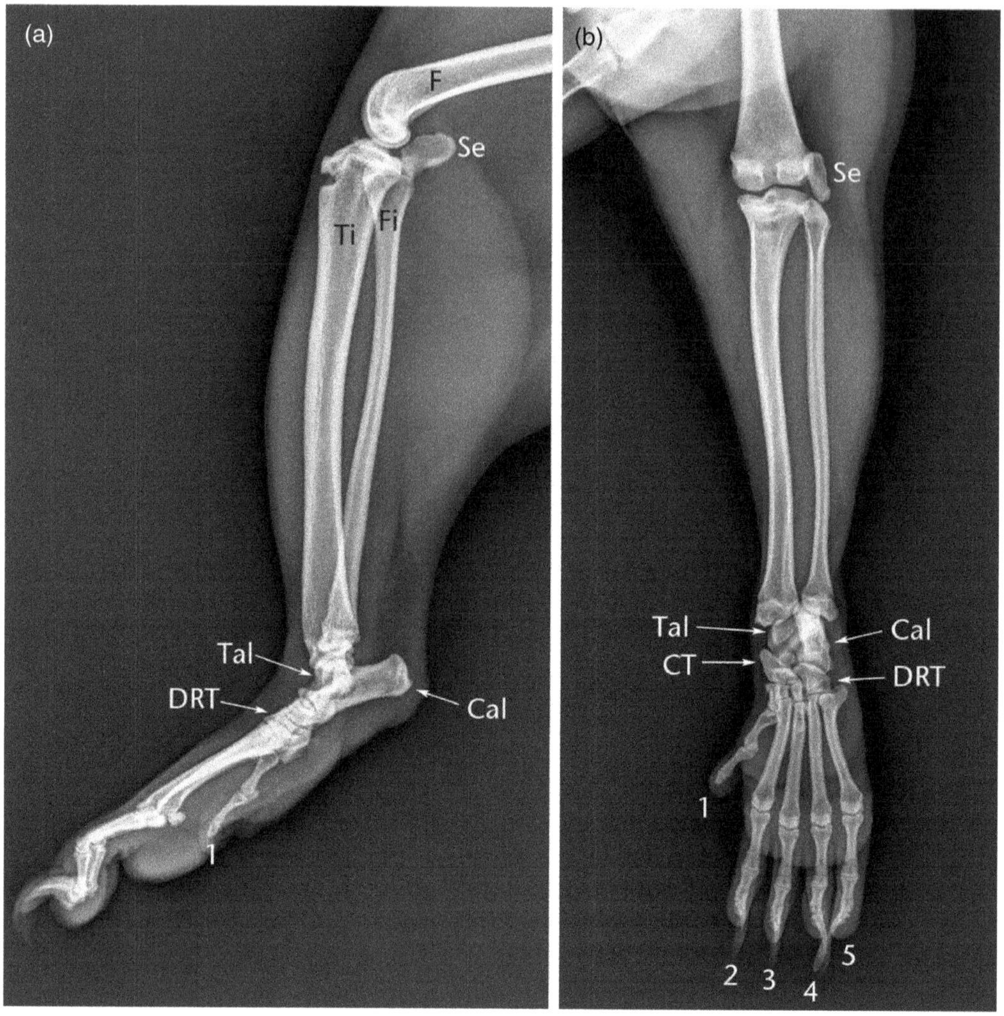

Figure 7.9: Stifle, crus and pes, spot-tailed quoll (*Dasyurus maculatus*): (a) mediolateral view, (b) craniocaudal view. The stifle joint comprises elements of the femur (F), tibia (Ti) and fibula (Fi) and an associated sesamoid bone (Se). The tarsocrural articulation is between the tibia and fibula proximally and the calcaneus (Cal) and talus (Tal) distally. The distal row of tarsal bones (DRT) is separated from the talus by a central tarsal bone (CT). The pes has five digits (1, 2, 3, 4, 5).

articulation of the second digit. In other species, the first digit of the pes is either short or absent, and when present is always clawless. The presence of the first digit is usually indicative of an arboreal habit. Arboreal species tend to have a broad pes and mobile first digit, whereas in those that are both terrestrial and arboreal, the pes is longer and the first digit is reduced (Morton *et al.* 1989). The second and third digits of the pes are not fused (i.e. they are didactylous). The third metatarsal is slightly longer that the second, third and fifth metatarsals. The phalangeal bones of digits 2–5 appear similar to one another in shape and dimensions (Fig. 7.10).

Axial skeleton

Skull

The skull structure of dasyurids varies significantly among species, ranging from the solid, broad, stout skull of the Tasmanian devil to the fine, elongated, narrow skulls of the smaller species (Fig. 7.11). The skull of the crest-tailed mulgara is short, broad and flattened with the bones of the

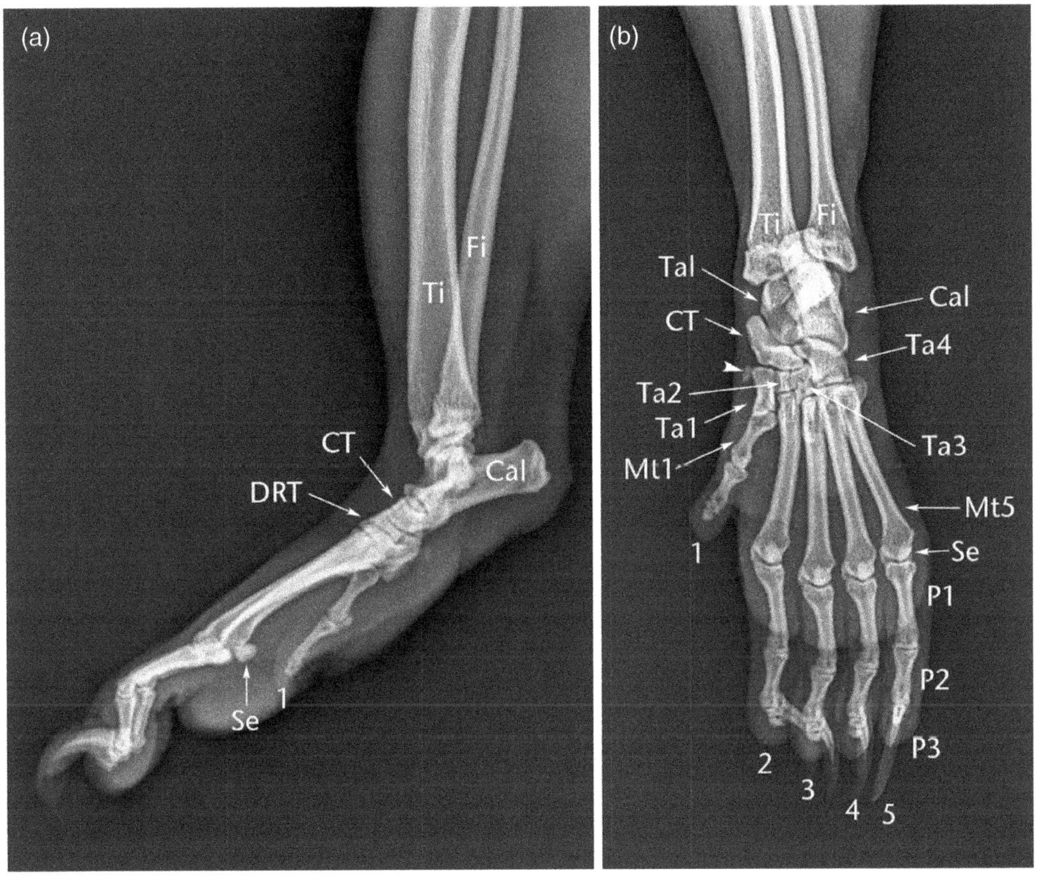

Figure 7.10: Pes, spot-tailed quoll (*Dasyurus maculatus*): (a) mediolateral view, (b) dorsopalmar view. The talus (Tal) and calcaneus (Cal) articulate proximally mainly with the tibia (Ti) and fibula (Fi), respectively. Distally, the talus is separated from the 1st three tarsal bones in the distal row (DRT, Ta1, Ta2, Ta3) by an angular central tarsal (CT) bone. These three tarsal bones articulate distally with their respective metatarsal bones. The 4th tarsal bone (Ta4) articulates proximally with the calcaneus and distally with the 4th and 5th metatarsal bones. There are five metatarsal bones (Mt1, Mt5 labelled). A small sesamoid bone (arrowhead) is present on the medial surface of the first tarsal bone. The 1st digit (1) has a well developed 1st phalanx and a poorly developed 2nd phalanx and is clawless. The remaining four digits (2, 3, 4, 5) each have three well developed phalanges (P1, P2, P3), the 3rd being clawed. A sesamoid bone is present on the caudal aspect of each metatarsophalangeal articulation (Se).

maxilla so thin that the tooth roots can be seen through them. In this species at least, there are paired rostral palatine vacuities between the canine teeth. Large palatine vacuities are present between the upper third and fourth molar teeth in the caudal palate in all dasyurids. In *Dasycercus* spp., there are two smaller vacuities just caudal to these (Jones 1949). The jugular process of the occipital bone is short or absent in dasyurids. The tympanic bullae vary in size among species. In the crest-tailed mulgara they are enormously inflated and occupy more than one-third of the total basal length of the skull (Jones 1949), whereas in the Tasmanian devil and spot-tailed quoll they are considerably smaller (Fig. 7.12). The glenoid cavity of the temporomandibular joint forms a concavity that is elongated transversely and shallower than that of eutherian carnivores. The joint also differs by not having inter-articular cartilage (Owen 1839). The coronoid process of the mandible is very tall. The angle of the mandible is very obtuse and the angular process is well developed.

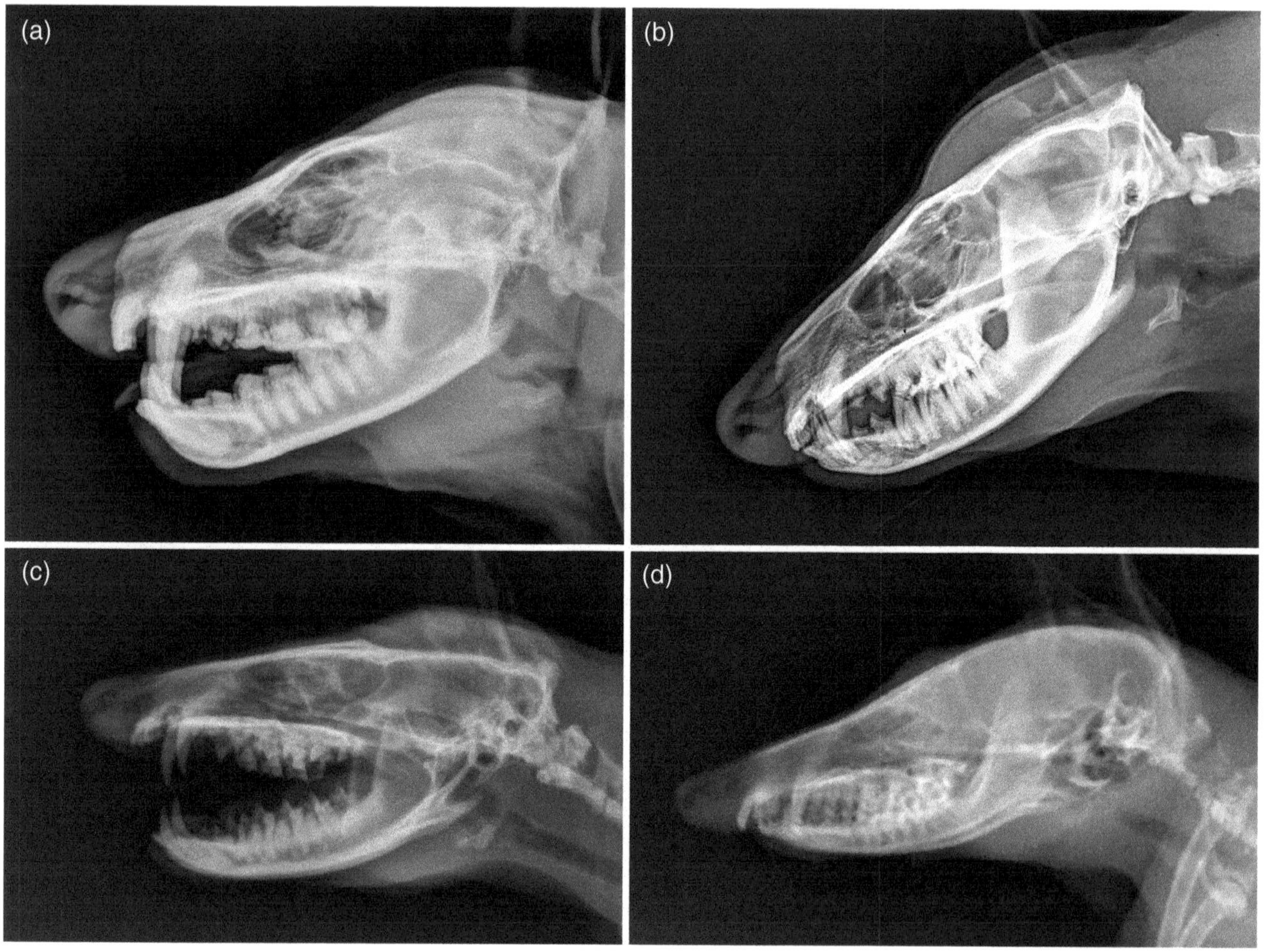

Figure 7.11: Skull, lateral view: (a) Tasmanian devil (*Sarcophilus harrisii*), (b) spot-tailed quoll (*Dasyurus maculatus*), (c) northern quoll (*Dasyurus hallucatus*), d) red-tailed phascogale (*Phascogale calura*). The profile of the skull of the Tasmanian devil is box-like, with a thick mandible. The profile is progressively more elongated and slender in the smaller quolls and the phascogale.

Dasyurids are polyprotodont marsupials because they have three pairs of lower incisors, as opposed to the diprotodonts, which have only a single pair. The dental characteristics of dasyurids reflect their insectivorous and carnivorous habits (Morton *et al.* 1989). Dentition and dental radiology is covered in Chapter 11.

Vertebrae

The vertebral column is made up of 7 cervical, 13 thoracic and 6 lumbar vertebrae, 2–3 sacral elements and up to or more than 20 caudal vertebrae (19 in the Tasmanian devil and 22 in the spot-tailed quoll) (Figs 7.13–7.15). In quolls, the two or three sacral vertebrae are ankylosed. However, the ilia only articulate with the expanded transverse processes of the first sacral element (Fig. 7.6). Varying numbers of haemal arches are located ventrally to the articulations of the caudal vertebrae (Owen 1839) (Figs 7.7 and 7.14). The tail of most species is shorter than the body length. Fat is stored in the tail of some species.

Ribs and sternum

The sternum is made up of six or seven elements. The xiphisternum has a long, styliform ossified

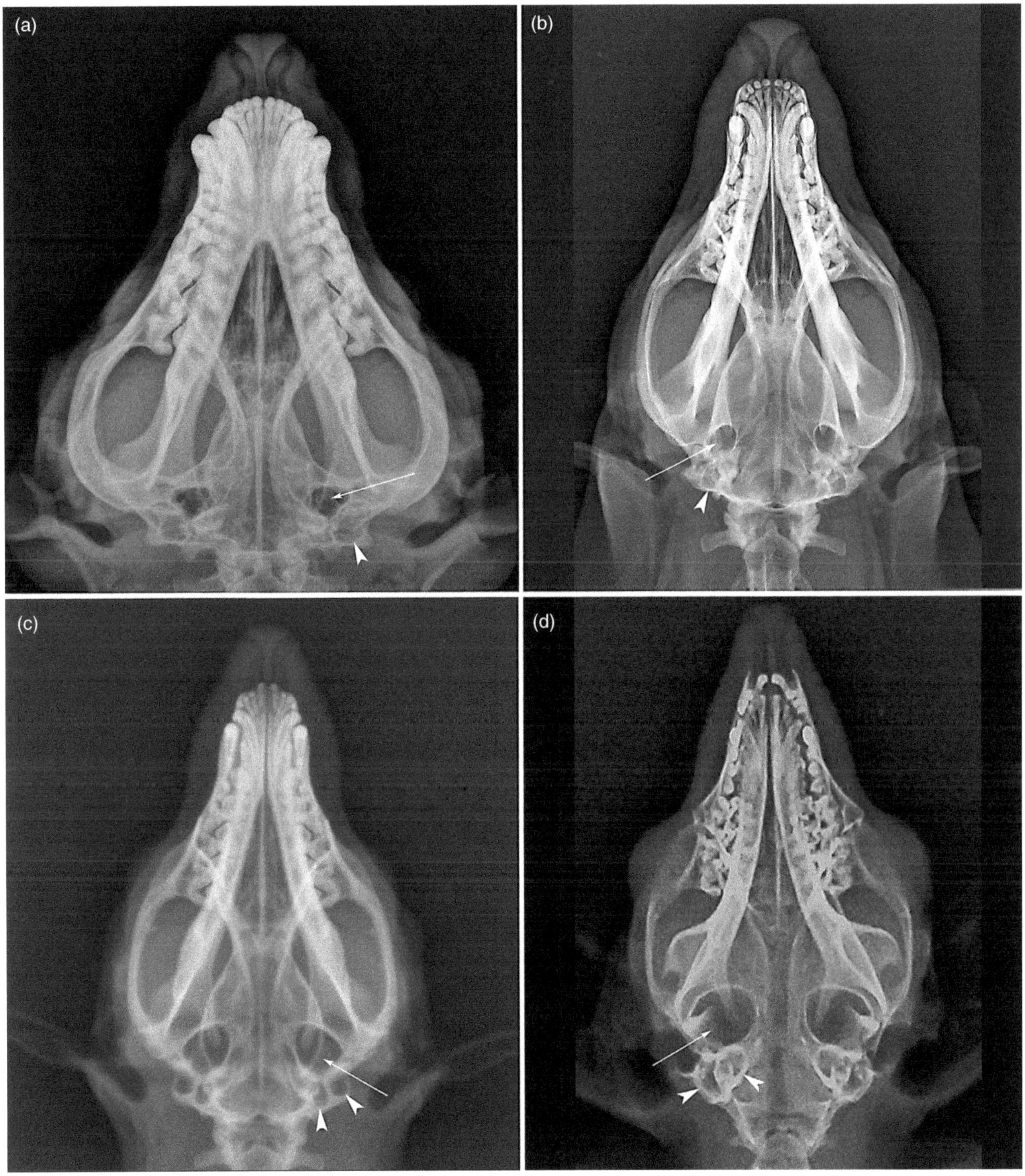

Figure 7.12: Skull, ventrodorsal view: (a) Tasmanian devil (*Sarcophilus harrisii*), (b) spot-tailed quoll (*Dasyurus maculatus*), (c) northern quoll (*Dasyurus hallucatus*), (d) red-tailed phascogale (*Phascogale calura*). The Tasmanian devil has a wide skull with flared zygomatic arches. This feature is less pronounced in the smaller dasyurids. Conversely, the tympanic bullae are proportionately larger in smaller species (arrows). There are small air-filled spaces caudal to the bullae (arrowheads) in each of the skulls.

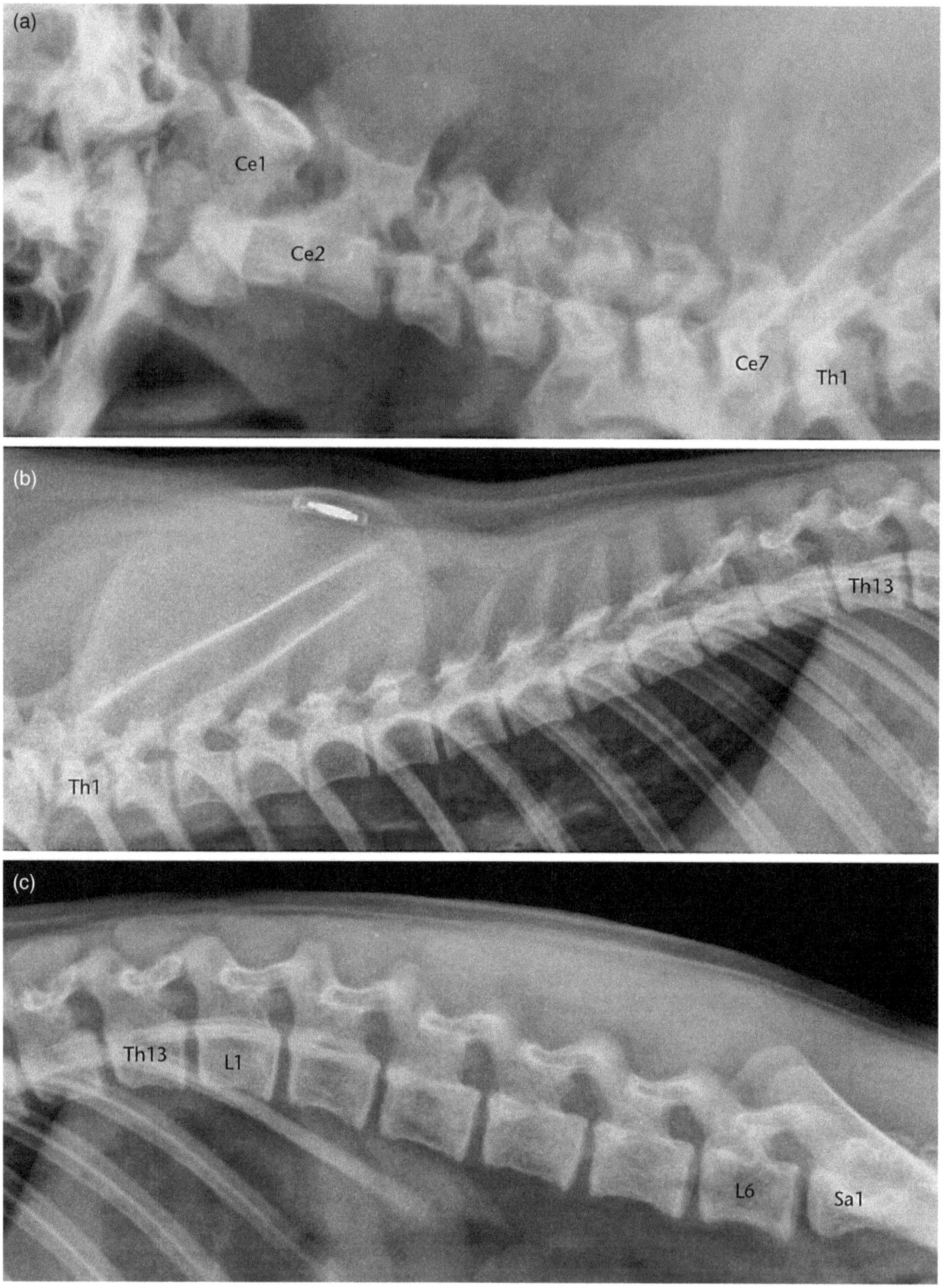

Figure 7.13: Vertebral column, Tasmanian devil (*Sarcophilus harrisii*), lateral view: (a) cervical vertebrae, (b) thoracic vertebrae, (c) lumbar vertebrae. There are 7 cervical (Ce1, Ce2, Ce7 labelled), 13 thoracic (Th1, Th13 labelled) and 6 lumbar (L1, L6 labelled) vertebrae. The 1st sacral element (Sa1) is included. Note: a transponder can be seen in (b).

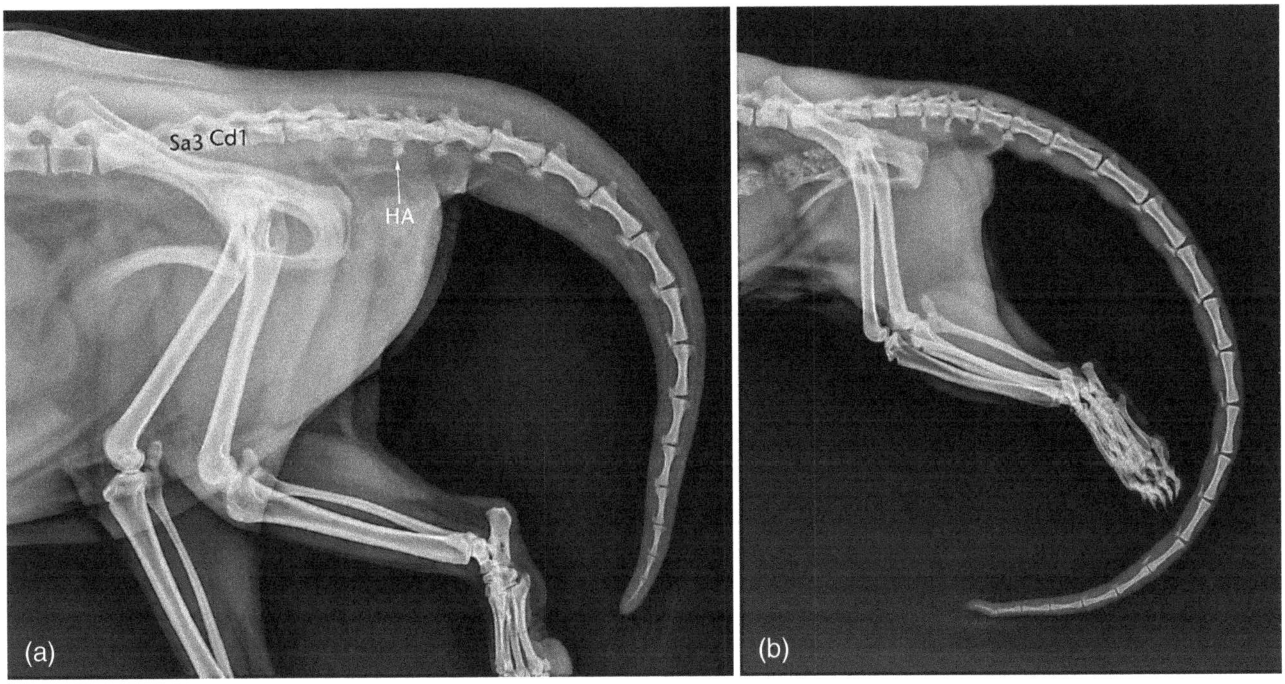

Figure 7.14: Caudal vertebrae, lateral view: (a) Tasmanian devil (*Sarcophilus harrisii*), (b) spot-tailed quoll (*Dasyurus maculatus*). Ventrally located haemal arches (HA) are present at the articulations of most caudal vertebrae. The 3rd sacral element (Sa3) and 1st caudal vertebra (Cd1) are labelled.

portion that expands caudally into a broad cartilaginous plate (Fig. 7.16). There are 13 pairs of ribs. Of these, the 7 cranial ribs articulate directly with the sternum (Figs 7.16 and 7.17). The 8th to 11th costal cartilages fail to reach the midline and articulate with each other in series. The 12th and 13th ribs terminate in free costal cartilages.

Soft tissue

Thorax

The cranial mediastinum is wider in Tasmanian devils than in other marsupials (Fig. 7.17), occupying 50–100% of the width of the cranial thorax. In other respects, the thorax, including the thoracic wall, diaphragm and pleural space is similar to that of the macropods and the koala.

Cardiovascular system

The radiographic appearance of the cardiovascular system is unremarkable and similar to that of other marsupials. The ovoid heart slopes ventrocaudally. Individual pulmonary vessels can be traced within the pulmonary parenchyma (Figs 7.16 and 7.17).

Respiratory system

Lobation of the right lung in dasyurids is identical to that of domestic carnivores. It is divided into cranial, middle, caudal and accessory lobes by interlobar fissures (Jones 1949; Cope *et al.* 2001). The left lung consists of cranial and caudal lobes that are not divided by an interlobar fissure (Cope *et al.* 2001). In the crest-tailed mulgara, the left lung is considerably smaller than the right and only shows a partial subdivision into cranial and caudal lobes (Jones 1949). Radiographically, the cranial parts of the lungs are less aerated than the caudal lobes, which is a function of size, but also an effect of the relatively wide cranial mediastinum in some species (Figs 7.16 and 7.17).

Abdomen

Dasyurids are carnivorous or insectivorous, although antechinus, phascogales and quolls also consume

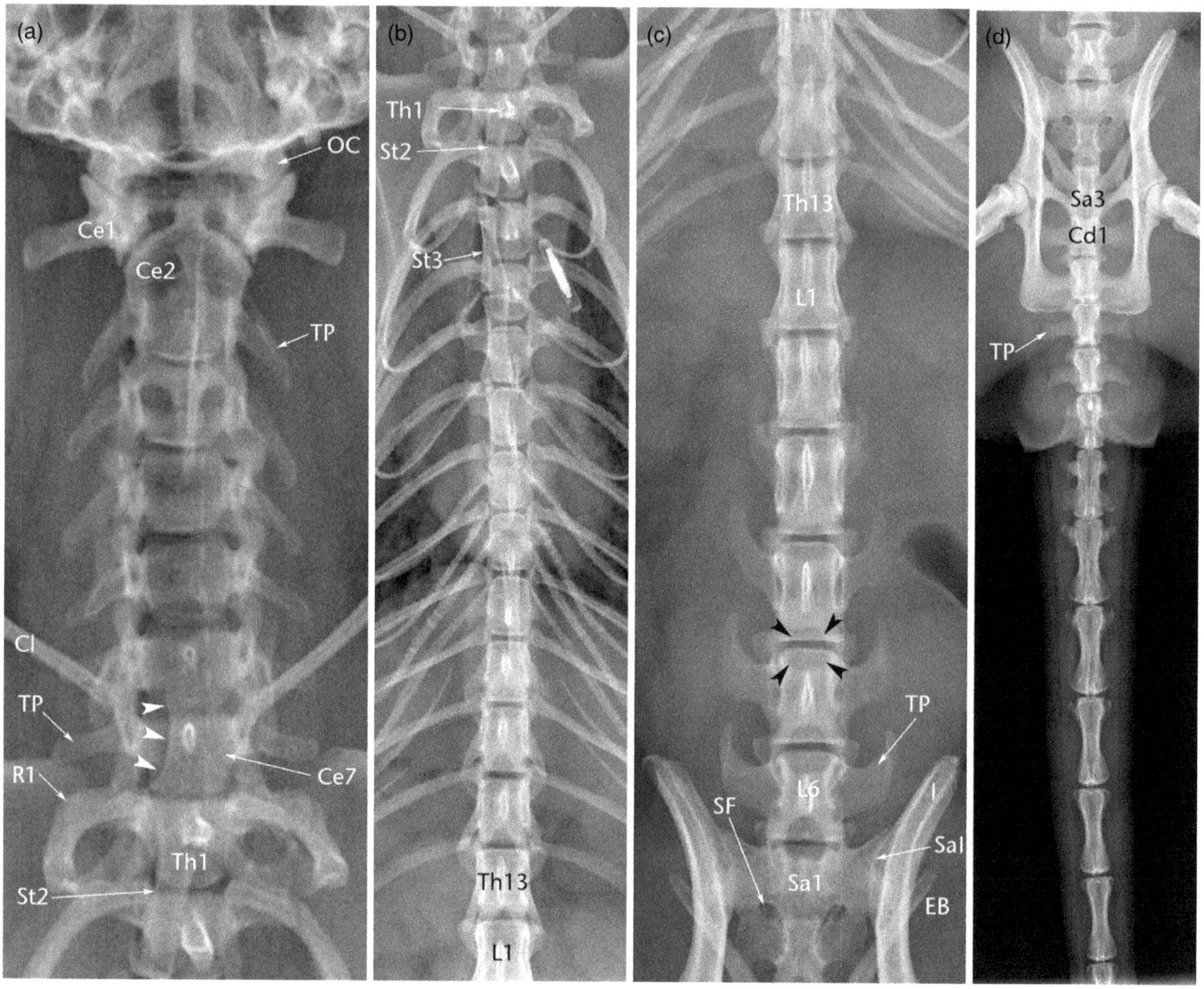

Figure 7.15: Vertebral column, spot-tailed quoll (*Dasyurus maculatus*), ventrodorsal view: (a) cervical vertebrae, (b) thoracic vertebrae, (c) lumbar vertebrae, d) sacrocaudal vertebral segments. The 1st cervical (Ce1) vertebra articulates cranially with the occipital condyles (OC) and has wide, flat lateral wings, while the remaining cervical vertebrae (Ce2 and Ce7 labelled) have caudally directed transverse processes (TP). The caudal cervical (Ce7 labelled) and cranial thoracic (Th1 labelled) vertebrae are superimposed on the sternum (sternal elements St2 and St3 labelled). The 1st sternal element has concave lateral margins (white arrowheads) that do not appear to have a close relationship with the distal ends of the clavicles (Cl). The caudal lumbar (L6 labelled) vertebrae have increasingly wide, curved transverse processes (TP). Ovoid, dorsally located intervertebral foraminae (black arrowheads) are present between the 10th thoracic vertebra and the sacrum (Sa1). Sacral foraminae (SF) are present between the 1st two sacral elements. The 1st five or six caudal (Cd1 labelled) vertebrae have transverse processes (TP). The 1st rib (R1), 13th thoracic vertebra (Th13), 1st lumbar vertebra (L1), epipubic bones (EB), ilium (I), 3rd sacral element (Sa3) and sacroiliac (SaI) articulation are labelled. Note: a transponder can be seen in (b).

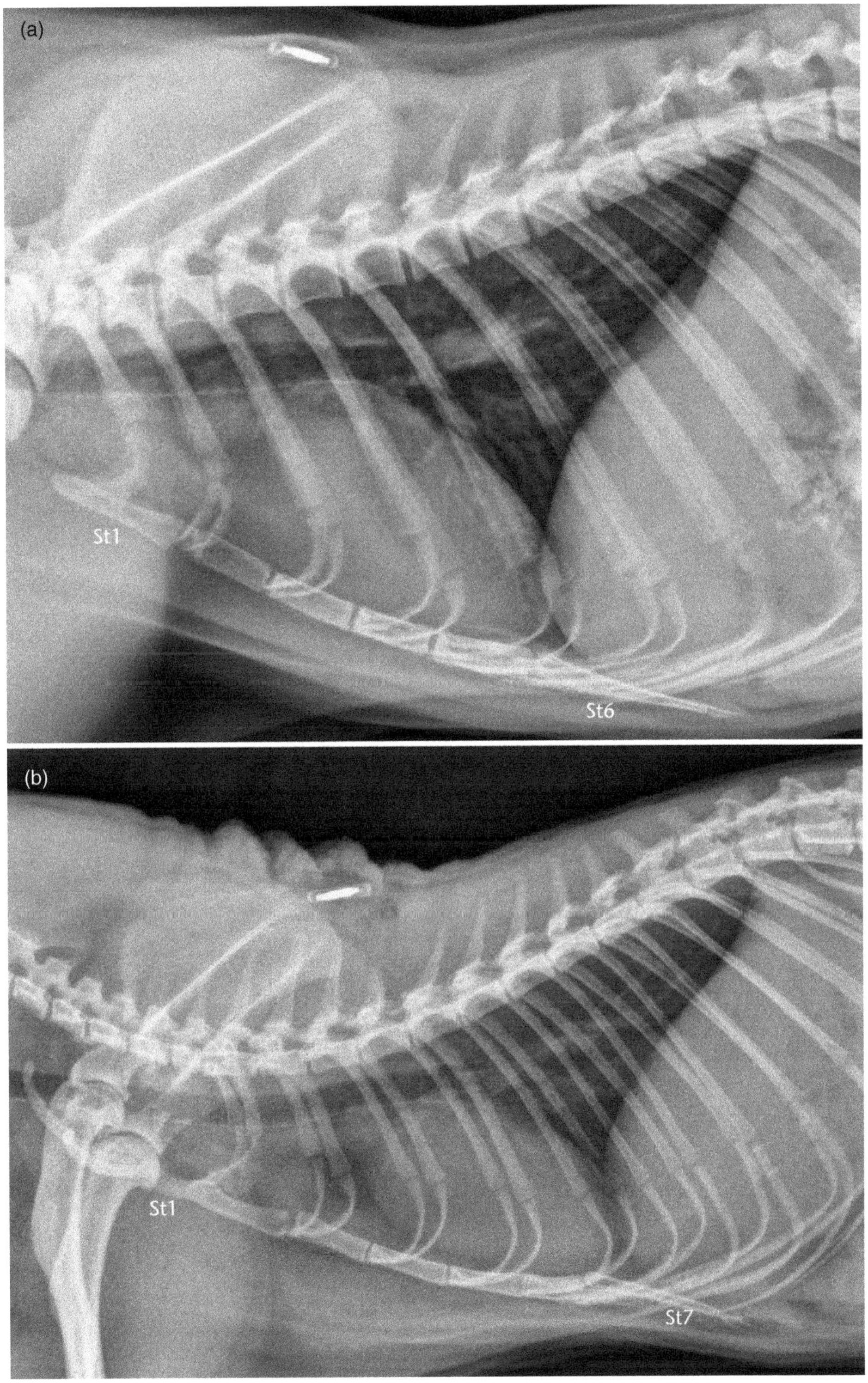

Figure 7.16: Thorax, lateral view: (a) Tasmanian devil (*Sarcophilus harrisii*), (b) spot-tailed quoll (*Dasyurus maculatus*). Poorer aeration of the cranial lung lobes in the Tasmanian devil may be related to the wide cranial mediastinum occupying a relatively greater proportion of the thorax than in the spot-tailed quoll. There are six sternebrae (St1, St6 labelled) in the Tasmanian devil and seven (St1, St7 labelled) in the spot-tailed quoll, although the first two appear fused. Note: transponders can be seen.

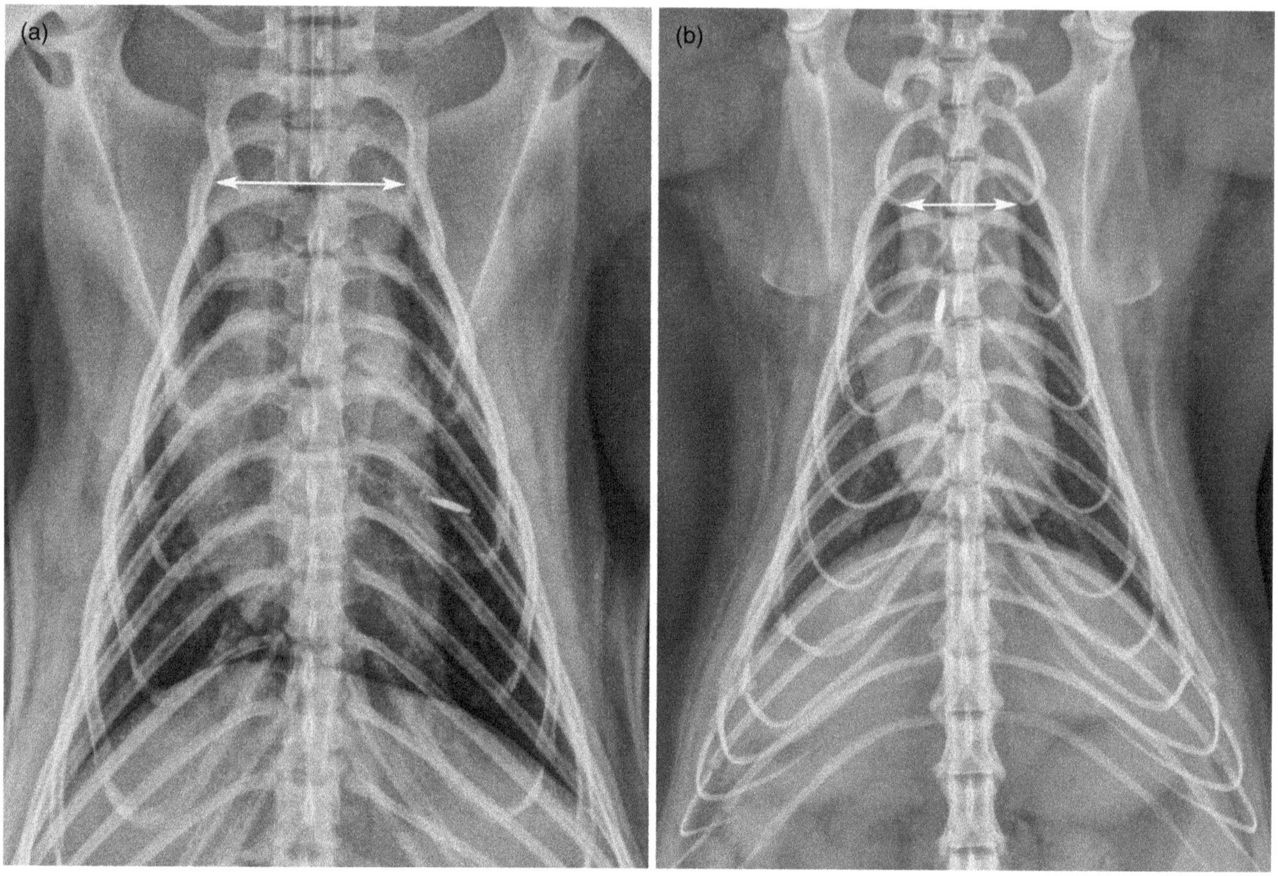

Figure 7.17: Thorax, ventrodorsal view: (a) Tasmanian devil (*Sarcophilus harrisii*), (b) spot-tailed quoll (*Dasyurus maculatus*). The relative width of the cranial mediastinum is defined with arrows. Note: transponders can be seen.

plant material and nectar, particularly in winter when invertebrates are scarce. Dasyurids have a short gastrointestinal tract consisting of a simple stomach, small intestine, large intestine and no caecum (Figs 7.18 and 7.19). Brunner's glands, which secrete an alkaline fluid that protects the proximal duodenal mucosa from the ulcerating effects of acid-pepsin secretions from the stomach, appear as a thick collar in the duodenum immediately distal to the pylorus. The glands form large lobes that drain into the duodenal lumen. Digesta transit times are variable, with 100% excretion occurring from 3–4 hours after ingestion in planigales and up to 13 hours in eastern quolls (*Dasyurus viverrinus*) (Hume 1999).

Female dasyurids have the typical marsupial reproductive tract of two uteri and two lateral vaginae. Between the vaginae is a mass of fibrous tissue called the urogenital strand. A birth canal forms within this strand during parturition. This canal becomes permanent after the first parturition in most macropods, but in dasyurids the cleft closes rapidly after birth so that no indication of the canal remains. Male dasyurids have a carrot-shaped prostate and paired bulbourethral glands: two pairs in Tasmanian devils and fat-tailed dunnarts and three pairs in eastern quolls and brown antechinus (*Antechinus stuartii*) (Holz 2008).

The pouch may be well developed or represented simply by raised ridges; its opening is directed either ventrally or caudally (Morton *et al.* 1989).

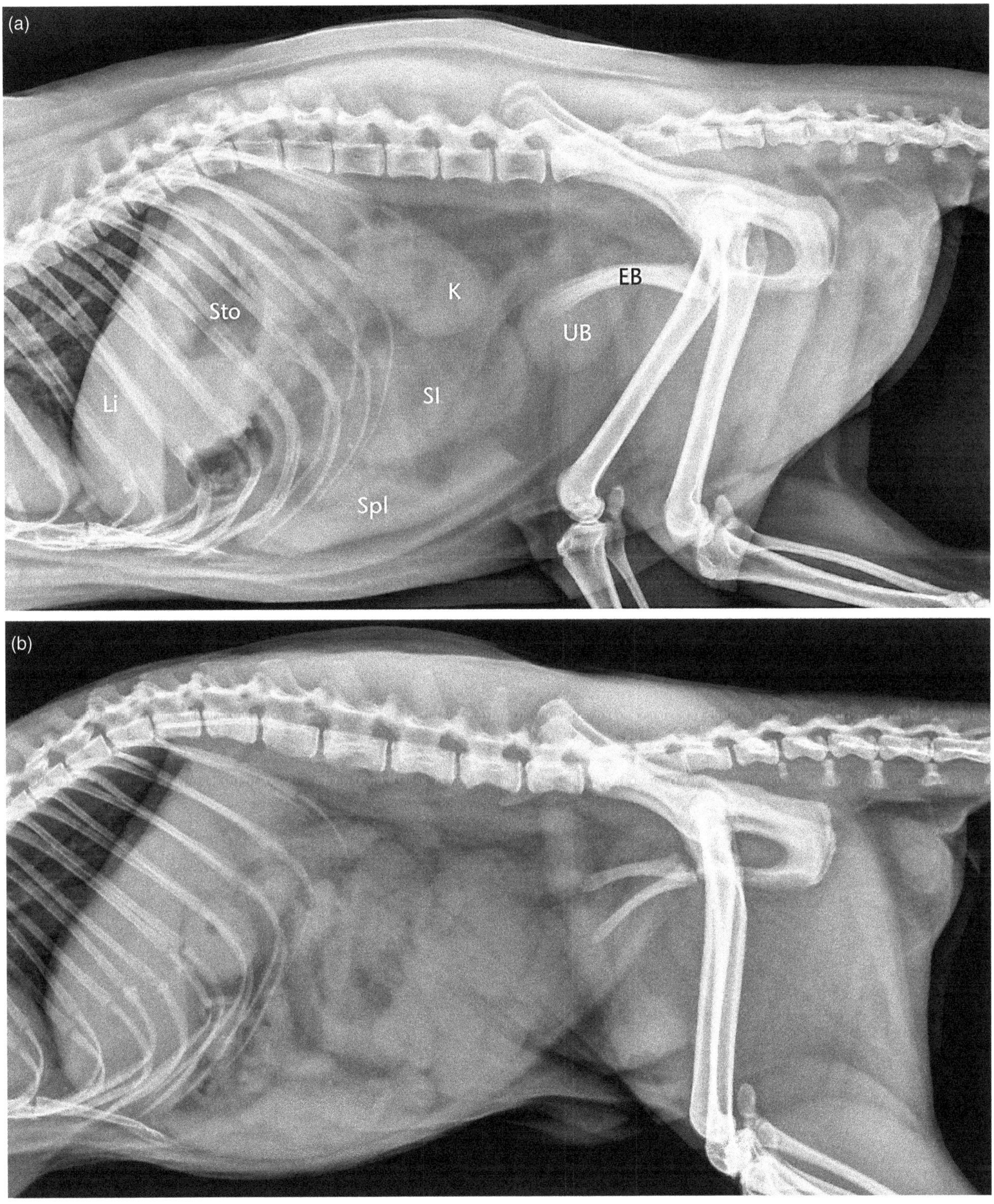

Figure 7.18: Abdomen, lateral view: (a) Tasmanian devil (*Sarcophilus harrisii*), (b) spot-tailed quoll (*Dasyurus maculatus*). With the exception of the epipubic bones (EB), the abdominal appearance is similar to most other species, with a simple stomach (Sto) and short intestinal tract. The liver (Li), kidneys (K, superimposed), small intestines (SI), spleen (Spl) and urinary bladder (UB) are labelled.

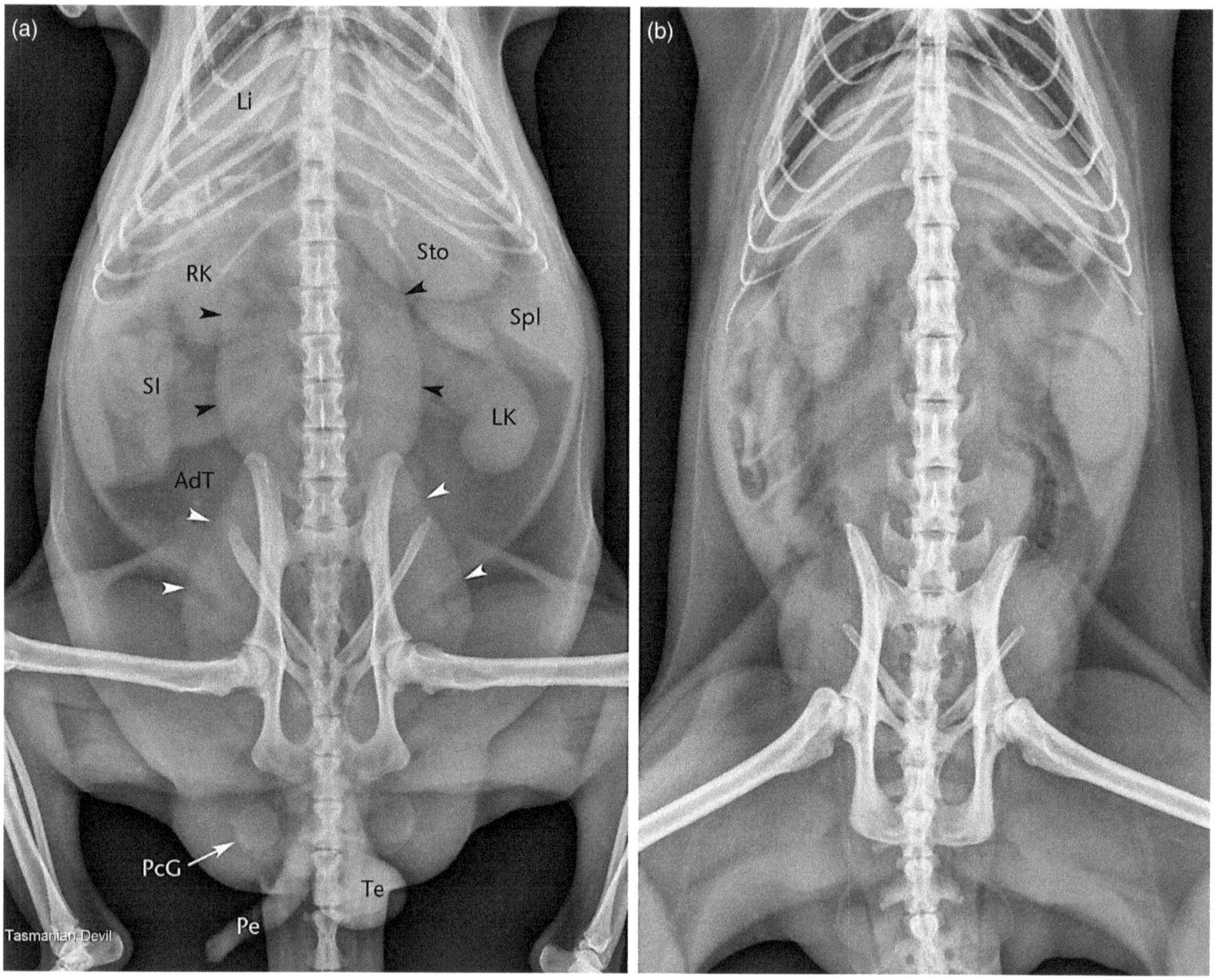

Figure 7.19: Abdomen, ventrodorsal view: (a) Tasmanian devil (*Sarcophilus harrisii*), (b) spot-tailed quoll (*Dasyurus maculatus*). The presence of adipose tissue (AdT) always enhances visibility of abdominal viscera. For this reason, the silhouettes of most of the abdominal viscera can be seen in the Tasmanian devil, but less so in the spot-tailed quoll. The liver (Li), stomach (Sto), spleen (Spl), right kidney (RK), left kidney (LK) and the small intestines (SI) are labelled. The edges of superimposed lumbar muscles (black arrowheads) and the gluteal muscles (white arrowheads) are defined. Caudal to the pelvis, the penis (Pe) and testes (Te) are seen. Paired paracloacal glands (PcG) are also visible.

References

Cope LA, Henry RW, Reed RB (2001) Macroscopic anatomy of the lower respiratory tract of the gray short-tailed opossum (*Monodelphis domestica*). *Brazilian Journal of Morphological Science* **18**, 111–117.

Holz PH (2008) Dasyurids. In *Medicine of Australian Mammals*. (Eds L Vogelnest and R Woods) pp. 359–382. CSIRO Publishing, Melbourne.

Hume ID (1999) Carnivorous marsupials. In *Marsupial Nutrition*. pp. 35–74. Cambridge University Press, Cambridge.

Jones FW (1949) The study of a generalised marsupial (*Dasycercus cristicauda* Kreft). *The Transactions of the Zoological Society of London* **26**(5), 409–501. doi:10.1111/j.1096-3642.1949.tb00225.x

Krajewski C, Westerman M (2003) Molecular systematics of Dasyuromorphia. In *Predators with*

Pouches: The Biology of Carnivorous Marsupials. (Eds M Jones, C Dickman and M Archer) pp. 3–20. CSIRO Publishing, Melbourne.

Morton SR (1980) Ecological correlates of caudal fat storage in small mammals. *Australian Mammalogy* **3**, 81–86.

Morton SR, Dickman CR, Fletcher TP (1989) Dasyuridae. In *Fauna of Australia, Volume 1B, Mammalia.* (Eds DW Walton and BJ Richardson) pp. 560–582. Australian Government Publishing Service, Canberra.

Owen R (1839) On the osteology of Marsupialia. *The Transactions of the Zoological Society of London* **2**, 379–408. doi:10.1111/j.1469-7998.1839.tb00030.x

Reese S, Pfuderer UR, Bragulla H, Loeffler K, Budras KD (2001) Topography, structure and function of the patella and the patelloid in marsupials. *Anatomia, Histologia, Embryologia* **30**, 289–294. doi:10.1046/j.1439-0264.2001.00334.x

Strahan R (1995) *The Australian Museum Complete Book of Australian Mammals.* Reed New Holland, Sydney.

8

Possums and gliders

There are 29 species of Australian possums and gliders within the order Diprotodontia and suborder Phalangeriformes. Several species also occur in New Guinea. There are six families: Burramyidae (pygmy possums); Phalangeridae (cuscuses and brush-tailed possums); Pseudocheiridae (ring-tailed possums and greater glider); Petauridae (possums and other gliders); Tarsipedidae (the honey possum [*Tarsipes rostratus*]); and Acrobatidae (the feather-tailed glider [*Acrobates pygmaeus*]) (see Appendix 1). The smallest species have an adult body size of less than 10 g, while the largest may weigh up to approximately 4.5 kg (Smith and Winter 1984). All gliders and nearly all possums are arboreal and live in forest or heathland, but some are semi-terrestrial, including the mountain pygmy possum (*Burramys parvus*), rock ring-tailed possum (*Petropseudes dahli*) and scaly-tailed possum (*Wyulda squamicaudata*) (Strahan 1995). Many species have a restricted range but some, such as the common brush-tailed possum (*Trichosurus vulpecula*), common ring-tailed possum (*Pseudocheirus peregrinus*) and sugar glider (*Petaurus breviceps*), have extensive ranges (Smith and Winter 1984). In addition, common brush-tailed and common ring-tailed possums often live in urban areas (How and Kerle 1995; McKay and Ong 1995). These two species are commonly seen in veterinary practice in Australia. Sugar gliders are common pets in other countries and the common brush-tailed possum was introduced to New Zealand in 1858, where it is now a major agricultural and ecological pest (Tyndale-Biscoe 2005).

The limbs and tails of possums and gliders reflect their generally arboreal niche. Gliders are distinguished from possums by the presence of a gliding membrane (patagium) extending from the hindlimb to the forelimb (Fig. 8.1). The gliding membrane consists of a fold of skin, connective tissue and muscle. In *Petaurus* spp. it extends from the fifth digit of the manus to the distal metatarsus or first digit of the pes. In the greater glider (*Petauroides volans*) and feather-tailed glider it extends from the elbow (Fig. 8.2) to the distal tibia and stifle respectively.

There is elongation of the long bones and vertebrae of the gliders (McKay 1989; Suckling 1995; Woodside 1995; Johnson and Hemsley 2008; Jackson 2012). Elongation of these bones increases the effective gliding surface area (Lindenmayer 2002). However, the feather-tailed glider does not show the marked elongation of the vertebrae or limb bones associated with the evolution of gliding. In such a small animal, the presence of the patagium alone is presumably sufficient to provide the necessary increase in surface area (Turner and McKay 1989; Jackson 2012). The tails of possums and gliders are long and moderately prehensile (McKay 1989; McKay and Winter 1989; Russell and Renfree 1989; Lindenmayer 2002).

In this chapter, anatomical descriptions and radiographic images focus primarily on readily accessible species. However, when radiographically

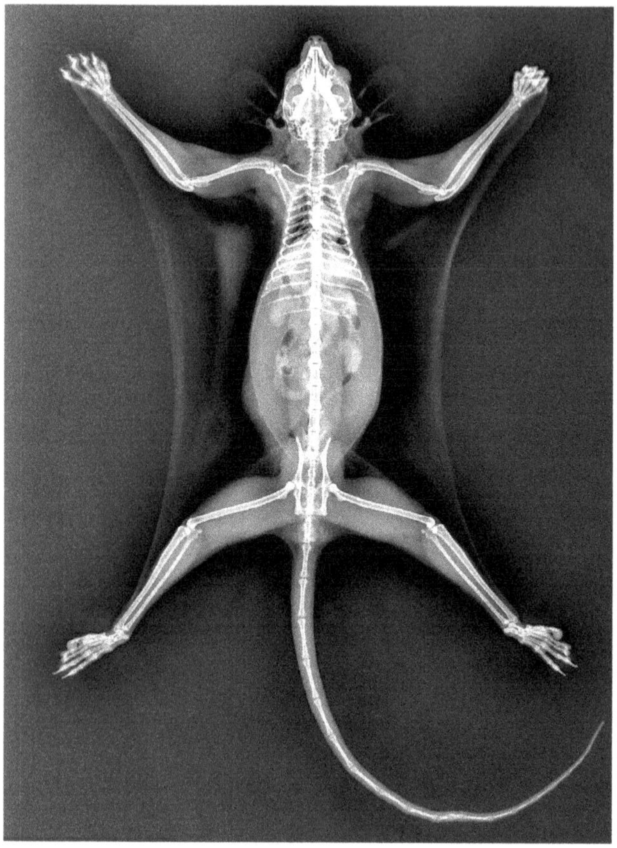

Figure 8.1: Whole body, ventrodorsal radiograph of a squirrel glider (*Petaurus norfolcensis*) demonstrating the gliding membrane (patagium). Note the displaced transverse fracture of a distal caudal vertebra.

available, relevant anatomical features of other species are included.

Skeletal system

Appendicular skeleton

Pectoral girdle and forelimb

As in most marsupials, a clavicle is present (Figs 8.3 and 8.4). The scapula is triangular. The greater glider is unique among gliding marsupials because it has a small accessory cartilaginous spur extending from the olecranon process of the elbow. The olecranon in this species is also slightly elongated, supporting the attachment of the patagium to the proximal antebrachium (Fig. 8.2b). The greater glider glides with its elbows flexed, rather than extended as in all other gliders, and with its forefeet positioned under the mandible (Jackson 2012).

The anatomy of the manus (Figs 8.5 and 8.6) is variable among species. In the Petauridae, the carpus is remarkable for the length of the accessory carpal bone. Some species, such as the ring-tailed possums, have a grasping manus with the first two digits opposable to the remaining three (McKay 1989). The striped possum (*Dactylopsila trivirgata*) uses an elongated fourth digit and long tongue to extract ants and termites (Smith 1982). The manus of gliders is elongated with large claws (McKay 1989).

Pelvis and hindlimb

The pelvis and bones of the hindlimb are unremarkable. A pair of epipubic bones, the dimensions of which vary among species, extend from the cranial border of the pubis (Figs. 8.7, 8.8 and 8.9). For example, the epipubic bones of gliders in the family Petauridae are significantly reduced, but the greater glider (family Pseudocheiridae) has moderately elongated epipubic bones. As in all marsupials other than bandicoots and the bilby, there is no ossified patella and it is represented only by a patelloid, which is a fibrocartilaginous induration of the quadriceps femoris extensor tendon (Reese *et al.* 2001) (Fig. 8.9). In the Phalangeridae, there is always a considerable range of movement between the tibia and fibula, comparable to some degree with the movements of pronation and supination of the radius and ulna (Parker and Haswell 1921).

Digit 1 of the pes is medially directed, clawless and opposes the remaining digits. Digits 2 and 3 are reduced in size, syndactylous and enclosed for most of their length in a sheath of skin (Smith and Winter 1984; McKay and Winter 1989). The first digit of brush-tailed possums is extremely robust and widely separated from the other digits to facilitate gripping branches in its arboreal habitat (Dawson *et al.* 1989) (Fig. 8.10).

Axial skeleton

Skull

The skulls of smaller species are fragile and have a narrow, pointed rostrum and slender zygomatic arch. In contrast, the skulls of Phalangeridae are broad and have a relatively short rostrum and

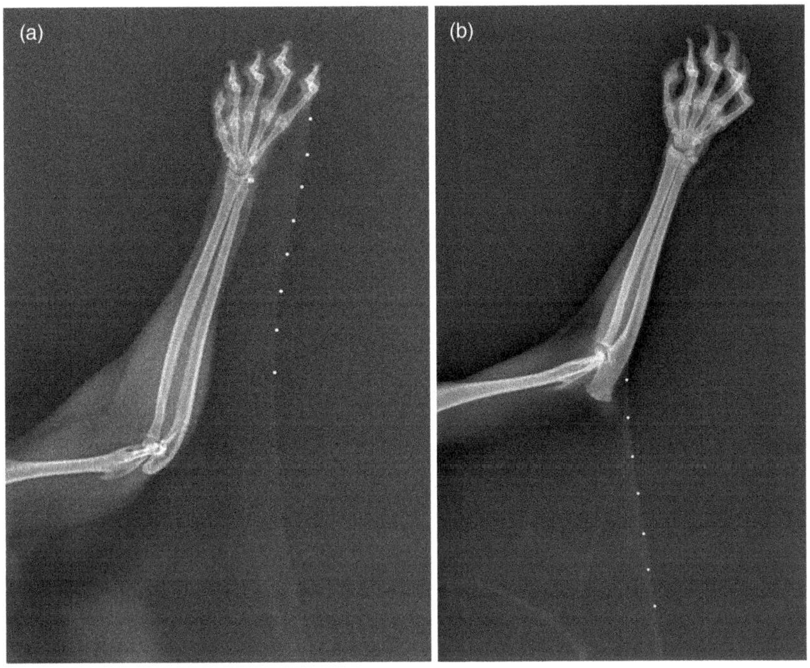

Figure 8.2: Forelimb and patagium: (a) squirrel glider (*Petaurus norfolcensis*), (b) greater glider (*Petauroides volans*). In the squirrel glider, the patagium inserts adjacent to digit 5 of the manus whereas in the greater glider the insertion is at the proximal antebrachium. The edge of the patagium is defined by the dotted line.

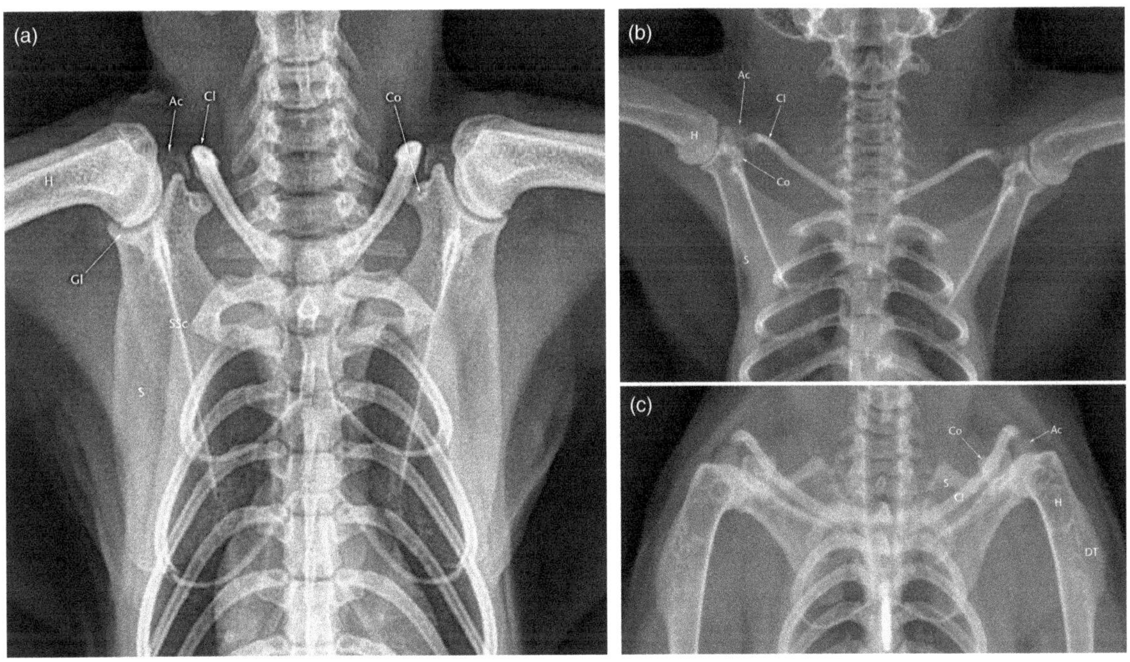

Figure 8.3: Pectoral girdle, ventrodorsal view: (a) common brush-tailed possum (*Trichosurus vulpecula*), (b) squirrel glider (*Petaurus norfolcensis*), (c) yellow-bellied glider (*Petaurus australis*). The acromio-clavicular articulation between the acromion (Ac) and the clavicle (Cl) is visible adjacent to the shoulder joint. The glenoid cavity (Gl) and spinous process (SSc) of the scapula (S) and the coracoid (Co) are visible in these views. The humerus (H) articulates with the glenoid cavity of the scapula. Note the well-developed deltoid tuberosity (DT). Note: a transponder can be seen in (c).

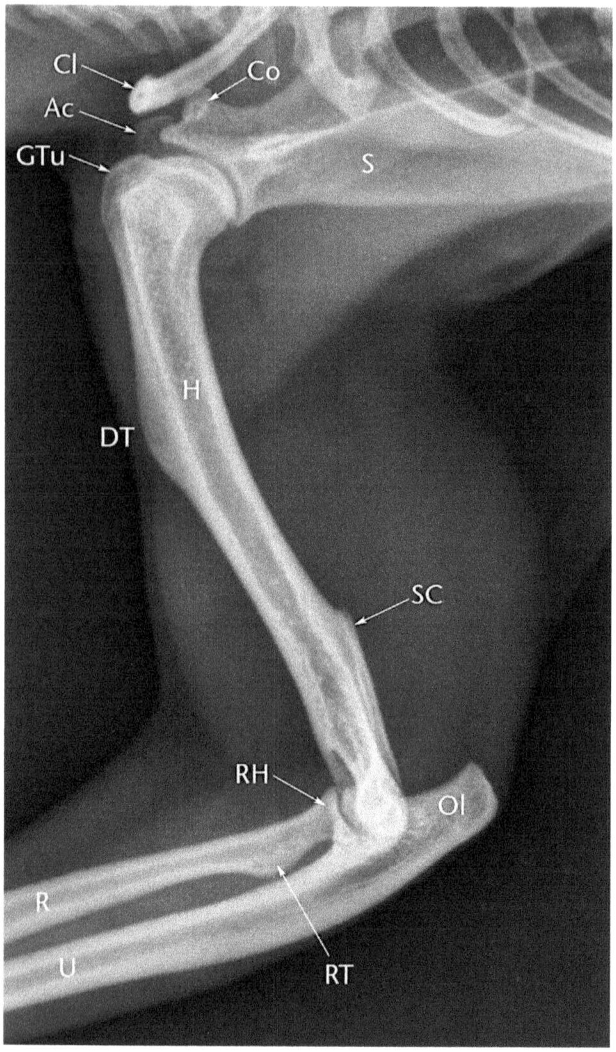

Figure 8.4: Forelimb, lateral view, common brush-tailed possum (*Trichosurus vulpecula*). The greater tubercle (GTu), deltoid tuberosity (DT) and supinator crest (SC) are features of the humerus (H). Proximally, the pectoral girdle is represented by the articulation of the clavicle (Cl) and acromion (Ac). The coracoid (Co) is visible on the cranial aspect of the distal end of the scapula (S). The olecranon (Ol) of the ulna (U) and the radial head (RH) and radial tuberosity (RT) of the radius (R) are features of the bones of the antebrachium.

strong zygomatic arch (Figs 8.11 and 8.12). Large caudal palatine vacuities are normally present (Turner and McKay 1989). With the exception of the honey possum, at the ventral edge of the caudomedial aspect of the mandible there is a medially directed shelf into which the internal pterygoid muscle inserts, as in most marsupials (Dawson *et al.* 1989).

In the common brush-tailed possum, the rostroventral part of the zygomatic arch extends ventrally to form a conical jugal process, the tip of which is at the level of the caudal maxillary molars (Owen 1839) (Fig. 8.13). This process is shorter and blunter than in macropods and does not extend beyond the margin of the maxillary bone (Gilmore and da Costa 1995). Jugular processes of the occipital bone are present but vary in dimensions among species (Owen 1839) (Fig. 8.12).

Pneumatisation of the cranial bones, which is radiographically visible (Figs 8.11, 8.12 and 8.14), is a feature of the skulls of possums and gliders. The tympanic bulla is augmented by a continuation of air-cells into the base of the zygomatic arch and the whole of the temporal bone, including its zygomatic process may be expanded to accommodate air-cells with very thin and smooth walls (Owen 1839; Macrini 2012). In possums, the coronoid and condyloid processes of the mandible are elongated and separated by a distinct notch. In gliders, the two processes are shorter and the notch less distinct. The angle of the mandible is very obtuse in possums and more acute in gliders. The degree of development, shape and length of the angular process varies among species.

The anatomy of the skull of the honey possum is unique within the suborder Phalangeriformes, exhibiting features reflecting its dietary specialisation. The teeth are vestigial and food collection is performed by means of the tongue. The parts of the skull that carry teeth and masticatory muscles are reduced. The skull is more elongated than in any other small diprotodont marsupial. The snout, with elongated nasal bones, maxillae and premaxillae, tapers to a narrow point. The frontal bones are small and the area of contact between the nasal and lacrimal bones is broad. The zygomatic arch is slender and the caudal margin of the palate is incomplete. The mandible is a splint-like bone with a reduced coronoid process and no inturned angular process (Russell and Renfree 1989).

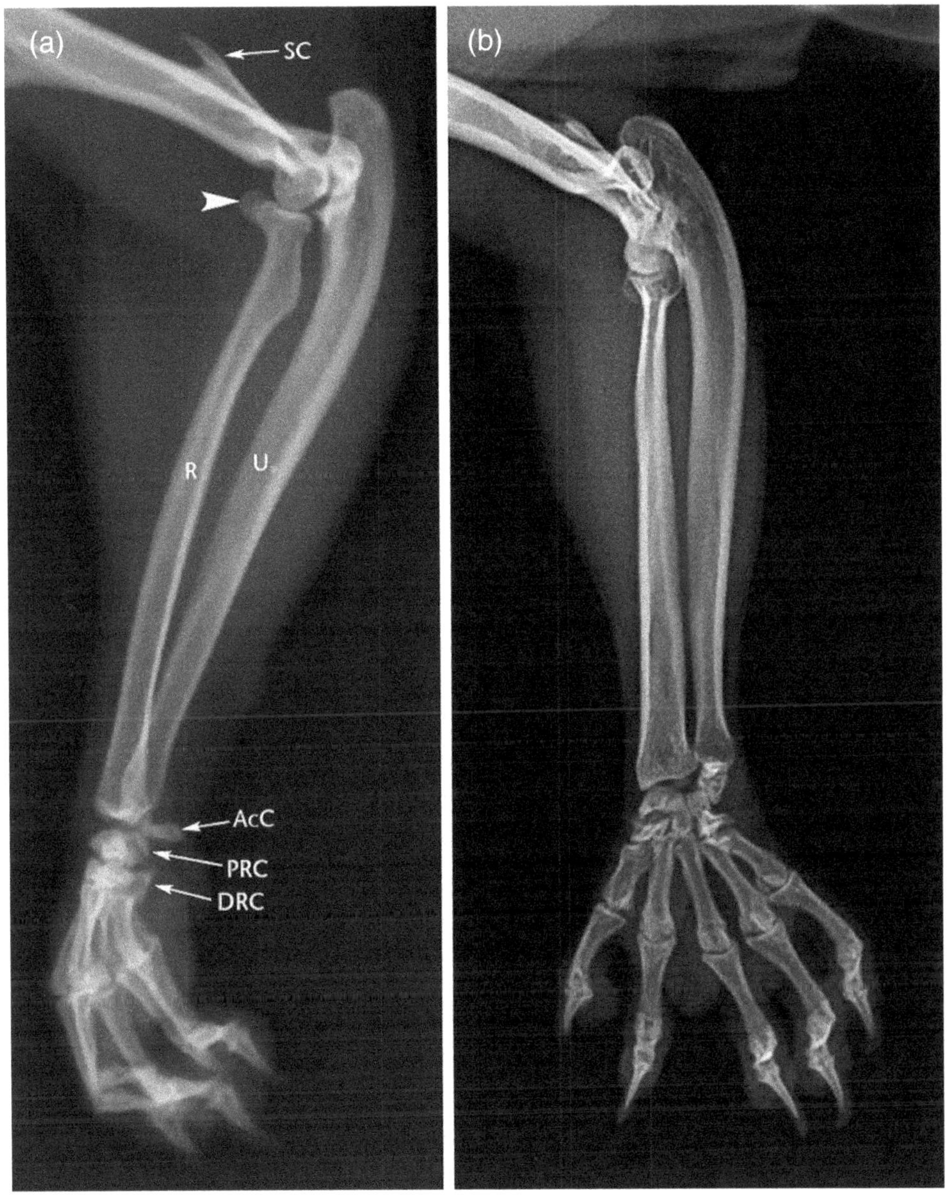

Figure 8.5: Antebrachium and manus, common ring-tailed possum (*Pseudocheirus peregrinus*): (a) mediolateral view, (b) oblique craniocaudal view of the antebrachium and dorsopalmar view of the manus. The supinator crest (SC) is present on the caudal aspect of the distal humerus. There is a single sesamoid bone (arrowhead) on the cranial aspect of the proximal end of the radius (R) adjacent to the elbow joint. The distal end of the radius and ulna (U) form the antebrachiocarpal articulation with the proximal row of carpal bones (PRC). The accessory carpal bone (AcC) projects palmar to the carpus. The distal row of carpal bones (DRC) articulates with the metacarpal bones.

Vertebrae

As in most marsupials, there are 7 cervical, 13 thoracic, 6 lumbar, 2–3 sacral and 20–29 caudal vertebrae in the Phalangeridae and Petauridae (Figs 8.15 and 8.16). Haemal arches are present adjacent to the ventral aspects of the articulations of the caudal vertebrae in the Phalangeridae and Petauridae, although they are smaller in the latter (Owen 1839) (Figs 8.16 and 8.17). The vertebrae, particularly the thoracic and lumbar vertebrae, of gliders within the families Pseudocheiridae and Petauridae are relatively much longer than those of the

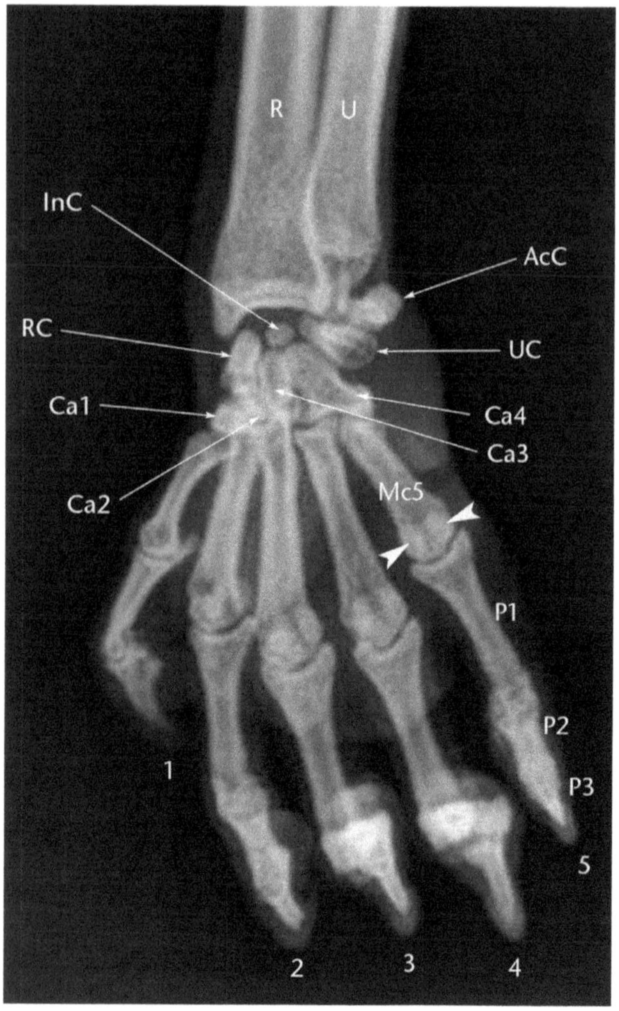

Figure 8.6: Manus, dorsopalmar view, common brush-tailed possum (*Trichosurus vulpecula*). The antebrachiocarpal articulation is between the radius (R) and ulna (U) and the proximal row of carpal bones, comprising the radial (RC), ulnar (UC), intermediate (InC) and accessory (AcC) carpal bones. The intermediate carpal bone is small. The distal row of carpal bones comprises the 1st, 2nd, 3rd and 4th carpal bones (Ca1, Ca2, Ca3, Ca4). At the distal end of each metacarpal bone (Mc5 labelled) is a pair of sesamoid bones (arrowheads). There are five digits (1, 2, 3, 4, 5). In four of them (2, 3, 4, 5) there are three phalangeal bones (P1, P2, P3). Digit 1 has two phalanges.

non-gliding species within these families (Fig. 8.18). The proximal caudal vertebrae are also elongated, resulting in a longer, but less flexible tail. Although this reduces the prehensility of the tail, it increases its efficacy as a rudder during flight (McKay 1989; Lindenmayer 2002).

Ribs and sternum

There are 13 pairs of ribs (Owen 1839) and 5–6 sternebrae (Figs 8.20 and 8.21).

Soft tissue

Head, neck and thorax

Pharynx and larynx

The common brush-tailed possum has a uniquely dilated thyroid cartilage, which forms a small, rigid-walled, spherical chamber. It opens through the floor of the larynx via a 1.5 mm circular aperture and possibly acts as a resonating chamber, which is consistent with this species being among the most vocal of marsupials (McKay and Winter 1989) (Figs 8.12d and 8.19).

Cardiovascular system

In the common brush-tailed possum (Figs 8.20 and 8.21), the heart is separated from the diaphragm by the accessory lobe of the lung and is attached to the ribs on the left side, rather than the diaphragm. The right atrium has two auricular appendages in this species (Runciman *et al.* 1992).

Respiratory system

In the common brush-tailed possum, the left lung is partially divided into two lobes of almost equal size. The right lung is divided into three lobes. A well-defined accessory lobe is separated from the right lung by the caudal *vena cava* and rests by its base on the dorsal surface of the diaphragm (Gilmore and da Costa 1995).

Abdomen

Most possums and gliders digest plant material by microbial fermentation in the caecum and colon and have a simple stomach. Many species also eat insects and some eat plant sap, nectar and pollen. An expanded caecum is the main site for fermentation of plant material in most species. Like the koala, the common ring-tailed possum and greater glider are folivorous and consequently, the caecum

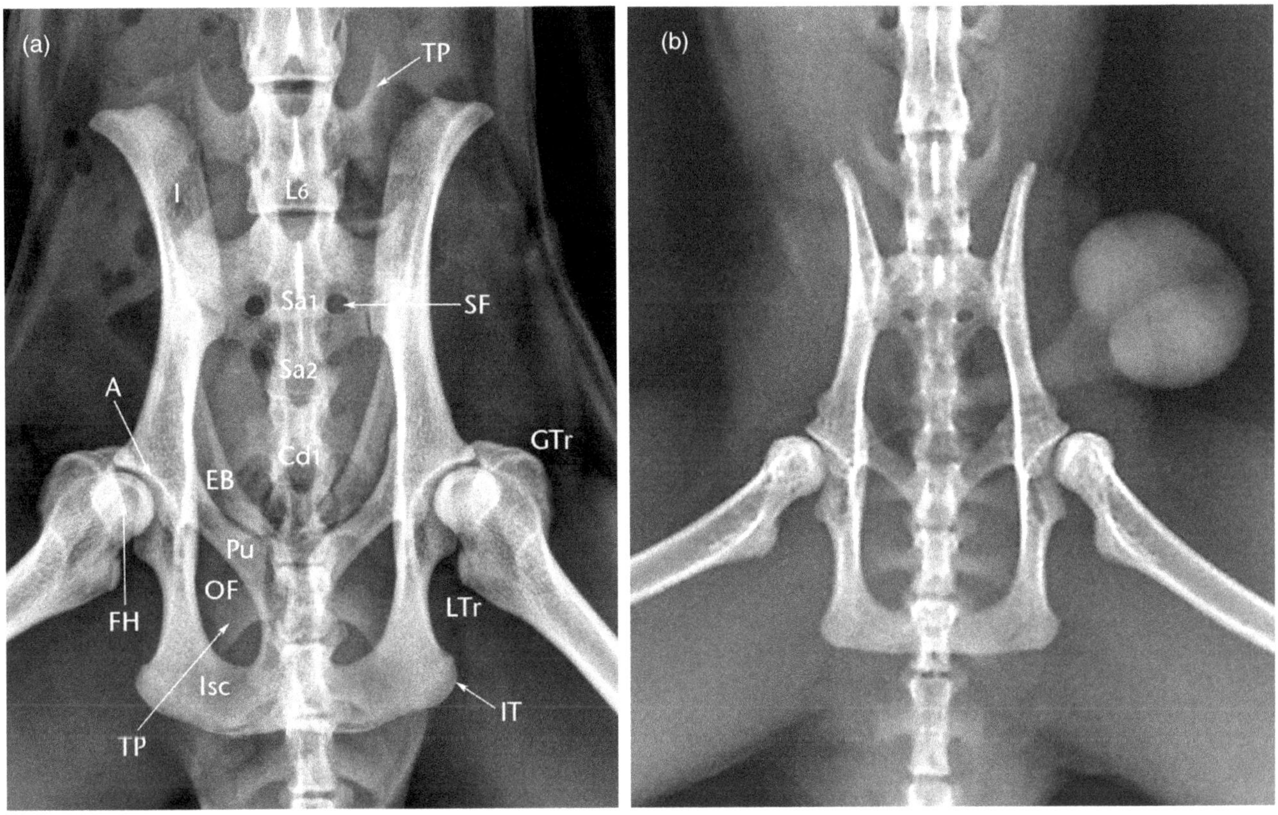

Figure 8.7: Pelvis, ventrodorsal view: (a) common brush-tailed possum (*Trichosurus vulpecula*), (b) squirrel glider (*Petaurus norfolcensis*). The lumbosacral articulation is between the 6th lumbar (L6) vertebra and the 1st sacral (Sa1) element. The 2nd sacral element (Sa2) articulates caudally with the 1st caudal element (Cd1). The two sacral elements are fused and feature a sacral foramen (SF). They articulate laterally with the ilia to form the sacroiliac joints. The ilium (I), ischium (Isc) and pubis (Pu) combine to form each half of the pelvis, which in turn provides an articulation between the acetabulum (A) and the femoral head (FH). The proximal femur features a greater (GTr) and lesser (LTr) trochanter. The paired epipubic bones (EB) articulate with the pubis (Pu) on each side. Transverse processes (TP) are present on the lumbar and caudal vertebrae. An ovoid obturator foramen (OF) is present on each side of the pelvis. The ischial tuberosity (IT) is in the form of a blunt protrusion on the caudolateral margin of the pelvis. In the squirrel glider (b), the acetabulae are shallow, epipubic bones are not evident and the large, pedunculated scrotum and testes are evident.

and colon are large, occupying the entire ventral portion of the abdomen. Intraluminal contents (gas and digesta) dominate the radiographic image of the abdomen, obscuring other abdominal organs (Figs 8.22 and 8.23). In addition, the viscera cannot be distinguished easily because the absence of abdominal adipose tissue results in positive silhouetting of adjacent viscera, camouflaging their serosal surfaces (Figs 8.22 and 8.23b, d). However, the intestines of the striped possum and honey possum are simple because of the lack of plant material in their diet. The stomach of the honey possum has a large diverticulum, which may have a role in nectar storage (Hume 1999; Smith 1982; Richardson *et al.* 1986).

The anatomy of the female reproductive tract of possums and gliders is typical of most marsupials. There are two lateral vaginae and a median vagina with a temporary central canal through which the young are born. Sugar gliders have elongated lateral vaginae compared with other species of possums and gliders (McKay 1989; Johnson and Hemsley 2008). The anatomy of the male reproductive tract is similar to that of most marsupials. Testes are located in a pre-penile scrotum. Accessory sex glands consist of only a prostate and one

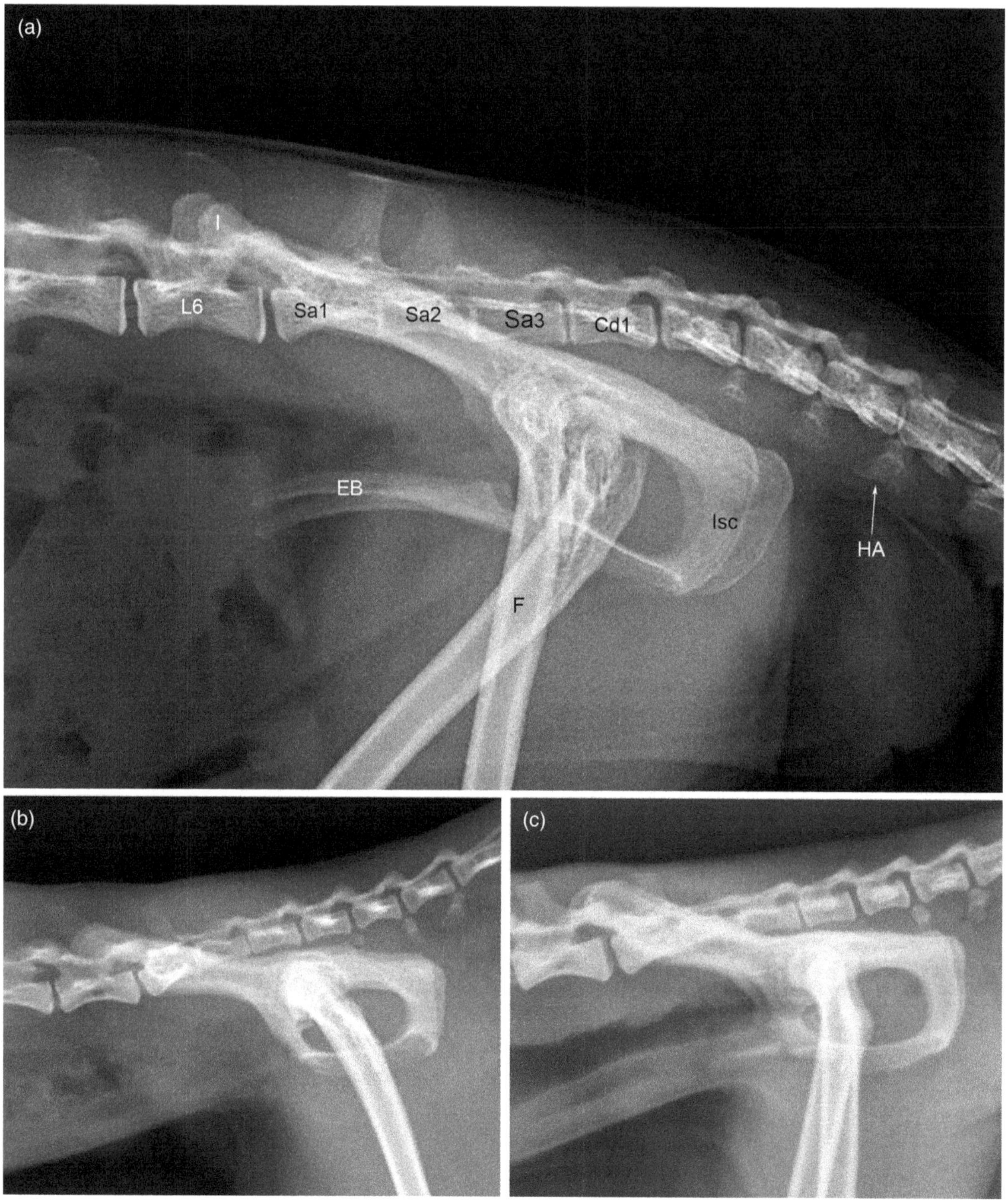

Figure 8.8: Pelvis, lateral view: (a) common ring-tailed possum (*Pseudocheirus peregrinus*), (b) squirrel glider (*Petaurus norfolcensis*), (c) yellow-bellied glider (*Petaurus australis*). The lumbosacral articulation is between the 6th lumbar (L6) vertebra and the 1st sacral (Sa1) element. In the common ring-tailed possum there are three sacral elements (Sa1, Sa2, Sa3), the 3rd of which articulates caudally with the 1st caudal element (Cd1). The ilium (I), ischium (Isc) and femurs (F) are labelled. The epipubic bones (EB) are well defined in the common ring-tailed possum but are greatly reduced in the squirrel and yellow-bellied gliders. Haemal arches (HA) are present ventral to each articulation between the caudal vertebrae.

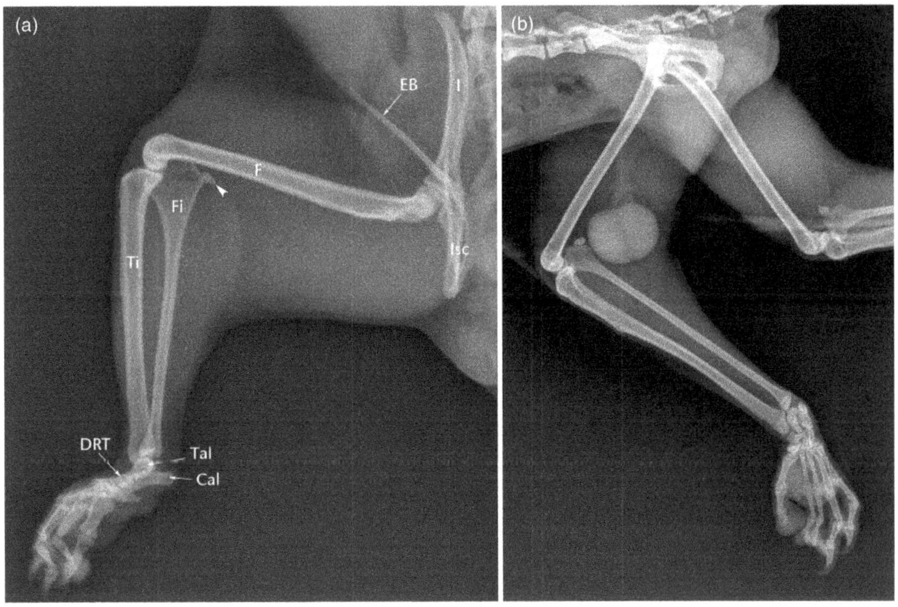

Figure 8.9: Hindlimb: (a) common ring-tailed possum (*Pseudocheirus peregrinus*), mediolateral view, (b) squirrel glider (*Petaurus norfolcensis*), lateromedial view. The ilium (I), ischium (Isc) and femur (F) are visible proximal to the stifle. An epipubic bone (EB) is visible in the common ring-tailed possum, but not in the squirrel glider. There is a sesamoid bone (arrowhead) adjacent to the proximal end of the fibula (Fi). The tibia (Ti) and fibula articulate distally with the talus (Tal) and calcaneus (Cal) of the tarsus. These two bones articulate distally with the distal row of tarsal bones (DRT).

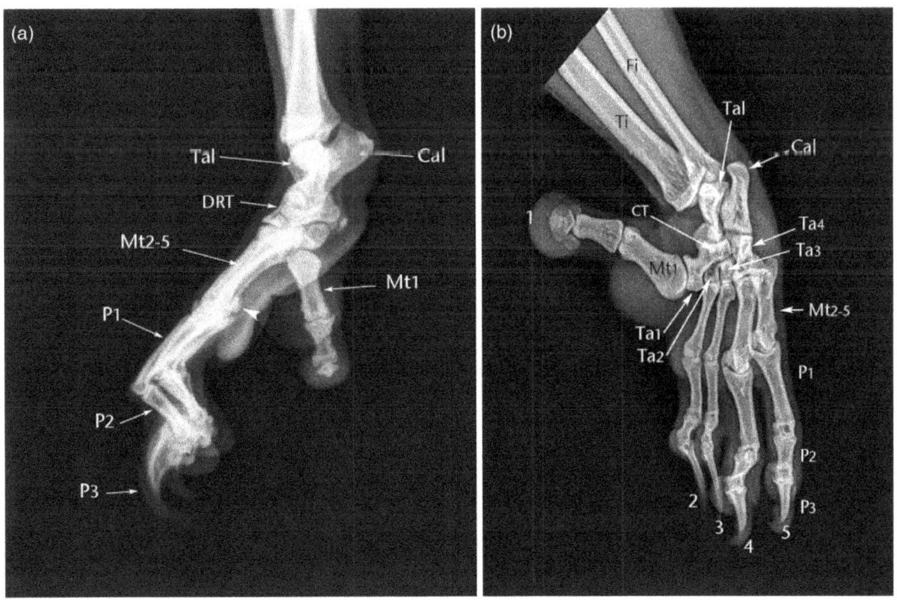

Figure 8.10: Pes: (a) common brush-tailed possum (*Trichosurus vulpecula*), lateromedial view, (b) common ring-tailed possum (*Pseudocheirus peregrinus*), dorsoplantar view. The talus (Tal) and calcaneus (Cal) contribute to the tarsocrural articulation with the tibia (Ti) and fibula (Fi). The distal row of tarsal (DRT) bones contains four cuboidal tarsal bones (Ta1, Ta2, Ta3, Ta4). A central (CT) tarsal bone is located between the talus and Ta2 and Ta3. Each digit articulates with a metatarsal (Mt1, Mt2–5) bone, each of which has three phalanges (P1, P2, P3), except for digit 1 (1), which has two. Digit 1 is opposable. Digits 2 and 3 (2, 3) are less well developed than digits 4 and 5 (4, 5). Paired metatarsophalangeal sesamoid bones (arrowhead) are present on the plantar surface of the metatarsophalangeal articulations.

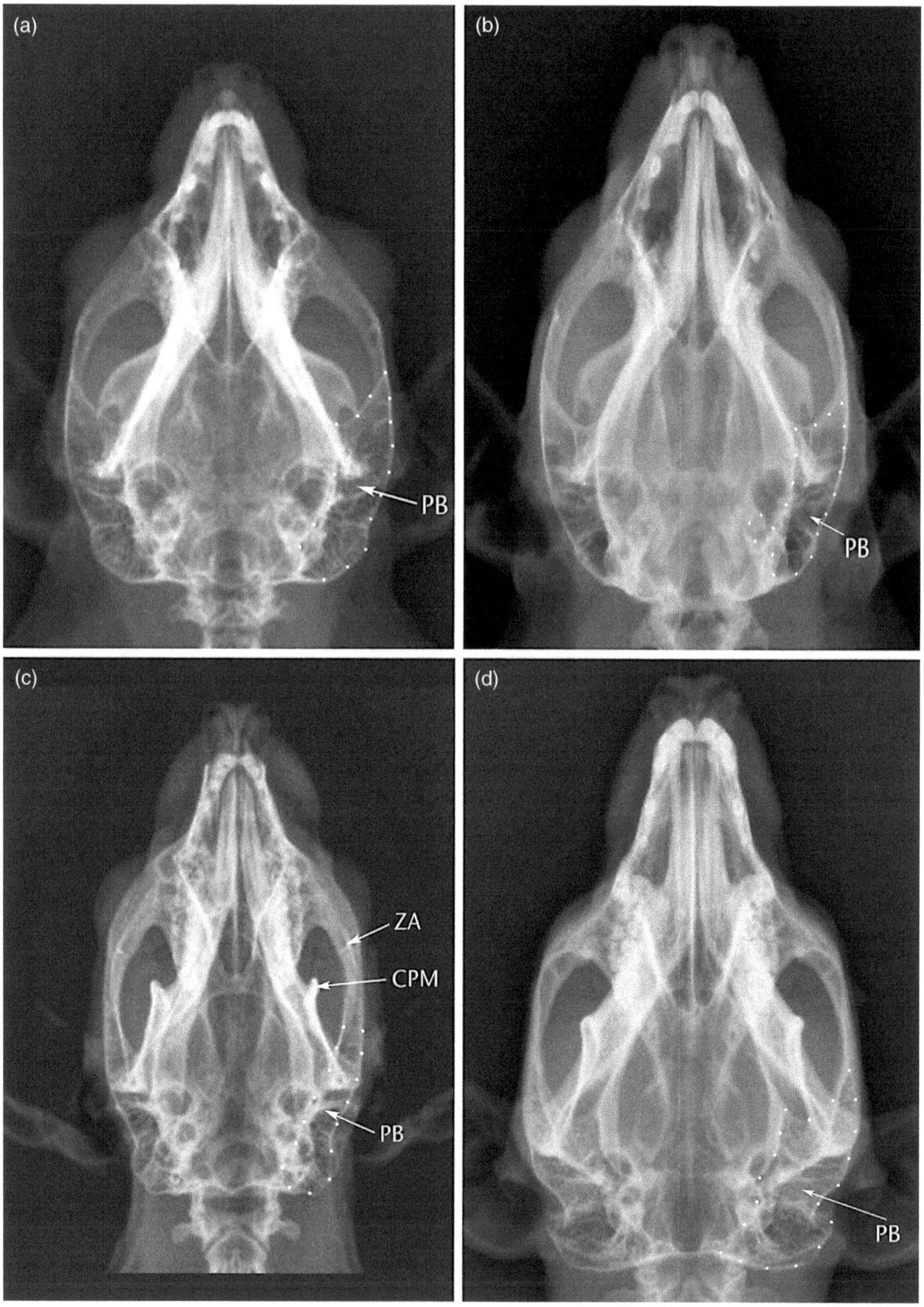

Figure 8.11: Skull, ventrodorsal view: (a) squirrel glider (*Petaurus norfolcensis*), (b) yellow-bellied glider (*Petaurus australis*), (c) common ring-tailed possum (*Pseudocheirus peregrinus*), (d) common brush-tailed possum (*Trichosurus vulpecula*). The extent of pneumaticised bone (PB) is marked with a dotted line. It occurs in the temporal and occipital bones to a varying extent. The coronoid process of the mandible (CPM) and the zygomatic arch (ZA) are also indicated.

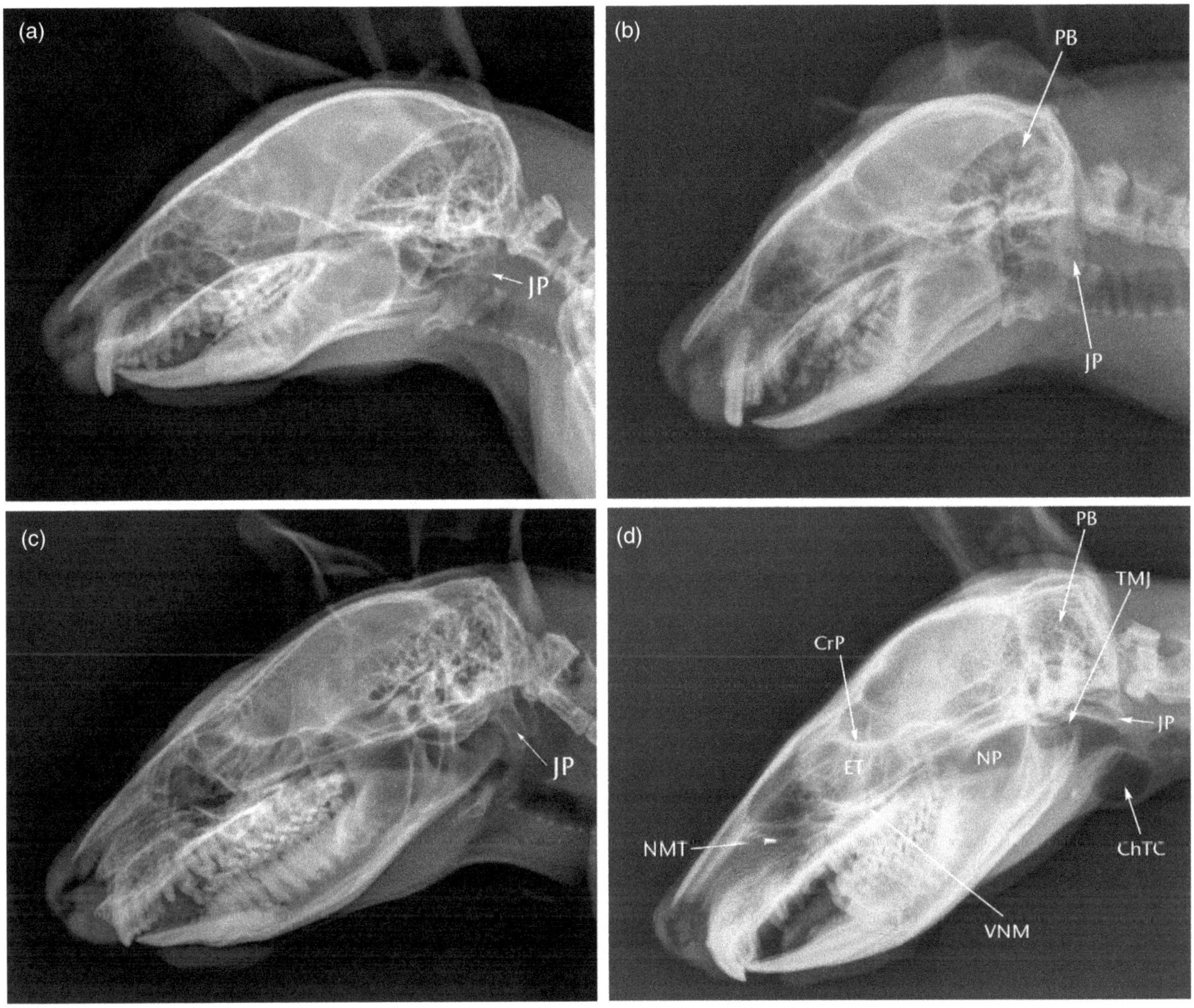

Figure 8.12: Skull, lateral view: (a) squirrel glider (*Petaurus norfolcensis*), (b) yellow-bellied glider (*Petaurus australis*), (c) common ring-tailed possum (*Pseudocheirus peregrinus*), (d) common brush-tailed possum (*Trichosurus vulpecula*). The nasomaxillary turbinates (NMT), the ethmoid turbinates (ET) and the cribiform plate (CrP) are labelled in the common brush-tailed possum. Pneumaticised bone (PB) in the caudal skull is visible in all four species. The ventral nasal meatus (VNM) is confluent with the nasopharynx (NP) caudally. The location of the temporomandibular joint (TMJ) is indicated in the common brush-tailed possum. The large air-filled chamber of the thyroid cartilage (ChTC), a feature of brush-tailed possums, can be seen in (d). The two large, ventrally directed jugular processes (JP) are more extensive in the yellow-bellied glider than in the common brush-tailed possum.

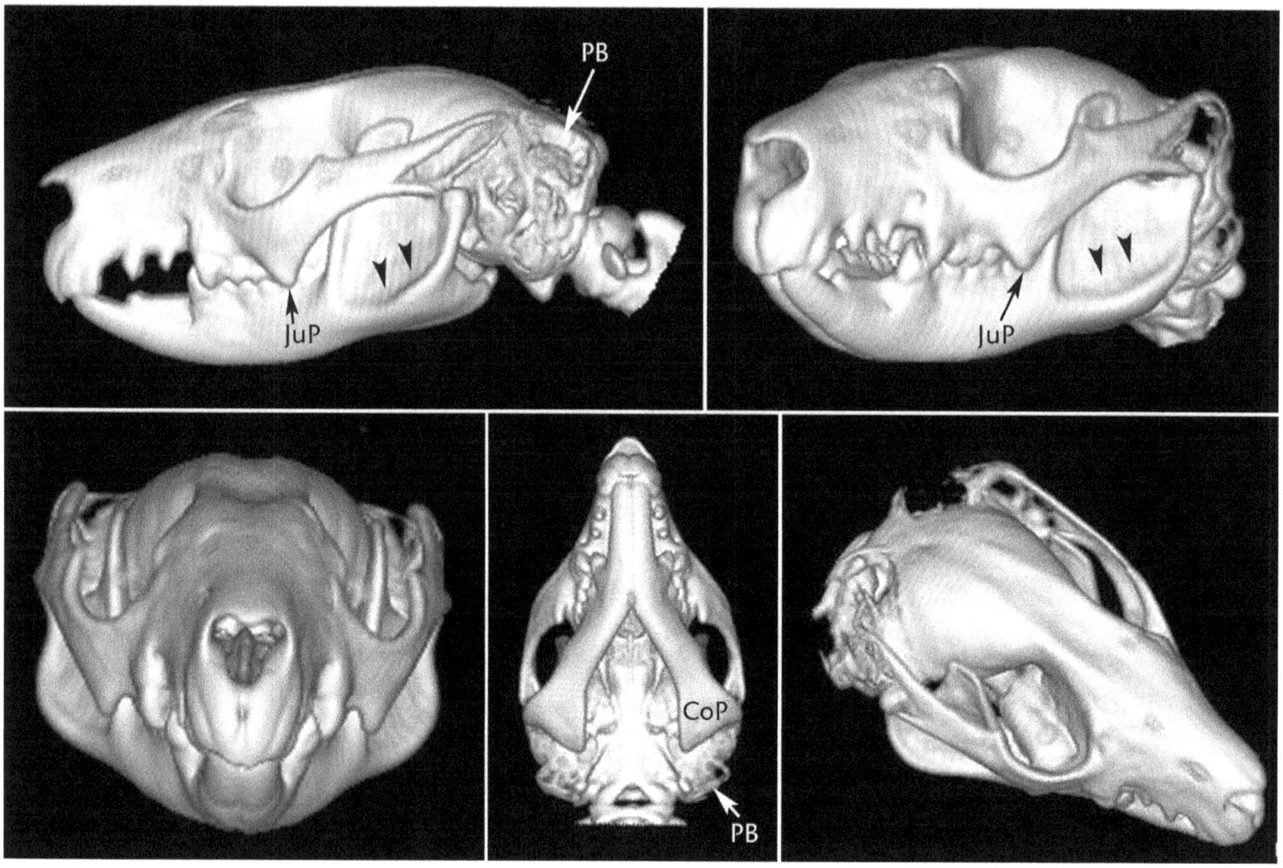

Figure 8.13: Skull, common brush-tailed possum (*Trichosurus vulpecula*), volume-rendered three-dimensional CT reconstruction. A large concavity in the lateral aspect of the mandible facilitates insertion of muscles of mastication (arrowheads). The cone-shaped, ventrally projecting jugal process (JuP) is visible at the rostral end of the zygomatic arch. The condyloid process (CoP) is wide. The pneumaticised temporal bone (PB) appears as a multi-cavitated bone on these reconstructed images.

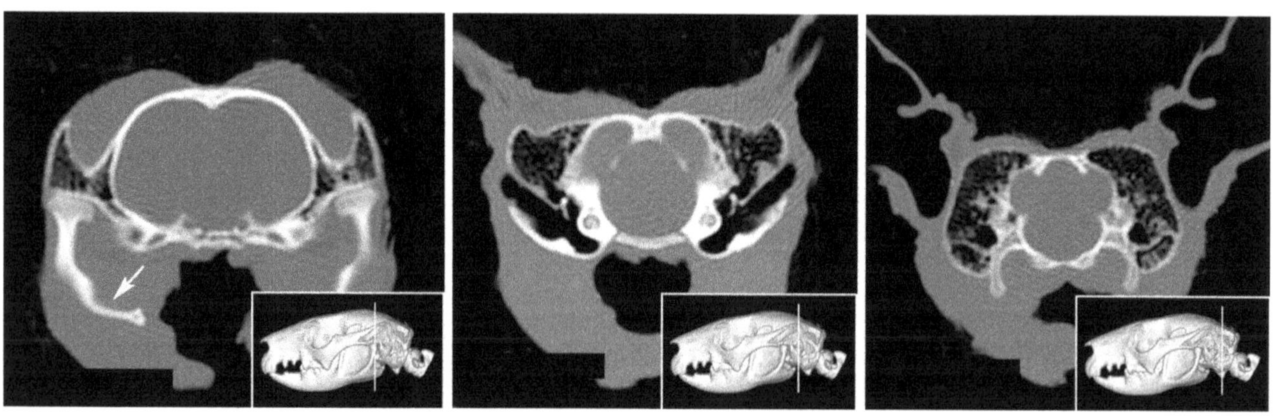

Figure 8.14: Skull, common brush-tailed possum (*Trichosurus vulpecula*), transverse reconstructed CT images. These images of the caudal aspect of the skull demonstrate extensive pneumatisation of the zygomatic arch and the temporal bone. Note the medial concavity (arrow) in the mandible that serves as an insertion for the pterygoid muscle.

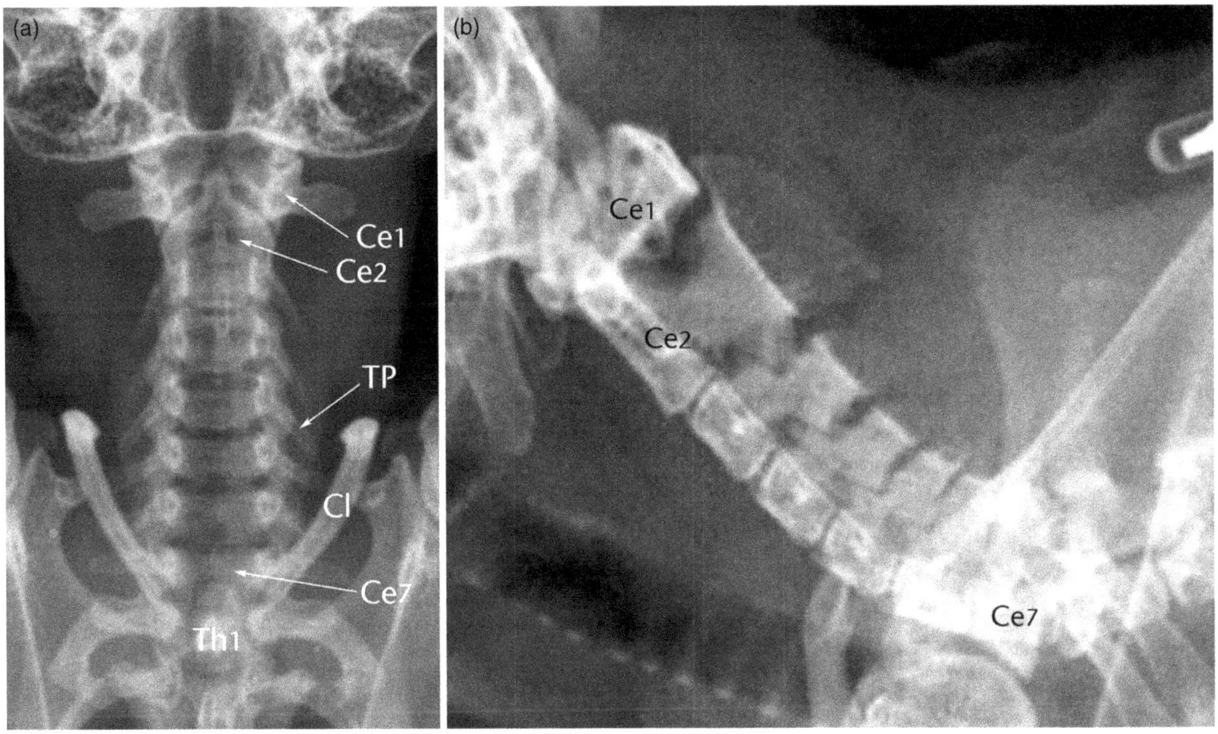

Figure 8.15: Cervical vertebral column, common ring-tailed possum (*Pseudocheirus peregrinus*): (a) dorsoventral view, (b) lateral view. There are 7 cervical vertebrae (Ce1, Ce2, Ce7 labelled). The vertebral canal is deep, giving the vertebrae a stretched appearance in the dorsoventral view. The 1st cervical vertebra (Ce1) has well defined lateral wings. The remaining cervical vertebrae have caudally directed transverse processes (TP). The clavicle (Cl) and 1st thoracic vertebra (Th1) are labelled. Note: a transponder can be seen in (b).

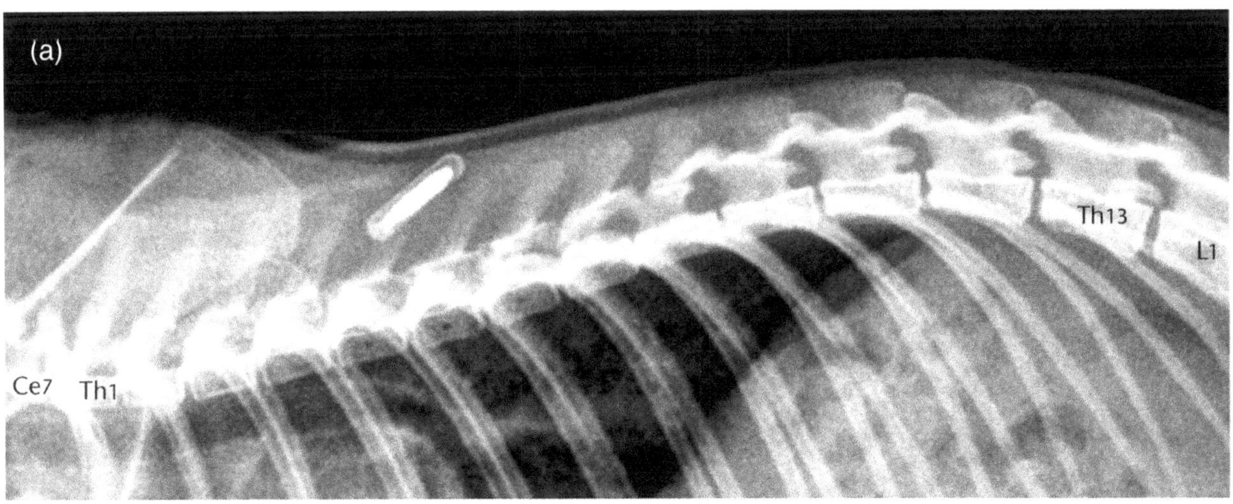

Figure 8.16: Vertebral column, common ring-tailed possum (*Pseudocheirus peregrinus*), lateral view: (a) thoracic vertebrae, (b) lumbar and sacral vertebrae, (c) caudal vertebrae.

(*Continued*)

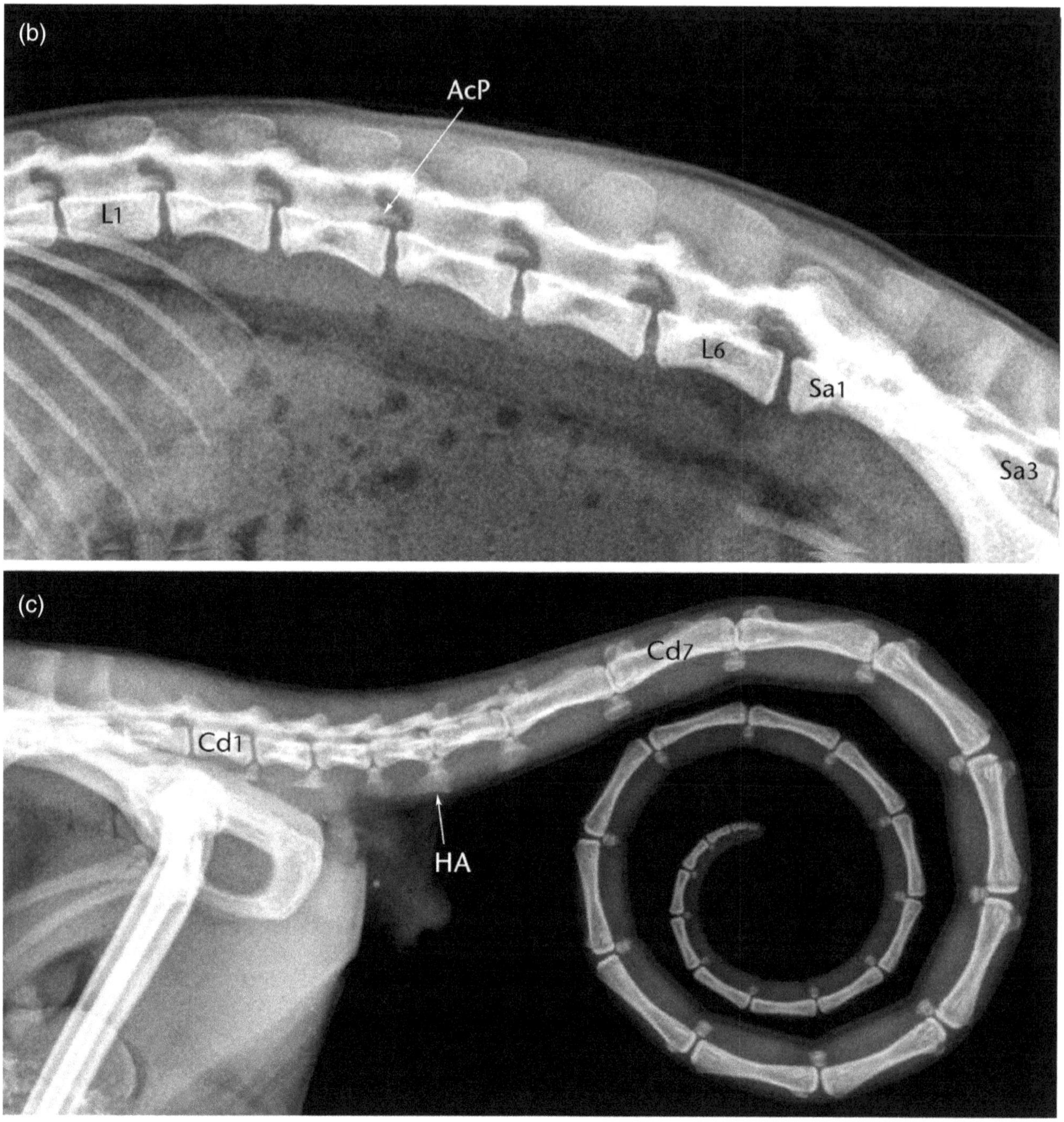

Figure 8.16: (*Continued*). There are 13 thoracic (Th1, Th13 labelled), 6 lumbar (L1, L6 labelled), 3 sacral (Sa1, Sa3 labelled) and 29 caudal (Cd1, Cd7 labelled) vertebrae. Accessory processes (AcP) arising from the caudal surface of the pedicles of the caudal thoracic and all of the lumbar vertebrae are present. Haemal arches (HA) are present adjacent to the ventral aspect of the articulations between the caudal vertebrae. The 7th cervical vertebra is labelled (Ce7). Note: a transponder can be seen in (a).

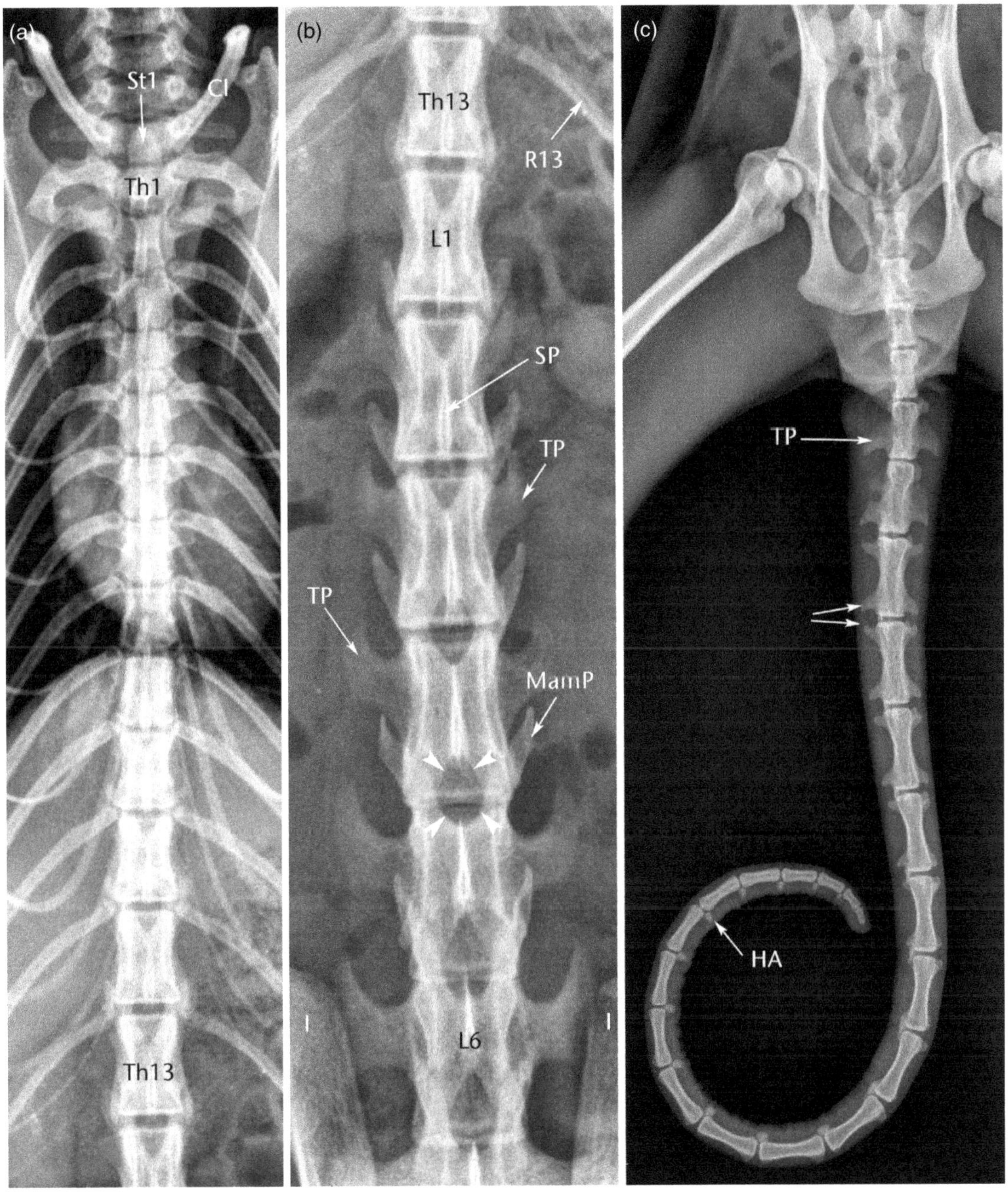

Figure 8.17: Vertebral column, common brush-tailed possum (*Trichosurus vulpecula*), ventrodorsal view: (a) thoracic vertebrae, (b) lumbar vertebrae, (c) caudal vertebrae. The articulation of the clavicle (Cl) and the sternum (St1) is visible superimposed on the junction of 7th cervical and the 1st thoracic vertebra (Th1). The 13th thoracic vertebra (Th13) is the most caudal thoracic vertebra. The 13th rib (R13) is labelled. Each vertebra between Th1 and the caudal vertebrae has a dorsal spinous process (SP). The lumbar vertebrae (L1, L6 labelled) carry a transverse process (TP), as do the cervical vertebrae, and a cranially directed mammillary process (MamP). The elongated ilia (I) project cranially adjacent to the 6th lumbar vertebra (L6). A well defined dorsal foramen (arrowheads) is located at the junction of the dorsal lamina of each lumbar vertebra and those of the first few caudal vertebrae. In the tail, transverse processes (TP) are visible, which become paired adjacent to the vertebral articulations of the mid-caudal vertebrae (paired arrows). Transverse processes are not visible caudally, although haemal arches (HA) persist to the end of the tail.

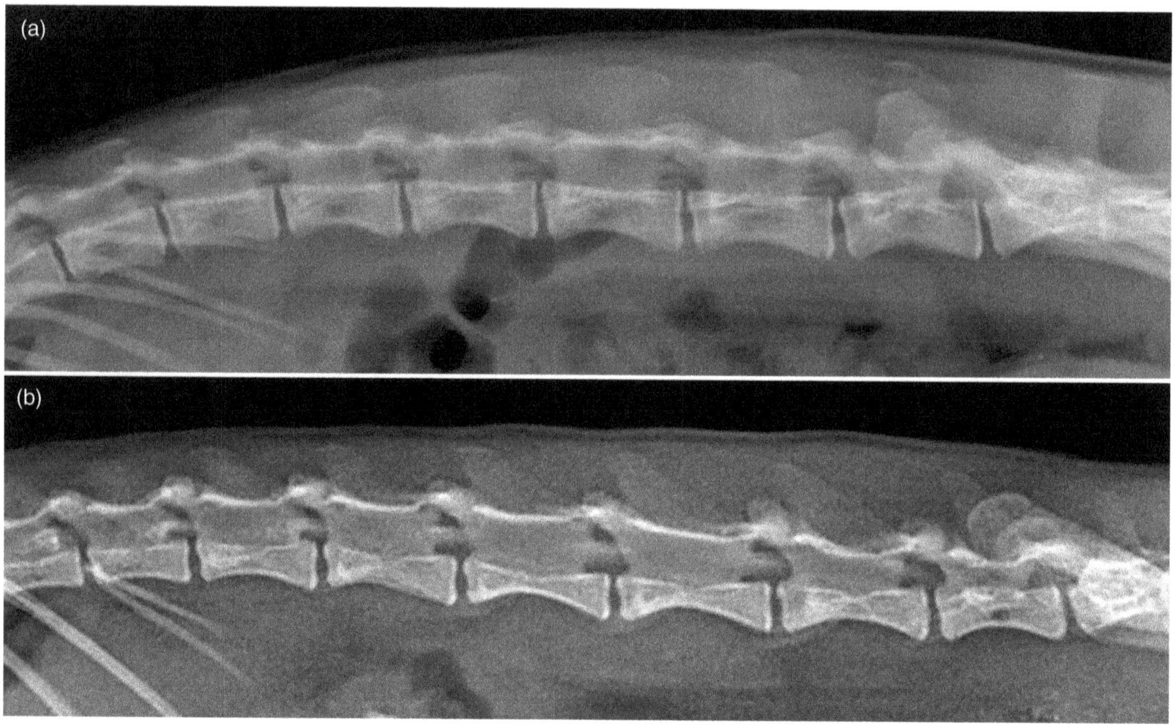

Figure 8.18: Lumbar vertebrae, lateral view: (a) common ring-tailed possum (*Pseudocheirus peregrinus*), (b) squirrel glider (*Petaurus norfolcensis*). The vertebral elements of the squirrel glider are longer and more slender that those of the common ring-tailed possum.

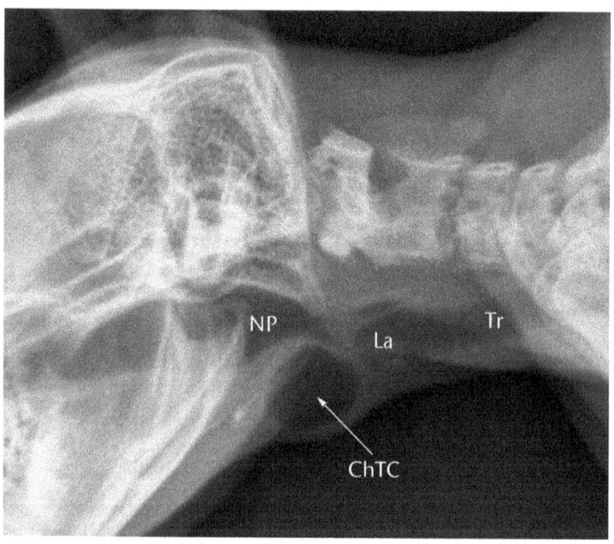

Figure 8.19: Neck, lateral view, common brush-tailed possum (*Trichosurus vulpecula*). The nasopharynx (NP) extends caudally to the larynx (La), which, in turn, is connected to the trachea (Tr). The aerated chamber of the thyroid cartilage (ChTC) connected to the larynx is located ventrally.

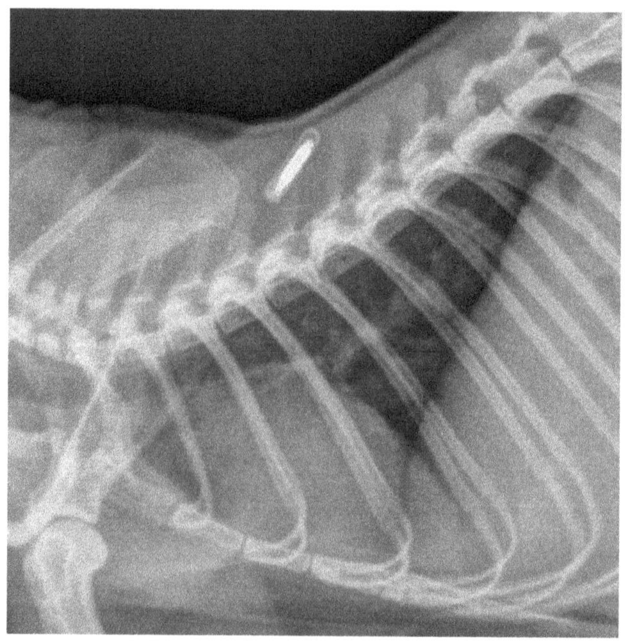

Figure 8.20: Thorax, common ring-tailed possum (*Pseudocheirus peregrinus*), lateral view. Note: a transponder can be seen.

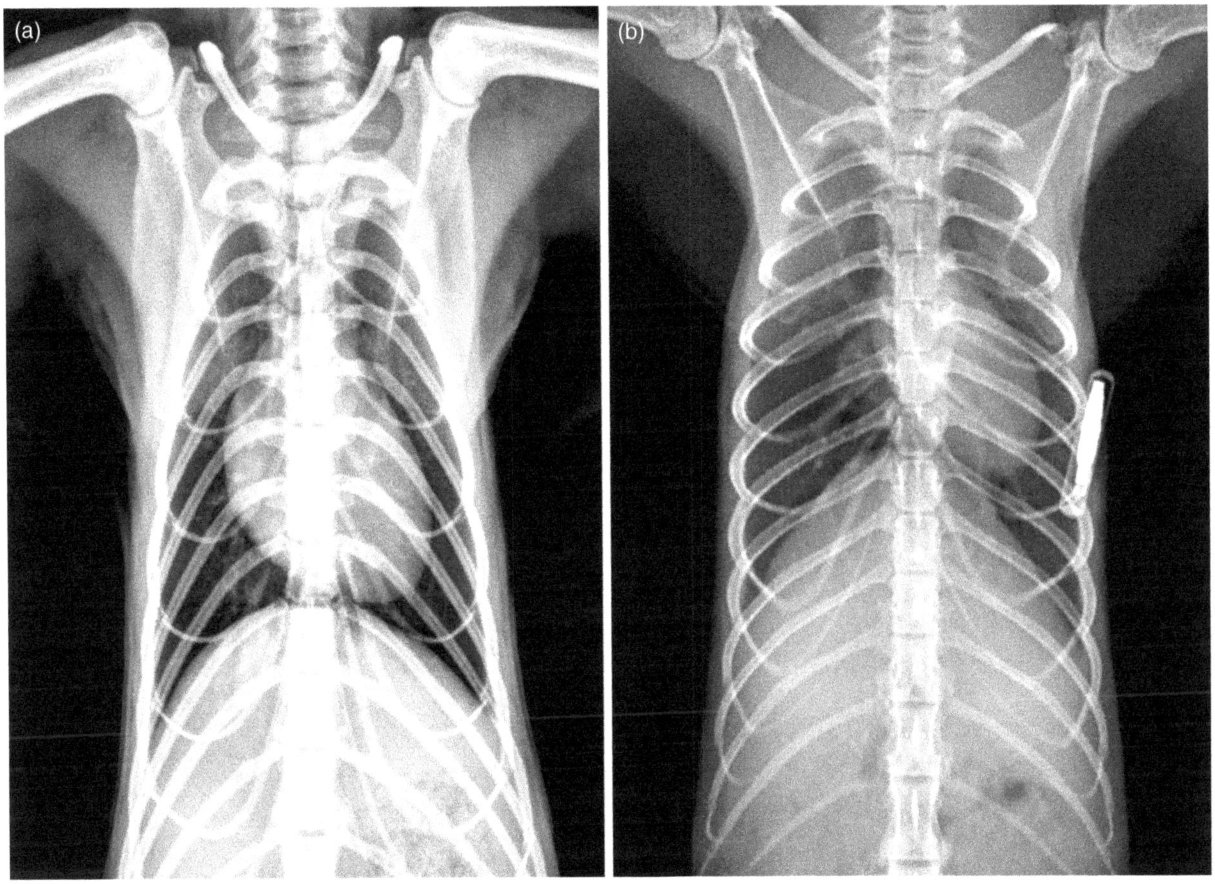

Figure 8.21: Thorax, ventrodorsal view: (a) common brush-tailed possum (*Trichosurus vulpecula*), (b) squirrel glider (*Petaurus norfolcensis*). Note: a transponder can be seen in (b).

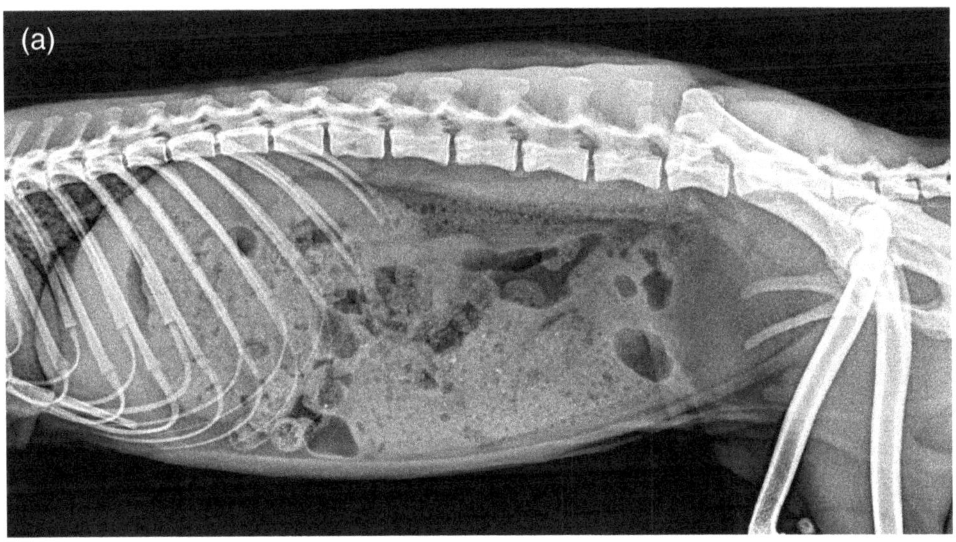

Figure 8.22: Abdomen, lateral view: (a) common brush-tailed possum (*Trichosurus vulpecula*), (b) common ring-tailed possum (*Pseudocheirus peregrinus*), (c) squirrel glider (*Petaurus norfolcensis*), (d) greater glider (*Petauroides volans*). Intraluminal contents (gas and digesta) of the caecum and colon dominate the radiographic image, obscuring other abdominal organs.

(*Continued*)

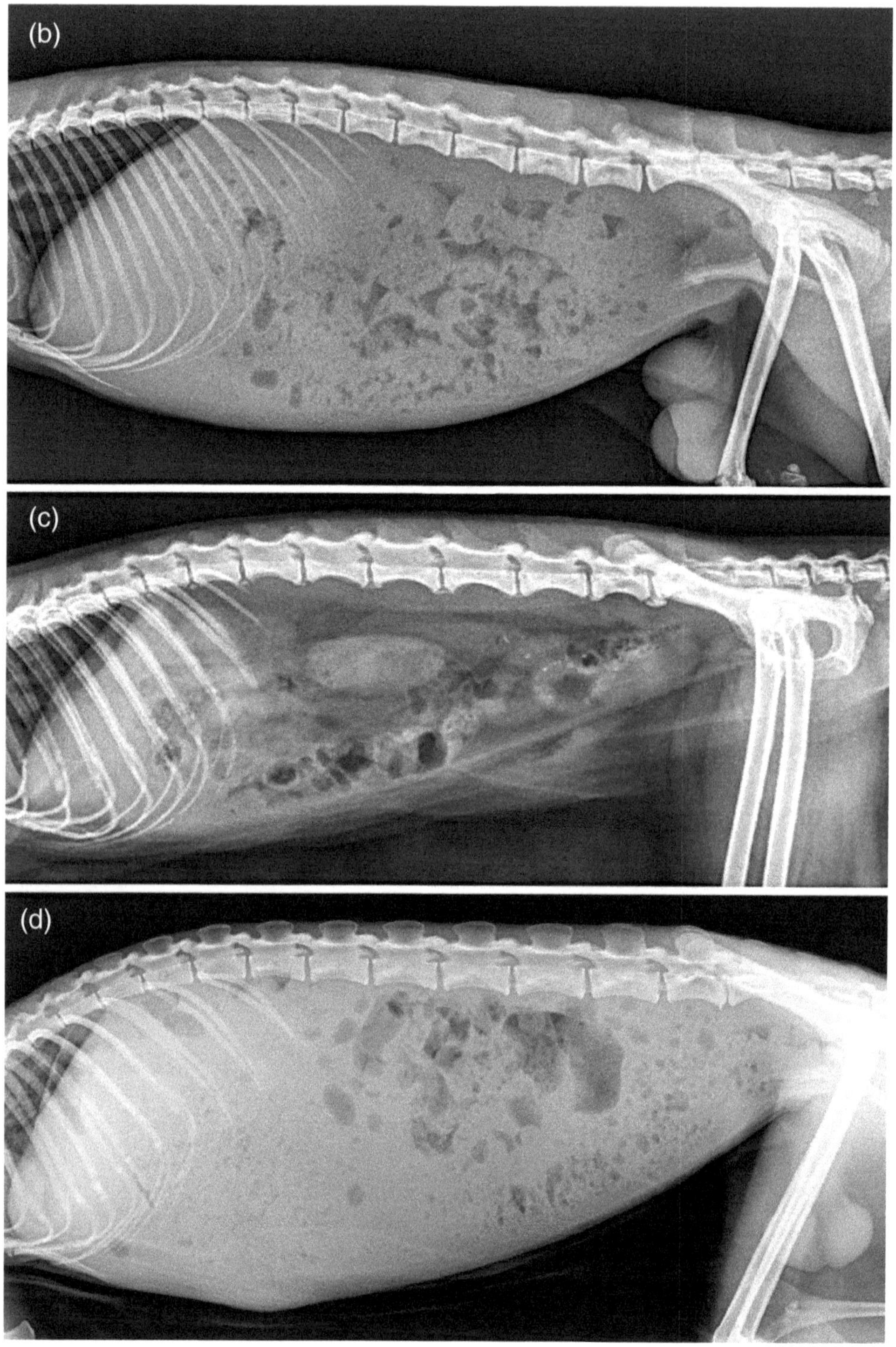

Figure 8.22: *(Continued)*.

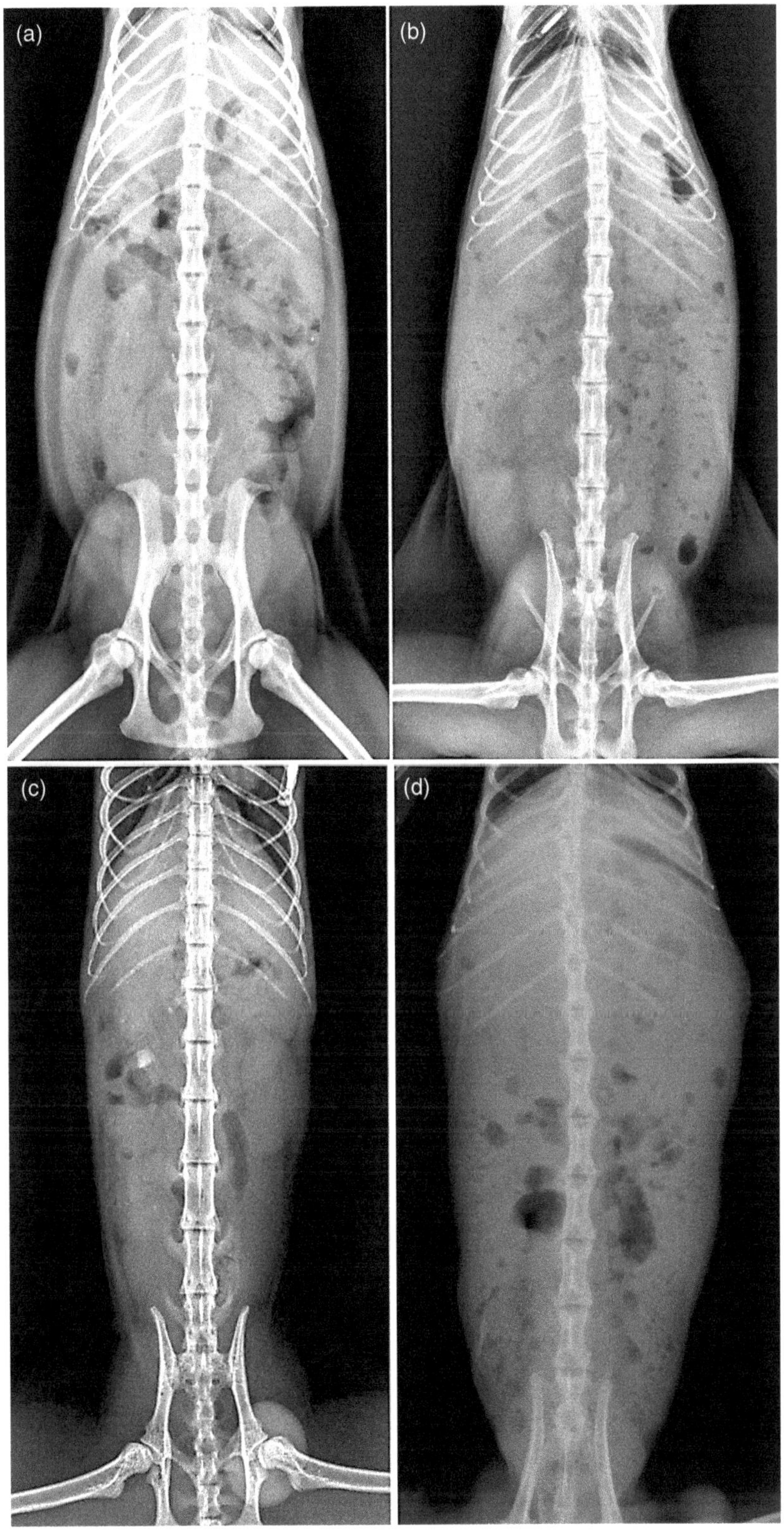

Figure 8.23: Abdomen, ventrodorsal view: (a) common brush-tailed possum (*Trichosurus vulpecula*), (b) common ring-tailed possum (*Pseudocheirus peregrinus*), (c) squirrel glider (*Petaurus norfolcensis*), (d) greater glider (*Petauroides volans*). Intraluminal contents (gas and digesta) of the caecum and colon dominate the radiographic image, obscuring other abdominal organs. Note: a transponder can be seen in (b) and (c).

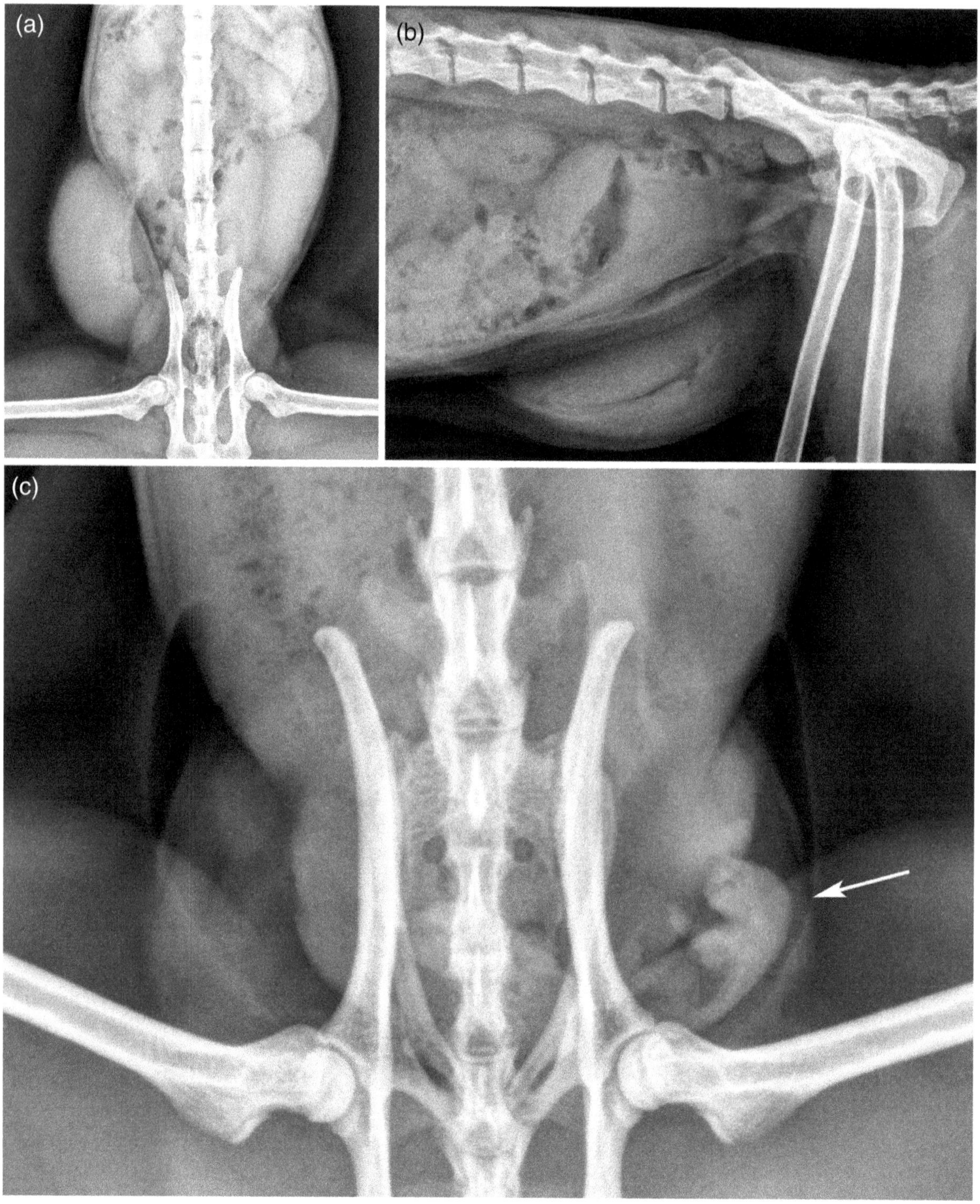

Figure 8.24: Caudal abdomen, yellow-bellied glider (*Petaurus australis*): (a) ventrodorsal view, (b) lateral view, (c) ventrodorsal view. (a, b) A lactating mammary gland can be seen. In this individual, the pouch young had emerged from the pouch but was still suckling. (c) A pouch young is present and attached to a nipple (arrow).

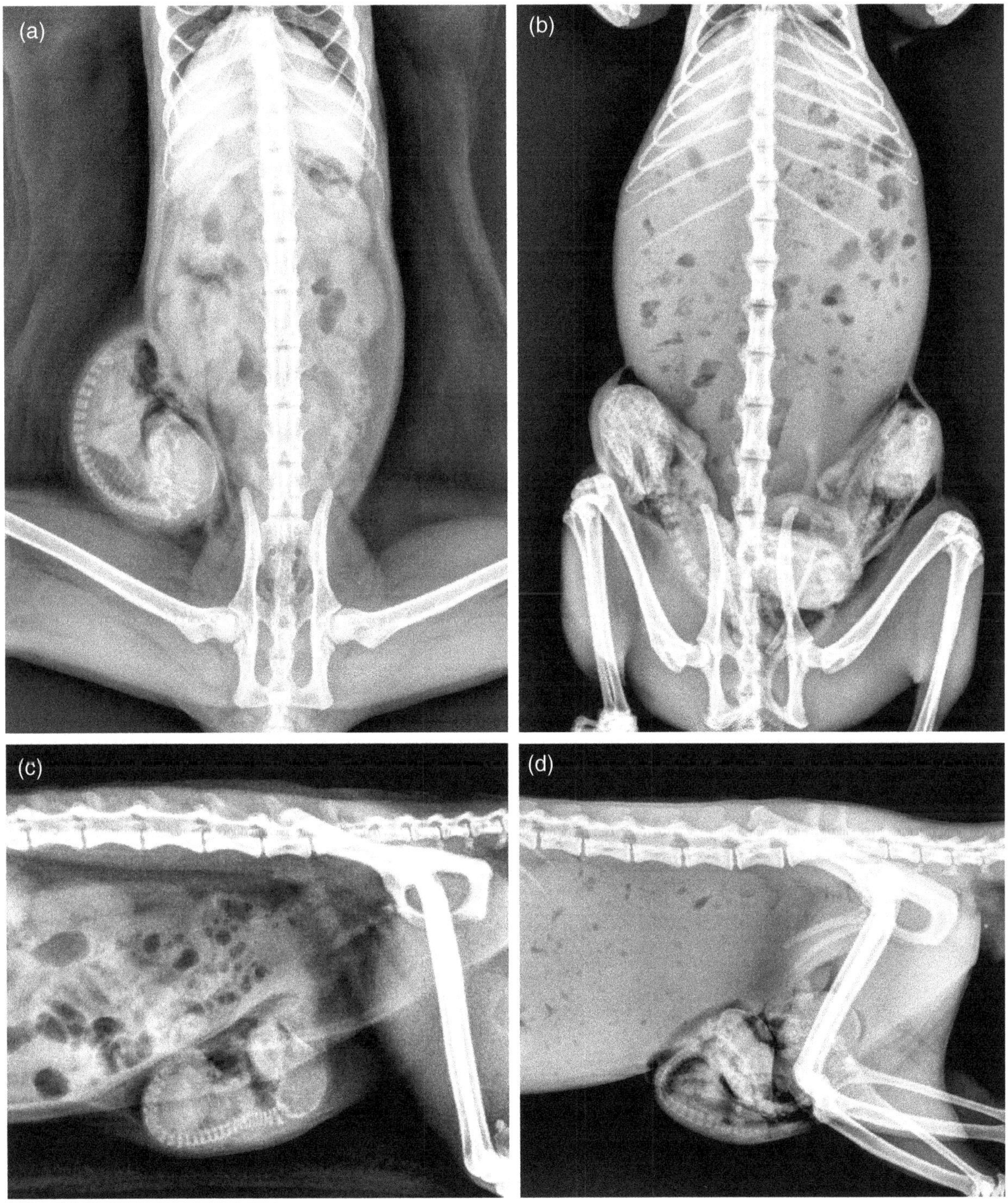

Figure 8.25: Pouch young clearly seen on radiographs: (a) Yellow-bellied glider (*Petaurus australis*) with a single pouch young, ventrodorsal view, (b) common ring-tailed possum (*Pseudocheirus peregrinus*) with three pouch young, ventrodorsal view, (c) squirrel glider (*Petaurus norfolcensis*) with a single pouch young attached to the nipple of the mammary gland, lateral view, (d) common ring-tailed possum (*Pseudocheirus peregrinus*) with two pouch young, lateral view.

or more pairs of bulbourethral (Cowper's) glands (Temple-Smith 1984).

The pouch opens cranially, has 2–6 teats and may have an incomplete median septum in *Petaurus* spp. and the Leadbeater's possum (*Gymnobelideus leadbeateri*) (McKay 1989; McKay and Winter 1989; Russell and Renfree 1989; Turner and McKay 1989). When lactating, the mammary gland may be large and can be seen on radiographs (Figs 8.24 and 8.25). If still present within the pouch, pouch young are clearly seen on radiographs (Fig. 8.25).

References

Dawson TJ, Finch E, Freedman L, Hume ID, Renfree MB, Temple-Smith PD (1989) Morphology and physiology of the metatheria. In *Fauna of Australia, Volume 1B, Mammalia.* (Eds DW Walton and BJ Richardson) pp. 451–504. Australian Government Publishing Service, Canberra.

Gilmore DP, da Costa CP (1995) *A Guide to Marsupial Dissection (Based Largely on the Australian Brush-tailed Possum* Trichosurus vulpecula*).* Departamento de Fisologica e Farmacologica, Universidade Federal de Pernambuco, Recife/PE.

How RA, Kerle JA (1995) Common brushtail possum *Trichosurus vulpecula*. In *The Mammals of Australia* (2nd edn). (Ed R Strahan) pp. 273–275. Reed New Holland, Sydney.

Hume ID (1999) *Marsupial Nutrition.* Cambridge University Press, Cambridge.

Jackson S (2012) *Gliding Mammals of the World.* CSIRO Publishing, Melbourne.

Johnson R, Hemsley S (2008) Possums and gliders. In *Medicine of Australian Mammals.* (Eds L Vogelnest and R Woods) pp. 395–437. CSIRO Publishing, Melbourne.

Lindenmayer D (2002) *Gliders of Australia: A Natural History.* University of New South Wales Press, Sydney.

Macrini TE (2012) Comparative morphology of the internal nasal skeleton of adult marsupials based on x-ray computed tomography. *Bulletin of the American Museum of Natural History* **365**, 1–91. doi:10.1206/365.1

McKay GM (1989) Petauridae. In *Fauna of Australia, Volume 1B, Mammalia.* (Eds DW Walton and BJ Richardson) pp. 665–678. Australian Government Publishing Service, Canberra.

McKay GM, Ong P (1995) Common ringtail possum *Pseudocheirus peregrinus.* In *The Mammals of Australia* (2nd edn). (Ed R Strahan) pp. 254–256. Reed New Holland, Sydney.

McKay GM, Winter JW (1989) Phalangeridae. In *Fauna of Australia, Volume 1B, Mammalia.* (Eds DW Walton and BJ Richardson) pp. 636–651. Australian Government Publishing Service, Canberra.

Owen R (1839) On the osteology of Marsupialia. *The Transactions of the Zoological Society of London* **2**(5), 379–408. doi:10.1111/j.1469-7998.1839.tb00030.x

Parker JT, Haswell WA (1921) Skeleton of metatheria. In *A Text-book of Zoology, Volume II.* pp 501–506. Macmillan and Co. Ltd, London.

Reese S, Pfuderer UR, Bragulla H, Loeffler K, Budras KD (2001) Topography, structure and function of the patella and the patelloid in marsupials. *Anatomia, Histologia, Embryologia* **30**, 289–294. doi:10.1046/j.1439-0264.2001.00334.x

Richardson KC, Wooller RD, Collins BG (1986) Adaptations to a diet of nectar and pollen in the marsupial *Tarsipes rostratus* (Marsupialia: Tarsipedidae). *Journal of Zoology* **208**, 285–297. doi:10.1111/j.1469-7998.1986.tb01515.x

Runciman SIC, Baudinette RV, Gannon BJ (1992) The anatomy of the adult marsupial heart: an historical review. *Australian Journal of Zoology* **40**, 21–34. doi:10.1071/ZO9920021

Russell EM, Renfree MB (1989) Tarsipedidae. In *Fauna of Australia, Volume 1B, Mammalia.* (Eds DW Walton and BJ Richardson) pp. 769–782. Australian Government Publishing Service, Canberra.

Smith AP (1982) Is the striped possum (*Dactylopsila trivirgata*; Marsupialia, Petauridae) an arboreal anteater? *Australian Mammalogy* **5**, 229–234.

Smith A, Winter J (1984) A key and field guide to the Australian possums, gliders and the koala. In *Possums and Gliders.* (Eds AP Smith and ID Hume) pp. 579–594. Surrey Beatty and Sons with the Australian Mammal Society, Sydney.

Strahan R (Ed) (1995) *The Mammals of Australia* (2nd edn). Reed New Holland, Sydney.

Suckling GC (1995) Sugar glider *Petaurus breviceps*. In *The Mammals of Australia* (2nd edn). (Ed R Strahan) pp. 229–231. Reed New Holland, Sydney.

Temple-Smith PD (1984) Reproductive structures and strategies in male possums and gliders. In *Possums and Gliders.* (Eds AP Smith and ID Humes) pp. 89–106. Surrey Beatty and Sons with the Australian Mammal Society, Sydney.

Turner V, McKay GM (1989) Burramyidae. In *Fauna of Australia, Volume 1B, Mammalia.* (Eds DW Walton and BJ Richardson) pp. 652–664. Australian Government Publishing Service, Canberra.

Tyndale-Biscoe H (2005) Life in the trees: koala, greater glider and possum. In *Life of Marsupials* (2nd edn). pp. 219–265. CSIRO Publishing, Melbourne.

Woodside DP (1995) Feather-tailed glider *Acrobates pygmaeus*. In *The Mammals of Australia* (2nd edn). (Ed R Strahan) pp. 262–264. Reed New Holland, Sydney.

9

Bandicoots and the bilby

The bandicoots and the bilby fall within the order Peramelemorphia. There are three families: Peramelidae, Chaeropodidae and Thylacomyidae. All species in the families Peramelidae and Chaeropodidae are bandicoots, and the bilby is in the family Thylacomyidae. They are found only in Australia, New Guinea and surrounding islands. There are 21 recognised species, of which 11 are found in Australia and three of these are now extinct. The family Peramelidae contains eight species in three genera: *Perameles* (long-nosed bandicoots), *Isoodon* (short-nosed bandicoots) and *Echymipera* (spiny bandicoots). The extinct pig-footed bandicoot (*Chaeropus ecaudatus*) is classified into a separate family, the Chaeropodidae (Tyndale-Biscoe 2005). The family Thylacomyidae comprises a single genus, *Macrotis*, containing two species: the lesser bilby (*Macrotis leucura*), now extinct, and the greater bilby (*Macrotis lagotis*) (Lynch 2008).

The bandicoots and bilby share anatomical characteristics with both major groups of Australian marsupials: the polyprotodonts and diprotodonts. Like the dasyurids, they have polyprotodont dentition, possessing three or more pairs of incisors in the maxilla and mandible. However, they also display syndactyly of the second and third digits of the pes, a characteristic shared with the diprotodont macropods, koala, wombats, possums and gliders. These characteristics collectively separate the bandicoots and bilby from all other marsupials (Gordon and Hulbert 1989). This separation has been supported by analysis of mitochondrial DNA. The divergence of the bandicoots and bilby from other Australian marsupials is estimated to have occurred approximately 60 million years ago (Tyndale-Biscoe 2005; Lynch 2008).

Bandicoots and the bilby are small to medium-sized omnivorous marsupials. The tail is one-quarter to one-half the length of the head and body and adults weigh 200–4700 g. They are usually nocturnal. They are stocky with short limbs and neck. The forelimbs are shorter than the hindlimbs and are designed for digging in soil to extract ground-living invertebrates and other food items. The elongated muzzle and powerful, flattened claws of the manus are used for probing crevices and rooting and digging in soil (Fig. 9.1). They are opportunistic feeders, exploiting a wide range of food items, have a rapid reproductive rate, grow quickly and are not long-lived (Gordon and Hulbert 1989; Lynch 2008).

In this chapter, anatomical descriptions and radiographic images focus primarily on readily accessible species. The greater bilby has been chosen as the primary representative of this taxonomic group for images in this chapter because they are anatomically very similar to most bandicoots and radiographic features are more readily identifiable because of their relatively large size. However, when radiographically available, relevant anatomical features of other species are included.

Skeletal system

Appendicular skeleton

Pectoral girdle and forelimb

Bandicoots and the bilby are the only marsupials that lack clavicles (Owen 1839; Jones 1924) (Fig. 9.2).

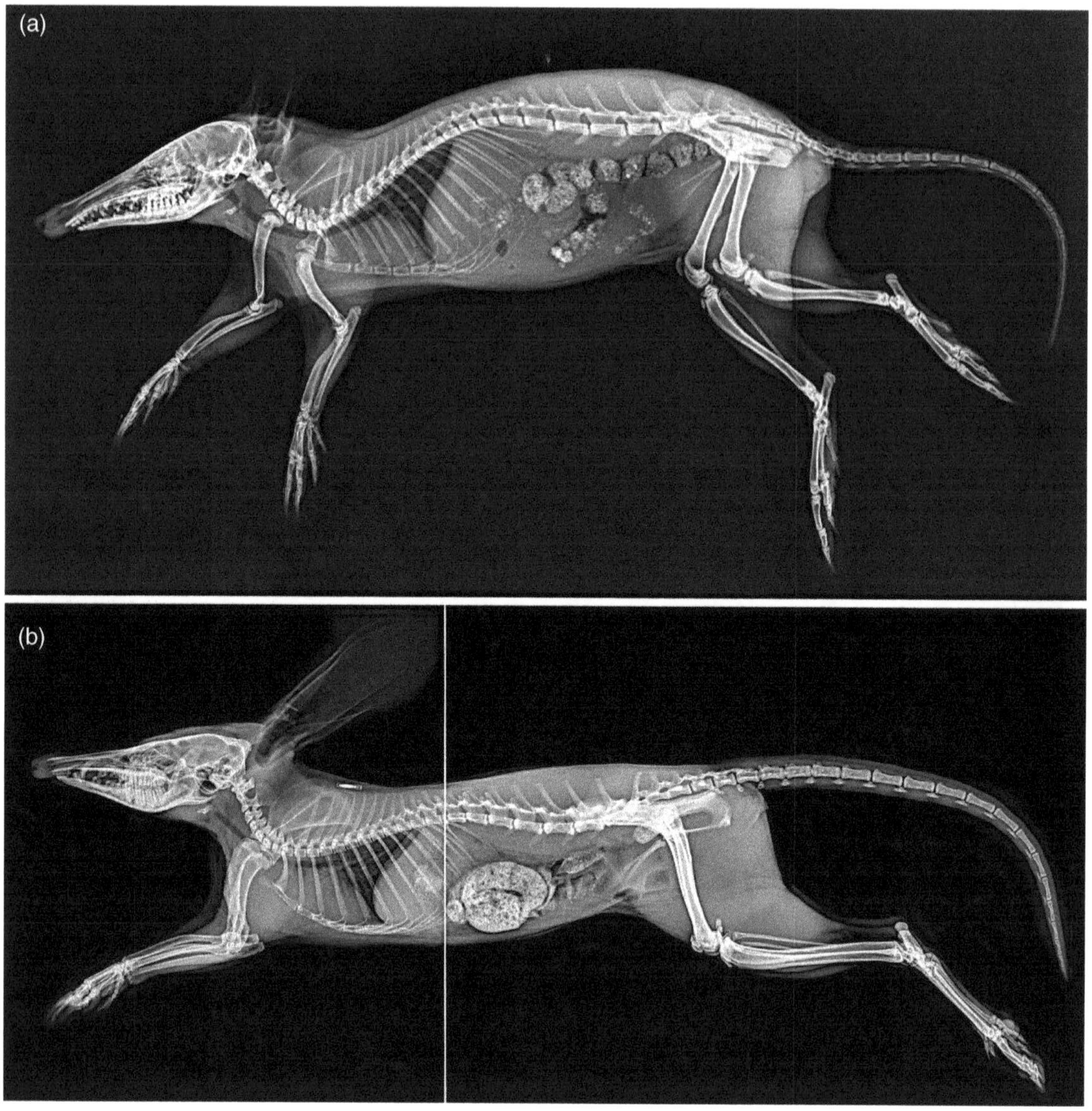

Figure 9.1: Whole body, lateral view: (a) southern long-nosed bandicoot (*Perameles nasuta*), (b) greater bilby (*Macrotis lagotis*), composite image. Note: a transponder can be seen in (b).

In the *Perameles* species (western barred bandicoot [*Perameles bougainville*], desert bandicoot [*P. eremiana*], eastern barred bandicoot [*P. gunnii*] and southern long-nosed bandicoot [*P. nasuta*]), the medial surface of the scapula is remarkably flat. The proportions of the supraspinous and infraspinous surfaces are reversed compared with other mammals, with the latter being larger. The entire scapular spine is bent downwards over the infraspinous surface (Owen 1839). The coracoid is represented by a slight projection of the dorsal margin of the glenoid cavity. The greater tubercle of the humerus appears smaller than the lesser tubercle and the humerus is relatively short and stocky. The deltoid tuberosity and the supinator crest are relatively small, smooth enlargements of the mid and distal humerus respectively (Fig. 9.3). They are greatly reduced in size when compared with the larger macropods and wombats. The radius and ulna are long, slender bones, although in adult male bandicoots the radius is often robust and bowed (Fig. 9.4).

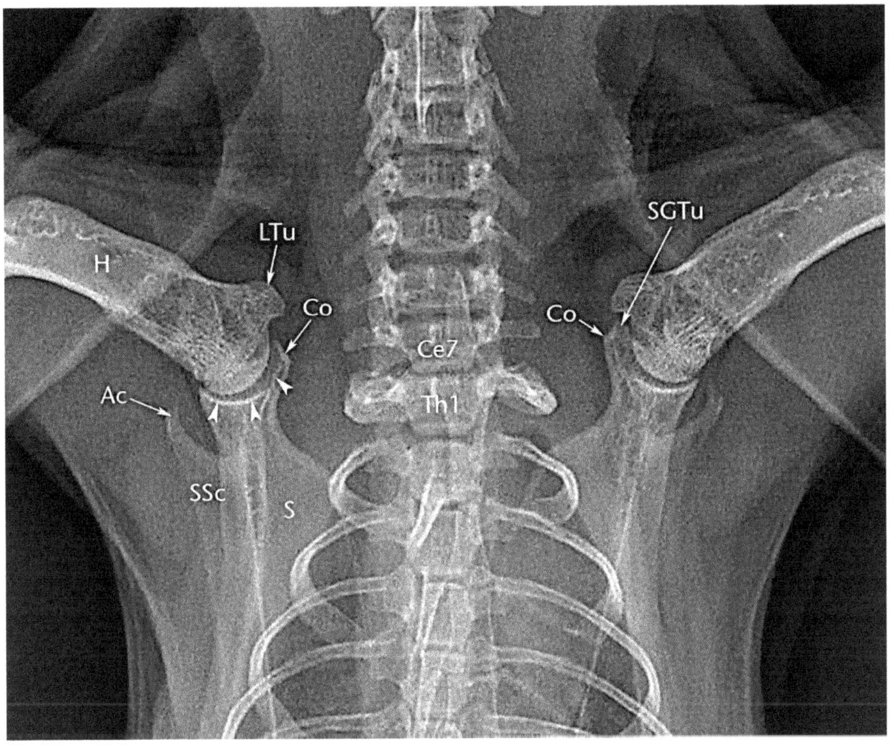

Figure 9.2: Pectoral girdle, greater bilby (*Macrotis lagotis*), ventrodorsal view. The humerus (H) articulates with the glenoid cavity (arrowheads) of the scapula (S). The spinous process (SSc) of the scapula curves distally and terminates at the acromion (Ac). Note the absence of a clavicle. The coracoid (Co) is reduced to a small protrusion from the supraglenoid tubercle (SGTu). In contrast, the lesser tubercle (LTu) of the humerus protrudes from the surface of the proximal humerus. The location of the 7th cervical (Ce7) and 1st thoracic (Th1) vertebrae is noted.

In bandicoots, there are four bones in the first row of carpal bones, including a slender accessory carpal bone and an intermediate carpal bone. In the second row of carpal bones, the first carpal is rudimentary and carpals three, four and five are well defined (Owen 1839) (Fig. 9.5). In bandicoots, digits 2, 3 and 4 of the manus are usually prominent with strong, flattened claws; digits 1 and 5 are reduced in size and clawless. In the pig-footed bandicoot (now extinct), only digits 2 and 3 are prominent and functional (hence the common name); digit 4 is a minute stub and the other digits are absent (Gordon and Hulbert 1989; Johnson 1989). In the greater bilby, digits 1 and 5 are clawless; the other digits are well developed with strong, curved claws (Figs 9.4 and 9.5). In bandicoots and the greater bilby, the distal phalanx of digits 2, 3 and 4 of the manus is split distally by a longitudinal fissure commencing at the upper part of its base (Owen 1839) (Fig. 9.5).

Pelvis and hindlimb

The ilia of the pelvis are broad and flat, the ischial tuberosities are curved slightly laterally and the epipubic bones are long, straight and slender (Gordon and Hulbert 1989) (Fig. 9.6). The femur, tibia, fibula and digits are elongated relative to the equivalent elements of the forelimbs and the structure of the pes is unique. The hindlimb of the greater bilby loosely resembles that of macropods, but bandicoots (Owen 1839; Jones 1924; Reese *et al.* 2001) and the bilby are the only marsupials with a well-developed patella (Fig. 9.7). In the greater bilby, the distal fibula is fused with the tibia, although a line of separation is apparent (Owen 1839) (Fig. 9.8).

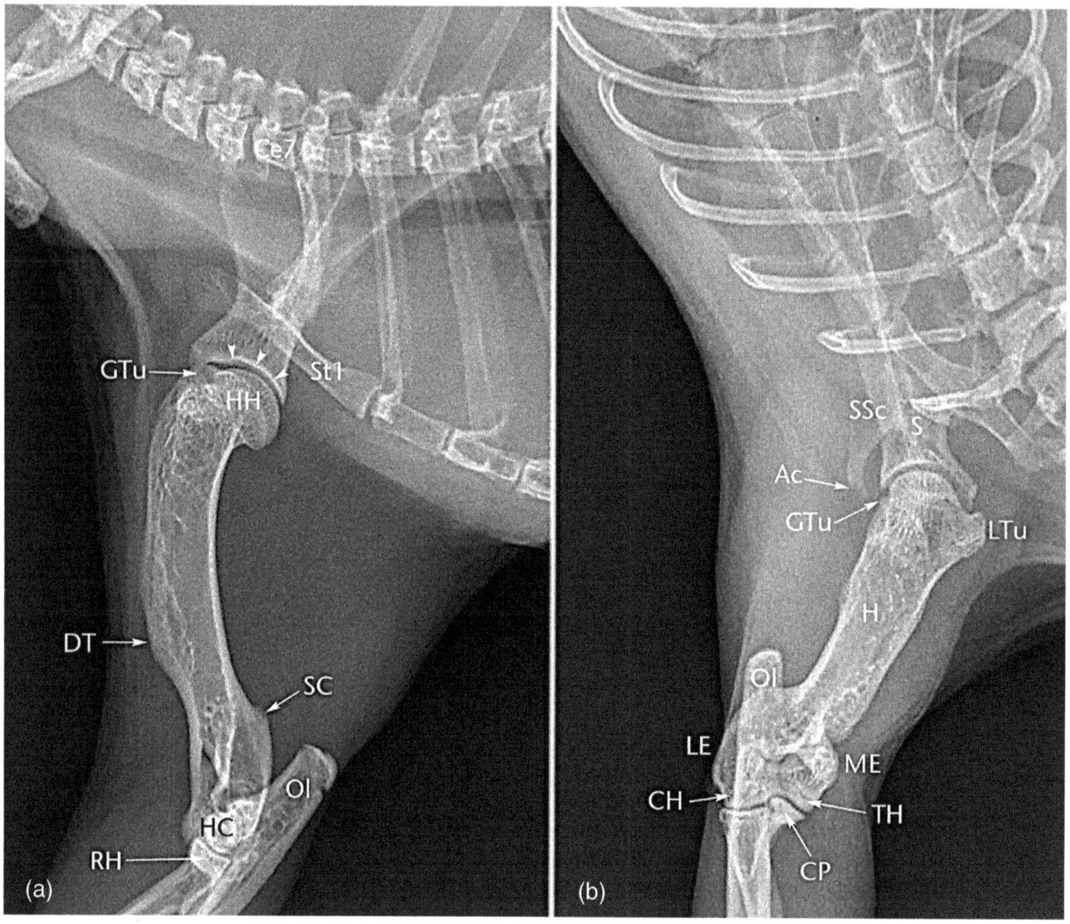

Figure 9.3: Pectoral girdle and brachium, greater bilby (*Macrotis lagotis*): (a) mediolateral view, (b) craniocaudal view. The glenoid cavity (arrowheads) and the humeral head (HH) create the ball and socket joint that is the shoulder joint. The greater tubercle (GTu) is relatively small and the lesser tubercle (LTu) is a large protrusion from the humeral neck. The spinous process (SSc) of the scapula (S) terminates at the acromion (Ac). There is no clavicle. The deltoid tuberosity (DT) is a smooth enlargement of the cranial surface of the diaphysis of the humerus (H). On the caudal surface of the distal humerus, the supinator crest (SC) is reduced to a smooth ridge. The humeral condyle (HC) at the distal end of the humerus is enlarged laterally by the lateral epicondyle (LE) and medially by the medial epicondyle (ME). The elbow joint is between the articular surfaces of the radial head (RH) and the *capitulum humeri* (CH) and between the trochlear notch and coronoid process (CP) of the ulna and the *trochlea humeri* (TH). The olecranon (Ol) of the ulna projects proximally to the elbow joint. The 1st sternal element is labelled (St1).

Bandicoots show only slight variation in the basic structure of the pes among species, however, there can be considerable variation in the relative lengths of the digits among the species. The pes is generally elongated in all species, with digit 4 larger than digit 5, which in turn is larger than the syndactylous digits 2 and 3. Digit 1 is reduced and clawless. Digit 1 is larger in the rufous spiny bandicoot (*Echymipera rufescens*) than other species and absent in the pig-footed bandicoot and greater bilby. Digit 5 in the pig-footed bandicoot is significantly reduced (Owen 1839; Gordon and Hulbert 1989; Johnson 1989). In the greater bilby, the pes is elongated, digit 1 is absent, digits 2 and 3 are syndactylous and slender, digit 4 is very large and digit 5 is short and slender. In bandicoots and the greater bilby, the distal phalanx of digits 4 and 5 is split distally by a longitudinal fissure commencing at the upper part of its base (Owen 1839) (Fig. 9.9).

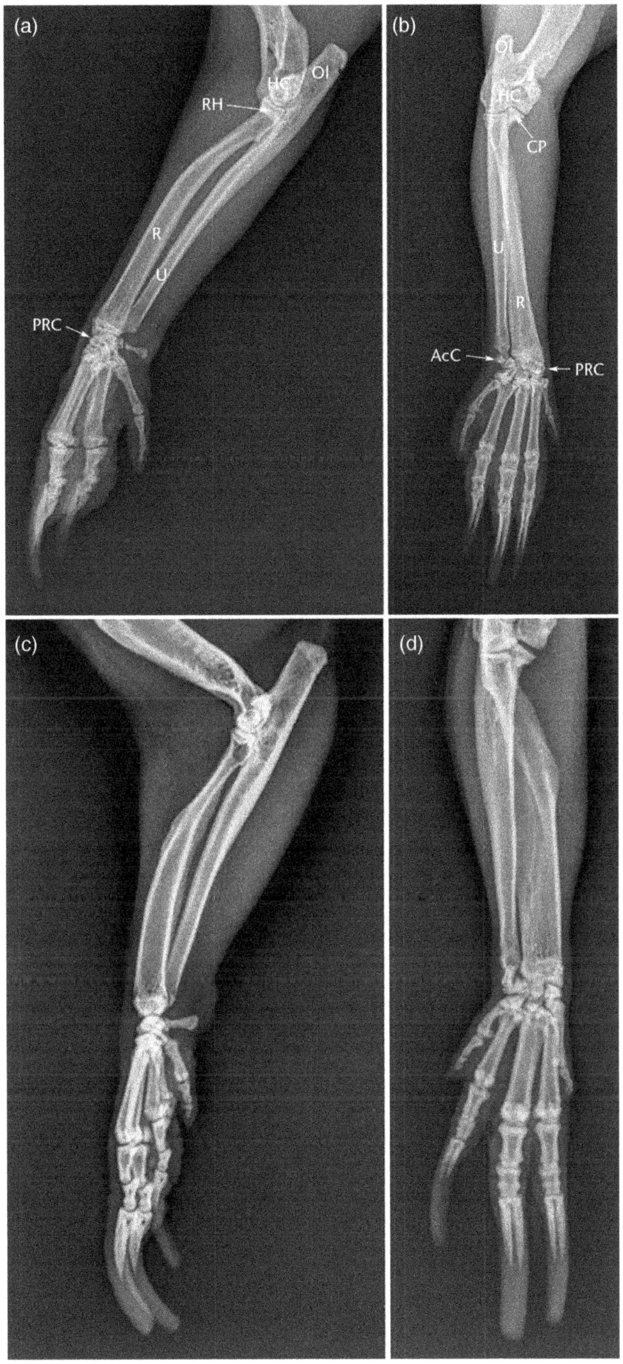

Figure 9.4: Antebrachium and manus: (a) greater bilby (*Macrotis lagotis*), mediolateral view, (b) craniocaudal and dorsopalmar view; (c) southern long-nosed bandicoot (*Perameles nasuta*), mediolateral view, (d) craniocaudal and dorsopalmar view. The antebrachium comprises the radius (R) and ulna (U), which are long and relatively slender in the greater bilby, whereas in the adult male southern long-nosed bandicoot they are more robust and the radius is bowed. The ulna extends proximal to the elbow joint as the olecranon (Ol). The elbow joint comprises the articular surfaces of the humeral condyle (HC), radial head (RH), coronoid process (CP) and trochlear notch of the olecranon. Distally, the antebrachiocarpal joint is mainly between the distal radius and the proximal row of carpal bones (PRC), with a smaller contribution from the styloid process of the ulna. The accessory carpal bone (AcC) is slender and projects caudal to the proximal row of carpal bones.

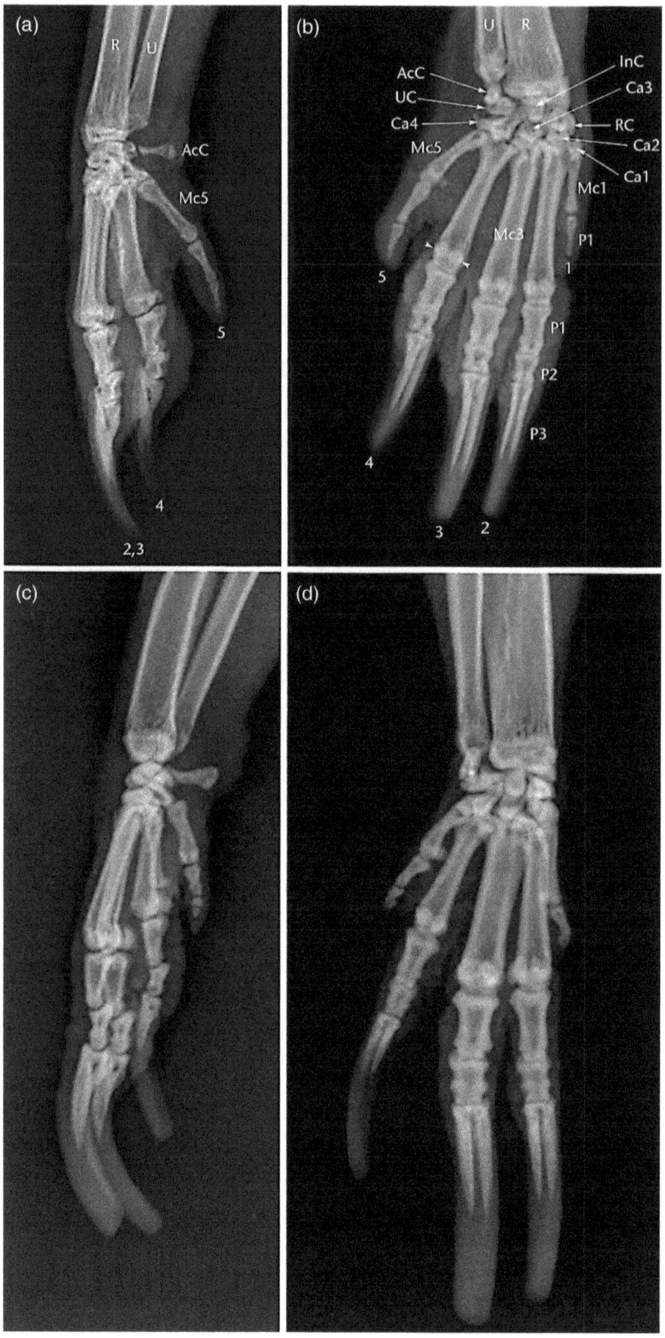

Figure 9.5: Manus: (a) greater bilby (*Macrotis lagotis*), mediolateral view, (b) dorsopalmar view; (c) southern long-nosed bandicoot (*Perameles nasuta*), mediolateral view, (d) dorsopalmar view. In the proximal row of carpal bones, an intermediate carpal bone (InC) separates the radial (RC) and ulnar (UC) carpal bones. The accessory carpal bone (AcC) is located on the lateral aspect of the carpus, articulating with the palmar surface of the ulnar carpal bone. There are four carpal bones in the distal row: the 4th (Ca4) is the largest, followed in size by the 3rd (Ca3), 2nd (Ca2) and 1st (Ca1). There is a small sesamoid bone adjacent to the lateral surface of Ca4. There are five metacarpal bones (Mc1, Mc3, Mc5 labelled). Paired sesamoid bones are located on the palmar surface of the manus proximal to the metacarpophalangeal articulations (arrow heads). Three digits (2, 3, 4) are well developed. They have three phalanges (P1, P2, P3) and the distal phalanges appearing bifid. Digit 1 (1) is clawless and may possess only P1. Digit 5 (5) is also clawless, although there is a rudimentary P3. The radius (R) and ulna (U) are labelled.

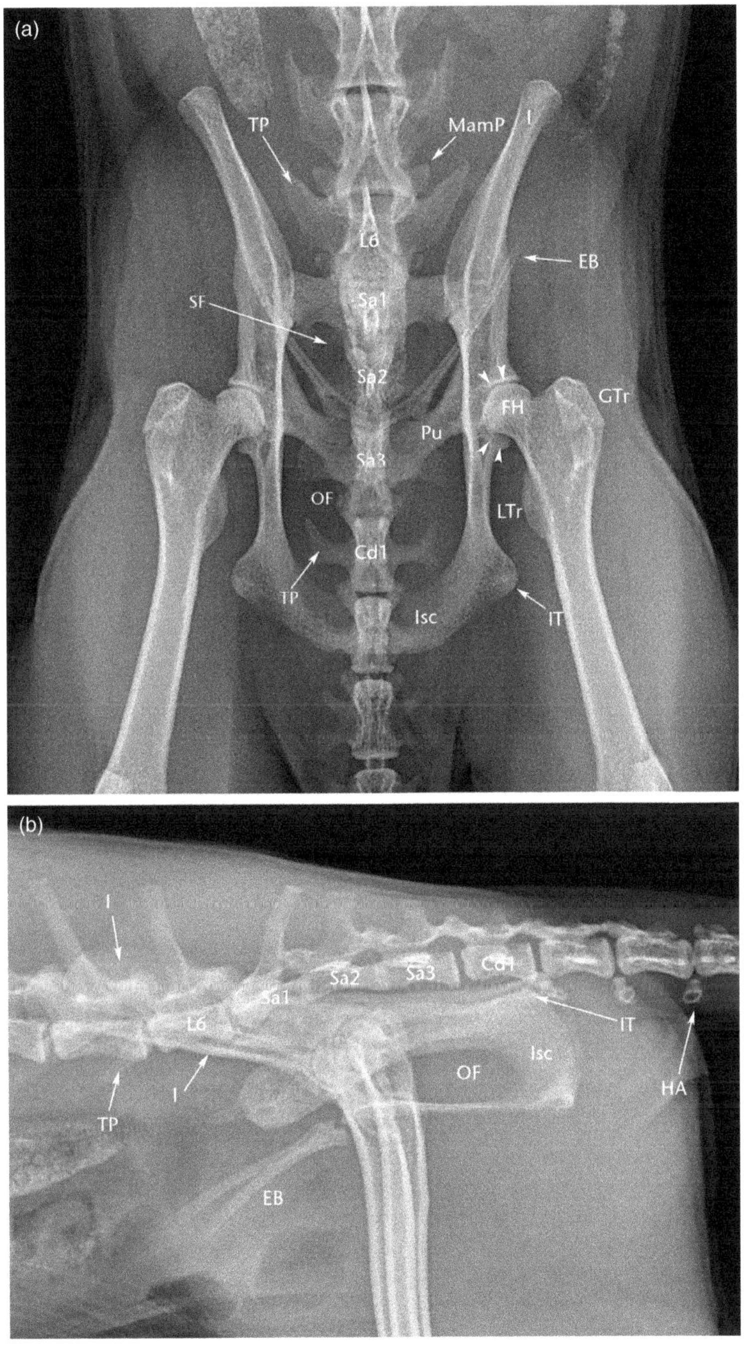

Figure 9.6: Pelvis, greater bilby (*Macrotis lagotis*): (a) ventrodorsal view, (b) lateral view. Of the three sacral vertebrae (Sa1, Sa2, Sa3), only the 1st contributes to the sacroiliac articulation. A large sacral foramen (SF) is present between the fused transverse processes of the sacral vertebrae. The foramen between Sa1 and Sa2 is visible but the foramen between Sa2 and Sa3 is obscured by the superimposed epipubic bones (EB). Each hemipelvis comprises three bones: the ilium (I), pubis (Pu) and ischium (Isc). They all contribute to the acetabulum (arrowheads) that articulates with the femoral head (FH). The pelvic bones are relatively slender and the obturator foramen (OF) is relatively large. The proximal femur possesses two large trochanters, the greater (GTr) on the lateral aspect and the lesser (LTr) on the medial aspect. Transverse processes (TP) are present on the lumbar vertebrae (L6 labelled) and the first few caudal vertebrae (Cd1 labelled). The lumbar vertebrae also have large, craniodorsally directed mammillary processes (MamP). Haemal arches (HA) are present ventral to the articulations between the caudal vertebrae. The ischial tuberosity (IT) has a blunt, rounded surface.

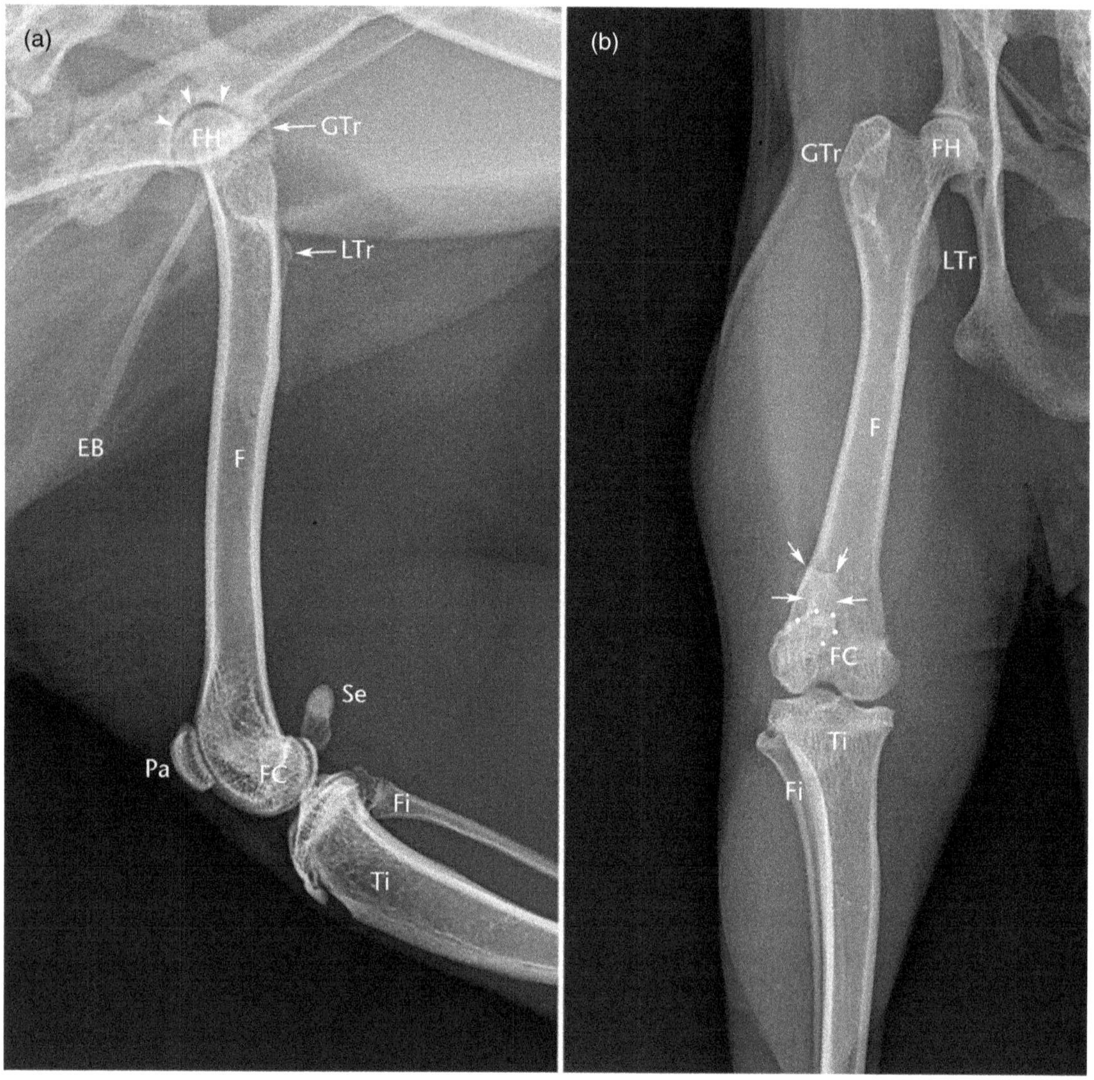

Figure 9.7: Femur and stifle, greater bilby (*Macrotis lagotis*): (a) mediolateral view, (b) craniocaudal view. The femoral head (FH) forms a congruent articulation with the acetabulum (arrowheads). The greater (GTr) and lesser (LTr) trochanters of the femur (F) are well developed. Distally, there is a patella (Pa) on the cranial aspect of the femoral condyle (FC) and a large sesamoid bone (Se) on the caudolateral aspect of the condyle. On the craniocaudal view, the patella (arrows) and sesamoid bone (dotted line) are only partially visible. The femur articulates with the tibia (Ti) and the fibula (Fi) appears lateral and slightly distal to the stifle joint. The epipubic bone (EB) is labelled.

The structure of the bandicoot and bilby pes is unique among marsupials. Bodyweight is transferred to the pes via a mechanism including contact between the third and fourth tarsal bones (Marshall 1972). This arrangement, convergent with that of the ungulate foot, is an adaptation for a cursorial life that permits bodyweight to be transferred to the distal tarsus. The third and fourth tarsal contact is the diagnostic character that truly sets bandicoots and the bilby apart from other marsupials (Gordon and Hulbert 1989).

Axial skeleton

Skull

There is some variation in the size, shape and structure of the skull of bandicoots and the greater bilby. Generally, the skull is elongated and narrow (Figs 9.10 and 9.11), an exception being the

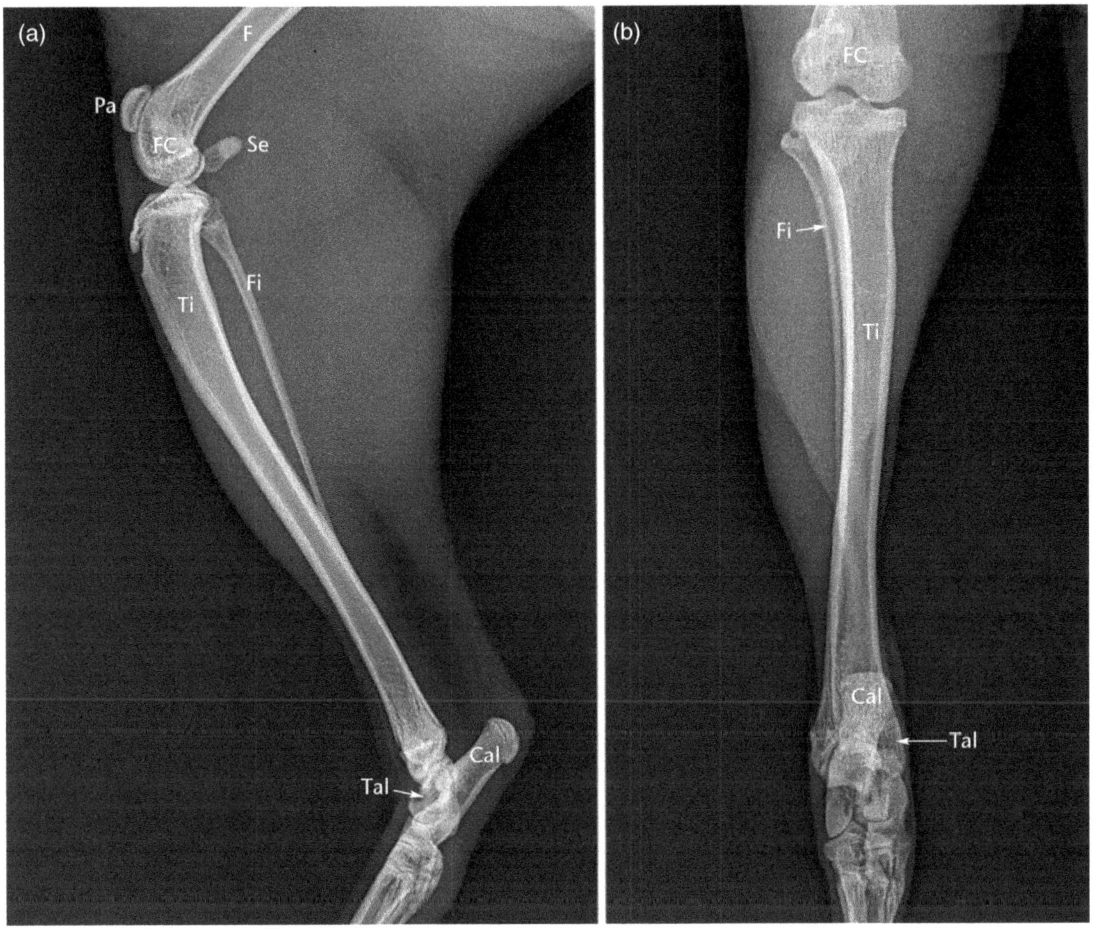

Figure 9.8: Crus, greater bilby (*Macrotis lagotis*): (a) mediolateral view, (b) craniocaudal view. The femur (F), femoral condyle (FC), patella (Pa) and stifle sesamoid bone (Se) are as described in Figure 9.7. The tibia (Ti) is the dominant bone of the crus. It is a solid, slightly curvilinear bone, while the fibula (Fi) is slender. The tibia's distal end contributes to the tarsocrural articulation, which is between the tibia, fibula, talus (Tal) and calcaneus (Cal).

pig-footed bandicoot, which has a shorter, broader skull. The snout is elongated by the forward growth of the maxilla, premaxilla and nasal bones. The cartilaginous nasal septum extends rostrally beyond the premaxillae, nasal bones and incisors (Freedman 1967). The skull of the greater bilby is distinguished radiographically by the cuboidal shaped molars and large maxillary canines.

The palate is long and narrows markedly rostral to the premolars. The paired rostral palatine vacuities are long and narrow, while the caudal pair are wider and have usually lost their thin median partition to form a larger single opening, the size of which is variable among species. At the level of the third and fourth molars the hard palate is very thin. In some species and individuals, additional areas of the palate may be fenestrated and this can be extensive. For example, one or two rostrocaudally elongated pairs of vacuities may be found between the rostral and caudal palatine vacuities and randomly distributed fenestrations sometimes occur in most of the palate caudal to the caudal palatine vacuities (Freedman 1967). In the greater bilby, the caudal palatine vacuities form a wide, oval space extending from the second premolars rostrally to the third molars caudally. Caudal to this there are six small perforations, in two transverse lines between the caudal palatine

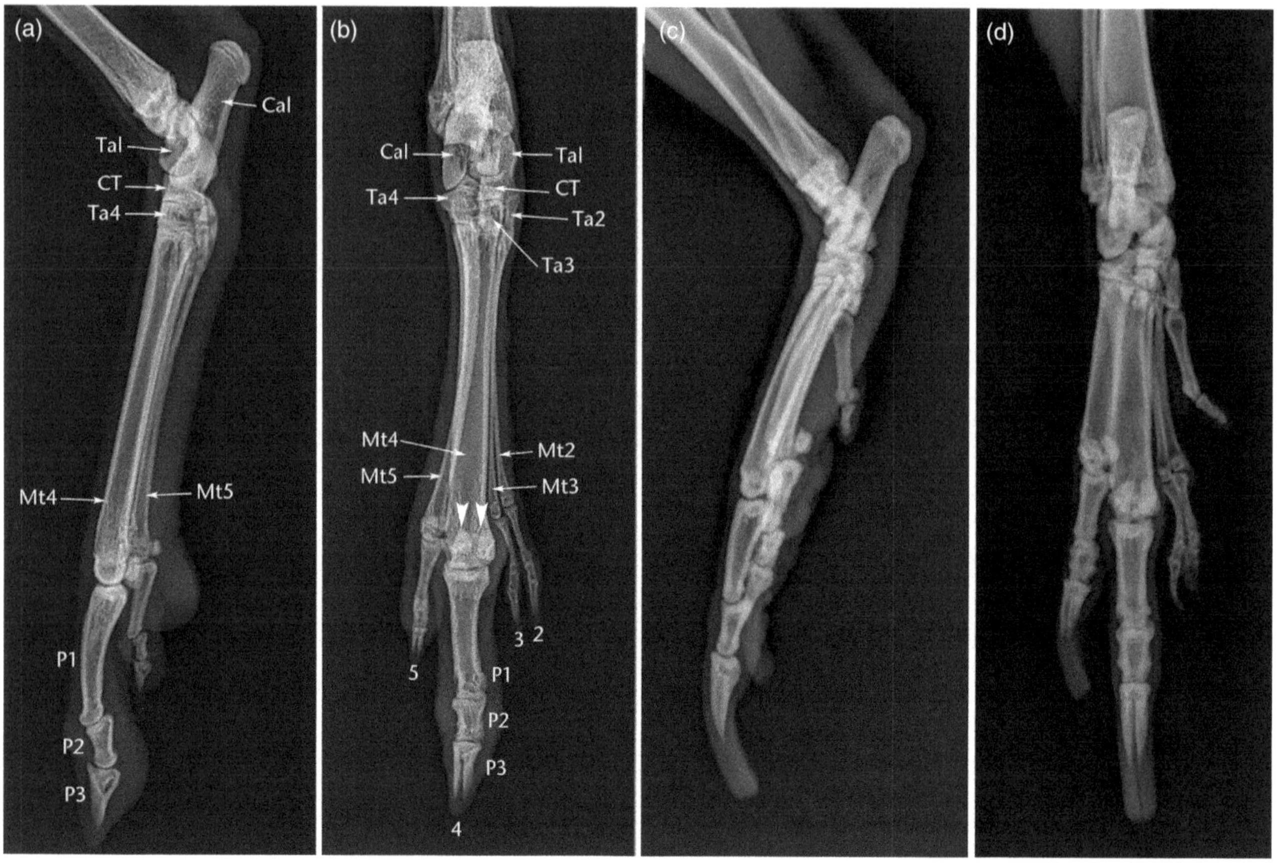

Figure 9.9: Pes: (a) greater bilby (*Macrotis lagotis*), mediolateral view, (b) dorsoplantar view; (c) southern long-nosed bandicoot (*Perameles nasuta*), mediolateral view, (d) dorsoplantar view. The tarsus has the usual arrangement of a talus (Tal) and calcaneus (Cal) that articulate proximally with the crus and distally with the central (CT) and 4th (Ta4) tarsal bones. The distal row contains the 2nd (Ta2), 3rd (Ta3) and 4th tarsal bones. The 2nd tarsal bone is located between the central tarsal bone and the syndactylous heads of the vestigial 2nd (Mt2) and 3rd (Mt3) metatarsal bones. The major bone of the pes is the 4th metatarsal bone (Mt4), which articulates proximally with the 3rd and 4th tarsal bones. The 5th metatarsal bone (Mt5) and digit 5 (5) are less developed and the former articulates with the 4th tarsal bone. Each of the four digits (2, 3, 4, 5) has three phalangeal bones (P1, P2, P3). The 3rd phalanges are unusual, because each is separated distally and fused proximally. Paired plantar metatarsophalangeal sesamoid bones (arrowheads) are visible on the plantar aspect of the metatarsophalangeal articulations of digits 4 and 5.

vacuities and the caudal margin of the palate (Owen 1839).

The tympanic bullae vary in size among the species. In *Perameles* species, the bullae are small to moderate in size, hemispherical and complete. They are relatively small in the southern long-nosed bandicoot and comparatively large in the western barred bandicoot. In *Isoodon* species, the bullae are large, complete and pear-shaped. The bullae are small, hemispherical and partly incomplete in the rufous spiny bandicoot (Freedman 1967). In at least the greater bilby and the southern long-nosed bandicoot, there are smaller accessory bullae caudal to the tympanic bullae (Owen 1839) (Figs 9.10 and 9.11).

The zygomatic arches are fine and arched laterally. In the greater bilby they are bifurcate at both extremities but more so rostrally than caudally (Owen 1839; Freedman 1967) (Fig. 9.11). A short jugular process extends from the exoccipital bones (Owen 1839) (Fig. 9.10).

The mandibles are slender. The coronoid process of the mandible is very tall and demarcated from the mandibular condyle by a deep notch. The

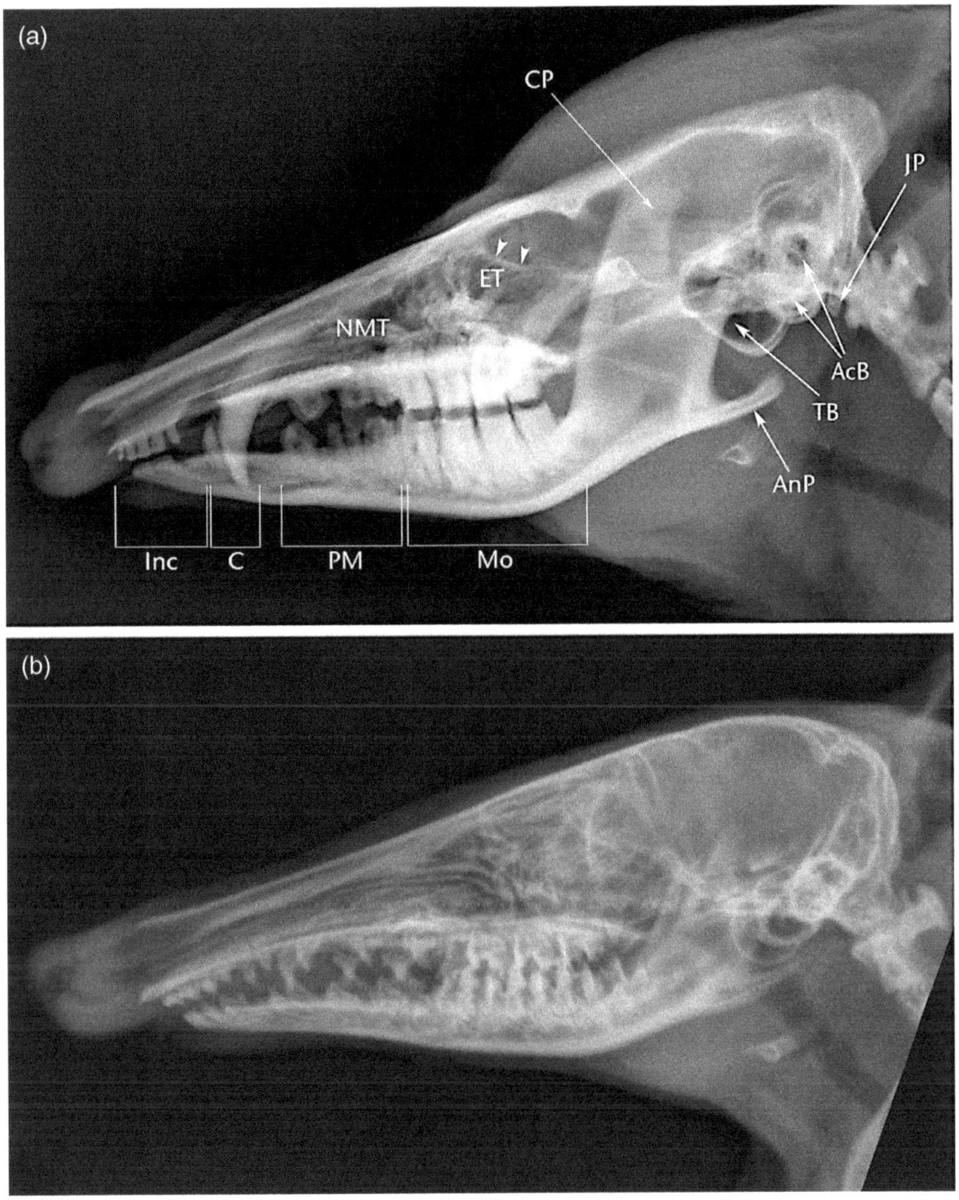

Figure 9.10: Skull, lateral view: (a) greater bilby (*Macrotis lagotis*), (b) southern long-nosed bandicoot (*Perameles nasuta*). The skull is long and slender. Dentition comprises incisor (Inc), canine (C), premolar (PM) and molar (Mo) teeth. The nasomaxillary turbinates (NMT) and ethmoid turbinates (ET) are radiographically visible. The cribriform plate (arrowheads) is almost horizontal. The tympanic bullae (TB) are relatively larger in the greater bilby and are supplemented by paired accessory bullae (AcB). The angular process (AnP) of the mandible is elongated and slender. The tall mandibular coronoid process (CP) is visible superimposed on the cranium. The jugular process (JP) of the occipital bone is small.

angle of the mandible is obtuse; the angular processes are well developed, flattened and curved slightly medially and dorsally (Owen 1839) (Fig. 9.10). The masseteric fossa is large and fairly deep. The mandibular symphysis is unfused.

Vertebrae

The vertebral formula of the greater bilby is 7 cervical, 13 thoracic, 6 lumbar, 3 sacral and 20 caudal vertebrae. Cervical ribs are present on the second cervical vertebra. The transverse processes of

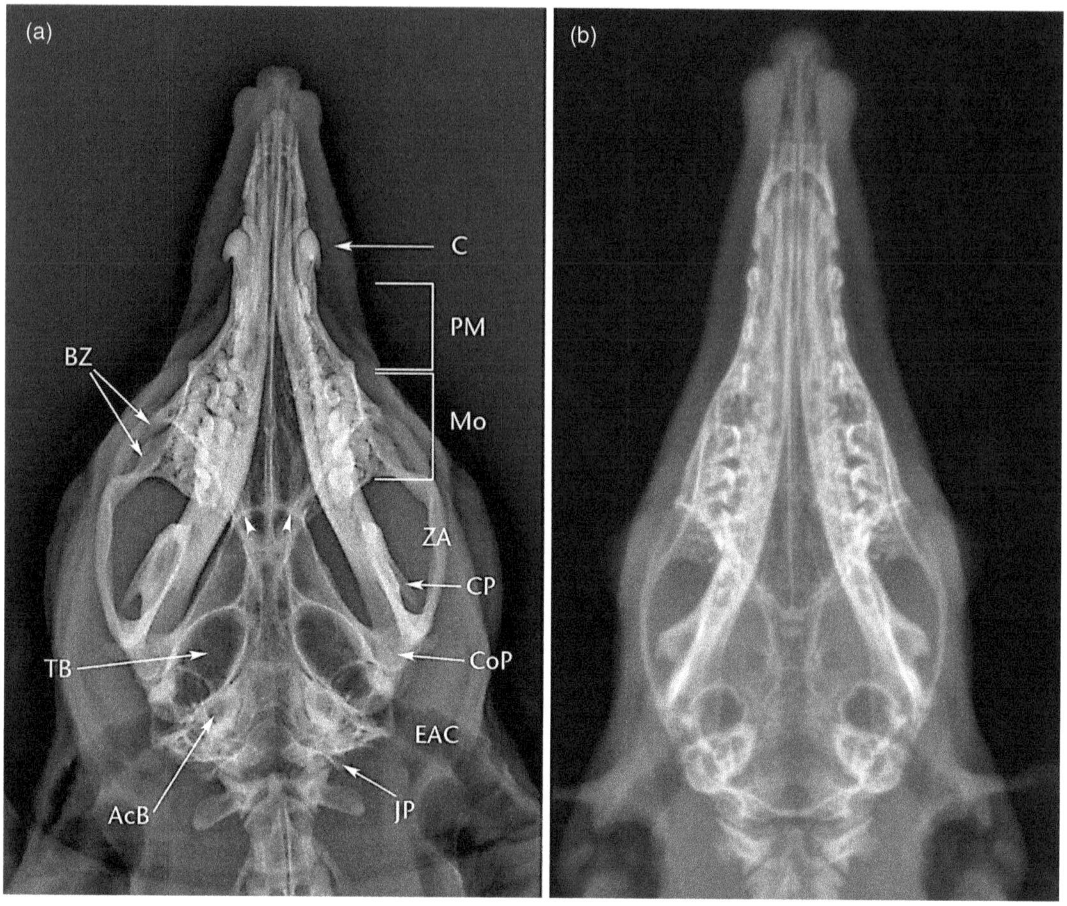

Figure 9.11: Skull, ventrodorsal view: (a) greater bilby (*Macrotis lagotis*), (b) southern long-nosed bandicoot (*Perameles nasuta*). The dental arcades are superimposed in this view, but the canine (C), premolar (PM) and molar (Mo) teeth are defined. The incisors are poorly visualised. The zygomatic arches (ZA) are curved. The coronoid (CP) and condyloid (CoP) processes of the mandibles can be distinguished. The external auricular canals (EAC) are seen leading to the tympanic bullae (TB), caudal to which the accessory bullae (AcB) are visible. The location of the cribriform plate is marked with arrowheads. The jugular processes of the occipital bones (JP) are small and the rostral ends of the zygomatic arches are bifid (BZ).

cervical vertebrae 3–7 are slender. The vertebral formula of bandicoots is 7 cervical, 13 thoracic, 6 lumbar, 3 sacral and 18–23 caudal vertebrae. The spinous processes of the cranial thoracic vertebrae are tall and slender. They gradually reduce in length caudally. The spinous processes of the sacral vertebrae are shorter and thicker than those of the lumbar vertebrae and are directed caudally (Figs 9.12–9.15). The tail of bandicoots is generally short and that of the greater bilby is long (50–66% of the body length). The tail is non-prehensile. Haemal arches are present on the ventral aspect of the articulations between the caudal vertebrae.

Ribs and sternum

In the greater bilby, there are 13 pairs of ribs, two of which are floating while the remainder articulate with the thoracic vertebrae (Johnson 1989). Bandicoots have 13 pairs of ribs, all of which have chondral elements, with the last two pairs floating (Figs 9.1 and 9.15).

Soft tissue

Thorax

There is a large thoracic thymus that lies dorsal to the cranial sternum and adjacent to the ventral

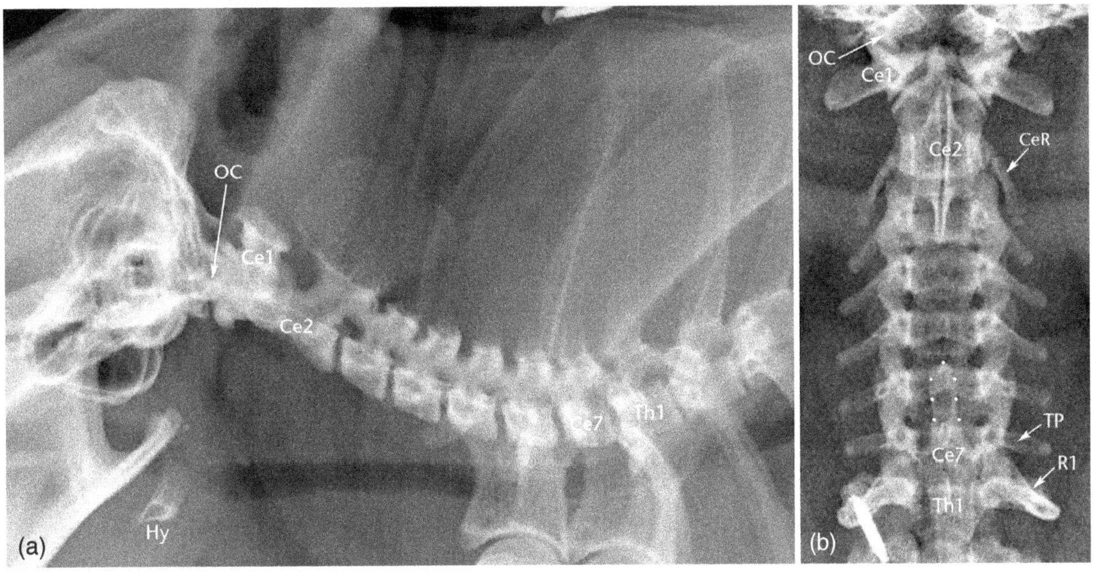

Figure 9.12: Cervical vertebral column, greater bilby (*Macrotis lagotis*): (a) lateral view, (b) ventrodorsal view. The occipital condyles (OC) and their articulation with the 1st cervical vertebra (Ce1) can be seen on both views. There are 7 cervical vertebrae. Cervical ribs (CeR) are present on the 2nd cervical vertebra (Ce2), which also possesses a large spatulate dorsal process. The following 5 vertebrae have caudally directed transverse processes (TP). The 7th cervical vertebra (Ce7) articulates with the 1st thoracic vertebra (Th1), which carries the 1st pair of ribs (R1). A hyoid bone (Hy) is visible on the lateral view. The cranial extent of the 1st sternal element (dotted line) is superimposed on the caudal cervical vertebrae in the ventrodorsal view. Note: transponders can be seen.

surface of the atria. Radiographically, the thymus silhouettes with the cranial cardiac silhouette, blurring out the cranial margin of the heart. There is no cervical thymus. The diaphragm arches deep into the thorax. Therefore, the heart and ventral parts of the lungs are restricted to the cranial half of the rib cage (Tedman 1990) (Figs 9.15 and 9.16).

Cardiovascular system

As in most marsupials, the heart is large relative to body size. The apex is less pointed than in other marsupials, but it is still formed by the left ventricle, although the tip of the right ventricle is also close to the apex in bandicoots. The more ventral of the two auricular appendages of the right atrium is small or absent in bandicoots. The 4–5 pulmonary veins enter the left atrium in bandicoots. In addition, as in most marsupials, the right atrium receives two cranial venae cavae and a single caudal *vena cava* (Tedman 1990; Runciman *et al.* 1992) (Figs 9.15 and 9.16).

Respiratory system

The right lung is divided into four lobes: cranial, middle, caudal and accessory. In the northern brown bandicoot (*Isoodon macrourus*) and possibly other species, the accessory lobe projects across the midline caudal to the heart and ventral to the oesophagus. The left lung is unilobate (Tedman 1990; Cope *et al.* 2001).

Abdomen

Bandicoots and the bilby have a simple, oval stomach and a small intestine that is not clearly demarcated into different regions. The size of the caecum varies among species. In most bandicoot species, the colon is short relative to the small intestine but in the greater bilby, the distal colon is much longer, presumably to maximise water resorption for this arid-zone species (Parsons 1903; Tedman 1990; Hume 1999). The presence of sand within the intestinal tract is common and visualised radiographically (Figs 9.17 and 9.18).

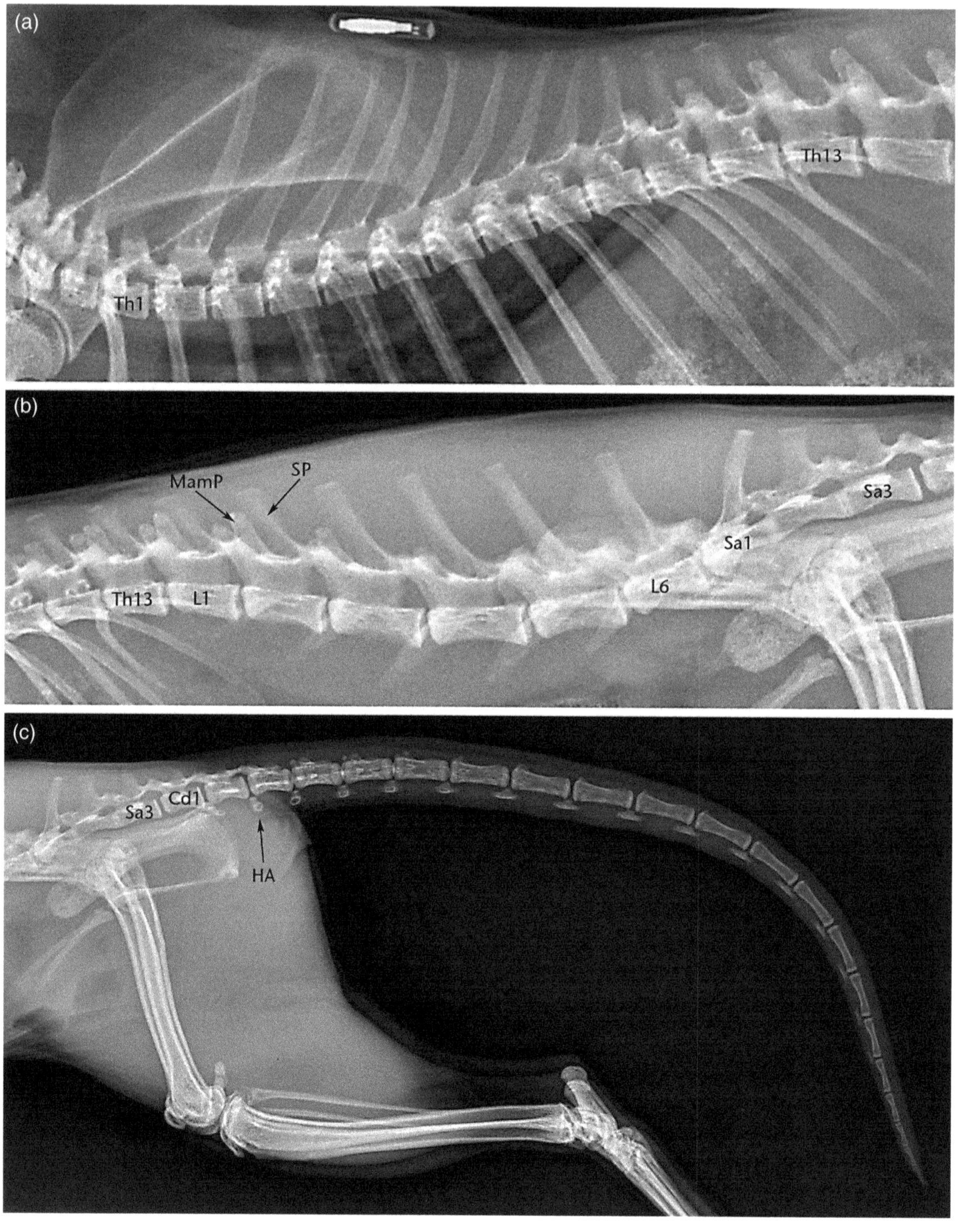

Figure 9.13: Vertebral column, greater bilby (*Macrotis lagotis*), lateral views: (a) thoracic, (b) lumbar, (c) sacral and caudal vertebrae. There are 13 thoracic (Th1, Th13 labelled), 6 lumbar (L1, L6 labelled), 3 sacral (Sa1, Sa3 labelled) and ~20 caudal vertebrae (Cd1 labelled). Well-developed dorsal spinous processes (SP) extend from the 2nd cervical to the 3rd sacral element. They continue caudally on the caudal vertebrae, diminishing gradually in size. There are craniodorsally directed mammillary processes (MamP) between the 11th thoracic and the 1st sacral vertebrae. Cranioventrally directed transverse processes are present on the lumbar vertebrae. Haemal arches (HA) are present ventral to the articulations between the caudal vertebrae. Note: a transponder can be seen in (a).

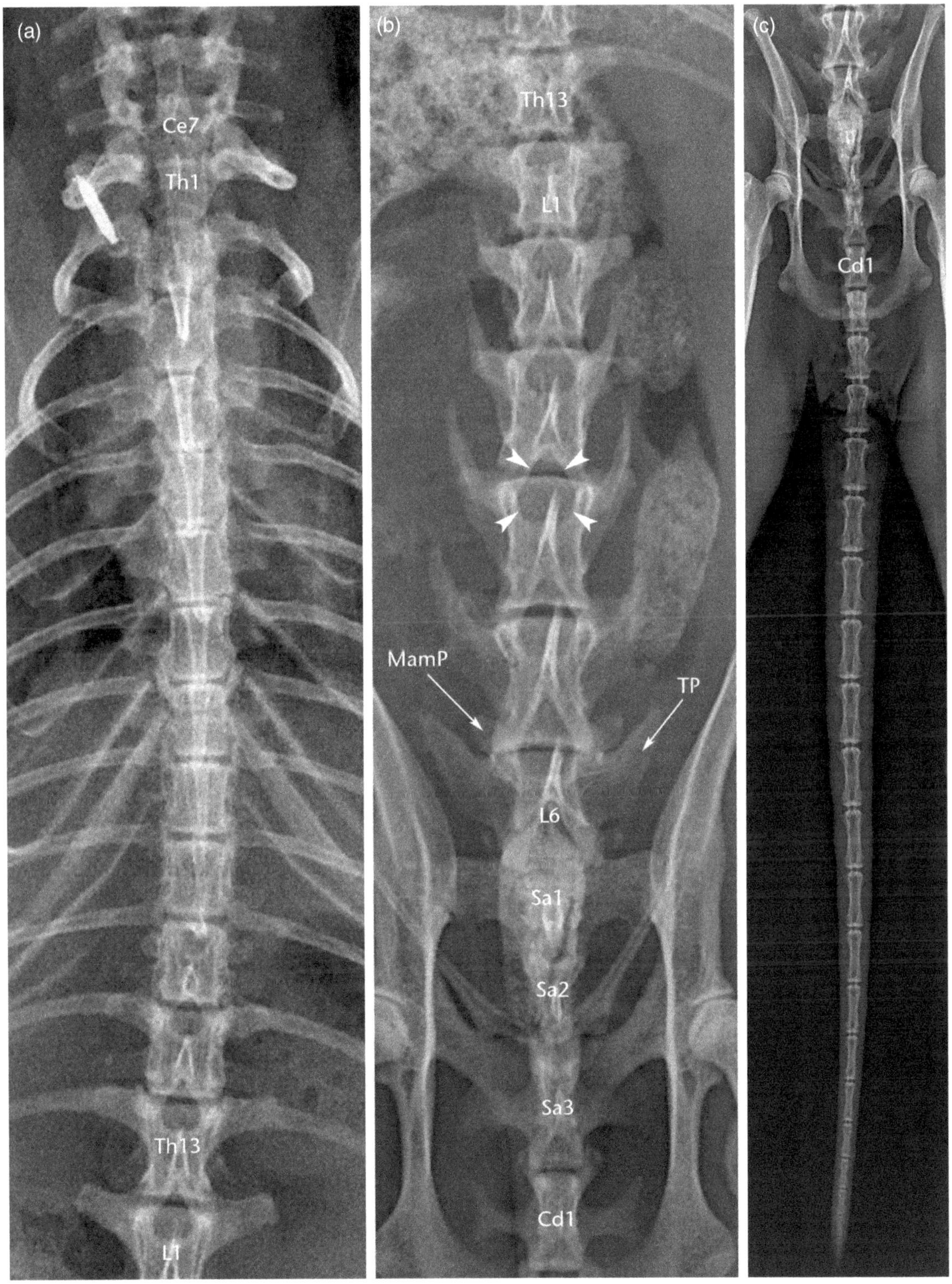

Figure 9.14: Vertebral column, greater bilby (*Macrotis lagotis*), ventrodorsal views: (a) thoracic, (b) lumbar, (c) sacral and caudal vertebrae. The corresponding regions of the vertebral column (Th1–Th13, L1–L6, Sa1–Sa3) and the caudal (Cd1) vertebrae are labelled. Associated features, such as well-defined transverse processes (TP) and mammillary processes (MamP), are visible. Dorsal intervertebral foramina are visible from the 11th thoracic vertebra to the sacrum and are defined at L3–4 by arrowheads. The 7th cervical vertebra is labelled (Ce7). Note: a transponder can be seen in (a).

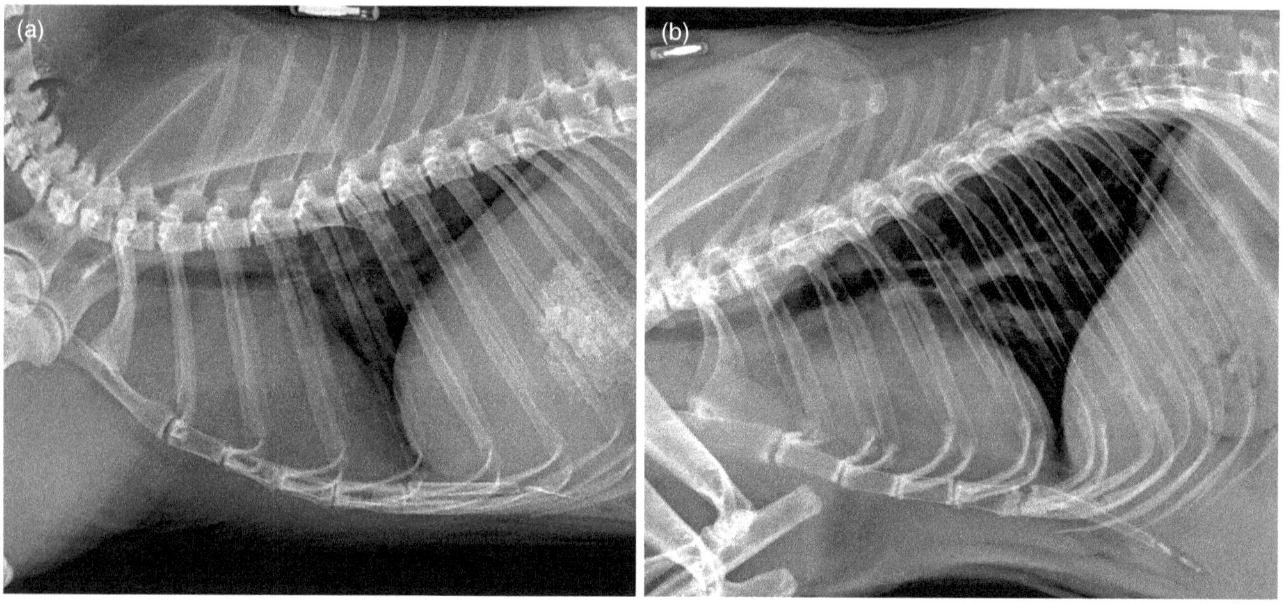

Figure 9.15: Thorax, lateral view: (a) greater bilby (*Macrotis lagotis*), (b) southern long-nosed bandicoot (*Perameles nasuta*). Note: transponders can be seen.

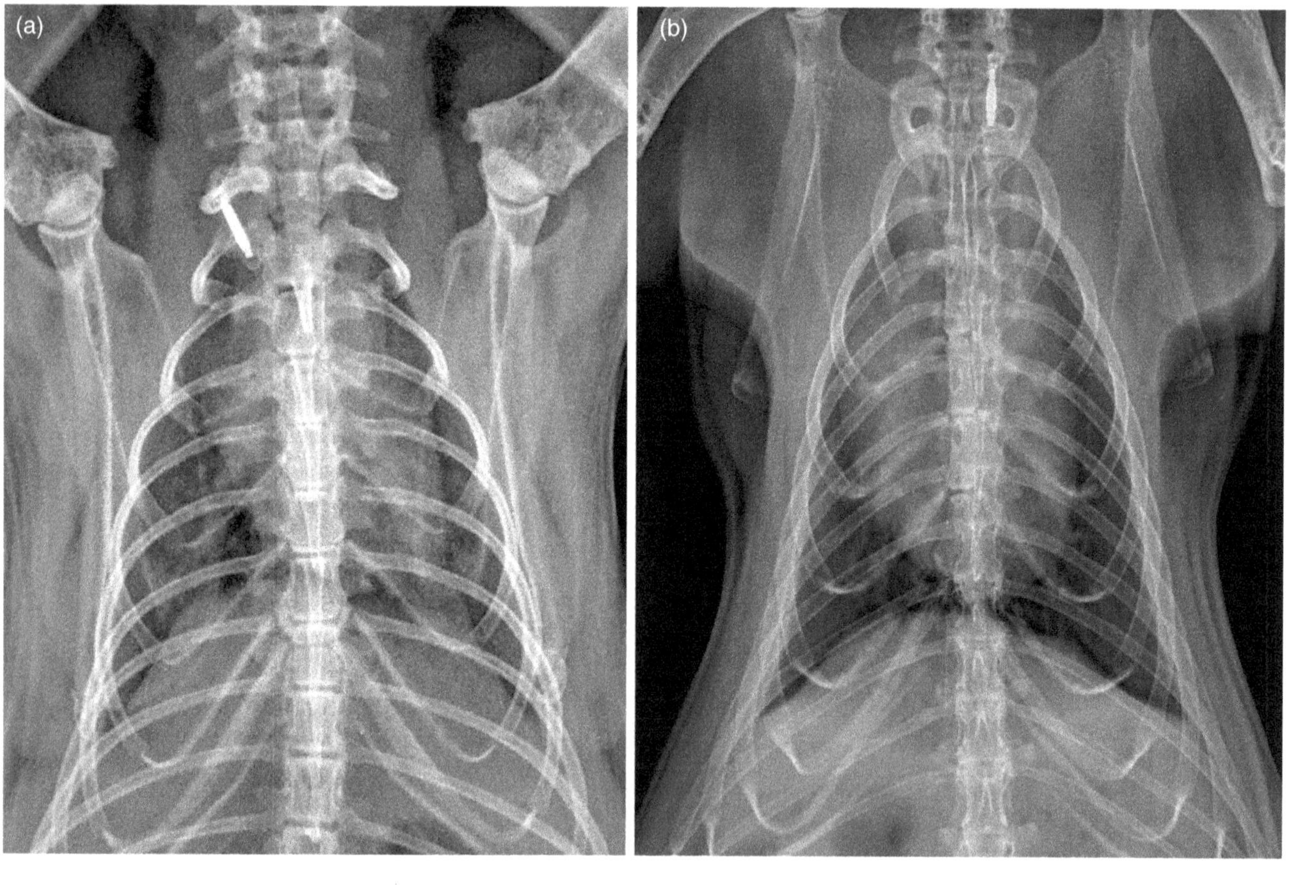

Figure 9.16: Thorax, ventrodorsal view: (a) greater bilby (*Macrotis lagotis*), (b) southern long-nosed bandicoot (*Perameles nasuta*). Note: transponders can be seen.

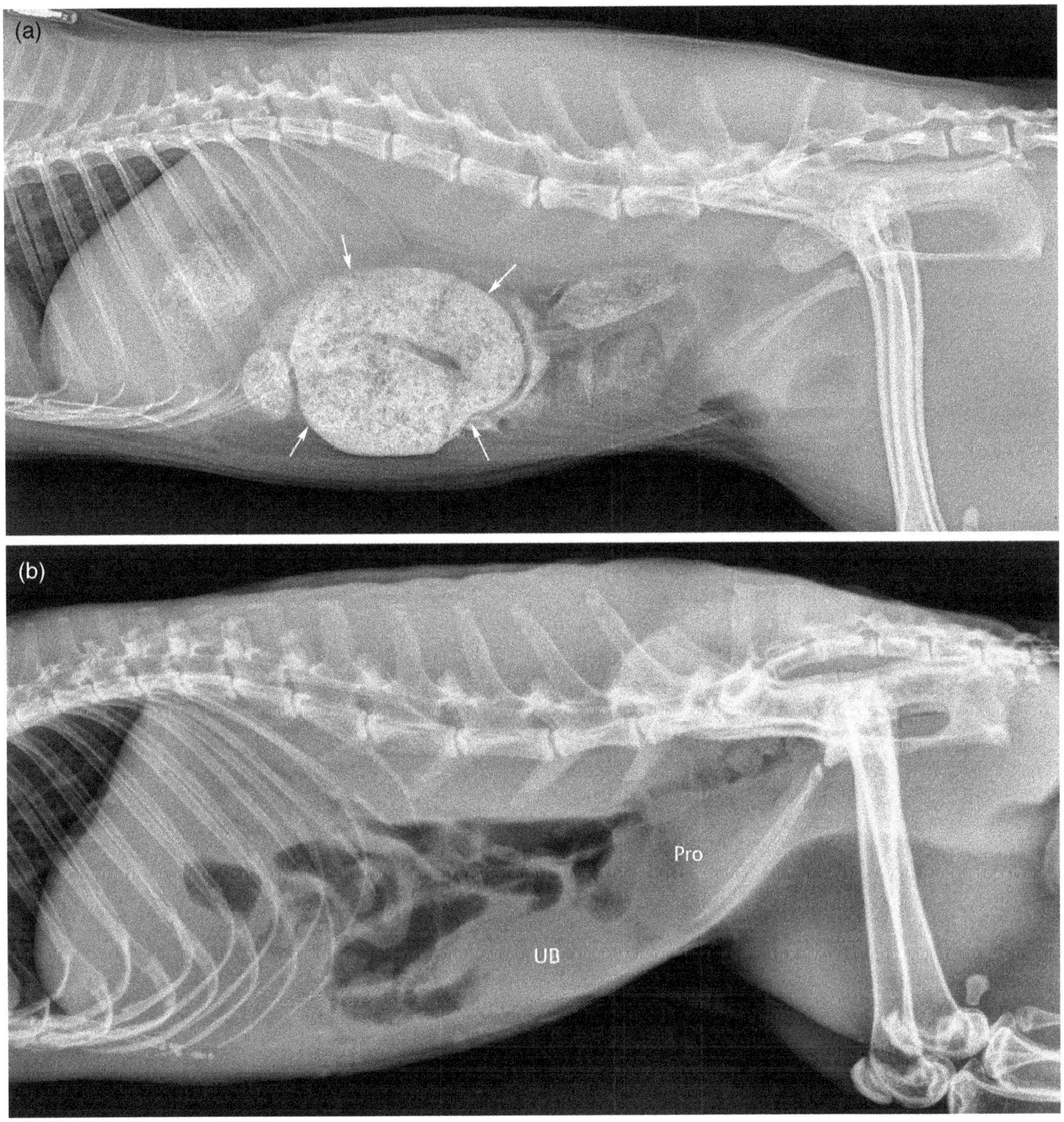

Figure 9.17: Abdomen, lateral view: (a) greater bilby (*Macrotis lagotis*), (b) male southern long-nosed bandicoot (*Perameles nasuta*). (a) Opaque material (arrows) in the intestines represents sand ingested during feeding. (b) The large prostate gland (Pro), which has displaced the urinary bladder (UB) cranially, is evident.

In the northern brown bandicoot and southern brown bandicoot (*Isoodon obesulus*), the spleen is large and triangular, with cranial and lateral processes that cover most of the greater curvature of the stomach, and extends to the right of the duodenum. In the southern long-nosed bandicoot, the spleen is narrow and lacks processes projecting from it, but in other *Perameles* species it has been described as small, flat and triangular. It is likely there is considerable variation in splenic morphology among bandicoot species (Tedman 1990).

Several anatomical features of the bandicoot and bilby urogenital tract are distinctive among the marsupials. The urinary bladder is situated quite cranially (Fig. 9.17b). Therefore, the urethra is long. The remarkable features of the reproductive tract are the expansive vaginal caeca, very long lateral vaginae and the short median

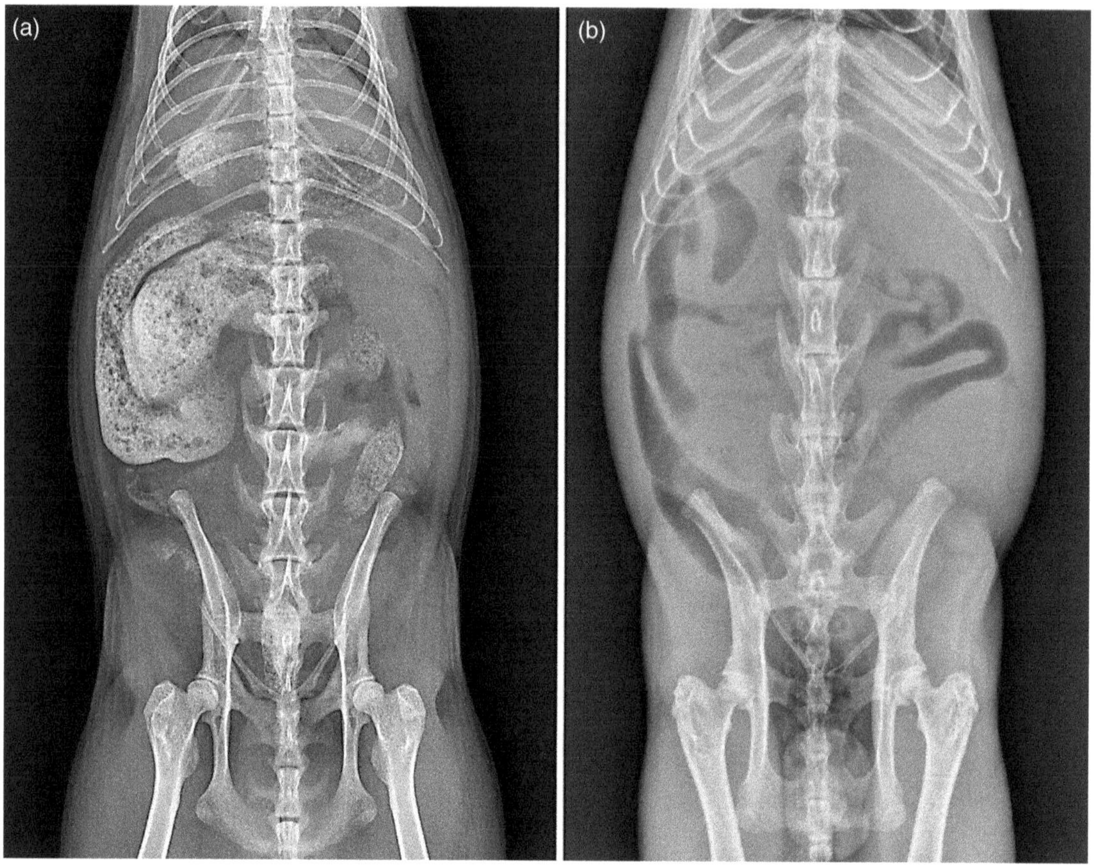

Figure 9.18: Abdomen, ventrodorsal view: (a) greater bilby (*Macrotis lagotis*), (b) southern long-nosed bandicoot (*Perameles nasuta*). Intestinal intraluminal sand, when present, may obscure the viscera.

vaginal cul-de-sac, which terminates a considerable distance from the cranial end of the short urogenital sinus. Dissection of the reproductive tract of a multiparous eastern barred bandicoot revealed that the lateral vaginae open into the vaginal caeca and each of the caeca connects with the median vagina. The uterine horns are separate and each has a single caudal opening into the median vagina, which is divided by a central septum. The anatomy of the urogenital tract has been described in three other bandicoot species (Hill 1899; Tedman 1990) and it differs from that of the bilby and eastern barred bandicoot in two respects. The vaginal caeca of the northern brown bandicoot, southern brown bandicoot and southern long-nosed bandicoot share a median partition wall and so externally have an undivided, bilobed appearance. Additionally, in the northern brown bandicoot the septum that divides the median vagina is reported to be present only in nulliparous individuals (Tedman 1990).

Male bandicoots have a large prostate and bulbourethral glands relative to their body size (Tedman 1990). These organs are usually palpated easily in the abdomen of breeding males and may be visible on radiographs (Fig. 9.17).

References

Cope LA, Henry RW, Reed RB (2001) Macroscopic anatomy of the lower respiratory tract of the gray short-tailed opossum (*Monodelphis domestica*). *Brazilian Journal of Morphological Science* **18**, 111–117.

Freedman L (1967) Skull and tooth variation in the genus *Perameles*. Part I: anatomical features. *Records of the Australian Museum* **27**, 147–165. doi: 10.3853/j.0067-1975.27.1967.441

Gordon G, Hulbert AJ (1989) Peramelidae. In *Fauna of Australia, Volume 1B, Mammalia* (Eds DW Walton and BJ Richardson) pp. 603–624. Australian Government Publishing Service, Canberra.

Hill JP (1899) Contributions to the morphology and development of the female urogenital organs in the Marsupialia 1. On the female urogenital organs in the *Perameles*, with an account of the phenomena of parturition. *Proceedings of the Linnean Society of New South Wales* **24**, 42–82.

Hume ID (1999) *Marsupial Nutrition*. Cambridge University Press, Cambridge.

Johnson KA (1989) Thylacomyidae. In *Fauna of Australia, Volume 1B, Mammalia* (Eds DW Walton and BJ Richardson) pp. 625–635. Australian Government Publishing Service, Canberra.

Jones FW (1924) *The Mammals of South Australia: Part II*. pp. 133–270. Government Printer, Adelaide.

Lynch M (2008) Bandicoots and bilbies. In *Medicine of Australian Mammals*. (Eds L Vogelnest and R Woods) pp. 439–464. CSIRO Publishing, Melbourne.

Marshall LG (1972) Evolution of the peramelid tarsus. *Proceedings of the Royal Society of Victoria* **85**, 51–60.

Owen R (1839) On the osteology of Marsupialia. *The Transactions of the Zoological Society of London* **2**, 379–408. doi:10.1111/j.1469-7998.1839.tb00030.x

Parsons FG (1903) On the anatomy of the pig-footed bandicoot (*Choeropus castanotis*). *Journal of the Linnean Society of London Zoology (Jena, Germany)* **29**, 64–80.

Reese S, Pfuderer UR, Bragulla H, Loeffler K, Budras KD (2001) Topography, structure and function of the patella and the patelloid in marsupials. *Anatomia, Histologia, Embryologia* **30**, 289–294. doi:10.1046/j.1439-0264.2001.00334.x

Runciman SIC, Baudinette RV, Gannon BJ (1992) The anatomy of the adult marsupial heart: an historical review. *Australian Journal of Zoology* **40**, 21–34. doi:10.1071/ZO9920021

Tedman RA (1990) Some observations on the visceral anatomy of the bandicoot *Isoodon macrourus* (Marsupialia: Peramelidae). In *Bandicoots and Bilbies*. (Eds JH Seebeck, PR Brown, RL Wallis and CM Kemper) pp. 107–116. Surrey Beatty and Sons, Sydney.

Tyndale-Biscoe CH (2005) *Life of Marsupials*. CSIRO Publishing, Melbourne.

10

Bats

Bats fall within the order Chiroptera (Greek: *cheir*, 'hand' and *pteron*, 'wing') and two suborders: Megachiroptera and Microchiroptera. There are 79 species of bats recognised in Australia. The megachiropterans are generally large bats, with some weighing up to 1 kg, but others, such as the tube-nosed fruit bats and blossom bats, are small, weighing only 30–65 g. The suborder Megachiroptera has a single family: the Pteropodidae. There are 5 genera and 13 species within this family in Australia, some of which also occur in New Guinea and surrounding islands. These include the blossom bats, tube-nosed bats, fruit-bats and flying-foxes. The terms 'pteropodids' or 'megabats' are often used to refer to these species. The Microchiropterans are small bats, varying in size from 3 to 120 g. In Australia, there are 6 families, 20 genera and 66 species (27 of which also occur in New Guinea or surrounding islands) within the suborder Microchiroptera. These include the ghost bats (Megadermatidae), horseshoe bats (Rhinolophidae), leaf-nosed bats (Hipposideridae), sheath-tailed bats (Emballonuridae), free-tailed bats (Molossidae) and ordinary bats (Vespertilionidae) (Hall and Richards 2000). These bats are often referred to as 'microbats'.

Bats are eutherian mammals and together with rodents (also eutherians) make up a significant proportion of Australia's native mammalian fauna.

The basic anatomy and external features of bats are similar to those of most other eutherian mammals, with one notable exception: they are the only mammals ever to have evolved true powered flight, having wings composed of a skin membrane stretched between elongated forelimb bones. Bats are well adapted for flight. Some of these adaptations are analogous to those seen in the other two groups of flying vertebrates: pterosaurs and birds. Adaptations related to the generation of the power flight stroke include an expanded bony sternum for anchoring flight muscles, a modified pectoral girdle, an enlarged deltopectoral crest on the humerus for the insertion of flight muscles and a shoulder articulation that limits forelimb movement to activities compatible with the flight stroke. None of these features is seen in gliding mammals. Unlike birds and pterosaurs, in which the wing is supported by the bones of the forelimb and one digit, a bat's wing membrane, or patagium, is supported by the forelimb and four highly elongated digits (hence the name Chiroptera, or 'hand-wing') (Fig. 10.1). In addition, the bones of bats are lightweight, slender and solid, the humerus and radius are large and the ulna and fibula are reduced (Padian 1985; Hutchinson 1996).

All bats have three flight membranes that form the main body of the wing. The membranes are essentially modified skin stretched around a framework of bones. They are the propatagium, the plagiopatagium and the dactylopatagium. In most Microchiropterans, the tail is connected to the legs by a fourth, broad membrane (the uropatagium), which is used during abrupt changes in flight direction and to facilitate capture of prey in some species (Hutchinson 1996; Hall and Richards 2000). The uropatagium in pteropodids is represented by

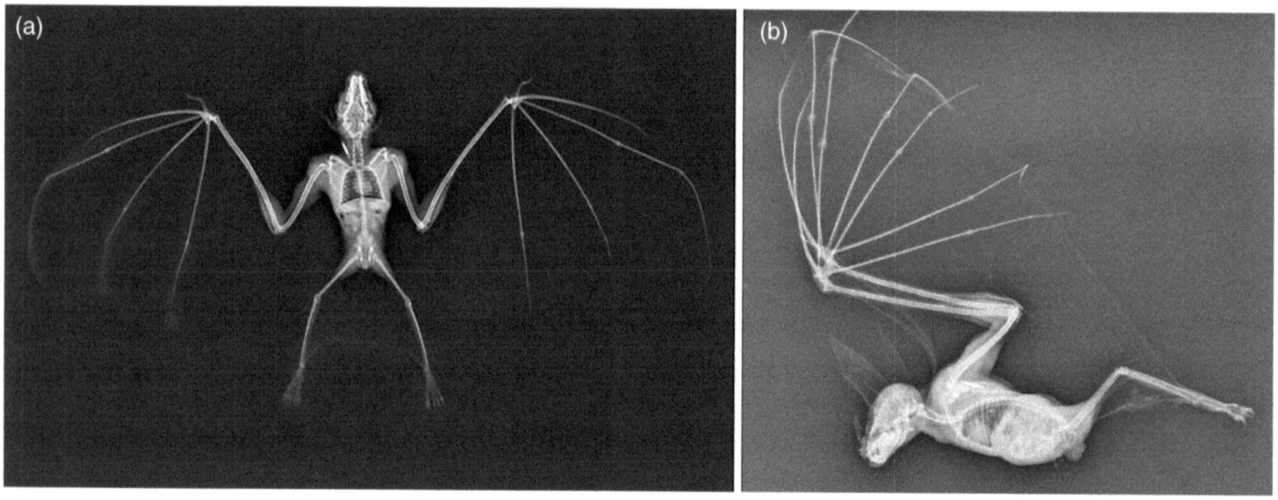

Figure 10.1: Whole body: (a) grey-headed flying-fox (*Pteropus poliocephalus*), ventrodorsal view, (b) ghost bat (*Macroderma gigas*), lateral view. Note: transponders can be seen.

a small flap of skin on the medial aspect of the hindlimbs. The membranes are very elastic. There are numerous parallel bands of elastin running through the plagiopatagium perpendicular to the humerus and radius, making the wing pliable and resilient (see Fig. 10.5) (Olsson and Woods 2008).

Pteropodids navigate primarily by sight and have large, forward-facing eyes that are well adapted for nocturnal vision while retaining a high degree of visual acuity even in broad daylight. All pteropodids feed on plant material. In contrast, Microchiropterans navigate by echolocation, having small eyes and poor vision (Hall and Richards 2000), and are primarily insectivorous or carnivorous. Their imaging system is made up of a transmitter (larynx) and a receiver (ears and associated neural systems) (Heard 2003). Microchiropterans have very acute hearing and most species have a well developed tragus (a prominent flap of skin-covered cartilage at the entrance to the ear) to assist in detecting sound outside the normal mammalian acoustic range. These bats often have large, and sometimes elaborate, ear pinnae. The nose varies from relatively unmodified to complex with elaborate dermal appendages. Thus, certain species are known as 'leaf-nosed bats', 'bulldog bats', 'flower-faced bats' or 'horseshoe bats'. These structures appear to function in directing and controlling sound pulses for navigation and hunting by echolocation (Hutchinson 1996; Olsson and Woods 2008).

Skeletal morphology shows considerable variation among the families of bats. This chapter will focus on general morphology, primarily using pteropodids and the largest Microchiropteran, the ghost bat (*Macroderma gigas*), as examples.

Skeletal system

Appendicular skeleton

Pectoral girdle and forelimb

The skeletal morphology of the pectoral girdle and forelimb is significantly modified to facilitate powered flight and form the structure of the wing. However, all the elements of the mammalian forelimb are present. With the exception of the first digit, the bones of the forelimb are elongated and slender and most are hollow. The morphology of the pectoral girdle and forelimb shows considerable variation among the families of bats, particularly those within the Microchiroptera.

The scapula is more cranially positioned on the thorax than in other vertebrates. It is large and oval and gives origin to some of the largest flight muscles (Parker and Haswell 1921). The position of the scapular spine and the proportions of the

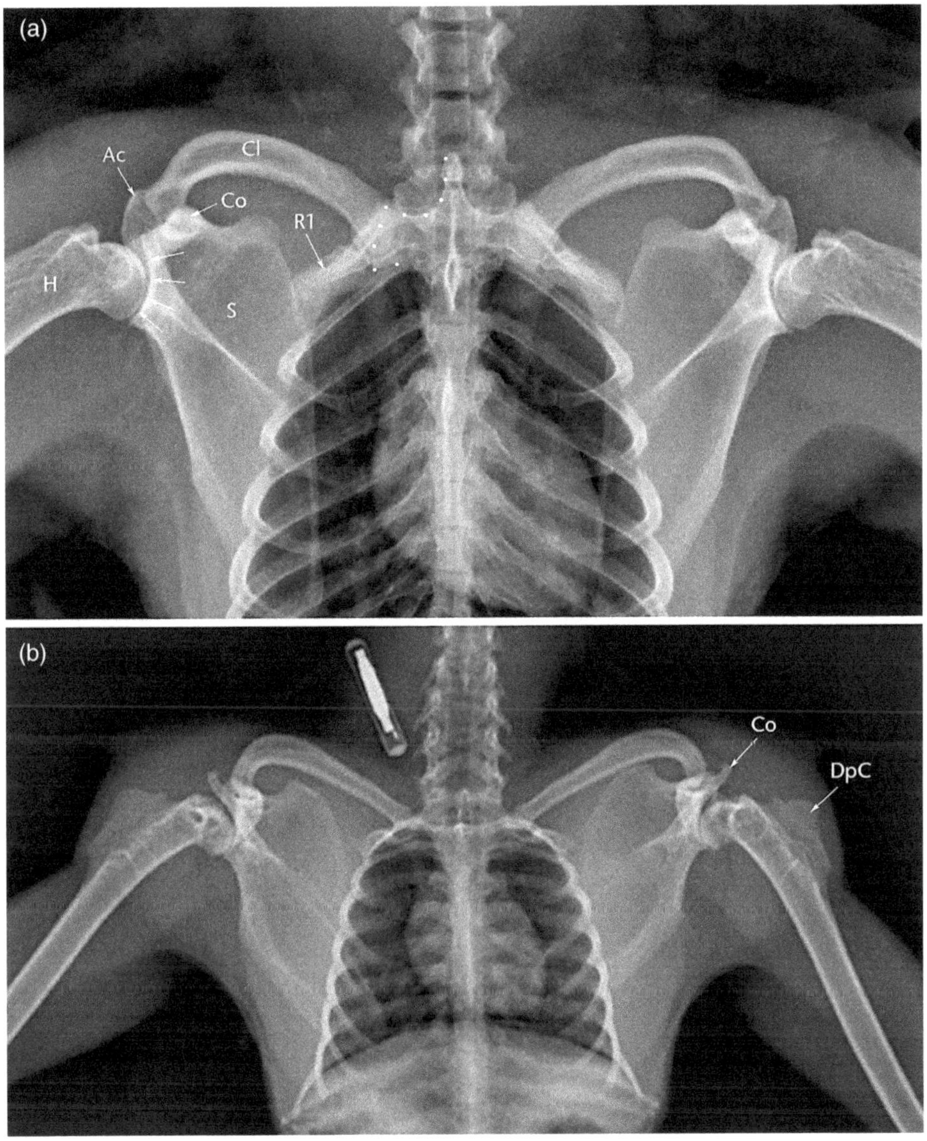

Figure 10.2: Pectoral girdle, ventrodorsal view: (a) grey-headed flying-fox (*Pteropus poliocephalus*), (b) ghost bat (*Macroderma gigas*). The clavicle (Cl) is a thick curvilinear bone that articulates closely with the acromion (Ac) and the 1st sternal element (dotted line). The 1st pair of ribs (R1) are modified, being thicker and more robust than subsequent ribs. The humerus (H) and scapula (S) articulate with the glenoid cavity (arrows). The coracoid (Co) and the deltopectoral crest (DpC) are more developed in the ghost bat than in the grey-headed flying-fox. Note: a transponder can be seen in (b).

supraspinous fossa and infraspinous fossa vary significantly among species. In most species, the infraspinous fossa is large and in some, it has numerous ridges for the insertion of muscles (Parker 1867). However, in *Pteropus* species, the scapula is narrow and the supraspinous fossa relatively large. The supraspinous fossa lies in the same plane as the infraspinous fossa, which has only a few ridges for muscular insertions. The scapular spine has a well-developed acromion (Figs 10.2 and 10.3). The length, shape and orientation of the coracoid process vary significantly among species and, in some, may be bifurcate (Parker 1867; Parker and Haswell 1921). The glenoid cavity is usually deep. Paired clavicles articulate with their ipsilateral scapular acromion abaxially and with the first

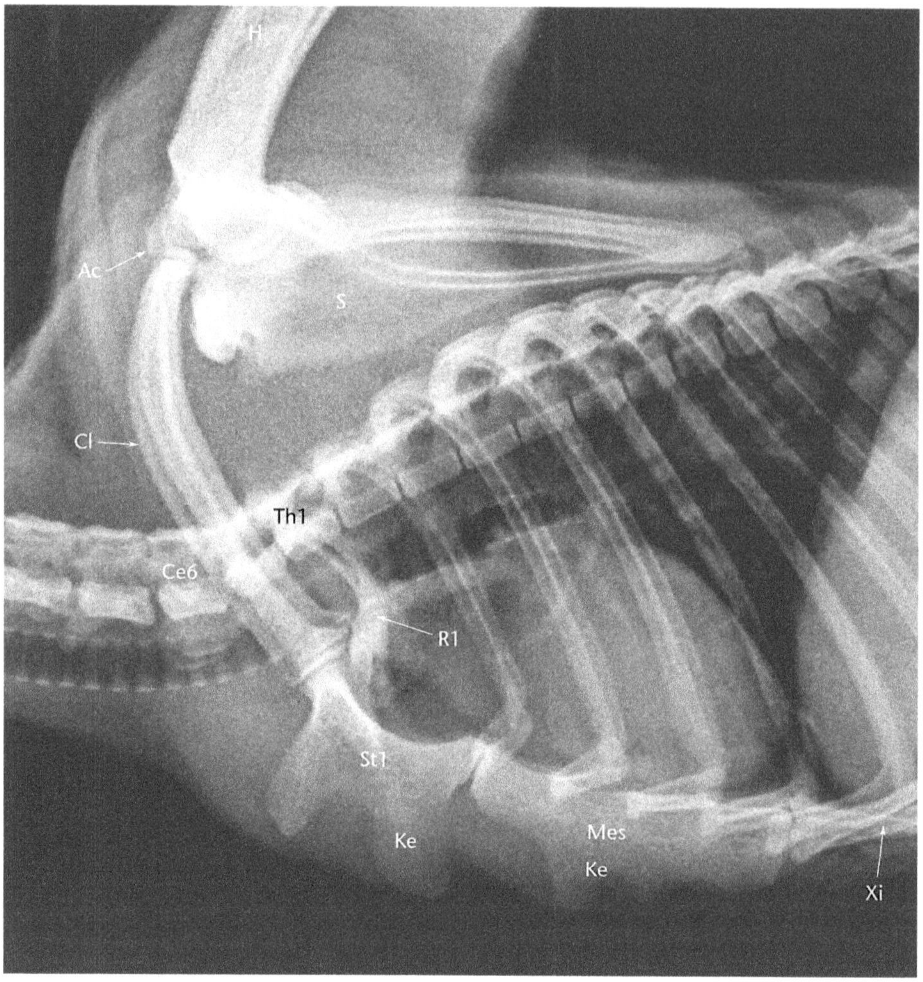

Figure 10.3: Pectoral girdle, grey-headed flying-fox (*Pteropus poliocephalus*), lateral view. The congruent articulations of the clavicle (Cl) with the acromion (Ac) and 1st sternal element (St1) can be seen in this view. Other adaptations to flight, such as the development of a keel (Ke) on the 1st sternal element, the fused subsequent three sternebrae (mesosternum, Mes), also with a keel, and the robust 1st pair of ribs (R1), are visible. The scapulae (S) are dorsally located and the humeri (H) are directed dorsally in this view. The 6th cervical (Ce6) and 1st thoracic (Th1) vertebrae and the xiphisternum (Xi) are labelled.

sternal element axially (Hall and Richards 2000). The clavicles are long and curved (Parker and Haswell 1921) (Figs 10.2 and 10.3). An epicoracoid between the clavicle and first rib may be present in some species (Parker 1867).

In pteropodids, the bony structure of the shoulder joint is simple but some features are specialised. One is the lateral displacement of the primitive insertion of the omohyoid muscle from the clavicle to the scapula and the other is an elaborate system of muscular slips that tie the tendon of the occipitopollicalis muscle to the skin over the thoracic and lower cervical regions (Nelson 1989).

The wing is essentially the mammalian antebrachium and manus highly modified for flight. The humerus is short relative to the radius and is slightly sigmoid shaped when viewed in the lateral projection. It has an enlarged deltopectoral crest for the insertion of flight muscles (Fig. 10.2). The radius is the main antebrachial bone and may be twice the length of the humerus. The ulna is fused to the radius for most of its length and is reduced to a

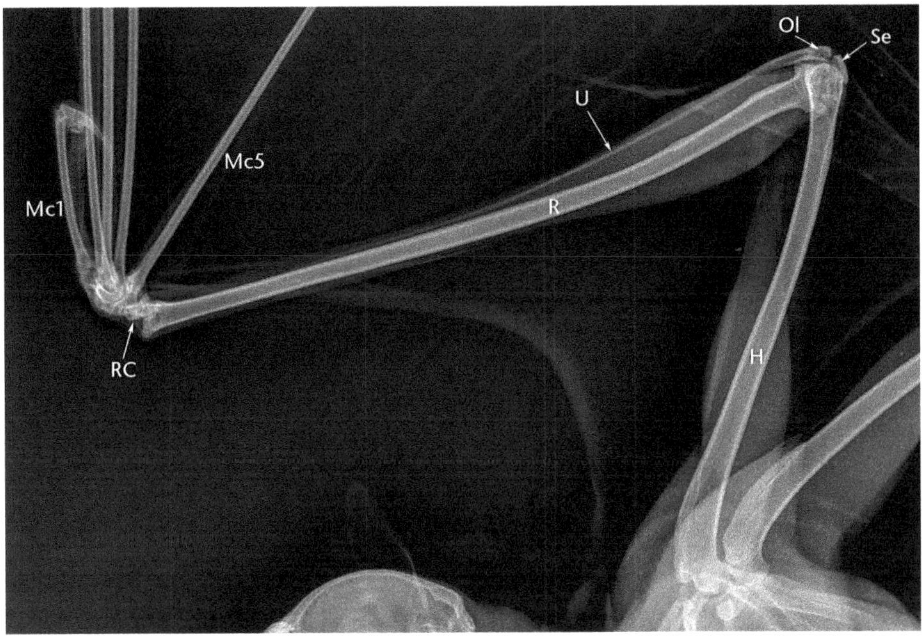

Figure 10.4: Brachium and antebrachium, grey-headed flying-fox (*Pteropus poliocephalus*), lateral view. The humerus (H) and radius (R) are long slender bones. They are described as being hollow and radiographically have a lucent medulla. The ulna (U) is filamentous in appearance and tapers distally to invisibility. At its proximal end is the olecranon (Ol), proximal to which is a relatively long, flat sesamoid bone (Se). In the carpus, a large radial carpal (RC) bone can be seen. The remaining carpal bones are indistinguishable. The metacarpal bones (Mc1, Mc5 labelled) fan out distally from the carpus.

fine, short bone, which in some species terminates within the proximal half of the radius (Figs 10.4 and 10.5). There is a large ulnar sesamoid bone within the tendon of the triceps muscle near the olecranon process of the ulna (Parker and Haswell 1921) (Figs 10.4 and 10.5).

The carpus is flexible, facilitating extreme extension during flight and tight flexion during rest. In the carpus, the radial and intermediate carpal bones are united and the accessory carpal bone is small (Parker and Haswell 1921) (Fig. 10.5).

The manus is made up of five digits, the longest segments of which are the metacarpal bones. Digit 1 is present in all species and is short relative to the other digits, although this difference in length is less extreme in pteropodids (Fig. 10.5). It is the only digit capable of free movement and it controls the shape of the propatagium during flight. In pteropodids, digit 1 is also an essential tool for manipulating branches and food items, hanging for defaecation and urination and fighting. In some species, it is used to assist climbing or walking on the ground. The remaining digits (2–5) are elongated, rigid spokes supporting the wing membranes. Each digit comprises a metacarpal bone and two phalanges, except in pteropodids where digit 2 has three phalanges, the distal phalanx being very small or rudimentary (Nelson 1989). The distal phalanx of digit 1 has a prominent claw (Hill and Smith 1985; Hall and Richards 2000; Heard 2003) (Fig. 10.5). With the exception of the bare-backed fruit-bat (*Dobsonia magna*), a distinguishing feature of pteropodids is the presence of a small claw at the end of digit 2 (Olsson and Woods 2008).

The wing membranes are composed of a double layer of epidermis between which is a thin layer of connective tissue. The connective tissue contains blood vessels, nerves, slips of skeletal muscle and a network of elastin fibres (Hall and Richards 2000)

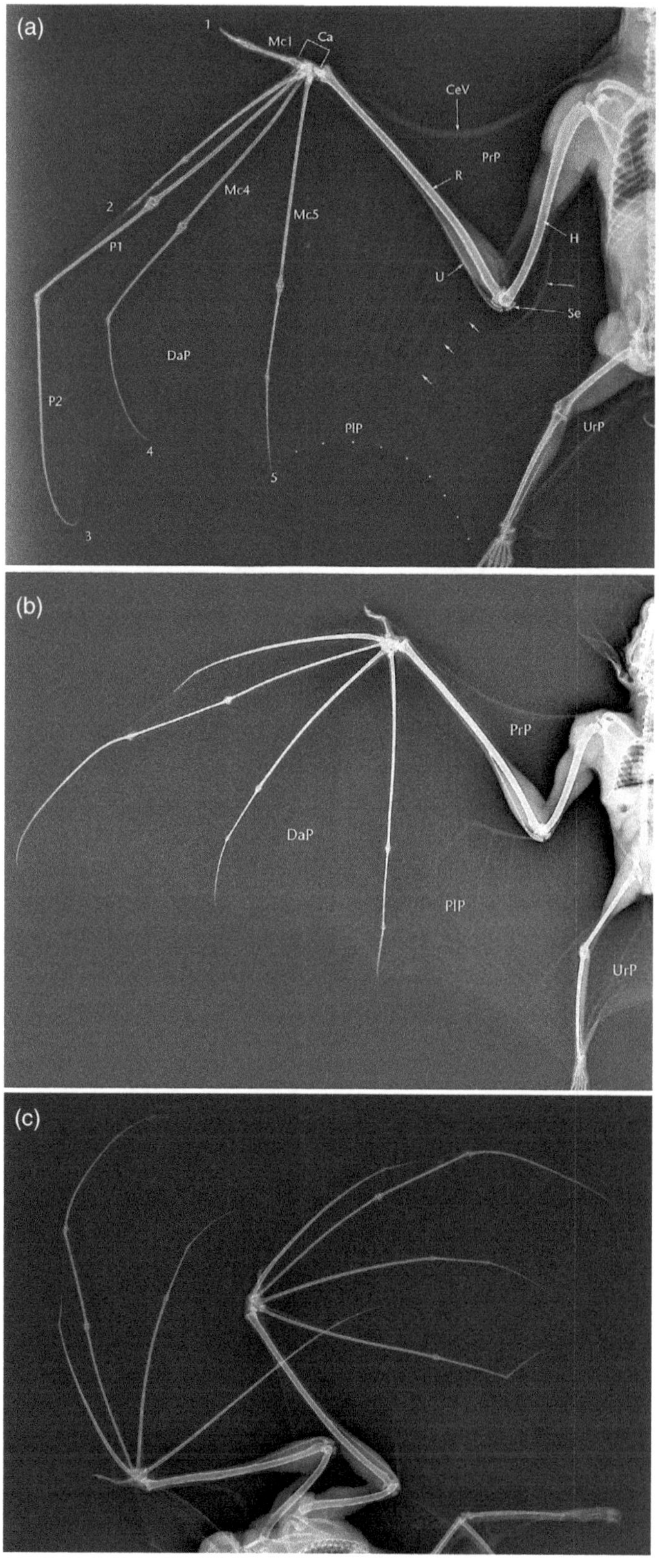
(a)
1
Mc1
Ca
CeV
PrP
R
2
Mc4
Mc5
P1
H
U
Se
DaP
P2
PlP
UrP
4
5
3
(b)
PrP
DaP
PlP
UrP
(c)

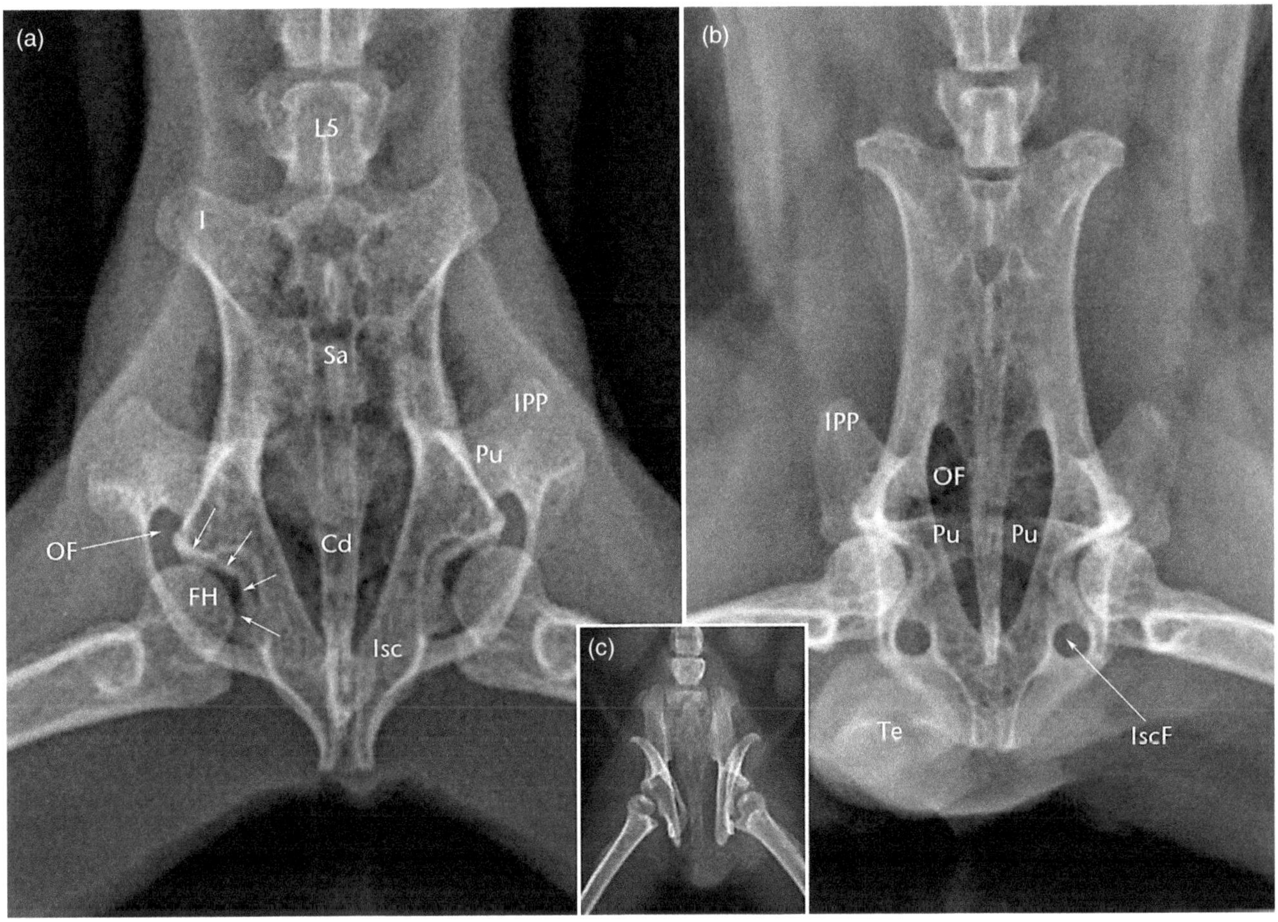

Figure 10.6: Pelvis, ventrodorsal view: (a) female grey-headed flying-fox (*Pteropus poliocephalus*), (b) male grey-headed flying-fox, (c) male ghost bat (*Macroderma gigas*). In the female flying-fox, the pubic bones (Pu) are flared abaxially, whereas in the male they are directed axially. The ilium (I) and sacrum (Sa) are fused and there is no sacroiliac joint. The caudal vertebrae (Cd) are also fused and there is no tail. The 5th lumbar vertebra (L5) is labelled. The shape of the obturator foramen (OF) of the sexes is different in this view, having a more open shape in the female and a narrower profile in the male. A cranially directed protuberance of the pubic bone, the iliopectineal process (IPP), is visible in both images. The ischia (Isc) are fused caudally and appear as a pointed caudal pelvic extremity. Ischial foraminae (IscF) are present in the ischium just caudal to the acetabulum in males. The acetabulum (arrows) houses the femoral head (FH) and is directed caudolaterally, a remodelling change associated with dorsocaudal rotation of the femurs. The testes (Te) are labelled in the male.

Figure 10.5: Wing: (a) grey-headed flying-fox (*Pteropus poliocephalus*), ventrodorsal view, (b) ghost bat (*Macroderma gigas*), ventrodorsal view, (c) ghost bat, lateral view. The humerus (H), radius (R), ulna (U) and its sesamoid bone (Se) are visible proximal to the carpus (Ca). The carpus articulates distally with five metacarpal bones (Mc1, Mc4, Mc5 labelled). There are five digits (1, 2, 3, 4, 5). Digit 1 has a small claw. The others digits have two phalanges (P1, P2) that provide osseous scaffolding for the flight membrane, or patagium. In the grey-headed flying-fox, digit 2 has three phalanges. The patagium is divided into the propatagium (PrP), dactylopatagium (DaP), plagiopatagium (PlP) and the uropatagium (UrP). The distal edge of the plagiopatagium between digit 5 and the ipsilateral pes is marked with a dotted line. Within the patagium, linear opacities (arrows) indicate the presence of elastin bands, which also support the patagium. The cephalic vein (CeV) runs along the leading edge of the propatagium. Note: a transponder can be seen in (c).

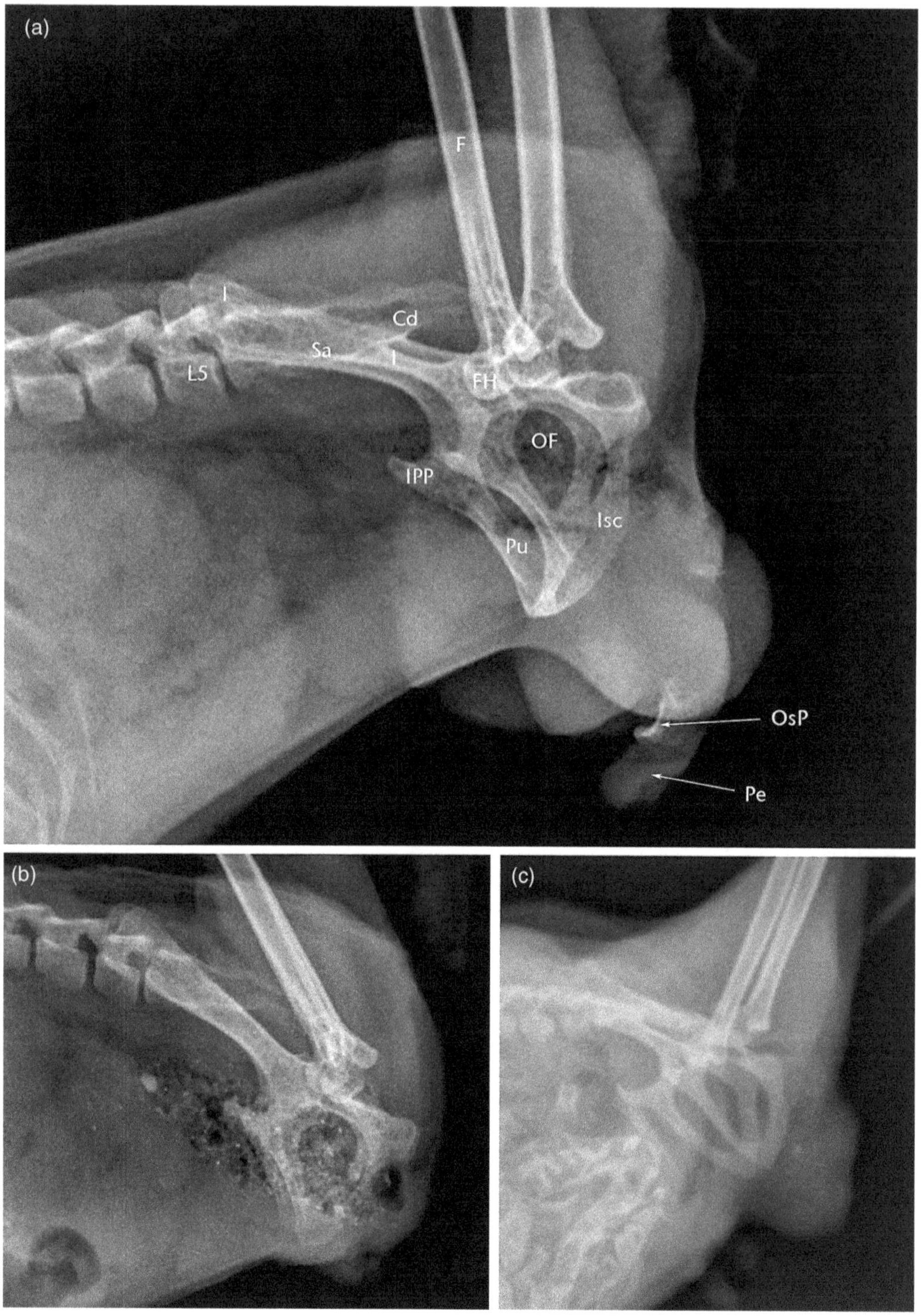

Figure 10.7: Pelvis, lateral view: (a) male grey-headed flying-fox (*Pteropus poliocephalus*), (b) female grey-headed flying-fox, (c) gravid female ghost bat (*Macroderma gigas*). The sacrum (Sa) and caudal vertebrae (Cd) are fused and appear as a continuous bone in the flying-foxes. They are superimposed on the ilia (I). The ovoid obturator foraminae (OF) are large and surrounded by the relatively slender ischia (Isc) and pubic bones (Pu). The cranially directed iliopectineal processes (IPP) are a feature of the cranial margins of the pubis. The femoral heads (FH) and dorsally directed femurs (F) are a departure from the usual mammalian structure. The 5th lumbar vertebra (L5) is labelled. The *os penis* (OsP) and penis (Pe) are visible in the male.

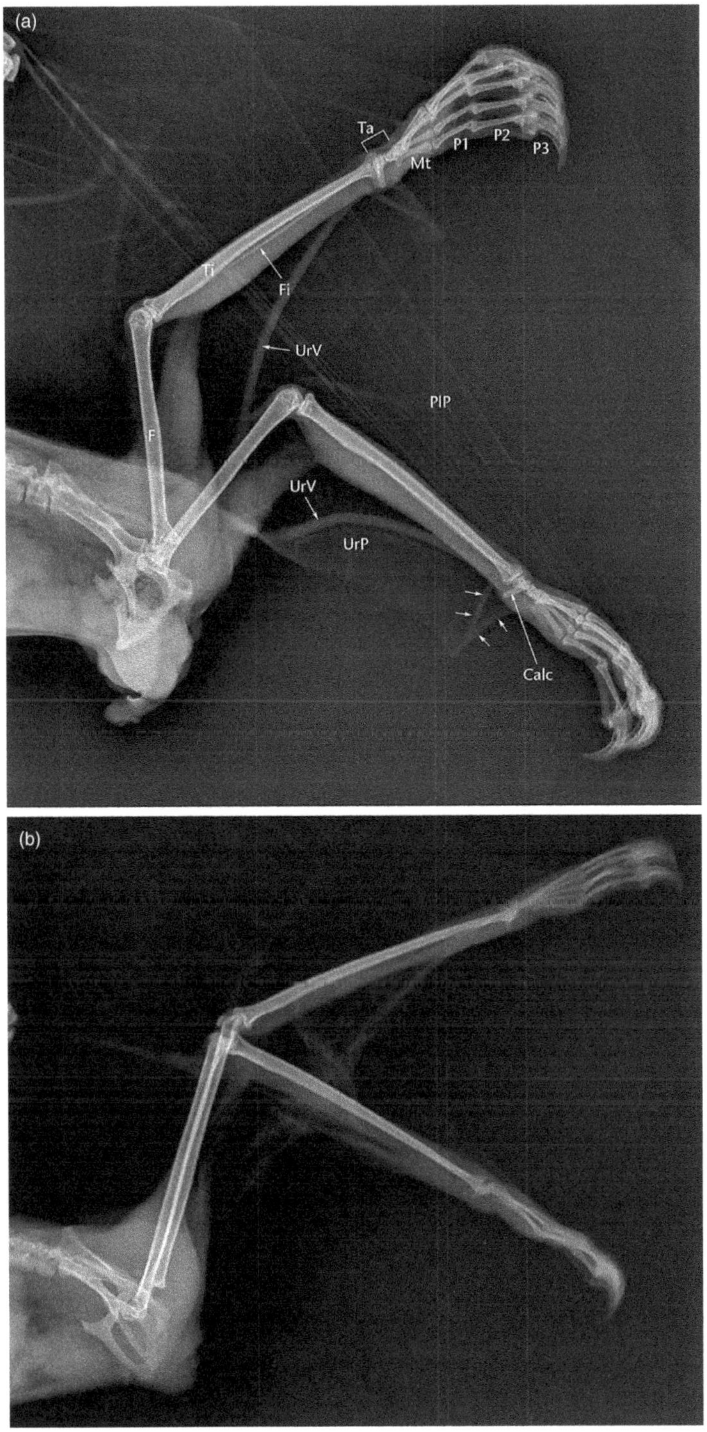

Figure 10.8: Hindlimbs, lateral view: (a) male grey-headed flying-fox (*Pteropus poliocephalus*), (b) female ghost bat (*Macroderma gigas*). The long bones, femur (F), tibia (Ti) and fibula (Fi), are directed dorsocaudally. The fibula is thin. The tarsus (Ta) carries a plantar extension of the calcaneus, the *calcar* (Calc), which supports the uropatagium (UrP). Additional support in the form of fibrocartilaginous bands (arrows) anchors the uropatagium to the *calcar*. The extension of the patagium from the wing to the foot, the plagiopatagium (PlP), is visible. Distal to the tarsus are five metatarsal (Mt) bones and five digits. Digit 1 has two phalanges and the remaining four digits have three (P1, P2, P3). Uropatagial blood vessels (UrV) are visible within the uropatagium.

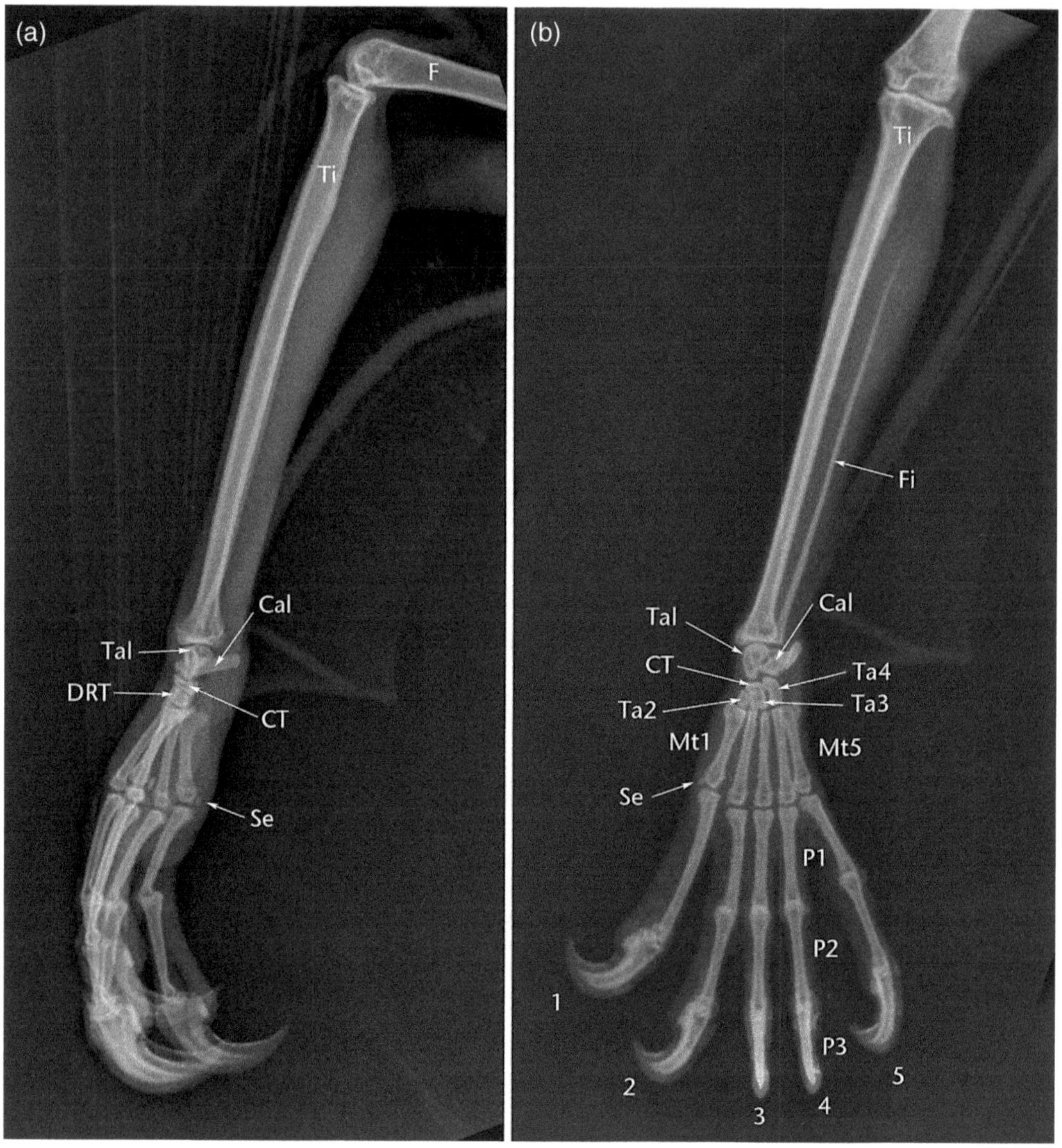

Figure 10.9: Crus, tarsus and pes, grey-headed flying-fox (*Pteropus poliocephalus*): (a) mediolateral view, (b) craniocaudal and dorsoplantar view. In the tarsus, the proximal row of tarsal bones comprises the talus (Tal) and calcaneus (Cal). The plantar extension of the calcaneus, which is modified to support the uropatagium, is called the *calcar*. There is a central tarsal bone (CT) between the talus and the 2nd tarsal bone (Ta2). In the distal row of tarsal bones (DRT), there is a 3rd (Ta3) and 4th (Ta4) tarsal bone in addition to the 2nd tarsal bone. The distal row of tarsal bones articulates with five metatarsal bones (Mt1, Mt5 labelled), which in turn articulate with the 1st phalanx (P1) of their respective digits (1, 2, 3, 4, 5). There is a small sesamoid bone (Se) at each metatarsophalangeal articulation. Digit 1 (1) has two phalanges and the remaining digits have three (P1, P2, P3). The terminal phalanx is a claw. The femur (F), tibia (Ti) and fibula (Fi) are labelled.

(Fig. 10.5). The three flight membranes forming the main body of the wing (the propatagium, plagiopatagium and dactylopatagium) and the uropatagium, which is poorly developed in pteropodids can be seen radiographically (Fig. 10.5).

Pelvis and hindlimb

There is considerable variation in the position, proportion and structure of the pelvis and hindlimbs among bats. This primarily relates to differences in feeding strategies, roosting habits and

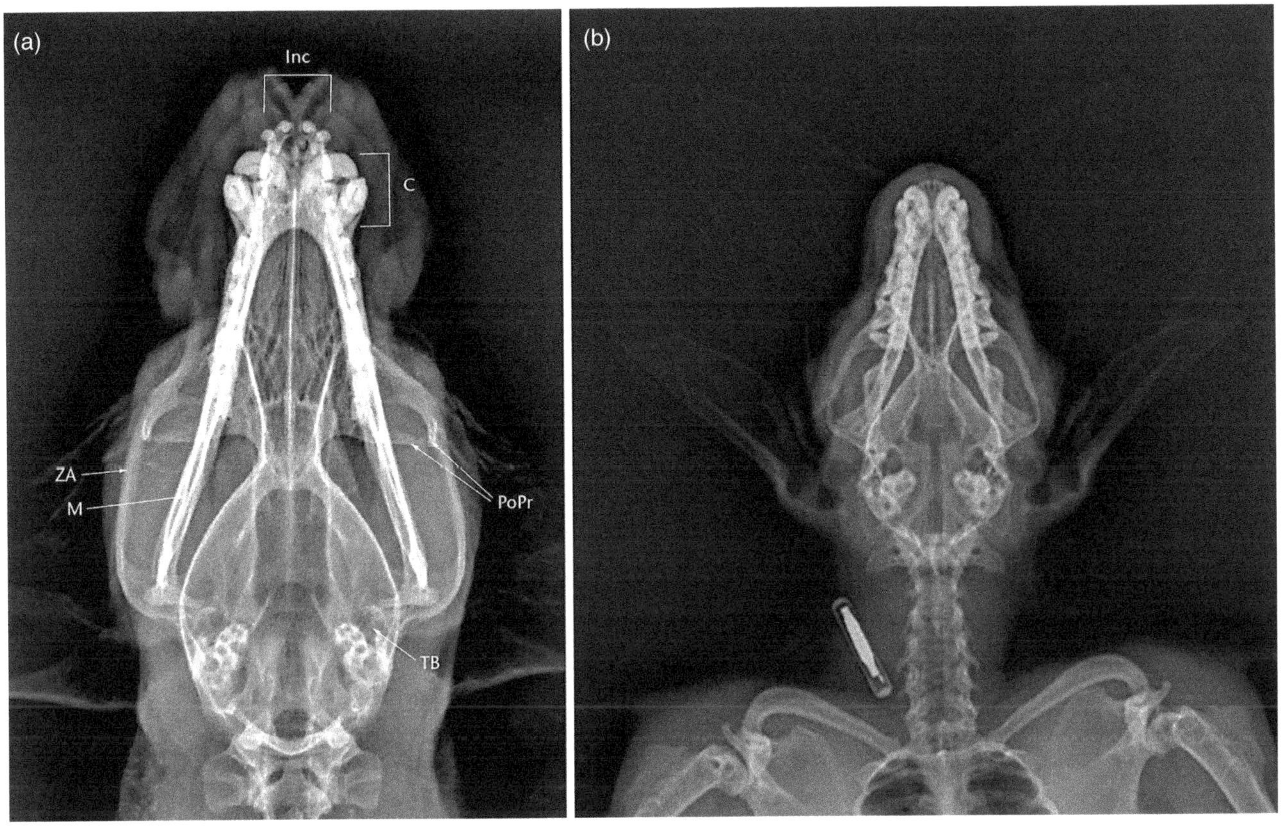

Figure 10.10: Skull, ventrodorsal view: (a) grey-headed flying-fox (*Pteropus poliocephalus*), (b) ghost bat (*Macroderma gigas*). The postorbital process (PoPr), a feature of the flying-fox, is not present in the ghost bat. The zygomatic arch (ZA) and mandible (M) are long and slender in the flying-fox. The tympanic bullae (TB) are relatively small in both species. The incisor (Inc) and canine (C) teeth appear well developed. Note: a transponder can be seen in (b).

the degree and mode of terrestrial locomotion. The pelvis of pteropodids also differs between males and females, the former being a closed 'O' shape and the latter a more open 'V' shape (Fig. 10.6) (Hall and Richards 2000). The pelvis is relatively small and the pubic symphysis in some species may be incomplete (Parker and Haswell 1921). In Microchiropterans, the sacrum and ilium are fused to the level of the acetabulum. There is no mobility at the sacroiliac joint. In pteropodids, the sacrum and ilium are entirely fused (Fig. 10.6) (Heard 2003). In most species, the hindlimbs are thin and lack significant muscle bulk. In all species, they are rotated through 180° at the coxofemoral joint so the hindlimbs can be directed cranially towards the ventrum. This allows bats to face outwards when hanging upside-down (Hall and Richards 2000). The stifles are projected upward during terrestrial locomotion. The femur is similar in length to, or shorter than, the tibia and is a straight, slender bone. The patella is absent in all species. The crus is composed almost entirely of the tibia, the fibula being vestigial and fused to the tibia in most species (Figs 10.8 and 10.9).

Most Microchiropterans, but none of the pteropodids, have a uropatagium, a membrane that extends between the hindlimbs and partially or totally covers the tail (Jackson 2003). A unique slender rod of bone or cartilage, the *calcar*, protrudes from the medial aspect of the tarsus (Figs 10.8 and 10.9). It has a ligamentous attachment to the *tuber calcanei* of the calcaneus and supports the uropatagium (Parker and Haswell 1921).

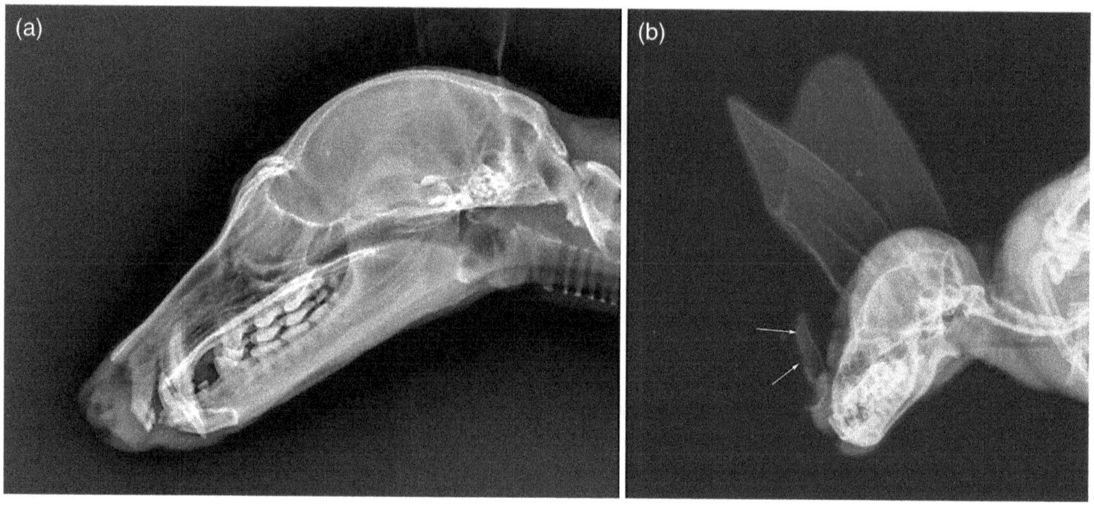

Figure 10.11: Skull, lateral view: (a) grey-headed flying-fox (*Pteropus poliocephalus*), (b) ghost bat (*Macroderma gigas*). The grey-headed flying-fox navigates by sight, so its ears and face are not modified for navigation. The ghost bat has large ears and a modified nasal skin fold (arrows), both of which are used in echolocation.

The digits of the pes have strong, laterally compressed claws to assist with suspension and grooming. Digit 1 has two phalanges and the remaining four have three (Figs 10.8 and 10.9). Pteropodids possess a locking mechanism for the claws that involves a tendon passing through a sheath with

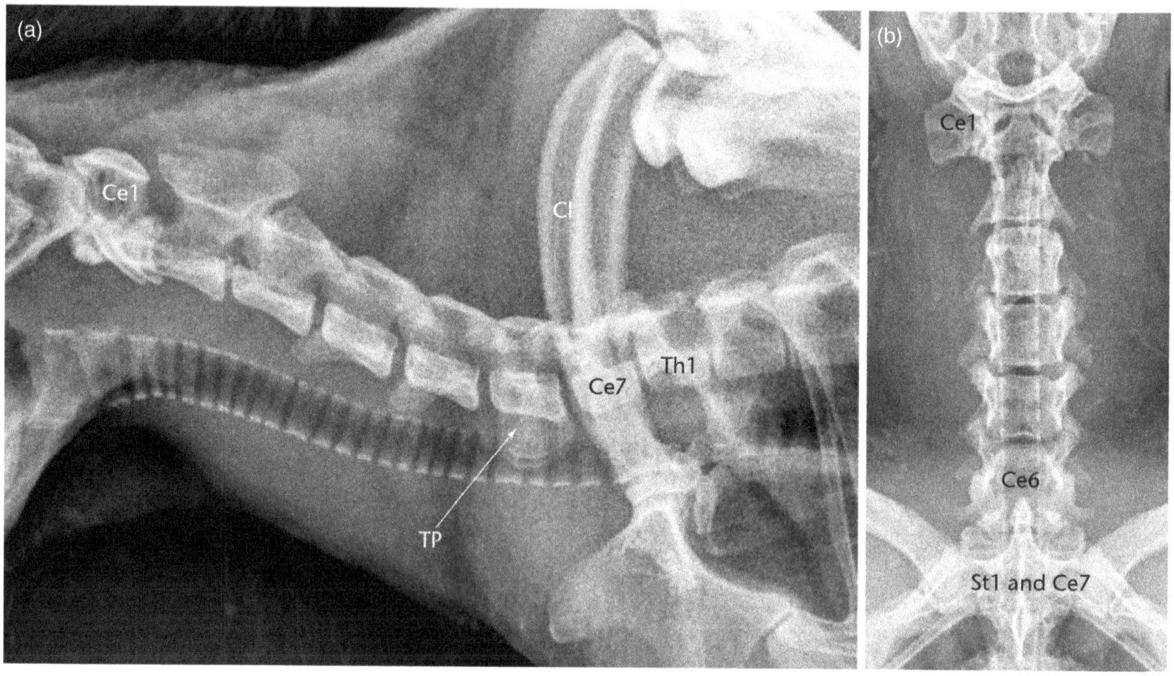

Figure 10.12: Cervical vertebrae, grey-headed flying-fox (*Pteropus poliocephalus*): (a) lateral view, (b) ventrodorsal view. There are 7 cervical vertebrae (Ce1, Ce6, Ce7 labelled). (b) The 1st sternal element (St1) and the 7th cervical vertebra (Ce7) are superimposed, so neither is clearly visible. The dorsal spinous processes are rudimentary. Transverse processes (TP) are visible on cervical vertebrae 4–6. If they are present on Ce7, they are obscured by the superimposed clavicles (Cl). The 1st thoracic vertebra is labelled.

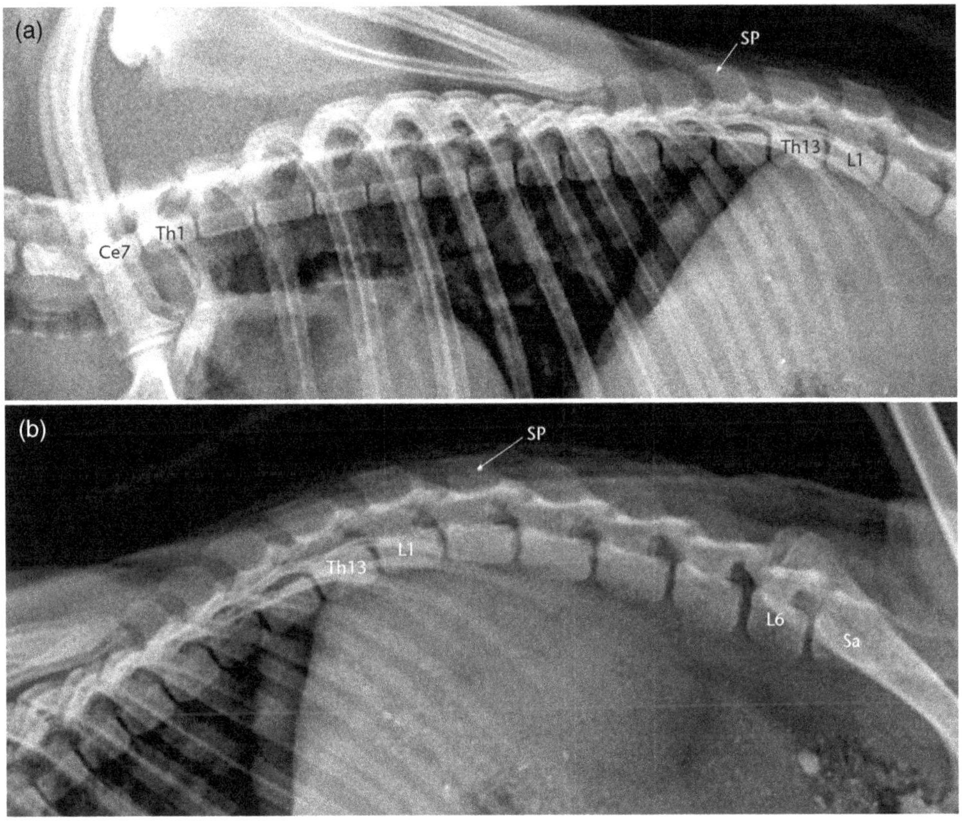

Figure 10.13: Thoracic and lumbar vertebrae, grey-headed flying-fox (*Pteropus poliocephalus*), lateral view: (a) thoracic vertebrae, (b) lumbar vertebrae. The grey-headed flying-fox has 13 thoracic (Th1, Th13 labelled) and 5–6 lumbar (L1, L6 labelled) vertebrae. The caudal thoracic and the lumbar vertebrae have short dorsal spinous processes (SP). The positions of the 7th cervical (Ce7) vertebra and the sacrum (Sa) are indicated.

cartilaginous ridges that is attached to the phalanges. The tendon has protrusions that lock into these ridges, constraining its movement. The bat's weight alone keeps the tendon taut and the claws locked in position. This allows the bat to rest and sleep without falling from its roost. To release the locking mechanism, the bat must flap its wings so the body becomes horizontal, thus removing the pressure of the bodyweight from the hindlimbs (Hall and Richards 2000).

Axial skeleton

Skull

The shape of the skull of bats is highly variable and dependent on diet. *Pteropus* species have long, narrow, almost fox- or dog-like skulls, *Nyctimene* species (tube-nosed bats) have blunter, more solid skulls with a prominent sagittal crest and the nectivorous blossom bats have long, slender skulls. Many Microchiropterans have relatively short, blunt skulls. Prominent features of the *Pteropus* skull are the large, rostrally directed orbits and a distinct postorbital process (Hall and Richards 2000). The zygomatic arch is long and slender. Sagittal and occipital crests are developed in the skull of some species. The tympanic bullae are small. The mandible has an angular process in Microchiropterans but not in pteropodids (Figs 10.10 and 10.11) (Parker and Haswell 1921).

Vertebrae

In general, bats have 7 cervical, 11 thoracic, 4 lumbar and between 0 and 10 caudal vertebrae. However, the grey-headed flying-fox has 13

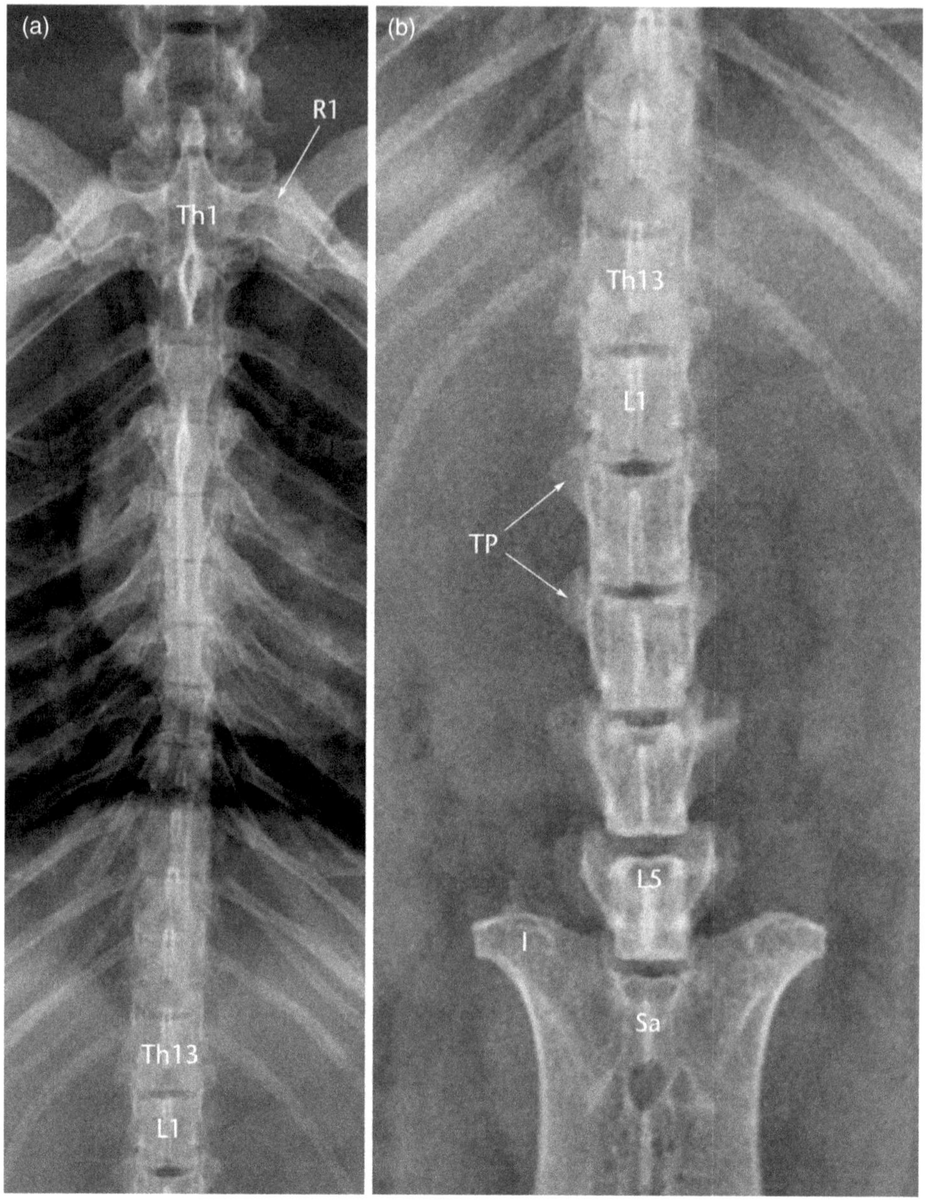

Figure 10.14: Thoracic and lumbar vertebrae, grey-headed flying-fox (*Pteropus poliocephalus*), ventrodorsal view: (a) thoracic vertebrae, (b) lumbar vertebrae. This grey-headed flying-fox has 13 thoracic (Th1, Th13 labelled) and 5 lumbar (L1, L5 labelled) vertebrae. The fused sacrum (Sa) and ilia (I) articulate with L5. There are rudimentary transverse processes (TP) on each lumbar vertebra. The location of the 1st thoracic vertebra is defined by the location of the 1st rib (R1).

thoracic and 5–6 lumbar vertebrae. In some species, the last cervical and first thoracic vertebrae are fused. In most species, the thoracic vertebrae fit tightly together to form a fairly rigid unit (Heard 1998). In pteropodids, the tail is short or absent (Nelson 1989). The cervical region is characterised by the absence of any distinct neural spines and the same holds true, but to a lesser extent, for the thoracic and lumbar vertebrae. The transverse processes of the lumbar region are also rudimentary. In bats with long tails, the caudal vertebrae are long, cylindrical and without processes (Figs 10.12–10.14) (Parker and Haswell 1921). The cervical segments of the spinal cord of bats are hypertrophied. This is

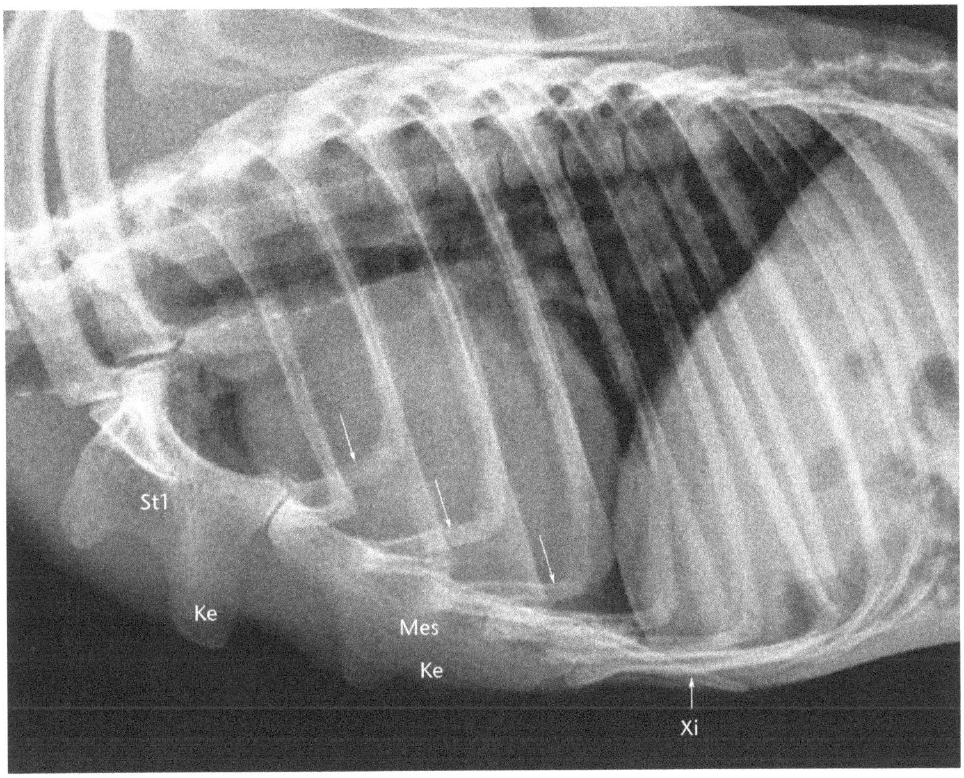

Figure 10.15: Thorax, grey-headed flying-fox (*Pteropus poliocephalus*), lateral view. The middle sternal elements are fused (the mesosternum, Mes) and enlarged by a ventral keel (Ke) for muscle attachment. The 1st sternal element (St1) also has a ventral keel and also provides a sturdy platform for insertion of the clavicles and 1st pair of ribs. The caudal sternal element, the xiphisternum (Xi), is long and narrow. The chondral elements of the ribs are ossified (arrows). The cardiac silhouette appears relatively large in this view but appears better proportioned in the ventrodorsal view (Fig. 10.16).

related to the innervation of the wings, which, like the forelimbs of all mammals, are innervated by the spinal nerves of the fourth cervical vertebra through to the first thoracic vertebra. In bats, the spinal cord is short, with the cervical region representing nearly half its entire length (Neuweiler 2000). This is related to the significantly greater function of the wings in locomotion as compared with the legs.

Ribs and sternum

The rib cage is large and usually more broad than deep (Heard 1998). The sternal ribs are ossified. The sternal elements, the number of which varies among species, may be distinct or fused (the mesosternum). The first sternal element may have a mesial keel or, in species where sternal elements are fused, the entire sternum may be keeled (Figs 10.3, 10.15 and 10.16). The pectoral muscles insert on the keel (Parker and Haswell 1921). The first sternal element may be greatly enlarged and T-shaped, with the lateral arms of the T articulating with the clavicles and first costal cartilages (Heard 1998).

Soft tissue

Thorax

Cardiovascular system

The cardiovascular system is typically mammalian. The heart is large relative to bodyweight, primarily in right atrial and ventricular muscle mass (Figs 10.15–10.18) (Hall and Richards 2000; Heard 2003). In large megachiropterans, the apex is

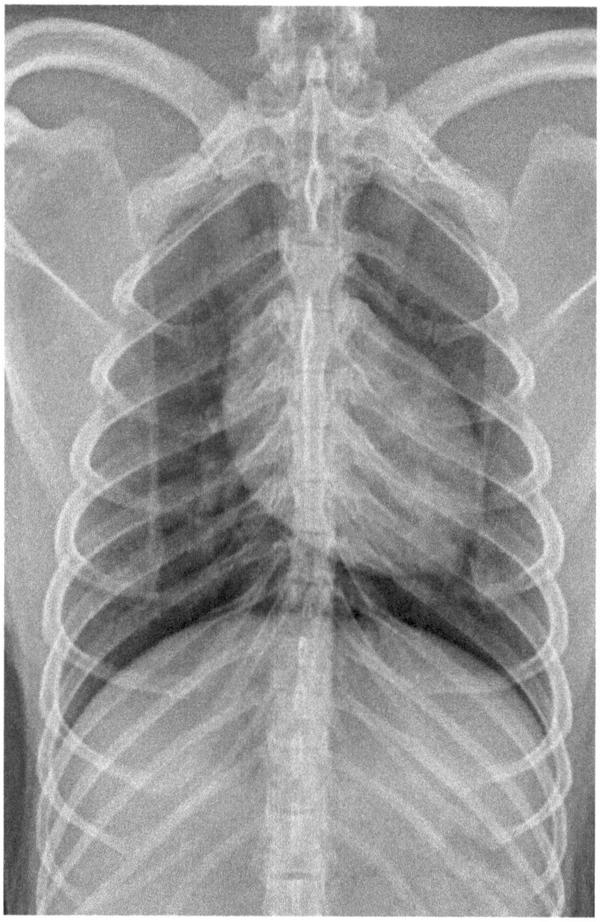

Figure 10.16: Thorax, grey-headed flying-fox (*Pteropus poliocephalus*), ventrodorsal view. The aerated lungs and heart appear more in proportion in this view compared with the lateral view.

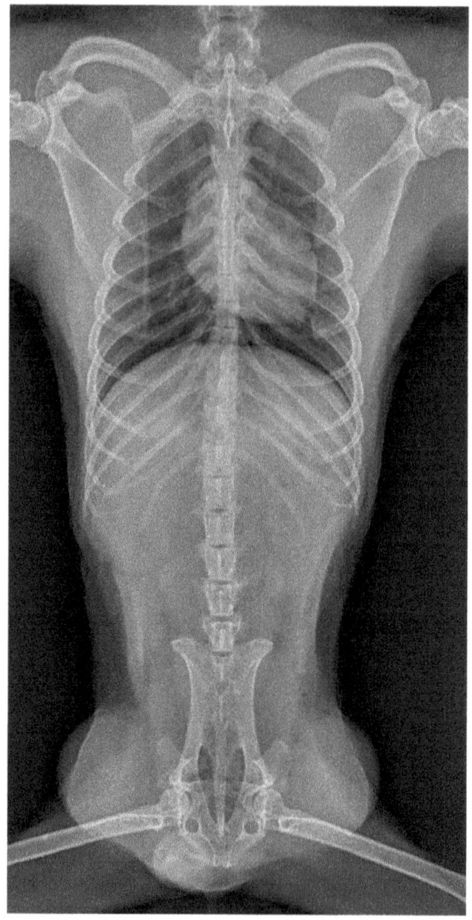

Figure 10.18: Thorax and abdomen, male grey-headed flying-fox (*Pteropus poliocephalus*), ventrodorsal view, showing the relative proportions of the thorax and abdomen.

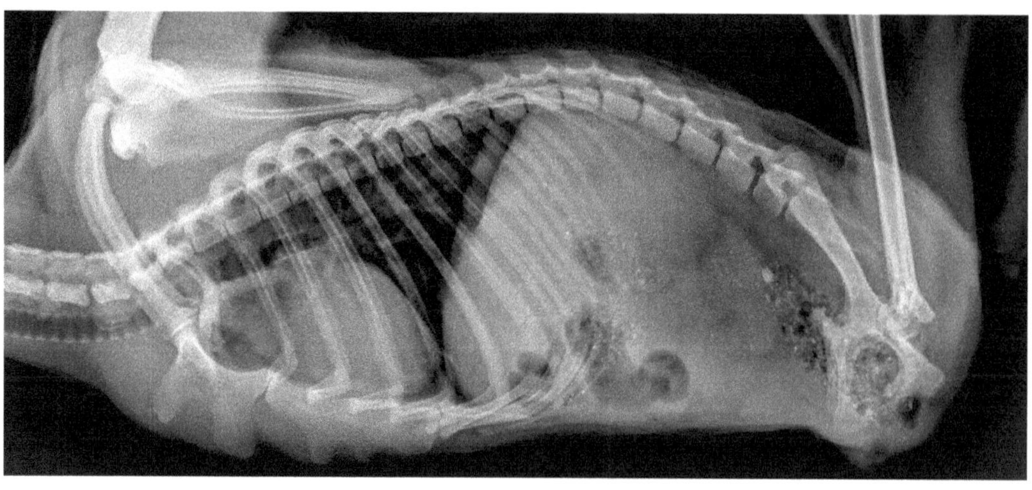

Figure 10.17: Thorax and abdomen, female grey-headed flying-fox (*Pteropus poliocephalus*), lateral view. The relative proportions of the thoracic and abdominal spaces are illustrated, with the thorax taking up nearly half of the combined space. The cardiac silhouette is large.

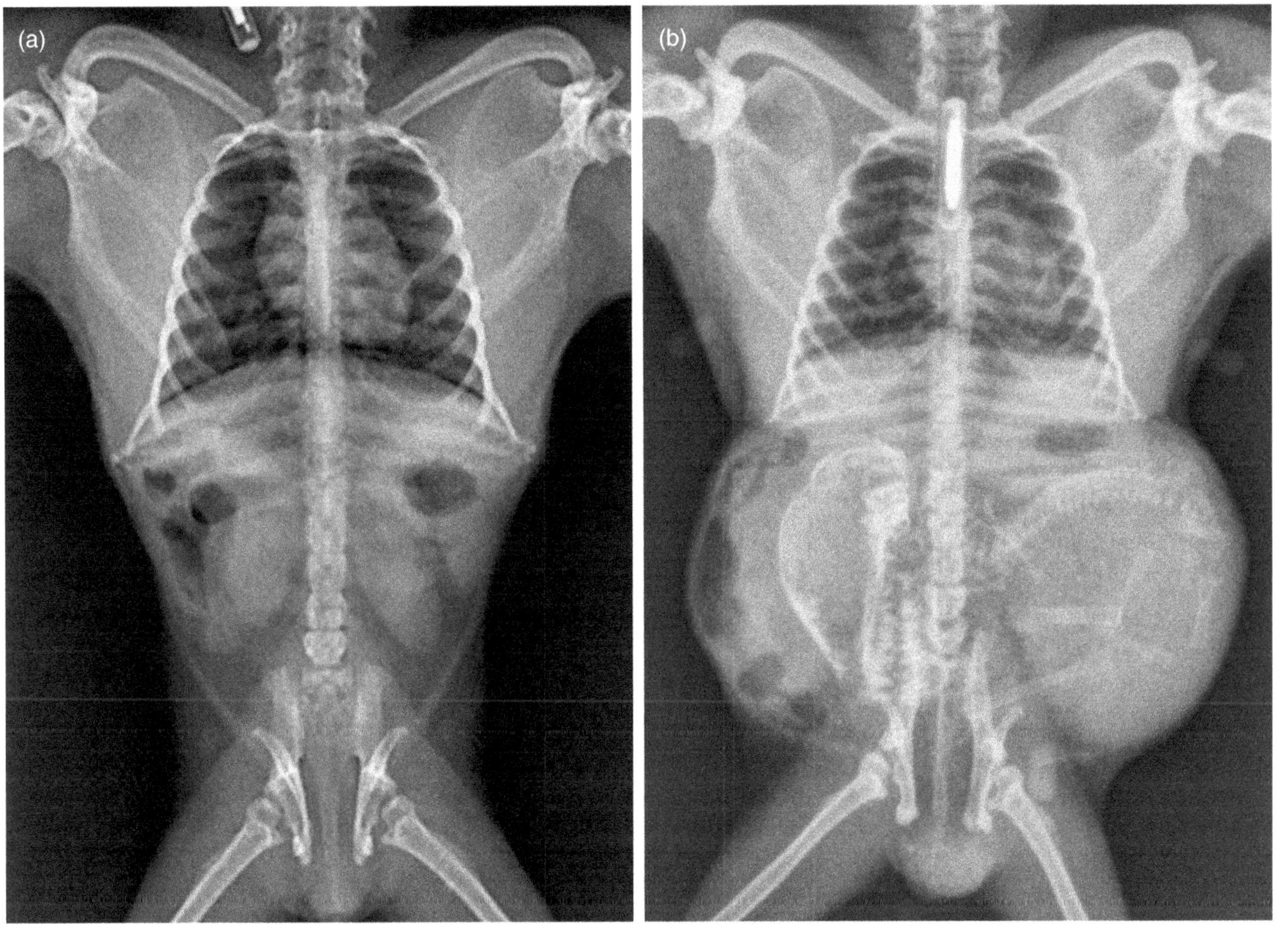

Figure 10.19: Thorax and abdomen, female ghost bat (*Macroderma gigas*), ventrodorsal view: (a) non-gravid, (b) gravid. The fetus is very large relative to the size of the dam. Note: transponders can be seen.

inclined to the left. Paired, large cranial venae cavae and a single caudal *vena cava* are present. The portion of the caudal *vena cava* cranial to the diaphragm is reduced in size, but caudal to it, the vessel is large and thin walled and is paired in megachiropterans (Heard 1998).

Respiratory system

The left lung is undivided, although a poorly defined notch may be present in some species. The right lung is divided into cranial, middle, caudal and accessory lobes.

Abdomen

In pteropodids, the stomach is simple, with large cardiac and fundic regions. There is no caecum, appendix or obvious demarcation between small and large intestines (Figs 10.17 and 10.18) (Nelson 1989). The large intestine is short. Lymph nodes are small, particularly in the mesentery, and the spleen is large and elongated (Heard 1998).

The gastrointestinal transit time is very short relative to other mammals. Insectivorous bats generally have longer transit times (35 minutes in fully active animals and up to 170 minutes in animals at rest) than *Pteropus* species (Jackson 2003). These anatomical and physiological modifications reduce the weight of gut contents while flying.

The female reproductive tract is typical of eutherian mammals. The anatomy, however, varies among the species: duplex (the two uteri are separate); bipartite (only the caudal end of the uteri are fused to form a single opening into the vagina); bicornuate (the caudal half of the uterus is fused); and simplex (where both uteri are fused to form a single organ). The duplex uterus is most common among *Pteropus* species (Neuweiler 2000). At birth, young are large relative to the size of the dam. A fetus, particularly late term, is readily seen on radiographs, taking up a large proportion of the abdominal space (Fig. 10.19). In the male, the penis and testes are large relative to body size. Almost 90% of the testicular volume consists of seminiferous tubules. The penis is pendant and an *os penis* (baculum) is present, the shape of which is characteristic for each species (Fig. 10.7) (Krutzsch 1959; Krutzsch 1962; Nelson 1989; Hall and Richards 2000; Neuweiler 2000).

References

Hall L, Richards G (2000) *Flying-foxes: Fruit and Blossom Bats of Australia.* University of New South Wales Press, Sydney.

Heard DJ (1998) Medical management of Megachiropterans. In *Zoo and Wild Animal Medicine, Current Therapy 4.* (Eds ME Fowler and RE Miller) pp. 344–353. Saunders, Pennsylvania.

Heard DJ (2003) Chiroptera. In *Zoo and Wild Animal Medicine.* (Eds ME Fowler and RE Miller) pp. 315–333. Saunders, Missouri.

Hill JE, Smith JD (1985) *Bats: A Natural History.* University of Texas Press, Texas.

Hutchinson JR (1996) Chiropteran flight. University of California, Museum of Palaeontology. <http://www.ucmp.berkeley.edu/vertebrates/flight/bats.html>

Jackson S (2003) Bats. In *Australian Mammals: Biology and Captive Management.* pp. 297–350. CSIRO Publishing, Melbourne.

Krutzsch PH (1959) Variation in the os penis of tropical fruit bats. *Journal of Mammalogy* **40**, 387–392. doi:10.2307/1376563

Krutzsch PH (1962) Additional data on the os penis of Megachiroptera. *Journal of Mammalogy* **43**, 34–42. doi:10.2307/1376878

Nelson JE (1989) Pteropodidae. In *Fauna of Australia, Volume 1B, Mammalia.* (Eds DW Walton and BJ Richardson) pp. 836–844. Australian Government Publishing Service, Canberra.

Neuweiler G (2000) *The Biology of Bats.* Oxford University Press, New York.

Olsson A, Woods R (2008). Bats. In *Medicine of Australian Mammals.* (Eds L Vogelnest and R Woods) pp. 465–502. CSIRO Publishing, Melbourne.

Padian K (1985) The origins of flight in extinct vertebrates. *Palaeontology* **28**, 413–433.

Parker WK (1867) *A Monograph on the Structure and Development of the Shoulder Girdle and Sternum in the Vertebrata.* Published for the Ray Society by Robert Hardwicke, Piccadilly, London.

Parker JT, Haswell WA (1921) Skeleton of Chiroptera. In *A Text-book of Zoology, Volume II.* pp. 534–535. Macmillan and Co. Ltd, London.

11
Dental radiology

Nadine Fiani

As the level of veterinary care of Australian mammals with dental pathology improves, a thorough understanding of dental anatomy and radiology is becoming more pertinent. Dental radiology is an inexpensive diagnostic modality that provides the clinician with an accurate understanding of pathology and thus the ability to plan treatment accordingly. The aim of this chapter is to discuss and demonstrate the normal dental anatomy of some representative Australian marsupials. A small section is dedicated to monotreme dentition.

Because of the limited availability of live animals of various ages with complete and non-diseased dentition, dental radiographs of high-quality skulls from the mammal collection at the Australian Museum were obtained. All efforts were made to radiograph the specimens in the same manner that would be used in a clinical setting.

Radiographic technique

Dental versus conventional X-ray generator

The radiographic signs of early dental diseases are subtle and can be missed if the positioning, exposure or processing is not optimal. Although diagnostic radiographs of dental conditions can be obtained using conventional radiographic equipment, the radiographic tube head lacks manoeuvrability, making the process inconvenient for the clinician. It is best to use either a wall-mounted dental radiography unit or a hand-held unit.

Dental radiographic film and digital systems

When compared with conventional skull radiographs, intraoral radiography has the advantage of eliminating superimposition from neighbouring anatomical structures. Three sizes of intraoral dental radiographic film are routinely used in veterinary dentistry: 0, 2 and 4. However, more film sizes and shapes are now becoming readily available, including ones that allow intraoral radiology in lagomorphs. These films may be readily applicable for use in many marsupial species.

Computed radiology (CR) systems using photo-stimulated phosphor (PSP) technology are popular. These systems typically use thin, reusable imaging plates without cables or sensors. The plates are placed intraorally and exposed using a dental X-ray unit. The plate is removed manually and placed in the processor, which will display a digital image on a computer screen.

Direct digital imaging systems are also available for dental radiography. The majority of these systems are based on charge-coupled device (CCD) technology. The disadvantages of CCD systems include the restrictions associated with the bulky sensor, limited sensor size (size 4 film not available), and high cost of the sensor. With these systems, the sensor is placed in the oral cavity and exposed using a dental X-ray unit. The image then appears on the connected computer screen.

Digital radiography eliminates the costs and complications of traditional film processing equipment, developer, fixer and cleaners, duplicating equipment, X-ray film and mounts.

Parallel versus bisecting angle technique

An in-depth discussion of dental radiographic technique is beyond the scope of this text, but is readily available in the veterinary literature (Lommer *et al.* 2000; Tsugawa and Verstraete 2000; Holmstrom *et al.* 2004; DuPont and DeBowes 2008). Two main techniques are used in dental radiography: the *parallel technique* and the *bisecting angle technique*.

In the parallel technique, the long axis of the tooth is parallel to the film and perpendicular to the X-ray beam. Although this is the ideal technique, it is difficult, if not impossible, to use in the majority of the oral cavity, because of the presence of the hard palate in the maxilla and the symphysis at the rostral aspect of the mandible. The only teeth that allow film placement parallel to their roots are the mandibular premolars and molars. Essentially, the film is placed upright in the intermandibular space (between the tongue and the medial aspect of the mandible).

The bisecting angle technique is used to obtain radiographs of all the other teeth. With this technique, the film is placed perpendicular to the tooth of interest. The X-ray beam is then directed perpendicular to the line that bisects the angle formed by the film and the long axis of the tooth. In doing so, the image obtained is neither elongated nor foreshortened.

All the radiographs shown in this chapter were obtained using these techniques.

Basic dental anatomy

Understanding the basic dental tissues and tooth-supporting structures is fundamental to the clinical practice of dentistry.

Enamel

Enamel is the hardest and most mineralized tissue in the body. It is the outermost layer of the crown.

Pulp–dentin complex

The bulk of the crown and root consists of dentin, a substance slightly harder than bone but weaker than enamel. Dentin is tubular in structure and is produced by highly specialised cells called *odontoblasts*. These are located at the periphery of the pulp and continuously make dentin throughout the life of a tooth, therefore causing the pulp cavity to narrow with time. The pulp is a very vascular structure containing nerves, blood vessels, loose connective tissue and odontoblasts. The pulp cavity in the crown is called the *pulp chamber* and in the root, the *root canal* (Fig. 11.1). If the pulp becomes non-vital, dentin production ceases and the pulp cavity will fail to narrow (see Chapter 12).

Cementum

Cementum is an avascular, highly mineralised and largely structureless tissue covering the root surface.

Periodontium

The periodontium consists of four tissues that help to hold a tooth within the jaw: the *gingiva, periodontal ligament, cementum* and *alveolar bone* (tooth socket). The periodontal ligament attaches between the alveolar bone and the cementum on the surface of the roots. Radiographically, the periodontal ligament space is visible as a distinct radiolucent line around the tooth roots (Fig. 11.1). In health, the alveolar bone margin should reach the cemento-enamel junction of the tooth (Fig. 11.1).

Important dental and oral terminology

When discussing dentition, the following abbreviations will be used throughout this chapter: I – incisor; C – canine; PM – premolar; M – molar.

The homology of the tooth will be further annotated with a number that will be superscript if the tooth is maxillary or subscript of the tooth is mandibular; for example, M^4 is the maxillary fourth molar tooth, whereas I_1 is the mandibular first incisor tooth. If the maxillary and mandibular

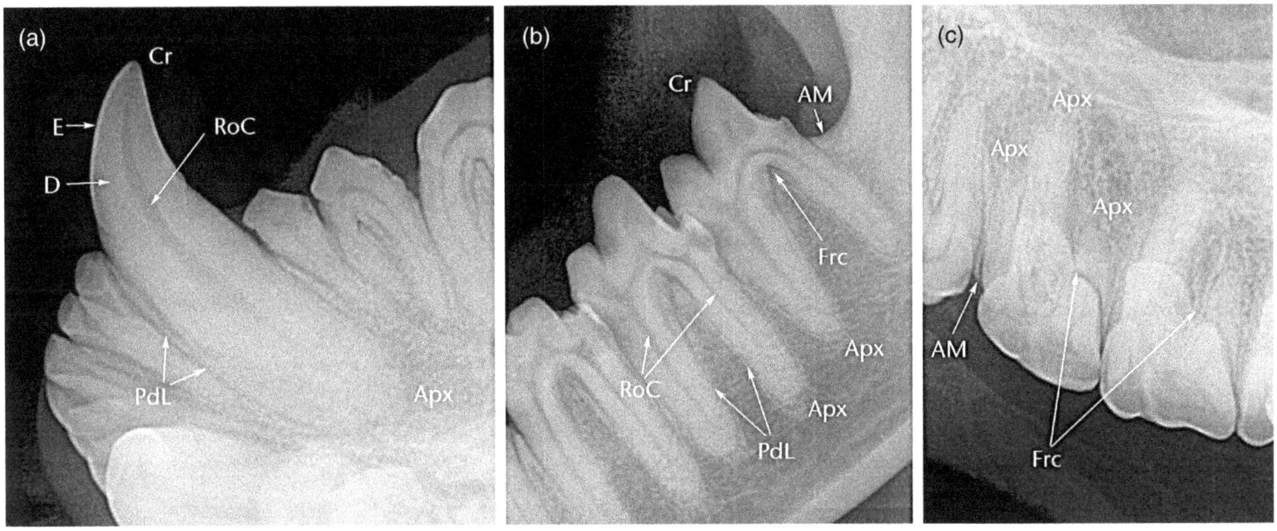

Figure 11.1: Normal dental anatomical features in (a) single-rooted tooth, (b) two-rooted teeth, and (c) three-rooted teeth. Crown (Cr), apex (Apx), enamel (E), dentin (D), root canal (RoC), periodontal ligament (PdL), furcation (Frc) and alveolar margin (AM) are indicated in this Tasmanian devil (*Sarcophilus harrisii*).

counterparts of the same tooth are both being discussed, the number is neither superscript nor subscript; for example, PM3 refers to the maxillary and mandibular third premolar teeth.

The oral cavity is divided into four *quadrants*; left and right, maxillary or mandibular. *Rostral* refers to a structure closer to, or a direction towards, the most forward aspect of the head. *Caudal* refers to a structure closer to, or a direction towards, the tail. The following positional terminology is used to describe the surfaces of the tooth and direction relative to the tooth or to the rest of the dentition:

Mesial: surface of a tooth that faces towards the first incisor (the midpoint of the dental arch)
Distal: surface of a tooth that faces away from the first incisor
Buccal: surface of a tooth that faces the cheek
Labial: surface of a tooth that faces the lips (limited to rostral dentition)
Lingual: surface of a tooth that faces the tongue
Palatal: surface of a maxillary tooth that face the palate
Coronal: direction towards the crown of the tooth (Fig. 11.1)
Apical: direction towards the apex of the root (Fig. 11.1)
Occlusal: interdigitation of the maxillary and mandibular teeth; more specifically, it refers to the chewing surface of the premolars and molars
Furcation: junction of roots of multirooted teeth (Fig. 11.1).

The word *diastema* is often used incorrectly in veterinary texts to refer to a normal space that exists between the rostral and caudal dentition. Diastema should only be used to describe a pathological space that has formed between teeth (e.g. rotated or malpositioned teeth) (Carmalt 2003). The normal anatomical space that exists between the rostral and caudal dentition (often encountered in herbivorous species) is simply referred to as an *interdental space*.

When discussing Australian marsupial dentition, animals are largely divided into two groups: *polyprotodont* or *diprotodont*. Polyprotodont species are those that have more than one incisor tooth in each mandibular quadrant. Diprotodont species are those that have one incisor tooth in each mandibular quadrant.

Marsupial deciduous dentition

Tooth replacement in marsupials differs from eutherian mammals. Broadly speaking, eutherian mammals typically possess two generations of incisors, canines, and most (or all) premolars. By definition, molar teeth are represented by only a single generation (Hillson 2005; van Nievelt and Smith 2005). The first generation of teeth, the deciduous dentition, is either absent or vestigial (minimally formed and unerupted) in all marsupials thus far examined (van Nievelt and Smith 2005; Smith 2006). The only exception to this is a single functional (erupted) deciduous tooth per quadrant. The tooth is generally thought to be the last premolar, termed deciduous premolar 3 (DP3) (Flower 1867; Freedman 1967; Hillson 2005; Smith 2006). The DP3 are replaced by the permanent premolar 3 (PM3).

The homology of the teeth in marsupial dentition is controversial and there are several alternate nomenclatures (Flower 1867; Thomas 1887; Wilson and Hill 1897; Archer 1976; Luckett and Wooley 1996). For example, Luckett and Wooley suggest that the PM1 and PM2, when present, are in fact deciduous premolar 1 and premolar 2 (DP1, DP2), as they are not replaced by a second generation. However, most other authors have continued to refer to the premolars as PM1 and PM2, with only the PM3 having a designated functional deciduous counterpart, DP3. The latter method of nomenclature is the one used in this chapter.

Polyprotodont marsupials

Dasyurids

Dental formula: I 4/3, C 1/1, PM 2–3/2–3, M 4/4

All dasyurids have four maxillary and three mandibular incisor teeth per quadrant (Archer 1976; Clarke 2003; Holz 2008) (Figs 11.2a, d and 11.3a, d). The first maxillary incisor tooth is often the last to

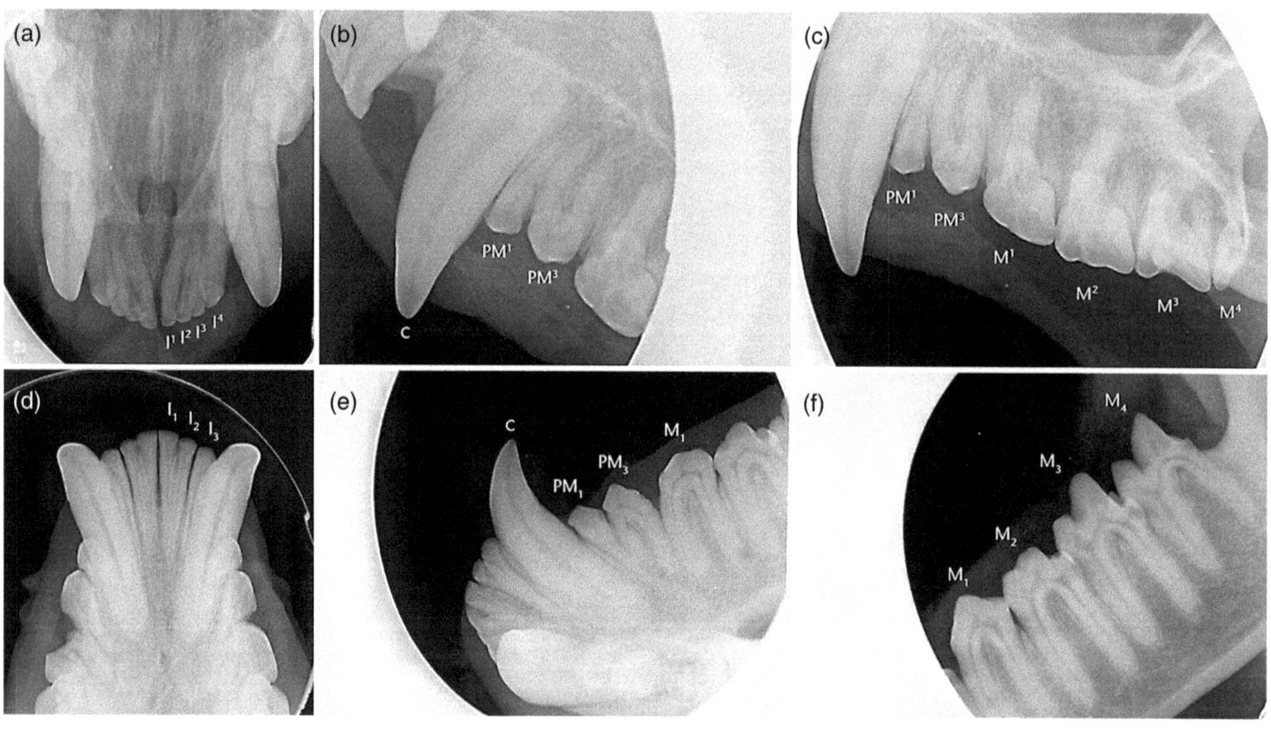

Figure 11.2: Tasmanian devil (*Sarcophilus harrisii*): (a) maxillary occlusal view, (b) left lateral maxillary canine and premolar view, (c) left caudal maxillary view, (d) mandibular occlusal view, (e) left lateral mandibular canine and premolar view, (f) left caudal mandibular view.

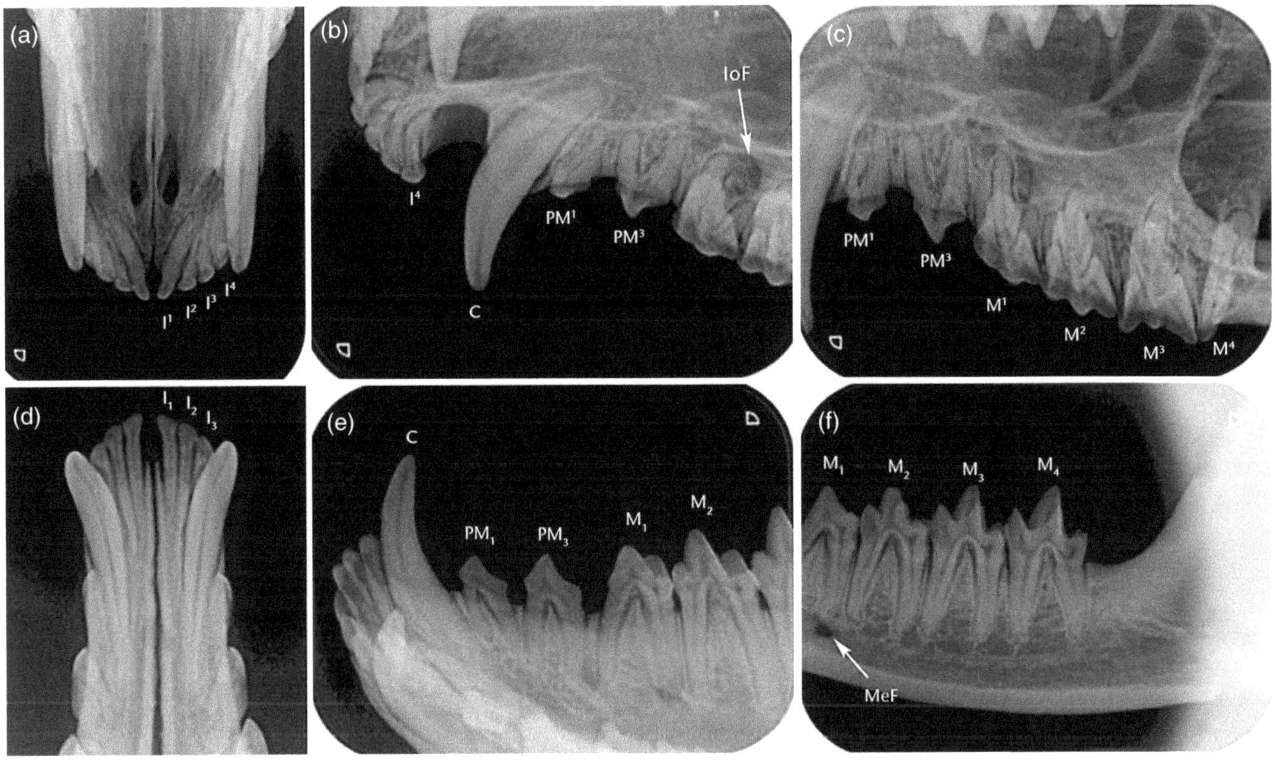

Figure 11.3: Spot-tailed quoll (*Dasyurus maculatus*): (a) maxillary occlusal view, (b) left lateral maxillary canine and premolar view, (c) left caudal maxillary view, (d) mandibular occlusal view, (e) left lateral mandibular canine and premolar view, (f) left caudal mandibular view. The infraorbital foramen (IoF) and mental foramen (MeF) are indicated.

erupt, resulting in a space between the left and right I^2. It has been suggested, that this prolonged eruption time may correlate with fixation of the young to the nipple (Guiler and Heddle 1974; Luckett and Wooley 1996).

There is one canine tooth per quadrant (Figs 11.2b, e and 11.3b, e). These can be greatly enlarged in some species, including the Tasmanian devil (*Sarcophilus harrisii*), but less prominent in others such as dunnarts (*Sminthopsis* spp.) (Archer 1976).

Dasyurids have two or three premolar teeth per quadrant (Thomas 1887). Species that have three premolar teeth include dunnarts and antechinus (*Antechinus* spp.). The premolar formula has been shown to be PM1, PM2 and PM3. Two premolar species include the Tasmanian devil and the quolls (*Dasyurus* spp.) (Figs 11.2b, e and 11.3b, e). It has been suggested that the missing premolar teeth are in fact the PM2, rendering their premolar sequence as PM1 and PM3 (Luckett and Wooley 1996).

There are four molars present in each quadrant, M1 M4 (Figs 11.2c, f and 11.3c, f). The maxillary molars are bunodont (low and rounded cusps distributed over a broad occlusal surface), making them ideal for crushing (Hillson 2005). The mandibular molars, however, have a relatively more prominent cusp structure.

Tasmanian devils have brachydont teeth (short crown and well-developed roots) (Hillson 2005). The crowns are covered with enamel, which extends to the gingival margin. The exception to enamel distribution is that of the incisor and canine teeth, where only the coronal two-thirds of the crown is covered.

The roots are distinct and well developed. All the incisor and canine teeth are single rooted. The maxillary and mandibular premolar teeth are two-rooted. It may be difficult to distinguish the two roots of the PM_1 in the Tasmanian devil because the roots converge and the tooth is often rotated, causing the roots to be superimposed on images. The maxillary molars have three roots. The M^1–M^3 have one palatal and two buccal roots. The M^4 is markedly reduced in size and all its roots converge, giving the radiographic impression of a single-rooted structure. All the mandibular molars are two-rooted.

Bandicoots and bilby

Dental formula: I 5/3, C 1/1, PM 3/3, M 4/4

Bandicoots and the bilby are omnivorous marsupials that exhibit characteristic elongation of the maxillary and incisive bones (Abbie 1937; Freedman 1967; Fosse and Risnes 1972; Flores *et al.* 2013). The elongated muzzle is thought to assist the animals in foraging for insects (Freedman 1967).

The maxilla holds five peg-like incisor teeth (Fig. 11.4a). The I^5 of the *Perameles* spp. has a distinctly caniniform shape and can be slightly distoverted (caudally pointing) (Freedman 1967). Interestingly, the I^5 of the eastern barred bandicoot (*P. gunnii*) appears to be two-rooted, unlike most incisors, which are normally single-rooted (Freedman 1967). The I^5 of bandicoots is often separated from the I^4 and canine teeth by an interdental space (Freedman 1967; Carmalt 2003; Flores *et al.* 2013). The rufous spiny bandicoot (*Echymipera rufescens*) differs from all other bandicoots by having four maxillary incisors (Gordon and Hulbert 1989). The mandible has three incisor teeth in all members of the order Peramelemorphia (Fig. 11.4d). The I_1 and I_2 have a thick enamel layer and sharp incisal edges. The I_3 has a bifid crown, with the mesial portion being about twice as big as the distal (Freedman 1967; Gordon and Hulbert 1989).

There is one large canine tooth per quadrant (Fig. 11.4b, e). Sexual dimorphism, with the exception of size, is not apparent in the cranium or

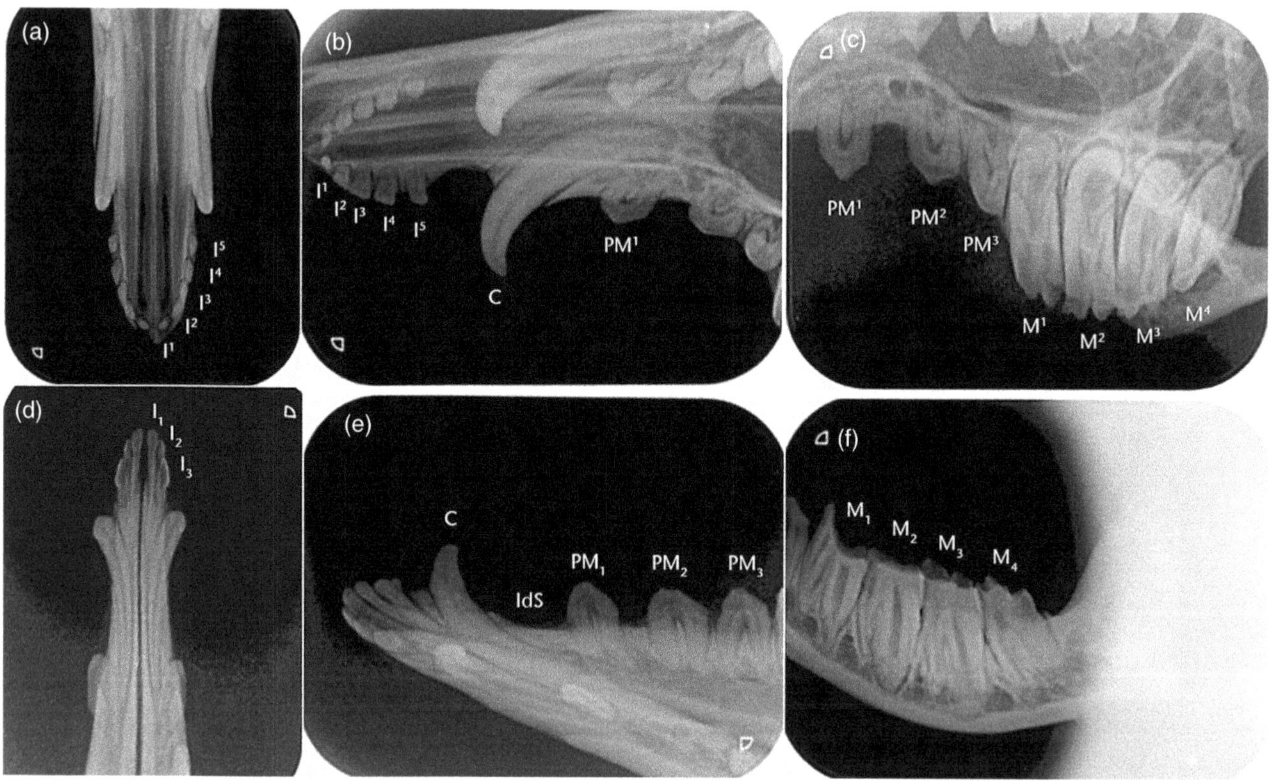

Figure 11.4: Greater bilby (*Macrotis lagotis*): (a) maxillary occlusal view, (b) left lateral maxillary canine and premolar view, (c) left caudal maxillary view, (d) mandibular occlusal view, (e) left lateral mandibular canine and premolar view, (f) left caudal mandibular view. The interdental space (IdS) is indicated.

mandible of bandicoots; however the canine teeth of the two sexes differ in both size and anatomical details (Freedman 1967; Lyne 1982; Johnson and Johnson 1983). In the male, the canine tooth is significantly longer than in the female (Freedman 1967; Freedman and Joffe 1967a, 1967b). Similar to the Tasmanian devil, only the coronal two-thirds of the canine teeth is enamel-covered (Freedman 1967).

Three premolar teeth are present in each quadrant (Fig. 11.4c, e): PM1, PM2 and PM3 (Freedman 1967). The premolar teeth are two-rooted and have a prominent central cusp, ideal for grasping and shearing prey (Hillson 2005). Prominent interdental spaces may exist between the C, PM1 and PM2 in both the maxilla and mandible.

Four block-like molar teeth exist in each quadrant: M1–M4 (Fig. 11.4c, f). The four mandibular molar teeth are relatively more elongated than their maxillary counterparts and they also have two distinct and equal-sized roots (Freedman 1967). The appearance of the crowns of the molar teeth is similar to that of the hypsodont dentition encountered in many herbivorous animals.

Diprotodont marsupials

Koala

Dental formula: I 3/1, C 1/0, PM 1/1, M 4/4

The first maxillary incisor tooth is well developed, but the second and third are relatively smaller and more slender (Fig. 11.5a) (Young 1881). The mandibular incisor teeth are typical of diprotodont species, with procumbent crowns and extensive root structure (Fig. 11.5d) (Eberhard 1972; Strahan 1978; Clarke 2003). In some of the earliest descriptions of koala dentition, the teeth, especially the mandibular incisors, were likened to those of the rabbit; however, unlike the rabbit, none of the koala teeth are elodont (always growing) (Cork and Sanson

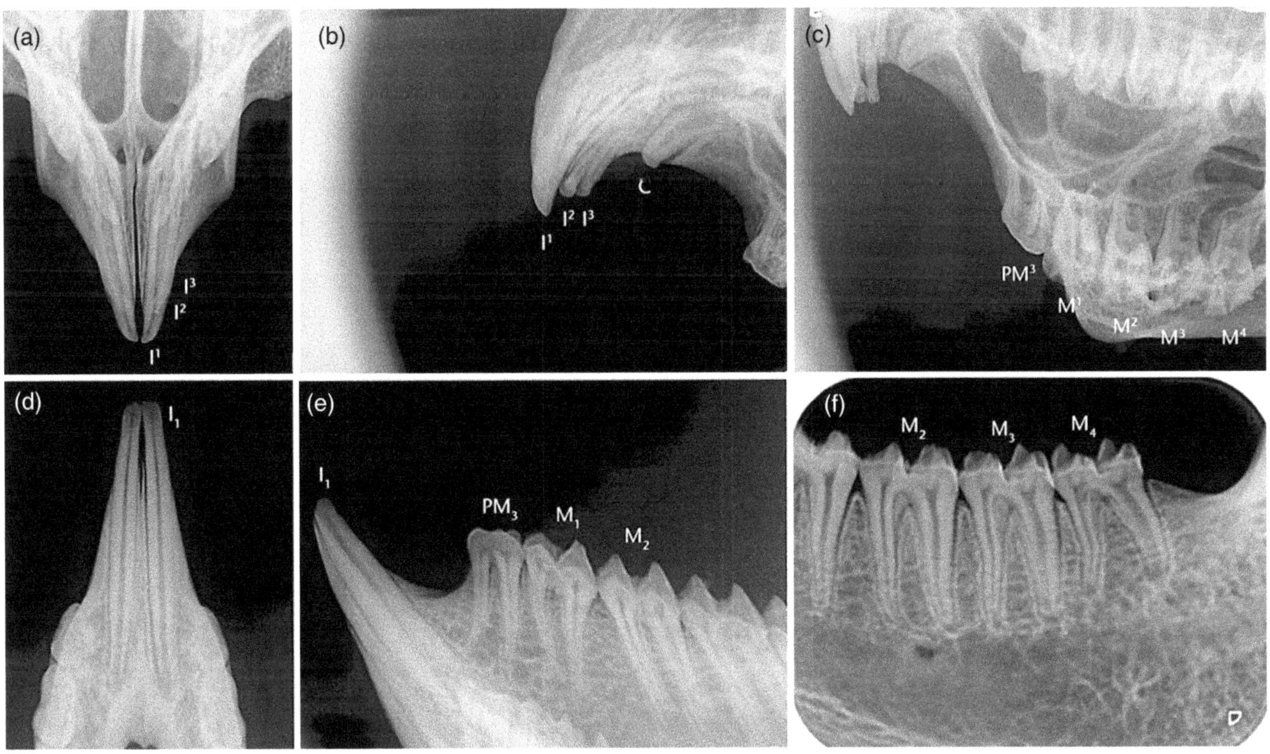

Figure 11.5: Koala (*Phascolarctos cinereus*): (a) maxillary occlusal view, (b) left lateral maxillary canine and premolar view, (c) left caudal maxillary view, (d) mandibular occlusal view, (e) left lateral mandibular premolar view, (f) left caudal mandibular view.

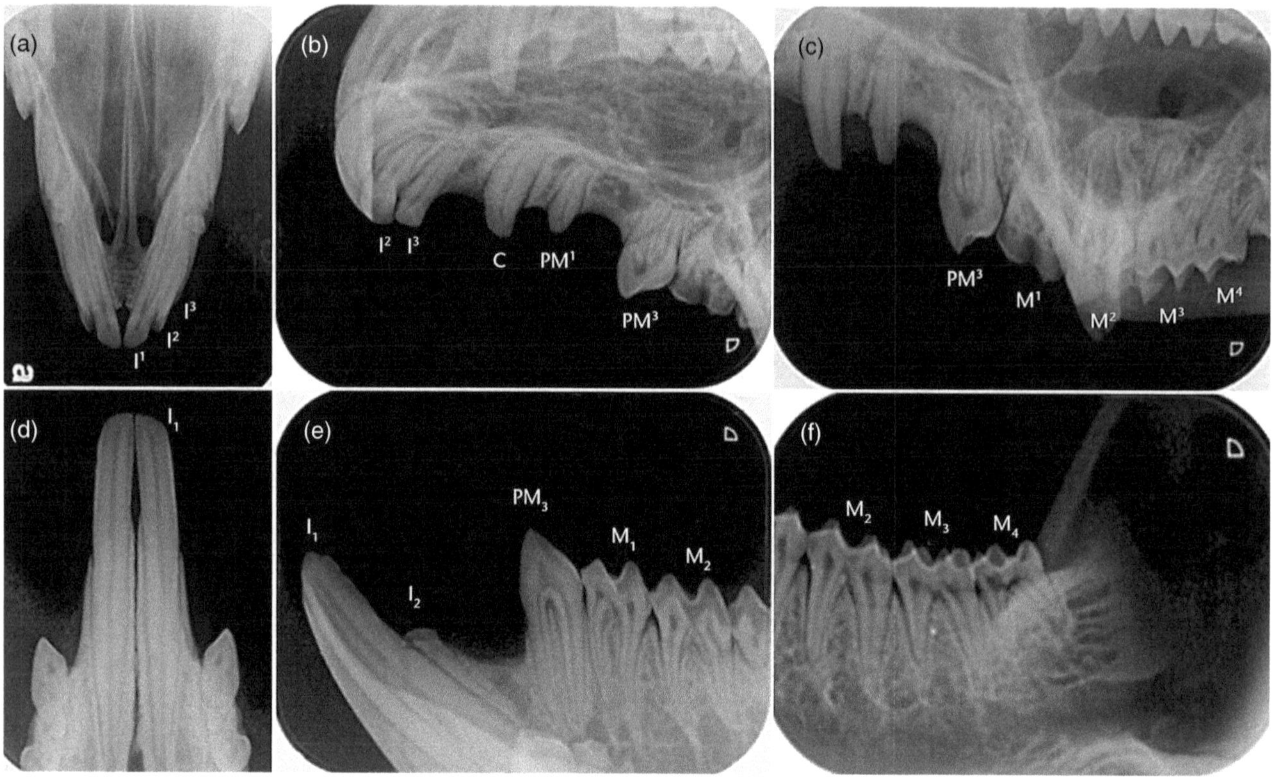

Figure 11.6: Common brush-tailed possum (*Trichosurus vulpecula*): (a) maxillary occlusal view, (b) left lateral maxillary canine and premolar view, (c) left caudal maxillary view, (d) mandibular occlusal view, (e) left lateral mandibular premolar view, (f) left caudal mandibular view.

1990; Hume 2006). There is no report of sexual dimorphism as far as dentition is concerned (Ellis and Bercovitch 2011).

One canine tooth is present in each maxillary quadrant and none in the mandibles. The canine tooth is small and peg-like (Fig. 11.5b) (Young 1881). It is located after a short interdental space caudal to the incisors.

A single premolar tooth, PM3, is present in each quadrant (Fig. 11.5c, e) (Lee and Martin 1988). It is the smallest of the cheek teeth and has considerably less prominent cusps relative to the molar teeth. The premolar teeth have two distinct roots.

There are four molar teeth per quadrant: M1–M4 (Fig. 11.5c, e, f). Each molar tooth has four pyramidal cusps that interdigitate with those on the opposing arch (Strahan 1978; Lanyon and Sanson 1986a). Throughout a koala's life, there is continual wear of the occlusal surfaces of the cheek teeth. The longitudinal cutting ridge of each premolar and the pyramidal cusps of each molar are gradually worn flat (Lanyon and Sanson 1986b). The molar teeth have two prominent roots.

Possums, gliders and cuscus

Dental formula: I 3/1, C 1/0, PM 1–3/1–3, M 4/4

The maxillary quadrant contains three incisor teeth (Figs 11.6a and 11.7a). The I^1 is considerably longer than the I^2 and I^3 (McKay 1989; McKay and Winter 1989; Turner and McKay 1989). The mandibular quadrant contains a single large procumbent incisor (Fig. 6d and 7c) (Johnson and Hemsley 2008). The shape of the incisor may be compressed and blade-like, mildly curved, or greatly enlarged and curved, depending on the species (McKay 1989). Lying directly behind the base of the crown of the procumbent incisor is sometimes a small, functionless tooth (Fig. 11.6e). The homology of

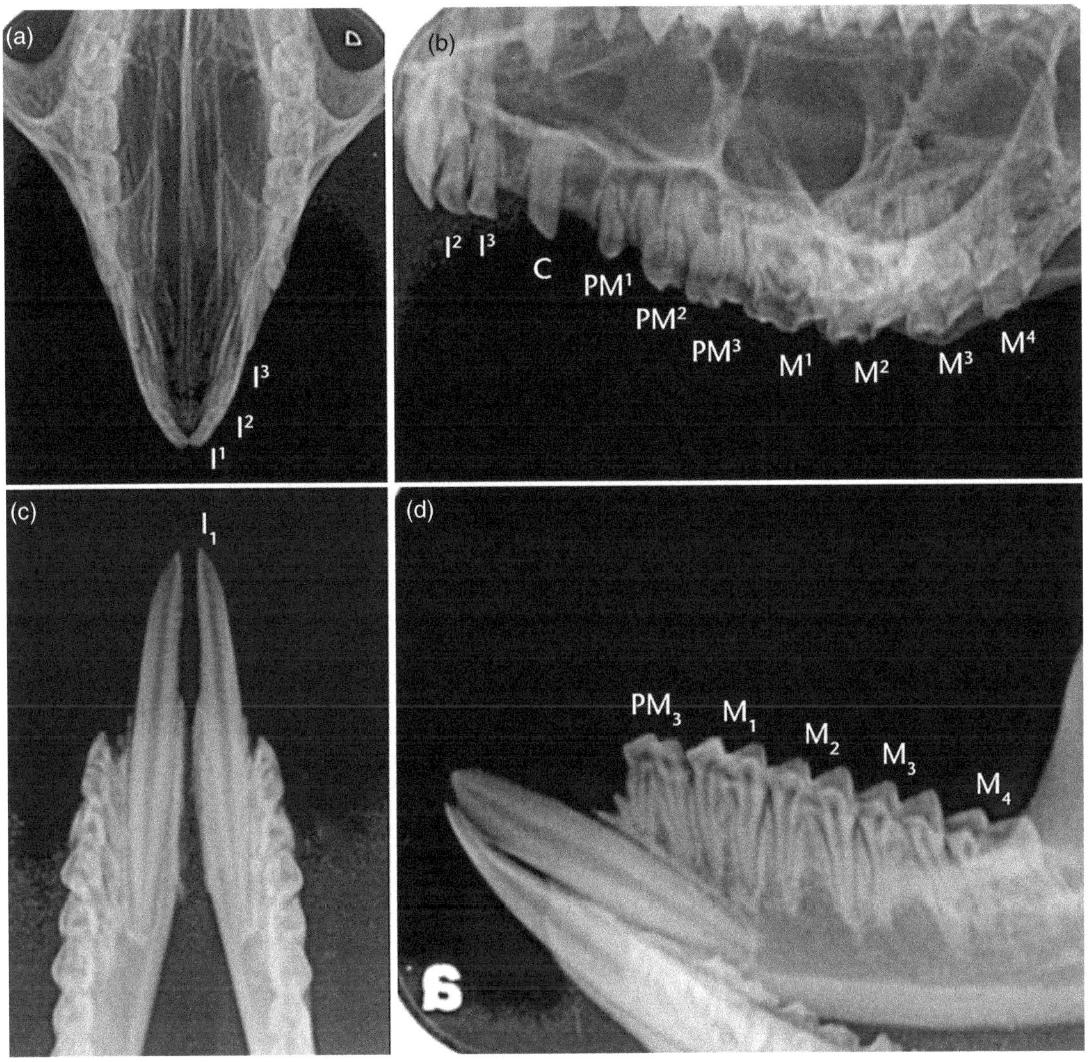

Figure 11.7: Common ring-tailed possum (*Pseudocheirus peregrinus*): (a) maxillary occlusal view, (b) left lateral maxillary canine, premolar and molar view, (c) mandibular occlusal view, (d) left lateral mandibular premolar and molar view.

this tooth remains unclear; however, it has been suggested that it is a vestigial second incisor (Kingsmill 1962; Berkovitz 1968; McKay 1989; McKay and Winter 1989; Turner and McKay 1989; Clarke 2003).

A single canine tooth is present in the maxilla but not in the mandible (Figs 11.6b and 11.7b). It is separated from the premolar teeth by an interdental space. The size of the canine tooth varies with the species. It is more prominent in the Phalangeridae than in the Pseudocheiridae (Kerle 2001). The canine tooth of the common brush-tailed possum (*Trichosurus vulpecula*) is connate and occasionally divided into larger mesial and smaller distal parts by a groove that separates the tips of the crowns (Clarke 2003).

The number of premolar teeth is variable. In the Phalangeridae, the maxillary first premolar tooth is normally caniniform but distinctly smaller than the canine tooth (Fig. 11.6b). The PM^2, if present, is reduced in size and has a single cusp. The PM^3 has a prominent crown, which may be conical or blade-like and set at an angle relative to the molar teeth behind it (Fig. 11.6c). The mandible may have 1–3 premolars (Fig. 11.6e). Generally the PM_1 and PM_2, if present, are vestigial. The mandibular PM_3 is similar in shape and size to its maxillary counterpart (McKay and Winter 1989; Clarke 2003).

In the Burramyidae, with the exception of the mountain pygmy possum (*Burramys parvus*), there are three maxillary premolar teeth. The mountain pygmy possum has lost the first premolar tooth. Three premolars exist in the mandible (Turner and McKay 1989).

In the Petauridae, three maxillary premolar teeth are always present and the first two may vary in size. The third premolar tooth is usually larger and has a variable number of cusps. Three small premolar teeth are often present in the mandible; however, the first and second may be either vestigial or absent in some species (McKay 1989).

There are four maxillary and mandibular molar teeth in all possums, gliders and cuscus (Figs 11.6c, e, f and 11.7b, d). These can vary in shape and size depending on the species. The molars generally have cusps, which may be bunodont (rounded cusps), lophodont (cusps that coalesce into buccolingual folds) or selodont (cusps that coalesce into mesiodistal folds) depending on the species and the diet (McKay 1989; McKay and Winter 1989; Turner and McKay 1989; Kerle 2001; Hillson 2005; Johnson and Hemsley 2008).

Macropods

Dental formula: I 3/1, C 0–1/0, PM 1–2/1–2, M 4/4

Macropod (kangaroos and wallabies) dentition is often characterised by the species' adaptation to a particular diet, either browsers, grazers or a combination (Sanson 1989; Hume 1999; Tyndale-Biscoe 2005). Sanson (1989) described three prominent variations between the three groups: (1) the size of the permanent third premolar tooth, (2) molar crown morphology and (3) the type of molar occlusion.

Browsers consume a heterogeneous diet consisting of a variety of plant materials such as shrubs and tree foliage. The plants vary in texture, thickness and levels of resistance to breakage (Sanson 1980, 1989). Members of this group include tree-kangaroos (*Dendrolagus* spp.), the swamp wallaby (*Wallabia bicolor*), the quokka (*Setonix brachyurus*), pademelons (*Thylogale* spp.), and several of the rock-wallabies (*Petrogale* spp.) (Sanson 1989; Tyndale-Biscoe 2005; Vogelnest and Portas 2008).

The intermediate browser–grazer diet consists largely of grasses, but animals will also browse (Hume 1982, 1999; Dawson 2012). Macropods in this group include the hare-wallabies (*Lagorchestes* spp.), nail-tailed wallabies (*Onychogalea* spp.), the banded hare-wallaby (*Lagostrophus fasciatus*), and some of the *Macropus* spp., including red-necked wallaby (*Macropus rufogriseus*) and black-striped wallaby (*M. dorsalis*) (Sanson 1989; Vogelnest and Portas 2008).

The grazers feed primarily on grass, which is largely homogeneous but abrasive in its physical properties (Hume 1982, 1999; Dawson 2012). The six largest species of macropods are strict grazers: the red kangaroo (*M. rufus*) the antilopine kangaroo (*M. antilopinus*), black wallaroo (*M. bernardus*), common wallaroo (*M. robustus*), western grey kangaroo (*M. fuliginosus*), and the eastern grey kangaroo (*M. giganteus*) (Sanson 1980; Sanson 1989; Tyndale-Biscoe 2005; Vogelnest and Portas 2008).

In browsing macropods the I^1 is larger and its crown is longer than that of the I^2 and I^3 (Figs 11.8a, b and 11.9a, b). The procumbent mandibular incisor teeth occlude immediately palatal to the I^1 and onto the incisive surface of the I^2 and I^3 (Figs 11.8d and 11.9d) (Sanson 1989). It is thought that this occlusion is similar to that of omnivorous animals and may reflect an emphasis on manipulation of individual items between specific upper and lower incisor teeth (Sanson 1989; Hume 1999). The crowns of the maxillary incisors of grazing macropods are all of the same height such that the mandibular incisor teeth occlude against the latter's incisive edges (Fig. 11.10a, e). The maxillary I^3 is considerably wider in grazers and has an evident labial groove. This increase in width, combined with relatively more procumbent lower incisor teeth and a flexible mandibular symphysis, results

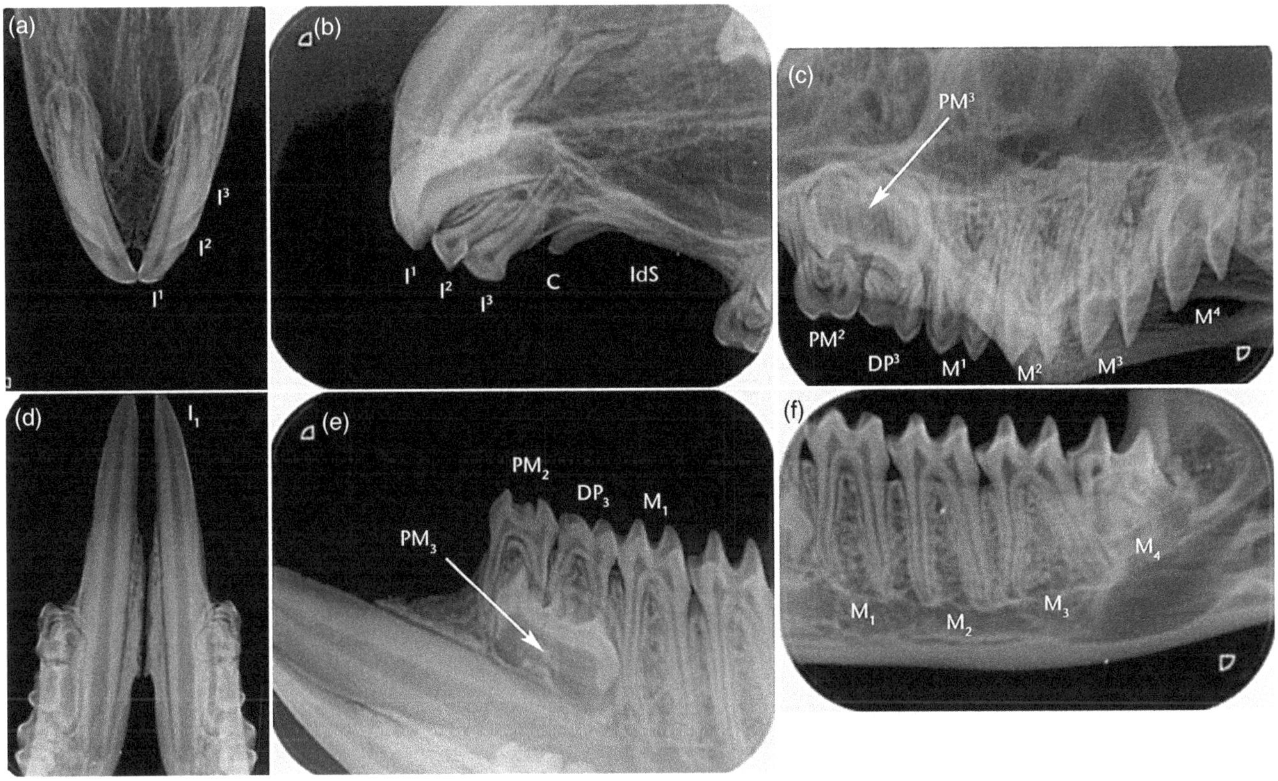

Figure 11.8: Sub-adult Lumholtz's tree-kangaroo (*Dendrolagus lumholtzi*): (a) maxillary occlusal view, (b) left lateral maxillary canine and premolar view, (c) left caudal maxillary view, (d) mandibular occlusal view, (e) left lateral mandibular premolar view, (f) left caudal mandibular view. The unerupted PM3 is clearly evident in (c) and (e). The interdental space (IdS) is indicated.

in a wide occlusal surface, which creates a more efficient cutting edge (Sanson 1989; Tyndale-Biscoe 2005; Vogelnest and Portas 2008). The mandibular incisors were once reported to be continuously growing, but this is not true (Clarke 2003). The incisors form a complete apex and the pulp chamber continues to narrow throughout the life of the tooth.

Canine teeth are rarely encountered in the macropod species. If they are present, they are usually vestigial, single-rooted and play no role in food prehension or preparation (Sanson 1989; Vogelnest and Portas 2008).

A large interdental space exists between the incisor and the premolar/molar teeth (Figs 11.8b, 11.9b, e and 11.10b). The first two cheek teeth in each quadrant of young macropods are a large, sectorial PM2 and molariform deciduous premolar 3 (DP3) (Fig. 11.8c, e). These premolars provide the young animal with a shearing tooth and a grinding tooth in each quadrant (Tyndale-Biscoe 2005; Vogelnest and Portas 2008). As the macropod ages and the M1 and M2 erupt, the PM2 and DP3 are shed and replaced by a single sectorial PM3 at the rostral aspect of each row (Figs 11.9c, e, 11.10c, e and 11.11a). In browsing macropods, the PM3 is large, has a well-developed shearing action similar to that of carnassial teeth in carnivores and is retained throughout life (Fig. 11.9c, f) (Sanson 1989; Vogelnest and Portas 2008). The smaller intermediate browser–grazer species have a PM3 that is reduced in size and sometimes lost. The PM3 is smallest in grazing macropods and is also rapidly lost by molar progression (Fig. 11.11a, b) (Sanson 1989). In some cases, the PM3 is pushed out even before it has come into wear, suggesting that its

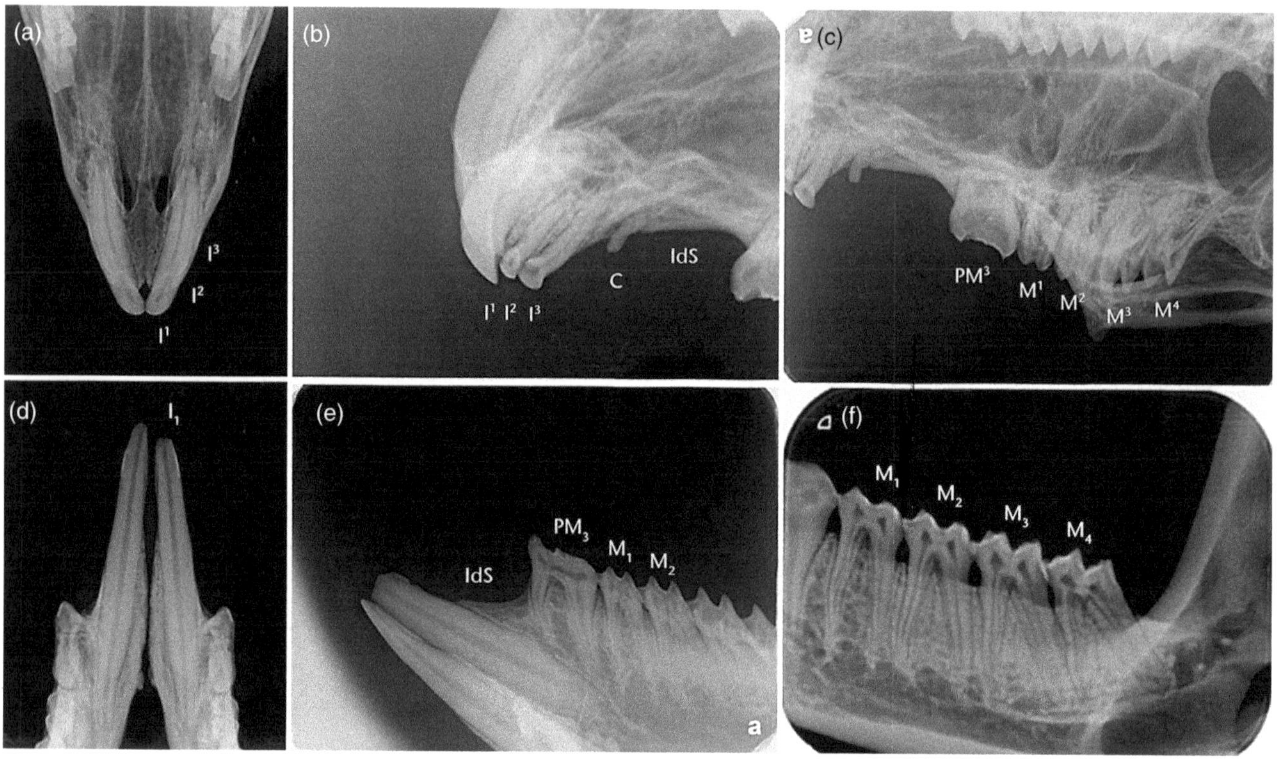

Figure 11.9: Adult Lumholtz's tree-kangaroo (*Dendrolagus lumholtzi*): (a) maxillary occlusal view, (b) left lateral maxillary canine and premolar view, (c) left caudal maxillary view, (d) mandibular occlusal view, (e) left lateral mandibular premolar view, (f) left caudal mandibular view. The interdental space (IdS) is indicated.

presence is largely vestigial in these species (Sanson 1980; Sanson 1989).

The molars of macropods are lophodont (Sanson 1980; Sanson 1989). In browsing species, the lophs, which are transverse ridges that form the occlusal surface of the crowns, have weak longitudinal ridges between them that are termed *links*. This morphology creates a large surface area of contact between the maxillary and mandibular teeth, leading to an efficient crushing action with some fine shearing and grinding capacity (Sanson 1980; Sanson 1989; Hume 1999). All of the cheek teeth of the browsing species are equal in height, so all the molars occlude simultaneously (Fig. 11.9c, f) (Hume 1982, 1999; Vogelnest and Portas 2008).

In grazing species, the molars are characterised by pronounced lophs and well-developed links (Fig. 11.10c, e) (Tyndale-Biscoe 2005). This results in a shearing action between the maxillary and mandibular dentition (Hume 1999). The molar rows are strongly curved, with the upper and lower rows curving in opposite directions such that only two pairs of molars occlude on each side at a time (Fig. 11.10c, e) (Tyndale-Biscoe 2005; Vogelnest and Portas 2008).

Intermediate browser–grazer species have similar molar morphology to the grazers, but the curvature of their molar rows is often less pronounced (Hume 1999; Tyndale-Biscoe 2005; Vogelnest and Portas 2008).

Molar progression

Molar progression is a phenomenon whereby erupted molars move rostrally within a jaw and the most worn premolar and molars are shed from the front of the row (Fig. 11.10e, f and 11.11a, b) (Vogelnest and Portas 2008; Lentle and Hume 2010). Molar progression occurs in all the grazing macropods and some of the intermediate browser–grazer species. The curvature of their cheek tooth rows means that only the rostral cheek teeth occlude initially. As these wear down and lose their function,

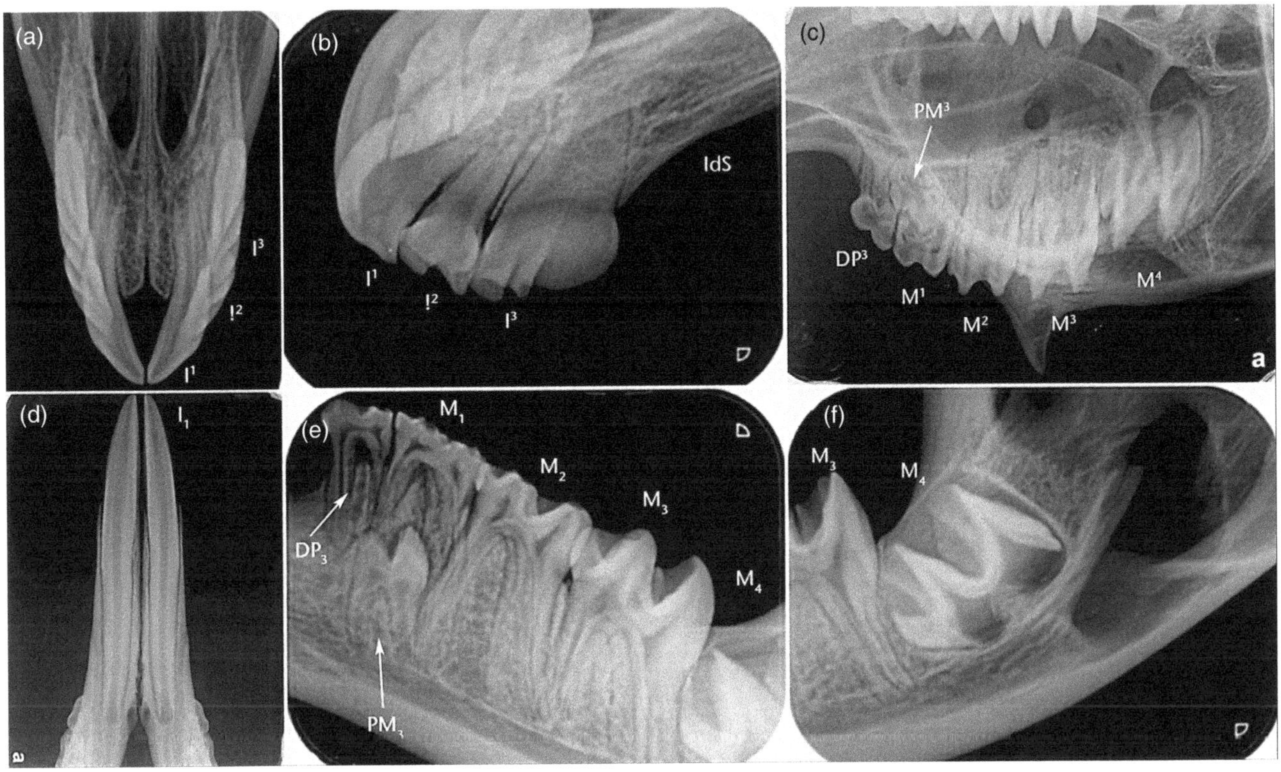

Figure 11.10: Sub-adult eastern grey kangaroo (*Macropus giganteus*): (a) maxillary occlusal view, (b) left lateral maxillary premolar view, (c) left caudal maxillary view, (d) mandibular occlusal view, (e) left lateral mandibular premolar view, (f) left caudal mandibular view. The unerupted PM3 and M4 are evident. The interdental space (IdS) is indicated.

the more caudal dentition must move rostrally within the jaw to take the place of the now inefficient rostral molars (Sanson 1980, 1989). Older animals will often only have a worn-down M3 and M4 remaining in each quadrant (Fig. 11.11b). Molar progression has been used to create a molar index, which may be helpful when estimating the age of an animal (Kirkpatrick 1964, 1965; Kirkpatrick and Johnson 1969). It is important to remember that molar progression appears to be, at least in part, driven by the rate of tooth wear, which in turn is related to the abrasiveness of the animal's diet (Sanson 1989).

Premolar and molar root anatomy

The root structure of the premolars and molars of macropods remains largely undescribed in the literature. Based on the radiographs presented here, browsing macropods appear to have two-rooted premolars and molars (Figs 11.8f and 11.9f). Grazing species, however, are considerably more difficult to interpret. Their cheek teeth appear to have two primary roots that further subdivide more apically (Figs 11.10e and 11.11a, b). The situation is further complicated by these teeth being in constant mesial progression and at various stages of resorption as they are in the process of being shed rostrally (Fig. 11.11b). Given that the necessity for extraction of premolar and molars often arises in the clinical setting, further investigation of the radicular structures is necessary.

Wombats

Dental formula: I 1/1, C 0/0, PM 1/1, M 4/4

Although wombats are in the order Diprotodontia, they differ by having only a single pair of maxillary incisor teeth (Fig. 11.12a) (Bryant and Reiss 2008). They are also especially unique among Australian marsupials as all their teeth are aradicular hypsodont (elodont), which means that none of the

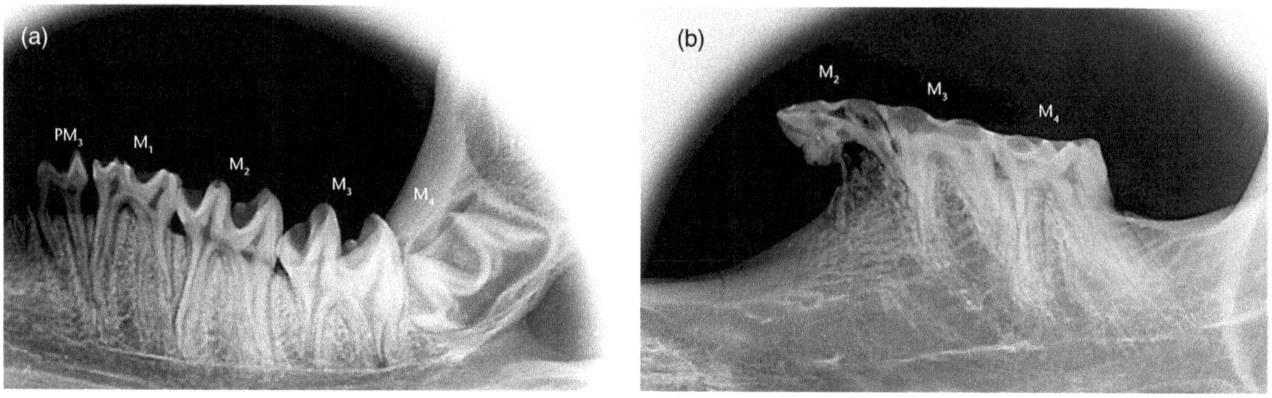

Figure 11.11: Eastern grey kangaroo (*Macropus giganteus*), left caudal mandibular view: (a) sub-adult, (b) aged adult. In the sub-adult animal, M_4 is yet to erupt and there is already evidence of resorption of the mesial root of PM_3. In the aged adult, all teeth have erupted and molar progression is advanced, with loss of PM_3 and M_1, and root resorption and advanced wear evident in the remaining molars.

teeth forms an apex and they continue to grow throughout life (Fig. 11.12b, e) (Clarke 2003; Bryant and Reiss 2008; Fagan and Ullrey 2008). Enamel only covers the labial aspect of the incisor teeth, the buccal aspect of the mandibular cheek teeth and the lingual aspect of the maxillary cheek teeth

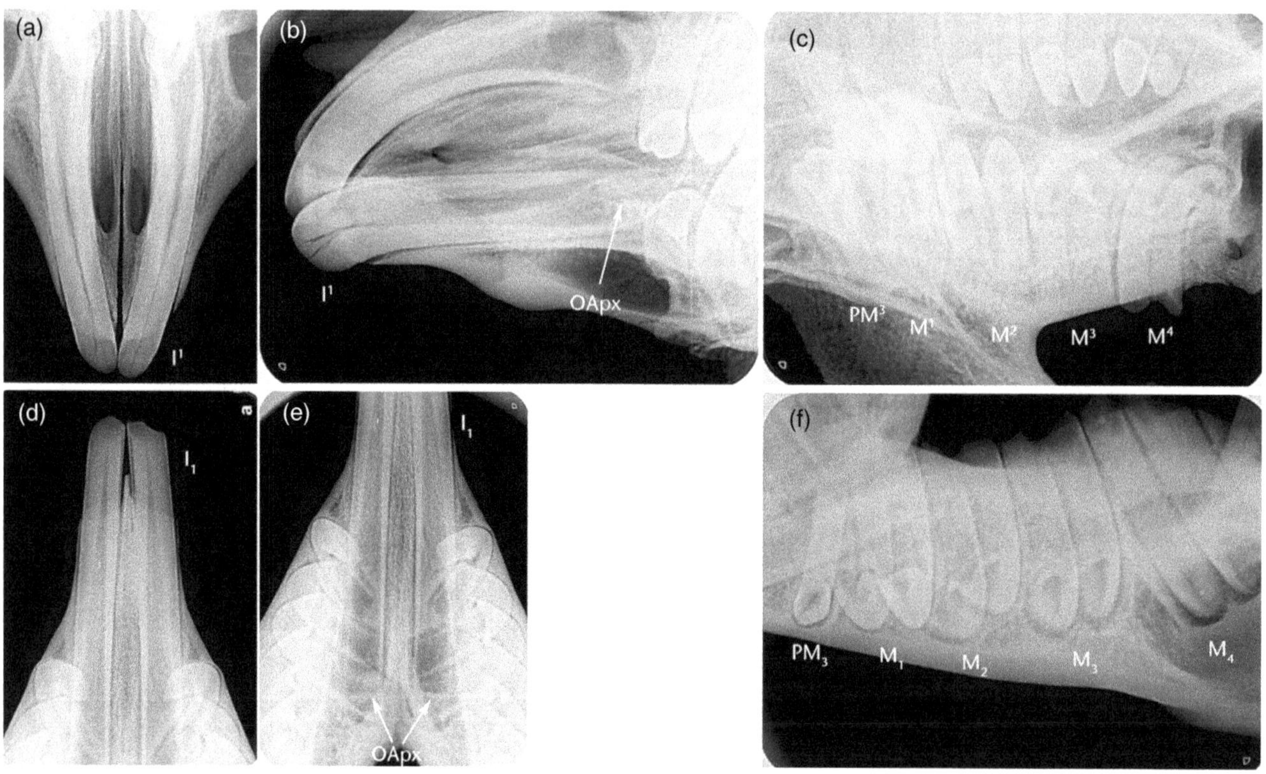

Figure 11.12: Common wombat (*Vombatus ursinus*): (a) maxillary occlusal view, (b) left lateral maxillary incisor view, (c) left caudal maxillary view, (d) mandibular occlusal view showing the coronal portion of the incisor teeth, (e) mandibular occlusal view showing the apical portion of the incisor teeth, (f) left caudal mandibular view. The open apex (OApx) of the incisor teeth is indicated.

(Bryant and Reiss 2008; Triggs 2009). The differential wear between the harder enamel and the softer dentin results in self-sharpening ridges through the life of the teeth (Ferreira *et al.* 1989).

There is a very long, single incisor in each quadrant (Fig. 11.12a, b, d, e). The clinical crown (the portion of the tooth above the gingival margin) of the maxillary incisors is relatively short and curved, whereas the mandibular counterpart is longer and protrudes horizontally. The maxillary and mandibular incisors meet together at the incisive margin, allowing the teeth to function in a sectorial fashion (Hillson 2005).

An interdental space exists between the incisors and the only premolar tooth in each quadrant, the PM3 (Fig. 11.12b, c, e, f). This tooth is mostly triangular in cross-section and has a single lobe (Merrilees 1967).

There are four molar teeth, M1–M4, in each quadrant (Fig. 11.12c, f). They are bilobate structures. There is a deeper groove on the lingual aspect of the maxillary teeth and the buccal aspect of the mandibular teeth (Merrilees 1967; Ferreira *et al.* 1989). The molar teeth are curved, with the convexity being lingual in the maxillary teeth and buccal in the mandibular teeth. Consequently, when the teeth are in occlusion, an upper molar and its mandibular counterpart form a sigmoid curve (Merrilees 1967) (see Chapter 6, Fig. 6.16b).

The oral cavity of wombats is very small and their gape is very narrow. Obtaining intraoral dental radiographs is enormously challenging. The best method for assessing both dental structure and occlusion is computed tomography.

Monotremes

Platypus

The platypus *(Ornithorhynchus anatinus)* has mammalian-like dentition as juveniles (Poulton 1887). There are three erupted teeth in the maxilla and the mandible (Wilson and Hill 1907; Simpson and Burrell 1929; Green 1937) (see Chapter 3, Fig. 3.10). The first maxillary tooth is small and has a single cusp, followed by two larger multicuspidate teeth. The two most rostral teeth in the mandible are large and multicuspidate, while the last tooth is much smaller (Thomas 1889; Wilson and Hill 1907; Green 1937). Following a lengthy review of numerous specimens, Green suggested that the erupted dental formula for the platypus should be I 0/0, C 0/0 PM 1/0 M 2/3 (Green 1937).

It has been noted that the first and second mandibular molars correspond to the second and third maxillary molar teeth. The third lower molar tooth and the remaining maxillary premolar tooth do not correspond to the erupted teeth. The dental occlusion is suggestive of a crushing action rather than a shearing or grinding one (Simpson and Burrell 1929).

The teeth are thought to erupt approximately 6–8 weeks after hatching and are shed close to 1 month after the young emerges from the nesting burrow (Green 1937; Musser and Archer 1998). Once shed, the teeth are replaced by keratinous grinding pads.

References

Abbie A (1937) Some observations on the major subdivisions of the marsupialia: with especial reference to the position of the peramelidae and caenolestidae. *Journal of Anatomy* **71**, 429–436.

Archer M (1976) The dasyurid dentition and its relationships to that of didelphids, thylacinids, borhyaenids (Marsupicarnivora) and peramelids (Peramelina: Marsupialia). *Australian Journal of Zoology* **24**, 1–34. doi:10.1071/AJZS039

Berkovitz B (1968) Some stages in the early development of the post-incisor dentition of Trichosurus vulpecula (Phalangeroidea: Marsupialia). *Journal of Zoology* **154**, 403–414. doi:10.1111/j.1469-7998.1968.tb01673.x

Bryant B, Reiss A (2008) Wombats. In *Medicine of Australian Mammals.* (Eds L Vogelnest and R Woods) pp. 329–358. CSIRO Publishing, Melbourne.

Carmalt J (2003) Understanding the equine diastema. *Equine Veterinary Education* **15**, 34–35. doi:10.1111/j.2042-3292.2003.tb00210.x

Clarke DE (2003) Oral biology and disorders of chiroptera, insectivores, monotremes, and marsupials. *The Veterinary Clinics of North America. Exotic Animal Practice* **6**, 523–564. doi:10.1016/S1094-9194(03)00033-1

Cork SJ, Sanson GD (1990) Digestion and nutrition in the koala: a review. In *Biology of the Koala.* (Eds AK Lee, KA Handasyde and GD Sanson) pp. 129–144. Surrey Beatty, Sydney.

Dawson TJ (2012) Living in the environment: feeding. In *Kangaroos.* (Ed. A Cuttler.) pp. 131–149. CSIRO Publishing, Melbourne.

DuPont GA, DeBowes LJ (2008) *Atlas of Dental Radiography in Dogs and Cats.* Saunders Elsevier, St Louis, Missouri.

Eberhard IH (1972) Ecology of the koala, *Phascolarctos cinereus* (Goldfuss) on Flinders Chase, Kangaroo Island. PhD thesis, The University of Adelaide, Australia.

Ellis WA, Bercovitch FB (2011) Body size and sexual selection in the koala. *Behavioral Ecology and Sociobiology* **65**, 1229–1235. doi:10.1007/s00265-010-1136-4

Fagan DA, Ullrey DE (2008) Dental malocclusion in a coarse-haired wombat (*Vombatus ursinus*). *Journal of Veterinary Dentistry* **25**, 240–244.

Ferreira JM, Phakey PP, Palamara J, Rachinger WA, Orams HJ (1989) Electron microscopic investigation relating the occlusal morphology to the underlying enamel structure of molar teeth of the wombat (*Vombatus ursinus*). *Journal of Morphology* **200**, 141–149. doi:10.1002/jmor.1052000204

Flores DA, Abdala F, Giannini NP (2013) Postweaning cranial ontogeny in two bandicoots (Mammalia, Peramelomorphia, Peramelida(e) and comparison with carnivorous marsupials. *Zoology (Jena, Germany)* **116**, 372–384. doi:10.1016/j.zool.2013.07.003

Flower WH (1867) On the development and succession of the teeth in the Marsupialia. *Philosophical Transactions of the Royal Society of London* **157**, 631–641. doi:10.1098/rstl.1867.0020

Fosse G, Risnes S (1972) Development of the teeth in a pouch-young specimen of Isoodon obesulus and one of Perameles gunnii (Peramelidae: Marsupialia). *Archives of Oral Biology* **17**, 829–838. doi:10.1016/0003-9969(72)90026-X

Freedman L (1967) Skull and tooth variation in the genus Perameles. Part 1: anatomical features. *Records of the Australian Museum* **27**, 147–165. doi:10.3853/j.0067-1975.27.1967.441

Freedman L, Joffe A (1967a) Skull and tooth variation in the genus Perameles. Part 2: metrical features of P. nasuta. *Records of the Australian Museum* **27**, 183–195. doi:10.3853/j.0067-1975.27.1967.444

Freedman L, Joffe A (1967b) Skull and tooth variation in the genus Perameles. Part 3: metrical features of P. gunnii and P. bougainville. *Records of the Australian Museum* **27**, 197–212. doi:10.3853/j.0067-1975.27.1967.517

Gordon G, Hulbert AJ (1989) Peramelidae. In *Fauna of Australia, Volume 1B, Mammalia* (Eds DW Walton and BJ Richardson) pp. 603–624. Australian Government Publishing Service, Canberra.

Green H (1937) The development and morphology of the teeth of Ornithorhynchus. *Philosophical Transactions of the Royal Society of London. Series B, Biological Sciences* **228**, 367–420. doi:10.1098/rstb.1937.0015

Guiler ER, Heddle R (1974) The eruption and growth of teeth in the Tasmanian devil, Sarcophilus harrisii (Marsupialia: Dasyurida(e). *Papers and Proceedings of the Royal Society of Tasmania* **108,** 137–140.

Hillson S (2005) *Teeth.* Cambridge University Press, New York.

Holmstrom SE, Fitch PF, Eisner ER (2004) *Veterinary Dental Techniques for the Small Animal Practitioner.* Saunders Elsevier, Philadelphia.

Holz P (2008) Dasyurids. In *Medicine of Australian Mammals.* (Eds L Vogelnest and R Woods) pp. 359–382. CSIRO Publishing, Melbourne.

Hume ID (1982) Diet and nutrition of kangaroos and wallabies. In *Digestive Physiology and Nutrition of Marsupials.* (Ed. ID Hume.) pp. 159–199. Press Syndicate of the University of Cambridge, Cambridge UK.

Hume ID (1999) Nutritional ecology of kanagaroos and wallabies. In *Marspuial Nutrition*. (Ed. ID Hume.) pp. 261–314. Cambridge University Press, Cambridge UK.

Hume ID (2006) Nutrition and digestion. In *Marsupials*. (Eds P Armati, C Dickmam and ID Hume) pp. 137–158. Cambridge University Press, Cambridge, UK.

Johnson R, Hemsley S (2008) Gliders and possums. In *Medicine of Australian Mammals*. (Eds L Vogelnest and R Woods) pp. 395–437. CSIRO Publishing, Melbourne.

Johnson C, Johnson K (1983) Behaviour of the bilby, Macrotis lagotis (Reid), (Marsupialia: Thylacomyida(e) in captivity. *Wildlife Research* **10**, 77–87. doi:10.1071/WR9830077

Kerle A (2001) Food and nutrition. In *Possums*. (Ed. PTJ Dawson) pp. 62–74. University of New South Wales Press Ltd, Sydney.

Kingsmill E (1962) An investigation of criteria for estimating age in the marsupials Trichosurus vulpecula Kerr and Perameles nasuta Geoffroy. *Australian Journal of Zoology* **10**, 597–616. doi:10.1071/ZO9620597

Kirkpatrick T (1964) Molar progression and macropod age. *Queensland Journal of Agricultural and Animal Sciences* **21**, 163–165.

Kirkpatrick T (1965) Studies of Macropodidae in Queensland. 2: age estimation in the grey kangaroo, the red kangaroo, the eastern wallaroo and the red-necked wallaby, with notes on dental abnormalities. *Queensland Journal of Agricultural and Animal Sciences* **22**, 301–317.

Kirkpatrick T, Johnson P (1969) Studies of Macropodidae in Queensland. 7: age estimation and reproduction in the agile wallaby (Wallabia agilis)(Gould). *Queensland Journal of Agricultural and Animal Sciences* **26**, 691–698.

Lanyon JM, Sanson G (1986a) Koala (*Phascolarctos cinereus*) dentition and nutrition. I: morphology and occlusion of cheekteeth. *Journal of Zoology* **209**, 155–168. doi:10.1111/j.1469-7998.1986.tb03572.x

Lanyon JM, Sanson G (1986b) Koala (*Phascolarctos cinereus*) dentition and nutrition. II: implications of tooth wear in nutrition. *Journal of Zoology* **209**, 169–181. doi:10.1111/j.1469-7998.1986.tb03573.x

Lee AK, Martin RW (1988) *The Koala: A Natural History*. New South Wales University Press, Sydney.

Lentle RG, Hume ID (2010) Mesial drift and mesial shift in the molars of four species of wallaby: the influence of chewing mechanics on tooth movement in a group of species with an unusual mode of jaw action. In *Macropods: The Biology of Kangaroos, Wallabies, and Rat-kangaroos*. (Eds G Coulson and M Eldridge) pp. 127–136. CSIRO Publishing, Melbourne.

Lommer MJ, Verstraete FJ, Terpak CH (2000) Dental radiographic technique in cats. *The Compendium on Continuing Education for the Practicing Veterinarian* **22**, 107–117.

Luckett WP, Wooley PA (1996) Ontogeny and homology of the dentition in dasyurid marsupials: development in Sminthopsis virginiae. *Journal of Mammalian Evolution* **3**, 327–364. doi:10.1007/BF02077449

Lyne A (1982) Observations on skull growth and eruption of teeth in the marsupial bandicoot Perameles nasuta (Marsupialia: Peramelidae). *Australian Mammalogy* **5**, 113–126.

McKay GM (1989) Petauridae. In *Fauna of Australia, Volume 1B, Mammalia*. (Eds DW Walton and BJ Richardson) pp. 665–678. Australian Government Publishing Service, Canberra.

McKay GM, Winter JW (1989) Phalangeridae. In *Fauna of Australia, Volume 1B, Mammalia*. (Eds DW Walton and BJ Richardson) pp. 636–651. Australian Government Publishing Service, Canberra.

Merrilees D (1967) Cranial and mandibular characters of modern mainland wombats (Marsupiala vombatida(e) from a palaeontological viewpoint, and their bearing on the on the fossils called Phascolomys Parvus by Owen (1872) *Records of the South Australian Museum* **15**, 399–418.

Musser A, Archer M (1998) New information about the skull and dentary of the Miocene platypus Obdurodon dicksoni, and a discussion of ornithorhynchid relationships. *Philosophical*

Transactions of the Royal Society of London Series B, Biological Sciences **353**, 1063–1079. doi:10.1098/rstb.1998.0266

Poulton EB (1887) True teeth in the young Ornithorhynchus paradoxus. *Proceedings of the Royal Society of London* **43**, 353–356. doi:10.1098/rspl.1887.0147

Sanson G (1980) The morphology and occlusion of the molariform cheek teeth in some Macropodinae (Marsupialia: Macropodidae). *Australian Journal of Zoology* **28**, 341–365. doi:10.1071/ZO9800341

Sanson GD (1989) Morphological adaptatios of teeth to diets and feeding in the Macropodoidea. In *Kangaroos, Wallabies and Rat-Kangaroos, Volume 1* (Eds G Grigg, P Jerman and ID Hume) pp. 151–168. Surry Beatty, Sydney.

Simpson GG, Burrell H (1929) The dentition of Ornithorhynchus as evidence of its affinities. *American Museum Novitates* **390**, 1–15.

Smith KK (2006) Craniofacial development in marsupial mammals: developmental origins of evolutionary change. *Developmental Dynamics* **235**, 1181–1193. doi:10.1002/dvdy.20676

Strahan R (1978) What is a koala? In *Proceedings of the Taronga Symposium on Koala Biology, Management and Medicine.* 11–12 March, Sydney. (Ed. TJ Bergin) p. 3. Zoological Parks Board of NSW, John Sands Pty Ltd, Sydney.

Thomas O (1887) On the homologies and succession of the teeth in the Dasyuridae, with an attempt to trace the history of the evolution of mammalian teeth in general. *Proceedings of the Royal Society of London* **42**, 310–312. doi:10.1098/rspl.1887.0059

Thomas O (1889) On the dentition of Ornithorhynchus. *Proceedings of the Royal Society of London* **46**, 126–131. doi:10.1098/rspl.1889.0019

Triggs B (2009) *Wombats.* CSIRO Publishing, Melbourne.

Tsugawa AJ, Verstraete FJ (2000) How to obtain and interpret periodontal radiographs in dogs. *Clinical Techniques in Small Animal Practice* **15**, 204–210. doi:10.1053/svms.2000.21042

Turner V, McKay GM (1989) Burramyidae. In *Fauna of Australia, Volume 1B, Mammalia.* (Eds DW Walton and BJ Richardson) pp. 652–664. Australian Government Publishing Service, Canberra.

Tyndale-Biscoe H (2005) Consummate kangaroo. In *Life of Marsupials* (Ed. H Tyndale-Biscoe) pp. 287–364. CSIRO Publishing, Melbourne.

van Nievelt AF, Smith KK (2005) To replace or not to replace: the significance of reduced functional tooth replacement in marsupial and placental mammals. *Paleobiology* **31**, 324–346. doi:10.1666/0094-8373(2005)031[0324:TRONTR]2.0.CO;2

Vogelnest L, Portas T (2008) Macropods. In *Medicine of Australian Mammals.* (Eds L Vogelnest and R Woods) pp. 133–226. CSIRO Publishing, Melbourne.

Wilson J, Hill J (1897) Observations upon the development and succession of teeth in Perameles; together with a contribution to the discussion of the homologies of teeth in marsupial animals. *The Quarterly Journal of Microscopical Science* **39**, 427–588.

Wilson J, Hill J (1907) Memoirs: observations on tooth-development in Ornithorhynchus. *The Quarterly Journal of Microscopical Science* **2**, 137–166.

Young AH (1881) Anatomy of the koala (*Phascolarctos cinereus*). *Journal of Anatomy and Physiology* **15**, 466–474.

12

Radiographic pathology case studies

Frances Hulst, Graeme Allan and Larry Vogelnest

Radiology is a fundamental tool in veterinary medicine, used for the diagnosis and monitoring of injuries and diseases. This chapter is not intended to provide a comprehensive review of the radiographic pathology of Australian mammals. Instead, cases have been selected to provide an appreciation of the radiographic findings seen in some common or noteworthy diseases of Australian mammals. The cases have been loosely organised along the anatomical regions affected.

Many Australian mammals have body shapes and anatomical features that do not conform to the more familiar domestic species. Recognition of radiographic abnormalities in species with which the clinician has little or no experience can be difficult and the normal radiographic images in the relevant chapters of this book should be used for comparison. Identification and interpretation of radiographic changes follow the same principles as those used in domestic animals. An orderly, logical and consistent approach to examination of images, without solely focusing on the area of interest, will ensure no abnormalities are missed. The roentgen sign approach to describing radiographic abnormalities is a useful method, as it involves describing abnormal radiographic morphology in terms of any change in *size, shape, location, number, margination* or *opacity*. Any radiographic abnormality can be described using one or more of these terms (Thrall 2013).

Radiographic changes are not necessarily specific for a particular aetiology or disease process. The radiographic appearance can provide support for a diagnosis suspected on the basis of signallment and history and can guide the choice of further diagnostics to reach a definitive diagnosis. Knowledge of common diseases of the species will assist in forming a list of differential diagnoses (Vogelnest and Woods 2008; Ladds 2009).

Many of the cases presented in this chapter involve animals in permanent care, such as those in zoological collections, but some are rescued wild animals. The cryptic nature and crepuscular or nocturnal habits of many Australian mammals, combined with the frequent need for chemical restraint in order to conduct a physical examination, makes survey radiographs a useful tool for reaching an early diagnosis and making timely decisions on treatment options based on welfare and prognosis for rehabilitation and release.

Unless otherwise indicated, cases in this chapter are from Taronga Wildlife Hospital and Taronga Zoo, Sydney, Australia.

Bones and joints

Fractures

Trauma resulting in fracture of bones is common in free-ranging Australian mammals, particularly where human factors such as fences, traps, motor vehicle accidents and predation by domestic pets are involved. Examples of some common fractures that may easily be missed without careful assessment are presented.

Fracture of the second cervical vertebra in macropods

Fracture of the second cervical vertebra (C2) is a relatively common injury in macropods, particularly after collision with fences or other obstacles. Animals will usually collapse immediately and become unconscious. Head trauma is often suspected and the neck fracture may be overlooked. The occipito-atlantal articulation is very flexible in macropods, allowing the head to remain level when hopping at high speed. This may be associated with altered forces at this articulation and other cervical joints, and may account for the high incidence of traumatic cervical fractures (Vogelnest and Portas 2008).

Agile wallaby, male, 9 years old

This agile wallaby (*Macropus agilis*) was found in its enclosure, recumbent and unresponsive to stimuli. It was hypothermic and both pupils were dilated with minimal light response although palpebral reflexes were present. Ventrodorsal and lateral neck radiographs (Fig. 12.1) revealed a fracture of C2. The wallaby was euthanased and post-mortem examination confirmed a fracture transecting the body of C2.

In juvenile macropods, the physis between the cranial and caudal buttress of C2 is clearly visible on radiographs and can be mistaken for a fracture (Fig. 12.2). It is therefore important to differentiate between the normal physis and a fracture by careful evaluation for displacement in both the lateral and ventrodorsal projections.

Although C2 is a common site for injury in macropods, cervical fractures do occur at other sites and a systematic examination of images is needed in every case.

Fractures in the short-beaked echidna

Motor vehicle accidents are the most common cause of traumatic injuries in the short-beaked echidna (*Tachyglossus aculeatus*). Fractures of the

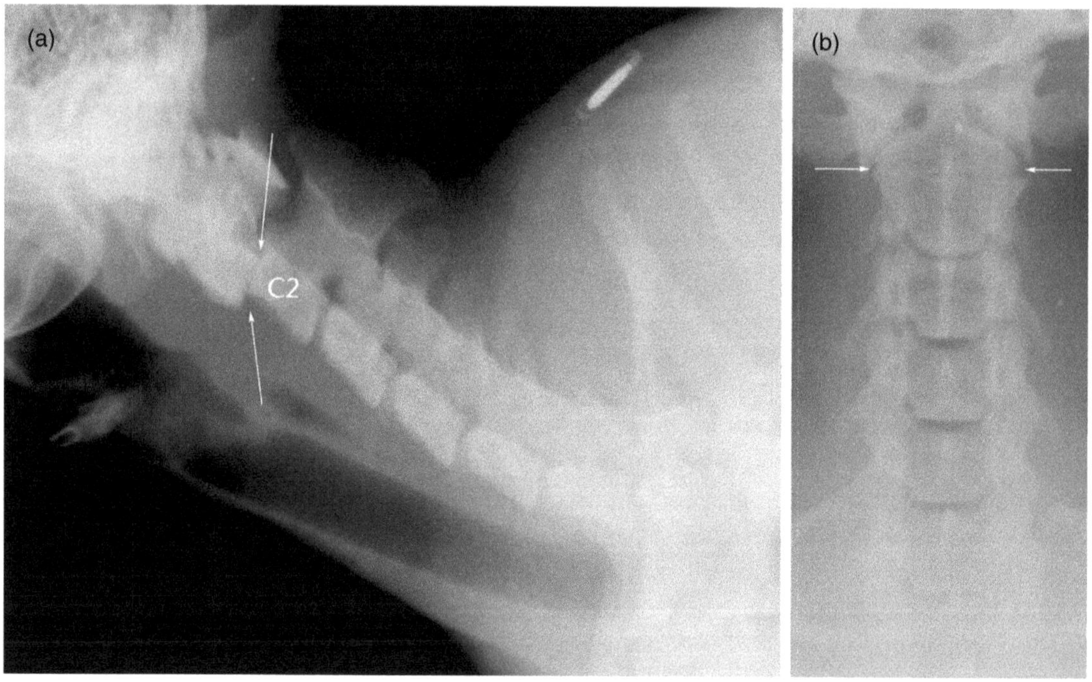

Figure 12.1: Cervical vertebrae, agile wallaby (*Macropus agilis*): (a) lateral view, (b) ventrodorsal view. The second cervical vertebra (C2) is fractured transversely through the body (arrows). (a) The caudal part of C2 is displaced dorsally, impinging on the vertebral canal. (b) There is minimal displacement at the fracture site in this view, reinforcing the need for two views to fully identify the injury. Note: a transponder can be seen in (a).

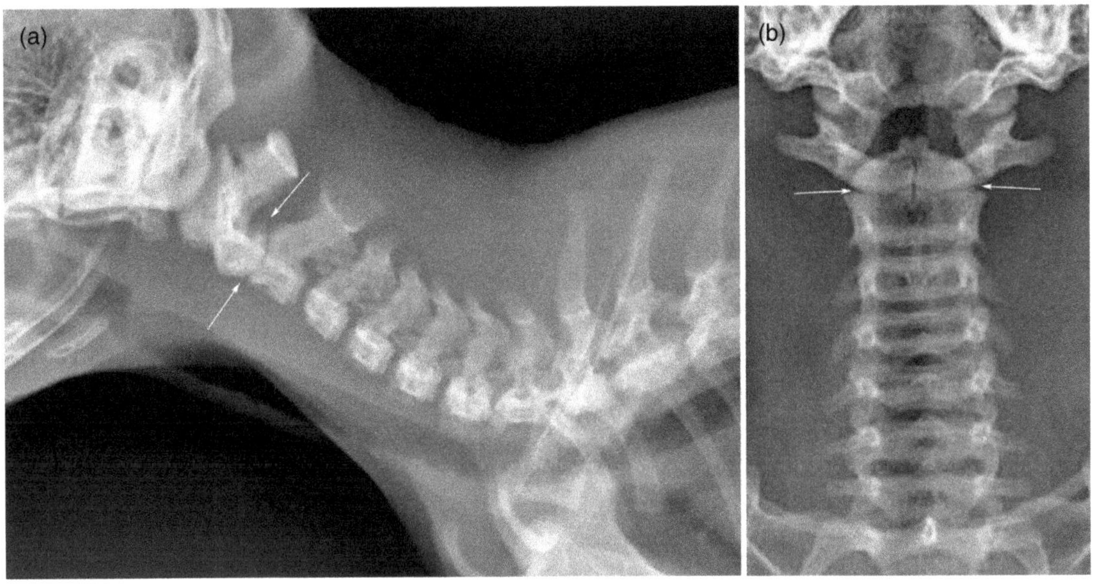

Figure 12.2: Cervical vertebrae, juvenile brush-tailed rock-wallaby (*Petrogale penicillata*): (a) lateral view, (b) ventrodorsal view. In this normal juvenile wallaby, there are several open physes in the cervical vertebral column, most notably between the odontoid process and the body of C2 (arrows). This could be mistaken for a fracture, but is a normal feature of growing macropods. In juveniles, it may be a weak point that is predisposed to traumatic separation.

delicate and relatively unprotected bones of the rostral beak are common and may affect the echidna's ability to feed or breathe (Middleton 2008). Fractures of other bones may also occur and are often difficult to recognise on physical examination because of the short-beaked echidna's unusual gait, compact body shape and the presence of spines. All injured short-beaked echidnas should be anaesthetised to allow correct positioning for whole-body survey radiographs using two views, in addition to beak radiographs, in order to evaluate the full extent of the animal's injuries (Fig. 12.3). Concurrent fractures will affect the prognosis for rehabilitation and release.

The set of images in Fig. 12.4 are from three wild short-beaked echidnas that were each presumed to have been hit by a car. The images demonstrate a range of common rostral beak fractures.

Short-beaked echidna, wild female, adult

This short-beaked echidna was found with a circular fitting from a garden watering hose embedded around its beak. The animal was thin with no other external wounds and had normal mobility. Whole-body survey radiographs (Fig. 12.5a, b) were taken under anaesthesia to assess the animal for non-evident injuries. On initial examination, the spine appeared normal, however, a fracture of the 15th thoracic vertebra (T15) was suspected when images were magnified (Fig. 12.5c, d). A CT scan confirmed a fracture of T15 (Fig. 12.5e). No beak fractures were evident after removal of the foreign body. The short-beaked echidna has a very short spinal cord, ending at the level of T7 (Ashwell and Zhang 1997). Fractures of the caudal thoracic and lumbar spine therefore may not be accompanied by neurological signs.

Dog predation injuries, platypus

The platypus (*Ornithorhynchus anatinus*) has a compact body shape, thick pelage and short limbs with an unusual gait that can make identification of bone fractures difficult without radiography. Motor vehicle trauma or predation by introduced species, especially dogs, is a common cause of trauma in the platypus (Connolly *et al.* 1998).

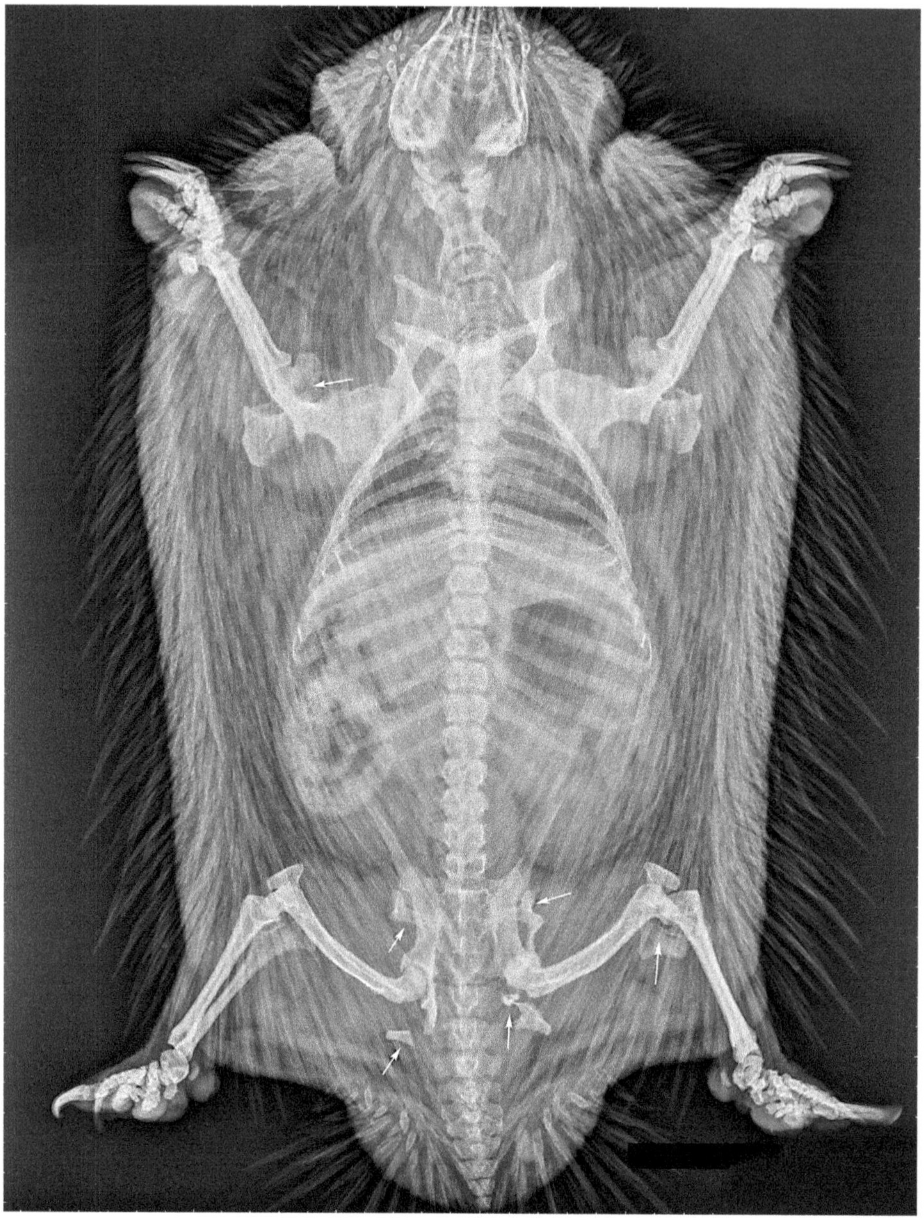

Figure 12.3: Whole body, short-beaked echidna (*Tachyglossus aculeatus*), ventrodorsal view. The beak fracture is not shown in this image, but there are multiple other fractures (arrows). The scapulae are distorted, there is a distal right humeral condylar fracture, multiple pelvic fractures and a displaced proximal left fibular fracture.

Platypus, female, 18 years old

This platypus had been dug out of its burrow by a dog as a 4–5-month-old juvenile and had sustained fractures to multiple ribs and left femur. The fractures were managed conservatively and the platypus was maintained in permanent care because of the poor prognosis for survival in the wild. Whole-body survey radiographs (Fig. 12.6) were taken as part of a health assessment 17 years later. At this time, the animal was considered to have small stature, but was able to walk and swim normally. This case is an example of bony injuries caused by the canine teeth of a large dog. The injuries could not be palpated when the animal was

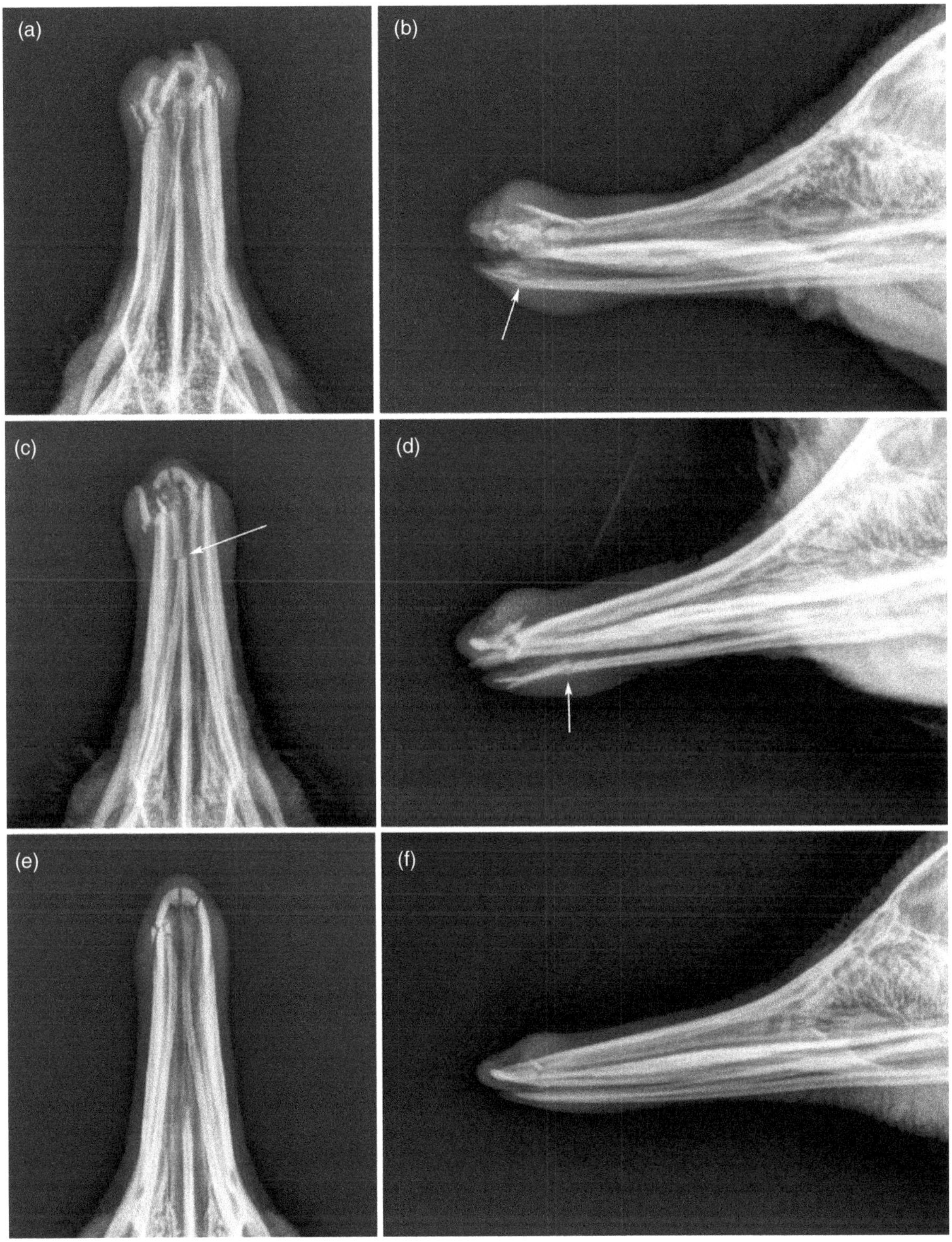

Figure 12.4: Skull, short-beaked echidna (*Tachyglossus aculeatus*): (a, c, e) dorsoventral views, (b, d, f) lateral views. (a–d) Displaced fractures of the beak, with peripheral soft tissue swelling. The rostral mandibles are also fractured (arrows). In both cases, the animals had occlusion of the nares, resulting in dyspnoea, and were unable to eat. (e, f) A non-displaced fracture of the left nasal bone, together with a minimally displaced comminuted fracture of the right maxilla with two small fragments. There is a small amount of soft tissue swelling. Such fractures may be managed conservatively with success.

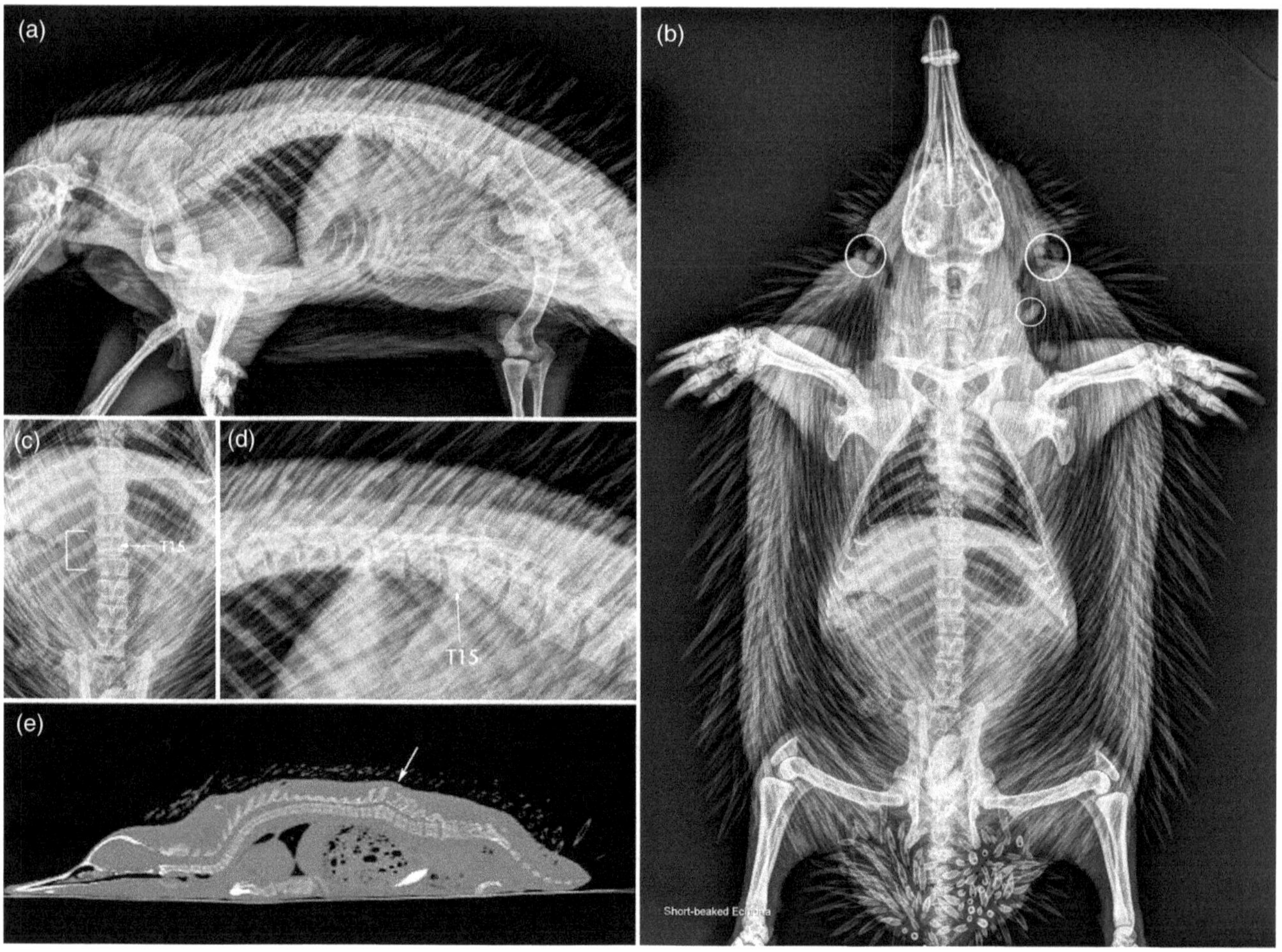

Figure 12.5: Short-beaked echidna (*Tachyglossus aculeatus*): (a) whole-body lateral view, (b) whole-body ventrodorsal view, (c) thoracic spine, magnified ventrodorsal view, (d) thoracic spine, magnified lateral view, (e) whole body CT reconstruction, mid-sagittal plane. There is a compression fracture of T15, with slight rotation and dorsal displacement of T16, which initially is not clear on the images in (a) and (b), because of the overlying spines, but is better defined (square bracket) on the same images magnified in (c) and (d). The location of the trauma (arrow) is clearly defined in the CT reconstruction in (e). Note the garden hose fitting around the beak and the engorged ticks around the neck (circled) in (b).

first presented, but the spacing of the fractures when the platypus was young corresponded with that of the canine teeth of the offending dog (a Labrador retriever).

Metabolic bone disease

Macropod pouch young

Metabolic bone disease (MBD) may occur in orphaned marsupials hand-reared on artificial milk formulas. Multiple factors may contribute to the development of MBD (McCracken 2008) and detection may be delayed because of the length of time the animal remains confined to an artificial pouch. The condition is also seen in pteropodid bats (Olsson and Woods 2008).

Swamp wallaby, female, pouch young, 1.04 kg

This orphaned wild swamp wallaby (*Wallabia bicolour*) joey had been rescued at 100 g bodyweight and hand-reared for 9 weeks on artificial milk formula. The joey was considered overweight for its size and was unable to stand. It had swelling and distortion

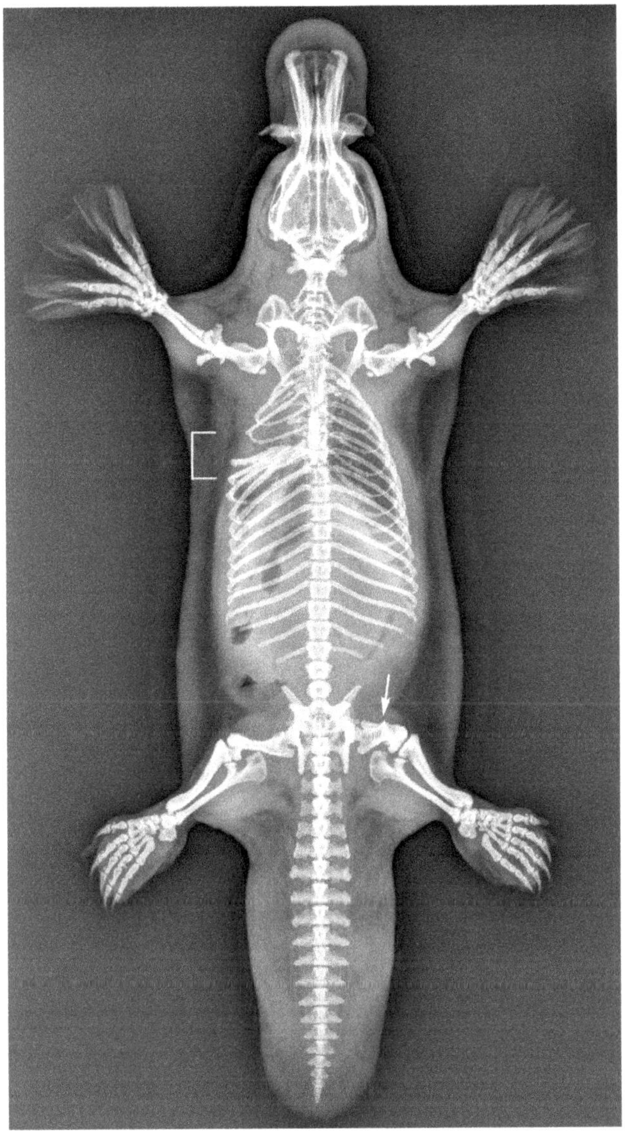

Figure 12.6: Whole body, platypus (*Ornithorhynchus anatinus*), ventrodorsal view. There is a row of fractured ribs (ribs 4–7) on the right side, leading to a chest wall deficit (square bracket). There is a healed, over-riding diaphyseal fracture of the left femur (arrow), resulting in shortening of the limb. Note: a transponder can be seen.

of the proximal tail. A series of whole-body and hindlimb radiographs were taken under anaesthesia (Figs 12.7 and 12.8). MBD was suspected. Oral vitamin D and calcium supplementation together with pouch rest and daily access to sunlight resulted in some improvement in limb strength; however, the joey was euthanased 5 weeks later because it was unable to walk normally. Histopathology confirmed increased osteoblast activity in affected bone.

Case submitted by: Dr Michelle Campbell-Ward, Taronga Western Plains Zoo, NSW, Australia.

Red kangaroo, male, pouch young, 2.07 kg

This orphaned wild red kangaroo (*Macropus rufus*) joey had been hand-reared for a period of 3 months and was presented with sudden onset of painful swelling of the mid-tail with no history of trauma. Radiographs (Fig. 12.9) showed a compression fracture of a caudal vertebra at the level of the swelling. Over the next month, the joey sustained two additional spontaneous fractures involving the left and right distal tibia and was euthanased. Histopathology of the tibia revealed osteoclastic resorption of the bone with fibrous replacement. Endochondral ossification appeared to be progressing normally at other sites. Renal glomerulosclerosis was also identified, suggesting that renal secondary hyperparathyroidism may have been the underlying disease process.

Case submitted by: Dr Michelle Campbell-Ward, Taronga Western Plains Zoo, NSW, Australia.

Koalas

Metabolic bone disease has been reported in parent-reared koalas (*Phascolarctos cinereus*) housed indoors with no UV exposure in zoos in the northern hemisphere (Figs 12.10, 12.11 and 12.12). No clinical signs were reported for joeys with confirmed MBD (Pye *et al.* 2013). All the joeys with MBD also had hip dysplasia, shoulder dysplasia or both (see cases under joint dysplasia).

Cases submitted by: Dr Geoff Pye, Disney's Animal Kingdom, FL, USA and images reproduced with permission from: Pye GW, Gait SC, Mulot B, Real de Asu MD, Martinez-Nevado E, Bonar CJ, Baines SJ, Baines EA (2013) Metabolic bone disease

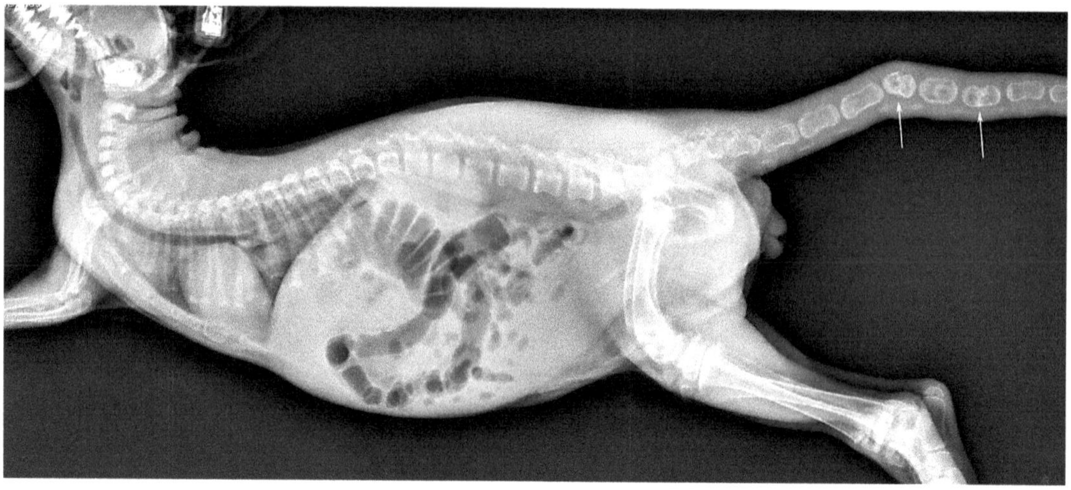

Figure 12.7: Whole body, juvenile swamp wallaby (*Wallabia bicolor*), lateral view. The skeleton is diffusely osteopenic and the cortices of the bones are thin. The skull shows loss of bone opacity. There are two pathological fractures of caudal vertebrae (arrows).

in juvenile koalas (*Phascolartcos cinereus*). *Journal of Zoo and Wildlife Medicine* 44, 273–279.

Spinal disease

Spondylosis

Spondylosis is a common incidental finding in older marsupials held in permanent care. As in domestic species, the cause is often idiopathic and the findings clinically insignificant. Figures 12.13 and 12.14 depict spondylosis as an incidental finding. Figures 12.15 and 12.16 are from the same long-nosed potoroo (*Potorous tridactylus*), which had a visibly deviated tail but was still able to ambulate well.

Spinal osteoarthrosis

Squirrel glider, male, 6 years old

This squirrel glider (*Petaurus norfolcensis*) moved with an abnormally hunched back and appeared to have restricted movement of the left shoulder when climbing. Whole-body radiographs (Fig. 12.17) revealed extensive degenerative changes in the caudal thoracic and lumbar spine. The glider failed to improve with treatment and was euthanased. On post-mortem examination the intervertebral discs appeared to be bulging slightly into the spinal canal, but there was no evidence of rupture or associated degeneration of the spinal cord. Discospondylitis was ruled out histologically.

Arthritis

Osteoarthritis

Southern hairy-nosed wombat, female, 21 years old

The radiograph in Fig. 12.18 was obtained as part of the routine health assessment of this southern hairy-nosed wombat (*Lasiorhinus latifrons*). Within weeks, the animal developed lameness in the right hindlimb that was unresponsive to medical treatment and a femoral head excision arthroplasty was performed. Histopathology on excised femoral head tissue was consistent with inflammatory degenerative joint disease with extensive loss of cartilage. The wombat recovered well from surgery, but was euthanased 5 months later because of congestive heart failure.

Case submitted by: Dr David McLelland, Zoos South Australia, SA, Australia.

Tasmanian devil, male, 3 years old

This Tasmanian devil (*Sarcophilus harrisii*) was part of a captive breeding program and had no history of previous illness. It presented with hindlimb lameness. Under anaesthesia, there was crepitus

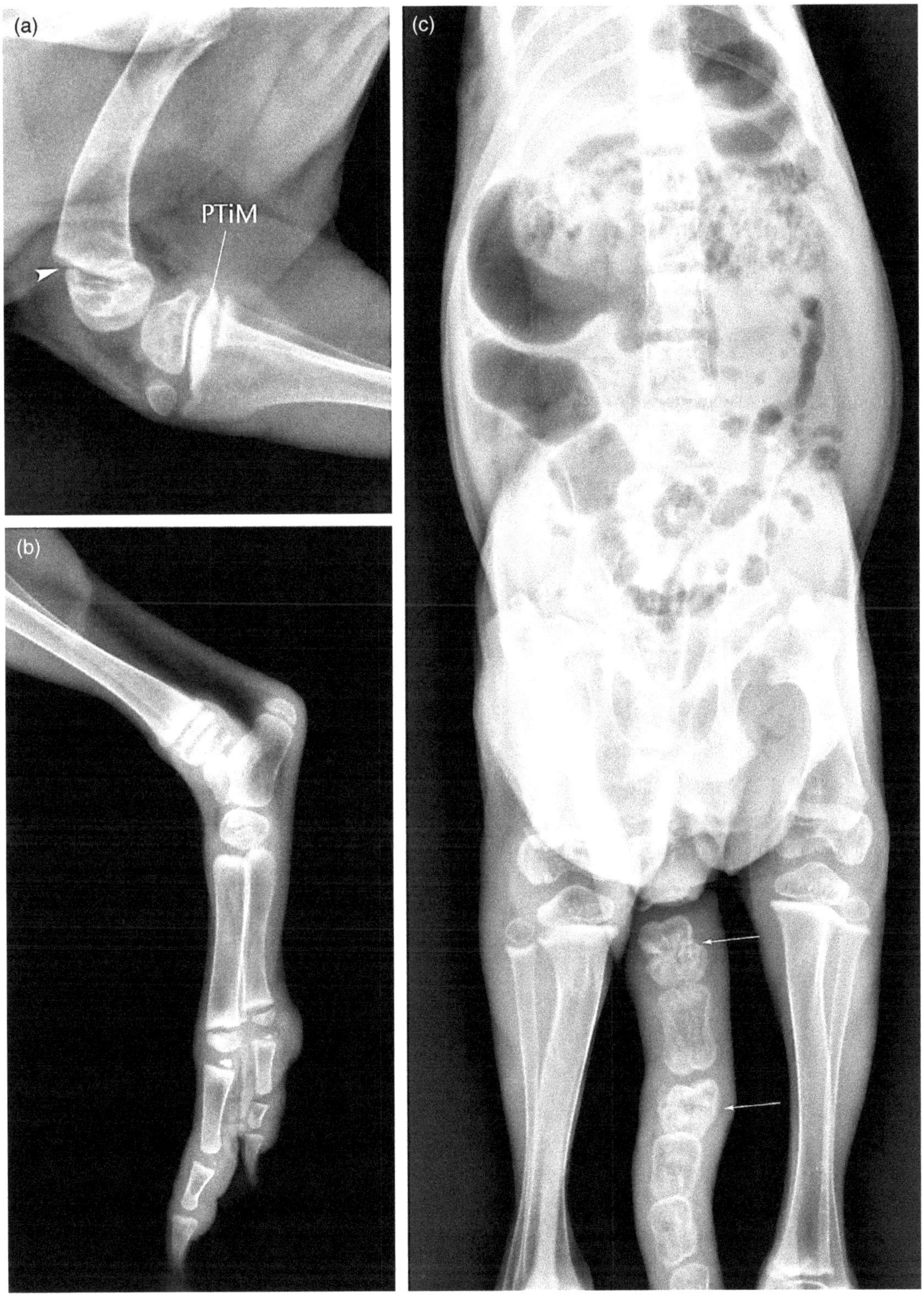

Figure 12.8: Juvenile swamp wallaby (*Wallabia bicolor*): (a) stifle, lateral view, (b) tarsus and pes, lateral view, (c) abdomen and hindlimbs, ventrodorsal view. The caudal vertebral fractures are indicated (arrows). The opacity of the lumbar vertebral column is reduced and the vertebrae are not distinct. The cortices of the long bones in the distal limb are thin. There is possible caudal displacement of the distal right femoral epiphysis (arrowhead). Proximal tibial metaphyses (PTiM) are increased in opacity adjacent to the growth plate, which indicates active endochondral ossification in the presence of bone resorption elsewhere.

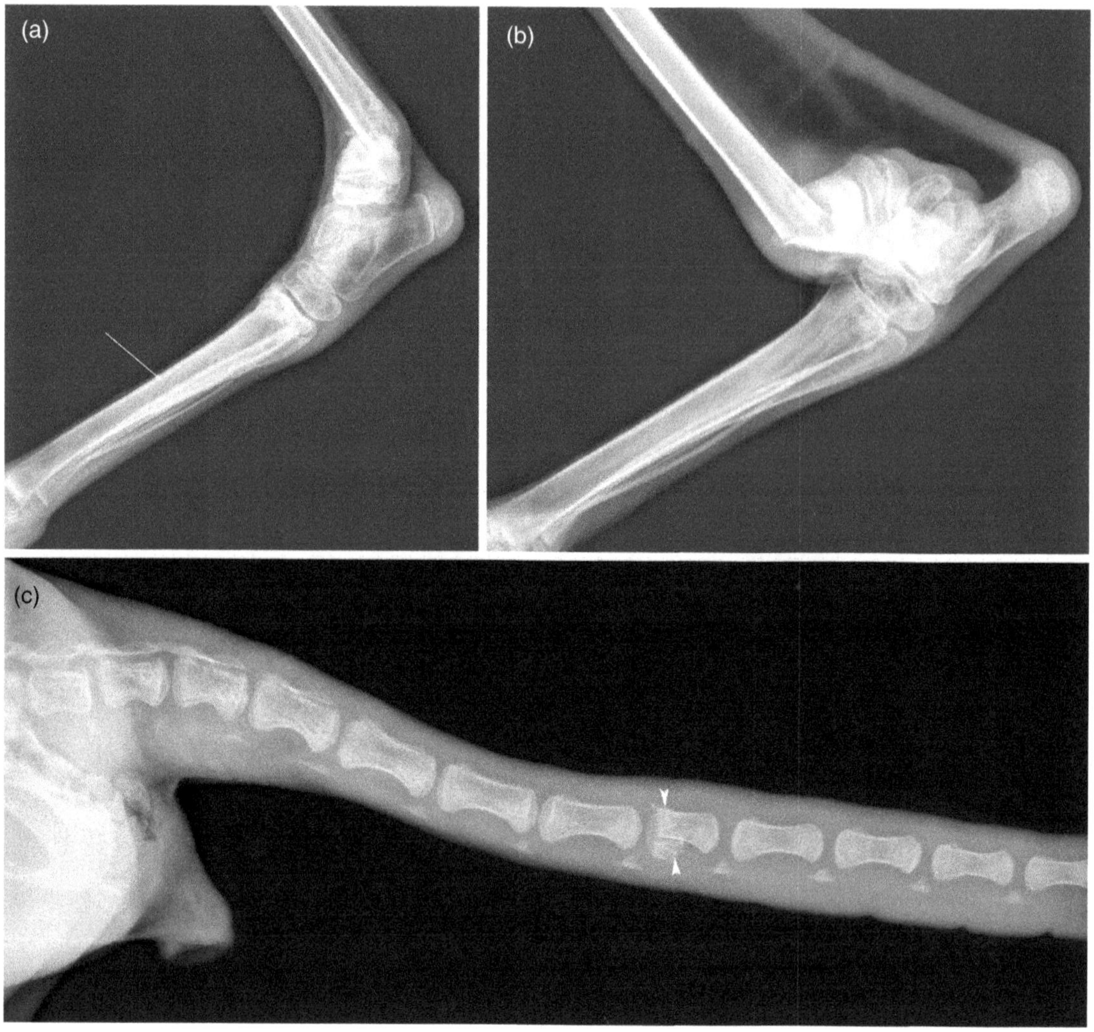

Figure 12.9: Juvenile red kangaroo (*Macropus rufus*): (a) left tarsus, mediolateral view, (b) right tarsus, mediolateral view, (c) tail, lateral view. The left and right distal tibias are fractured transversely at the junction of the diaphyses and metaphyses. Although the fractures appear minimally displaced, the tibial diaphyses are rotated, cranially in the left and caudally in the right limb. A non-displaced spiral fracture is present in the left 4th metatarsal bone (arrow) and there is a dorsally displaced transverse fracture of a caudal vertebra (arrowheads). The bones of the hock regions and the tail appear poorly mineralised and the cortices appear thinner than expected at this stage of bone development, consistent with secondary hyperparathyroidism.

and a reduced range of movement in the right hip. Radiography of the pelvis revealed the probable cause (Fig. 12.19). There were no other abnormal findings on clinical examination.

Case submitted by: Dr Franciscus Scheelings, Healesville Sanctuary, VIC, Australia.

Southern hairy-nosed wombat, female, 12 years old

This southern hairy-nosed wombat had a 29-month history of shifting and progressive hindlimb and right forelimb lameness. Initial radiographs showed osteoarthritis of both hips and tarsi, and the right carpus with digital degenerative arthropathies (Fig. 12.20). Both tarsi developed moderate soft tissue swelling over the calcaneus and the right carpus was also variably hot and swollen. The wombat was at times profoundly lame. Medical management and bodyweight reduction improved quality of life for 2 years, after which its condition deteriorated and the wombat was euthanased.

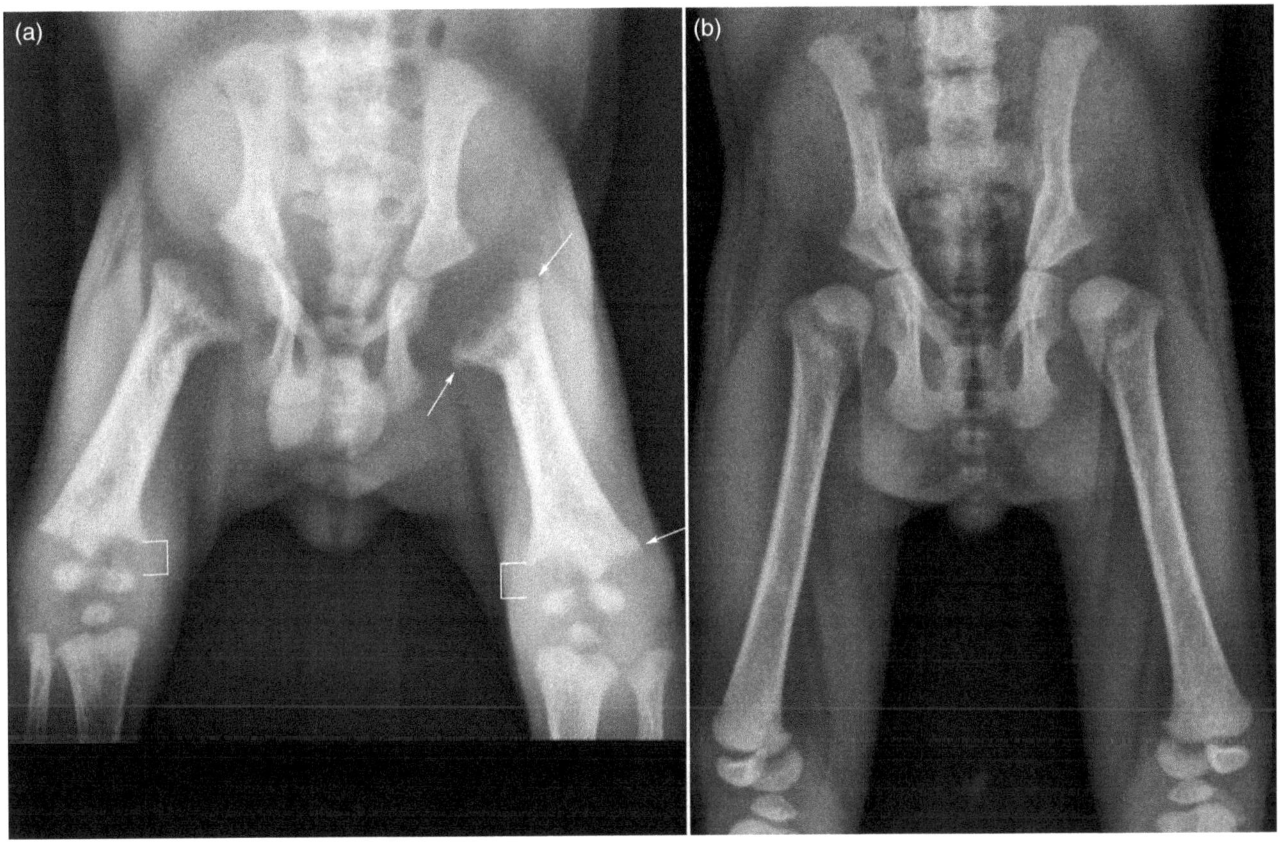

Figure 12.10: Pelvis, 34-week-old koala (*Phascolarctos cinereus*), ventrodorsal view: (a) affected koala, (b) normal koala. In the affected koala, the proximal and distal femoral metaphyses are flared (arrows). The femoral head and neck epiphyses are absent and the distal femoral physes (brackets) are wide. These changes are most characteristic of rickets, although most animals affected by rickets have concurrent secondary hyperparathyroidism.

Osteomyelitis

The radiographic appearance of osteomyelitis in Australian mammals does not differ to that in other animals. Disseminated cryptococcosis involving bone distant to the nasal cavity is a possible cause of osteomyelitis that should be considered in species prone to cryptococcal infection.

Cryptococcosis

Koala, male, ~3 years old

This wild koala was found near a road, lame in the hindquarters and suspected to have been hit by a motor vehicle. The koala was euthanased after radiographs revealed advanced lytic changes in the left humerus and the right wing of the ilium (Fig. 12.21). Post-mortem examination and histopathology confirmed cryptococcal osteomyelitis in both sites. The cryptococcal species was not identified. There was no gross evidence of respiratory cryptococcal granulomas.

Case submitted by: Dr Amber Gillett, Australia Zoo Wildlife Hospital, QLD, Australia.

Cryptococcosis occurs more frequently in koalas than in domestic animals in Australia, with most cases caused by *Cryptococcus gattii*. This increased frequency is thought to be related to both the organism and the koala's close association with eucalyptus trees (Blanshard and Bodley 2008). The most likely route of cryptococcal infection in koalas is through inhalation of airborne organisms from the environment. The most common areas affected are thus the respiratory tract and, by direct extension from the nasal sinuses, the central nervous

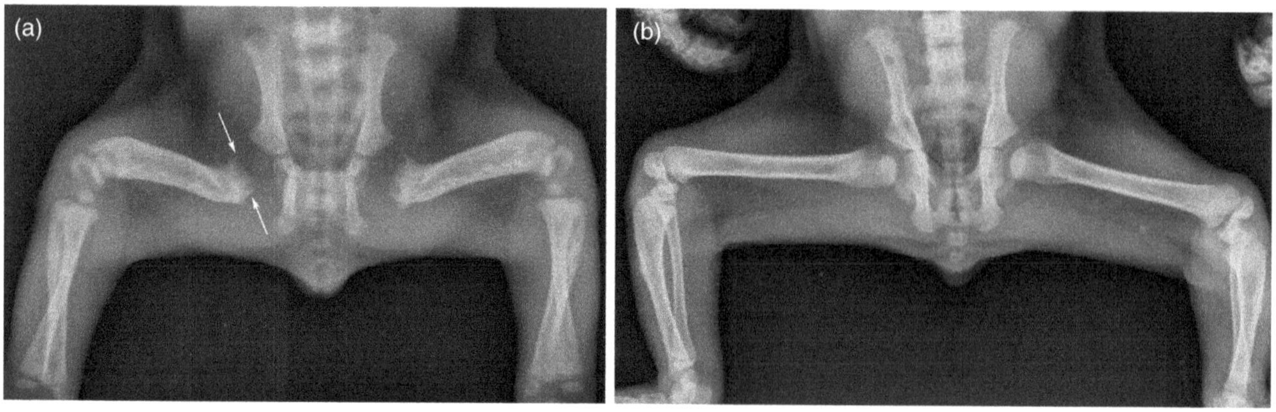

Figure 12.11: Pelvis, 34-week-old koala (*Phascolarctos cinereus*), frog-leg ventrodorsal view: (a) affected koala, (b) normal koala. In the affected koala, the proximal metaphyses are flared (arrows) and the femoral head and neck have not yet developed.

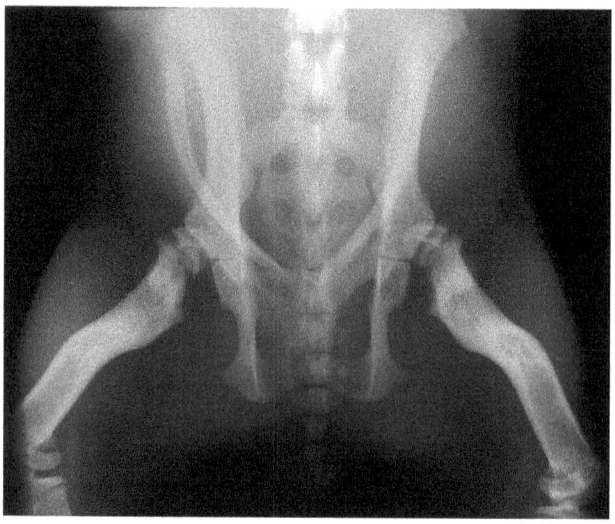

Figure 12.12: Pelvis, koala (*Phascolarctos cinereus*), ventrodorsal view. There is a long bone (femoral) deformity resulting from metabolic bone disease. The femur has remodelled with a sigmoid diaphyseal curvature.

system. Dissemination to other sites appears common (Krockenberger *et al.* 2003), however, bone involvement appears rare and spread is likely to be haematogenous.

Bone tumours in the koala

Tumours of mixed cartilage and bone have been reported in both wild koalas and those in permanent care (Canfield *et al.* 1987), most commonly affecting the bones of the head, but also the pelvis, ribs, clavicle and long bones (Hanger *et al.* 2003). The tumours are slow-growing and expansive rather than erosive. It is unusual for tumours to develop in more than one site in affected koalas.

Osteochondroma

Koala, male, 5 years old

This koala had a 6-month history of right forelimb lameness, although its appetite and demeanour remained normal. No abnormalities were noted on physical examination or on radiographs at the time of initial presentation (Fig. 12.22a), although subtle bony changes were evident on retrospective review. The koala developed atrophy of muscles of the right shoulder, together with paresis of the limb. Bony swelling of the proximal antebrachium developed over several weeks. Radiographs taken 6.5 months (Fig. 12.22b) and 10.5 months (Fig. 12.22c) from the onset of lameness showed marked bony proliferation. A biopsy confirmed an osteochondroma, considered to show low-grade malignancy.

Case submitted by: Dr Robert Johnson, South Penrith Veterinary Clinic, NSW, Australia.

Koala, male, 6 years old

This koala developed a slow-growing bony swelling on the left zygomatic arch. A bone biopsy collected under anaesthesia was not diagnostic, but radiographs showed bone proliferation consistent

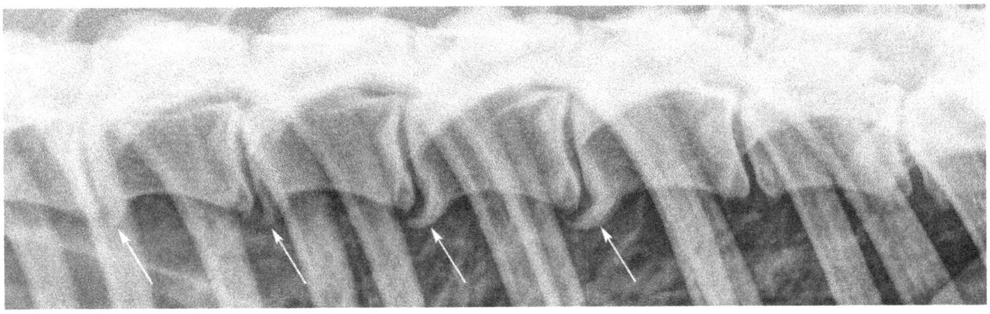

Figure 12.13: Thoracic spine, red kangaroo (*Macropus rufus*), lateral view. Large, curving spurs of bone on the ventral aspect of the cranial end of several thoracic vertebrae (arrows) represent spondylosis.

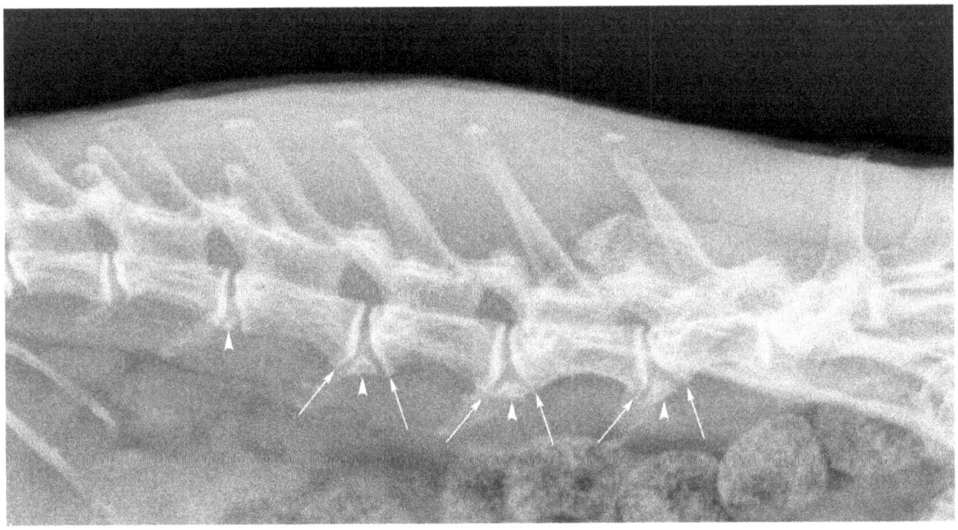

Figure 12.14: Lumbar spine, 5-year-old greater bilby (*Macrotis lagotis*), lateral view. Mineralisation of the annulus or ventral longitudinal ligament (arrowheads) is producing a bridge between adjacent spondylosing changes (arrows).

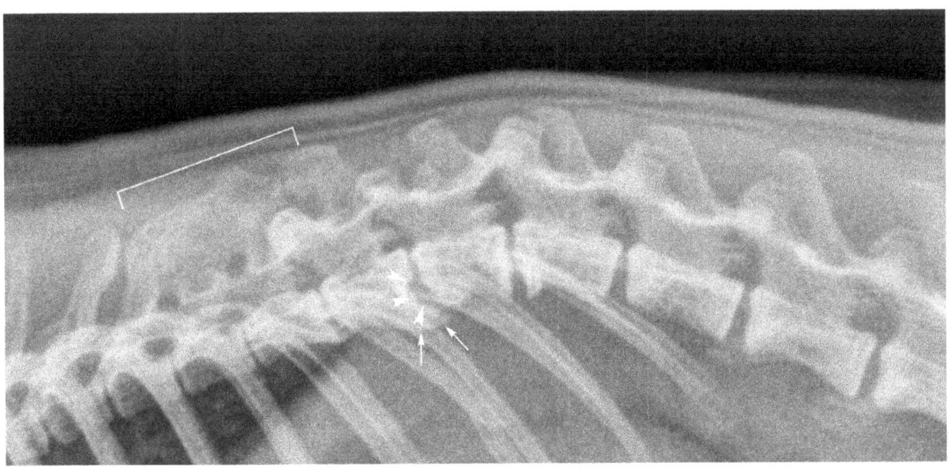

Figure 12.15: Thoracolumbar spine, 7-year-old long-nosed potoroo (*Potorous tridactylus*), lateral view. Spondylopathy is tending to ankylose adjacent vertebrae (arrows). Note the signs of end-plate erosion (arrowheads), which may be indicative of present or previous infection (discospondylitis). Several dorsal spinous processes are fused (square bracket). This may be a congenital anomaly or could be secondary to trauma.

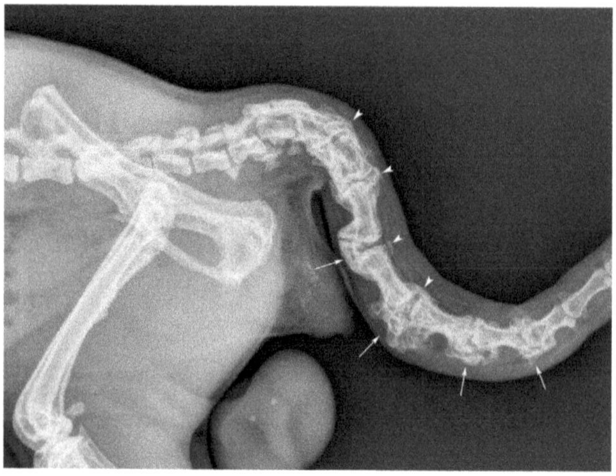

Figure 12.16: Pelvis and tail, 7-year-old long-nosed potoroo (*Potorous tridactylus*), lateral view. Extensive ankylosing spondylopathy is fusing many of the caudal vertebrae, both ventrally (arrows) and dorsally (arrowheads).

with an osteochondroma (Fig. 12.23). Approximately 2.5 years later, the tumour was considerably larger and a second bony growth was present on the left caudal mandible, causing considerable facial distortion. Another bony mass was present on the sacrum (Fig. 12.24). The koala was euthanased. At post-mortem examination, extensive proliferative osteochondroma of the left mandible, left zygomatic arch and left sacrum were identified.

Case submitted by: Dr Robert Johnson, South Penrith Veterinary Clinic, NSW, Australia.

Joint dysplasia in the koala

Hip dysplasia

Hip dysplasia has been diagnosed in captive-bred koalas held in a North American zoo. Dysplastic changes were readily apparent by the time the growth plates were closed at approximately 2 years of age. Affected koalas were mostly asymptomatic, unless secondary degenerative joint disease ensued (Pye *et al.* 2008). To evaluate the hip joints fully, three standard radiographic views should be used, as described in Chapter 5.

Figure 12.25 demonstrates three views of the increasing degree of severity of coxofemoral malformation encountered in koalas with hip dysplasia. These changes include coxofemoral joint incongruity, flattening of the femoral head and acetabulum, and development of periarticular

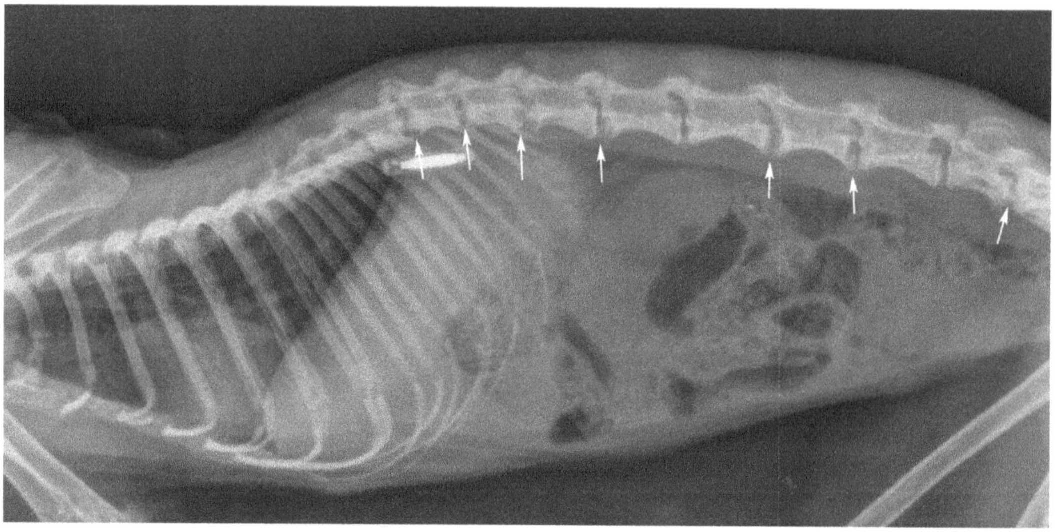

Figure 12.17: Thoracolumbar spine, squirrel glider (*Petaurus norfolcensis*), lateral view. Many intervertebral disc spaces are wider than normal and there is reduced opacity in the adjacent vertebral end-plates (arrows). These changes are strongly suggestive of an inflammatory disorder in the intervertebral disc extending into the adjacent vertebral end-plates. Discospondylitis is the usual cause of these changes. The histology in this case indicated degenerative arthrosis, which may be secondary to earlier discospondylitis. Note: a transponder can be seen.

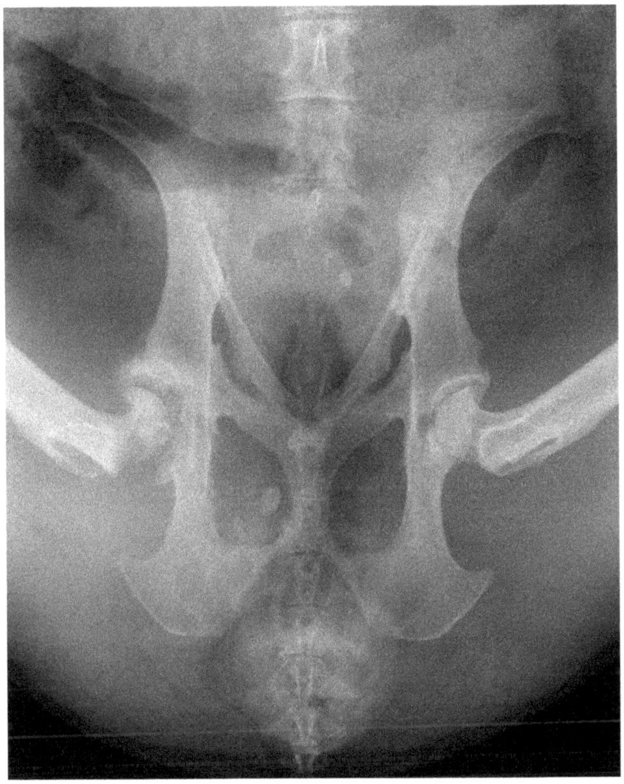

Figure 12.18: Pelvis, southern hairy-nosed wombat (*Lasiorhinus latifrons*), ventrodorsal view. The right coxofemoral joint has poorly defined margins, a result of a mixture of erosive and proliferative articular changes. The left coxofemoral joint appears normal. The changes in the right coxofemoral joint are consistent with osteoarthritis.

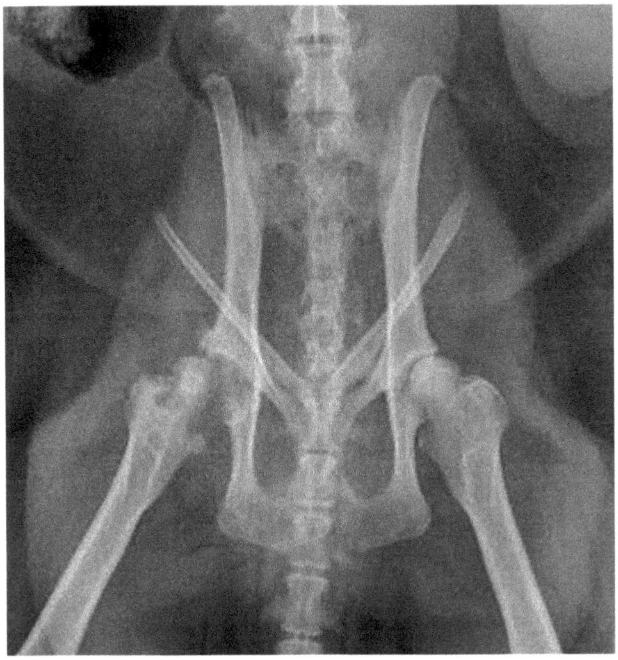

Figure 12.19: Pelvis, Tasmanian devil (*Sarcophilus harrisii*), ventrodorsal view. The right coxofemoral joint appears subluxated. Periosteal proliferative new bone surrounds the acetabulum and the proximal femur. There appear to be lytic or erosive changes within the acetabulum and the adjacent femoral head. The left coxofemoral joint appears normal. The changes in the right coxofemoral joint indicate an erosive arthropathy. An infectious cause (septic arthritis) or severe degenerative joint disease in association with synovial hyperplasia might cause these radiographic changes.

osteophytes. The images range from apparently normal to the worst outcome with total remodelling of the acetabulum and femoral head such that the coxofemoral joint no longer exists.

Cases submitted by: Dr Geoff Pye, Disney's Animal Kingdom, FL, USA and images reproduced with permission from: Pye GW, Hamlin-Andrus C, Moll J (2008) Hip dysplasia in koalas (*Phascolarctos cinereus*) at the San Diego Zoo. *Journal of Zoo and Wildlife Medicine* **39**, 64–68.

Shoulder dysplasia

A high proportion of koalas in the same North American zoo had radiographic evidence of bilateral shoulder dysplasia affecting the scapulae and humeri. Affected koalas were typically asymptomatic, unless secondary degenerative joint disease occurred. The cause of the dysplasia remains unknown, although a polygenic aetiology is suspected. There was a positive correlation between the degree of hip and shoulder dysplasia in this population of koalas (Pye 2009). To evaluate the shoulder joints fully, three standard radiographic views should be used, as described in Chapter 5.

The collages in Figures 12.26 and 12.27 show the variation in the extent and severity of changes seen in the shoulder joint of koalas affected by shoulder dysplasia. Changes include malformation of the supraglenoid and infraglenoid tubercles and the coracoid process, shallowing or loss of the glenoid cavity, flattening or loss of the humeral head, malformation of the greater and lesser tubercles, loss of

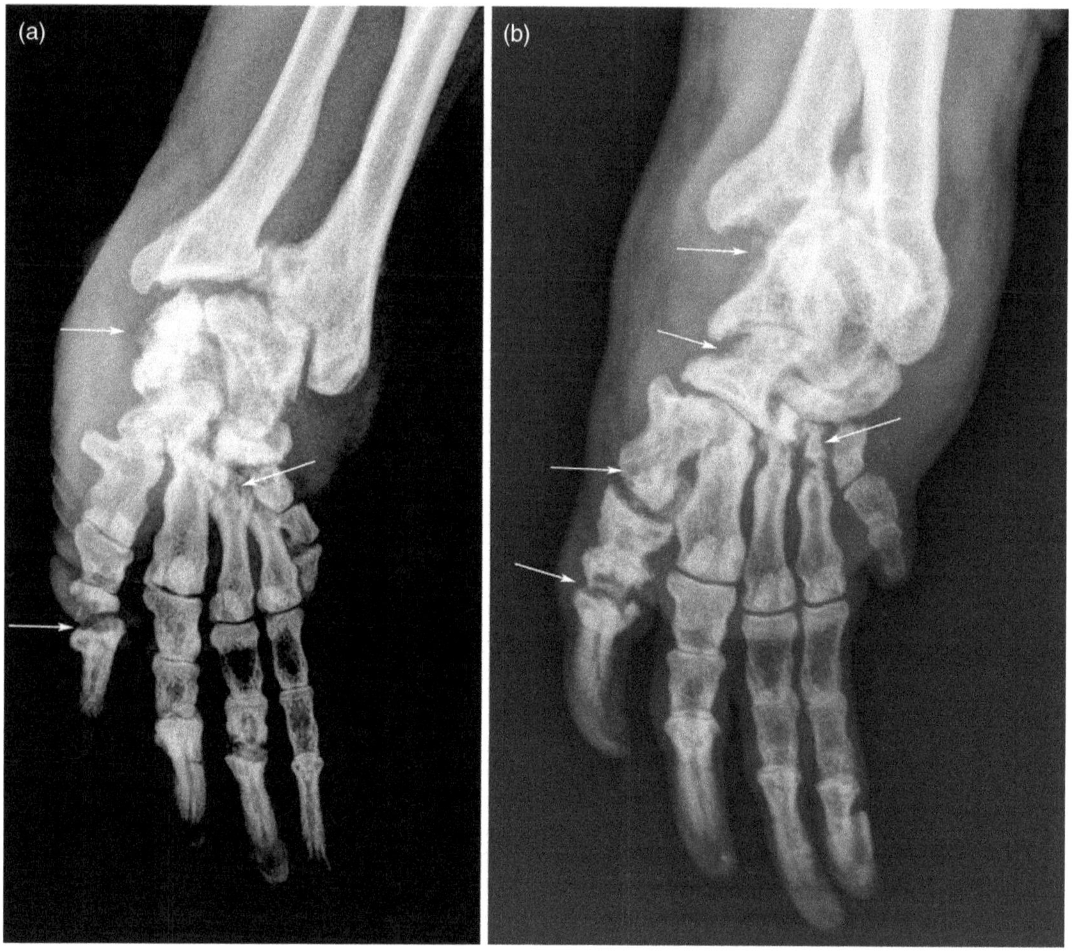

Figure 12.20: Pes, southern hairy-nosed wombat (*Lasiorhinus latifrons*), plantarodorsal view. There are many foci of bone erosion and bone proliferation (arrows). Most of the erosive changes are intra-articular and the proliferative changes are periarticular. The radiographic appearance is consistent with an erosive polyarthropathy. The image in (b) was taken 10 months after the image in (a). The changes in (b) are slightly more pronounced than in (a).

the intertubercular groove and humeral diaphyseal abnormalities.

Cases submitted by: Dr Geoff Pye, Disney's Animal Kingdom, FL, USA and images reproduced with permission from: Pye GW (2009) Shoulder dysplasia in koalas (*Phascolarctos cinereus*) at San Diego Zoo. *Journal of Zoo and Wildlife Medicine* **40**, 453–457.

Neither shoulder nor hip dysplasia has been reported in free-ranging koalas or those in permanent care in Australia. Osteoarthrosis of the hip joint has been reported and damage to one or both round ligaments of aged free-ranging koalas appears relatively common (Canfield and Spencer 1993), which could indicate an inherent laxity in the hip joints of koalas, perhaps as an adaption to an arboreal existence.

Special structures of the skull

Dental disease

Chapter 11 describes the normal dental radiographic appearance of a range of Australian mammals. Dental disease can occur in all species of Australian mammals, both free-ranging and those in permanent care; however, some taxonomic groups appear to be particularly susceptible to

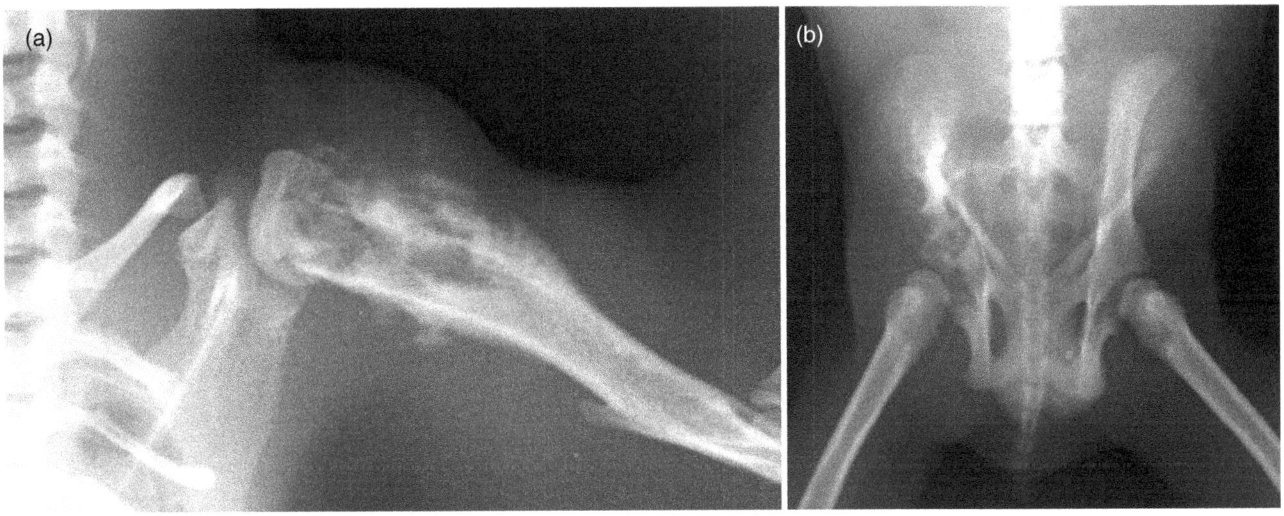

Figure 12.21: Koala (*Phascolarctos cinereus*): (a) shoulder, lateral view, (b) pelvis, ventrodorsal view. Poorly marginated loss of bone opacity is present in the left proximal humerus in (a) and the right ilium in (b) where it is encroaching on the dorsal aspect of the acetabulum. There is soft tissue mineralisation around each affected bone. The patchy, poorly demarcated lesions, together with the unstructured bone changes, indicate an osteo-aggressive disease. Because there are lesions in multiple locations, metastatic neoplasia and osteomyelitis are the most likely differential diagnoses.

developing dental disease. Because of the wide variation in anatomy and function of teeth across taxonomic groups, commonly occurring dental pathology will also vary between species. Dental malocclusion and fractures, caries and tartar involving the crown of the tooth will usually be obvious on clinical examination. Radiography is a useful tool for imaging the portion of the tooth that

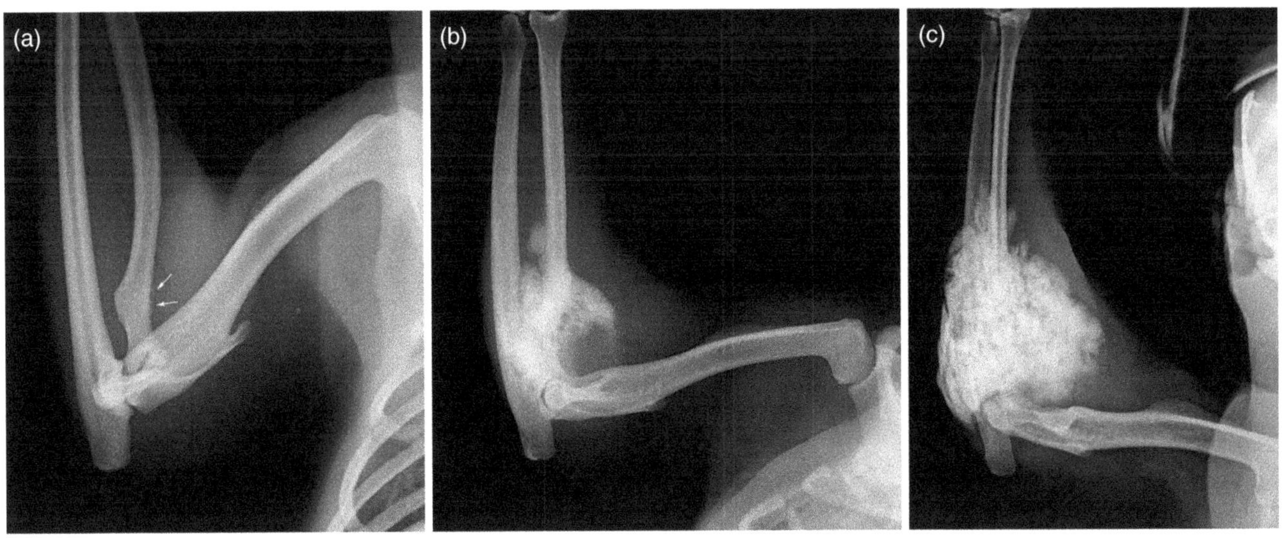

Figure 12.22: Antebrachium, koala (*Phascolarctos cinereus*), lateral view: (a) initial image, (b) 6.5 months after initial image, (c) 10.5 months after initial image. (a) Subtle periosteal proliferation (arrows) and adjacent soft tissue swelling. The subsequent image (b) reveals a large bone mass in the right proximal radius. There is an associated pathological fracture of the proximal radius and extensive bone proliferation into adjacent soft tissue and impingement on the ulna. Pathological fracture and irregular, extensive periosteal reaction suggests the lesion is aggressive indicating neoplasia or infection. (c) Extensive bony proliferation incorporates the proximal radius and ulna.

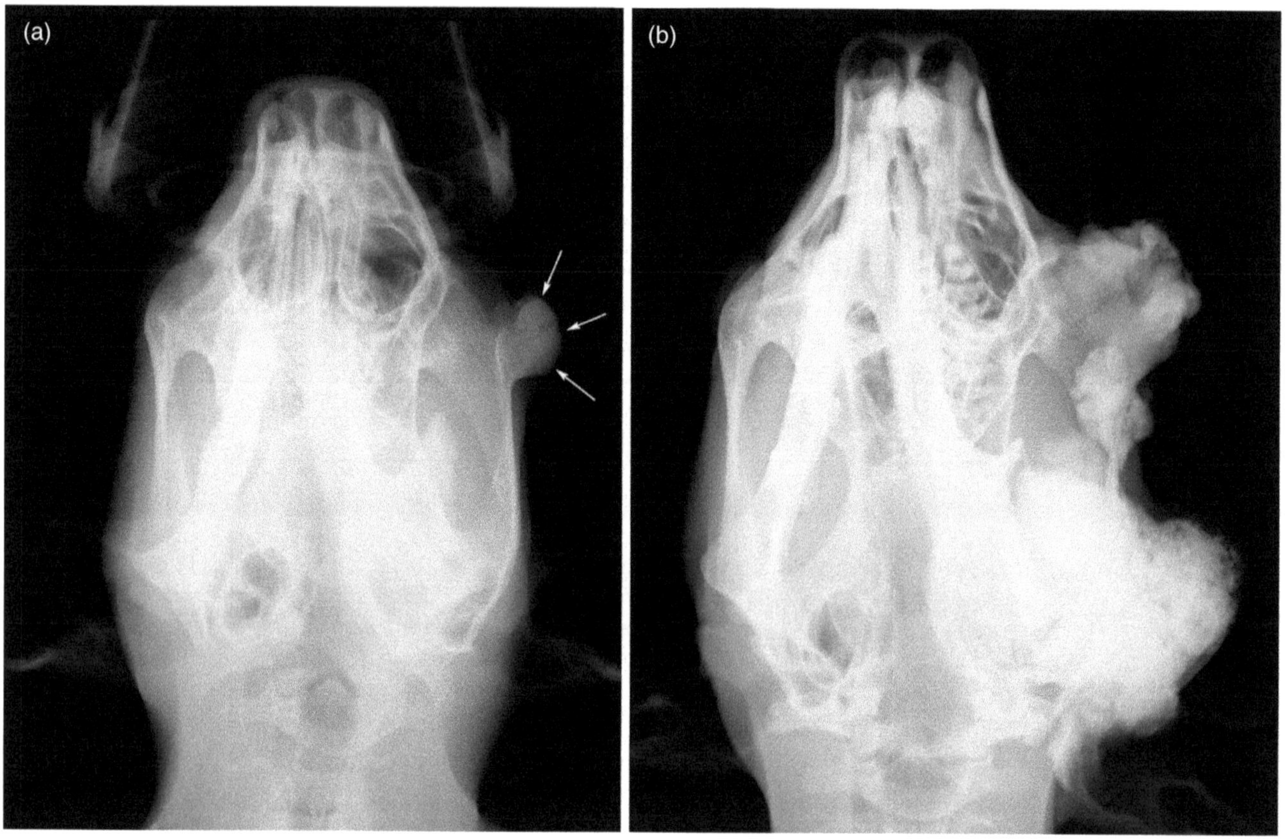

Figure 12.23: Skull, koala (*Phascolarctos cinereus*): (a) initial ventrodorsal oblique view, (b) ventrodorsal oblique view taken 28 months later. Images show progressive enlargement of a bone nodule arising from the rostral end of the zygomatic arch (arrows). (b) The zygomatic nodule has progressed to become a florid mass. A second mass impinges on the coronoid process of the mandible and zygomatic arch.

lies below the gum line where significant pathology may otherwise be overlooked. Diagnostic extra-oral radiographs can be obtained in most cases with careful positioning and the use of oblique projections (see Chapter 1). When additional information is required, intra-oral views using dental X-ray equipment are preferable. CT scans can accurately depict the extent of involvement of neighbouring structures in three dimensions, which may aid planning for surgical intervention.

Oral necrobacillosis in macropods

Oral necrobacillosis or 'lumpy jaw' is one of the most significant causes of illness and death in macropods, primarily affecting those in human care, but also occurring occasionally in free-ranging animals. Numerous published texts describe the pathology, epidemiology, treatment and management of this disease (Jackson 2007; Vogelnest and Portas 2008; Ladds 2009). *Fusobacterium necrophorum* is considered the main infectious agent. The lesions likely begin with a breach in the gingiva from trauma, molar progression or sharp food items. Disease commences as periodontitis and, if not contained, may progress to osteomyelitis of the alveolar bone and subsequently to necrosis of the surrounding mandible or maxilla. Extensive subperiosteal new bone growth and associated soft tissue swelling occurs in chronic cases, giving the condition its common name of 'lumpy jaw'.

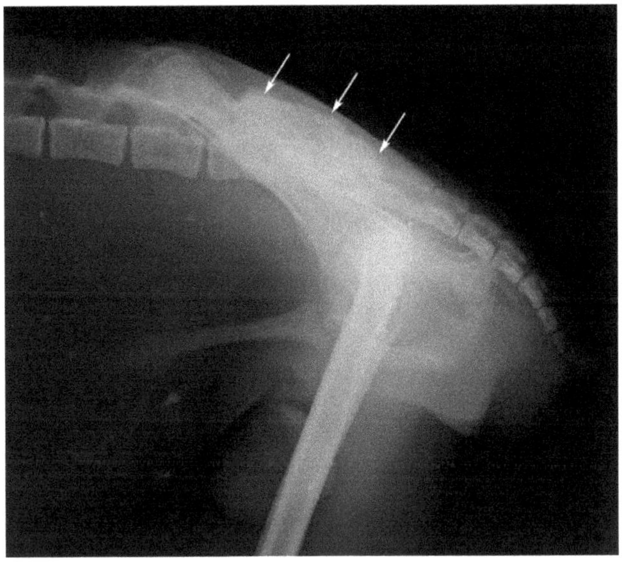

Figure 12.24: Pelvis, koala (*Phascolarctos cinereus*), lateral view. A bone growth (arrows) has developed on the sacrum of the same koala shown in Fig. 12.23.

Red kangaroo, female, 7 years old

This red kangaroo was first presented for lethargy and inappetence. Oral examination was unremarkable, but on extra-oral oblique skull radiographs (Fig. 12.28a) there was lysis of the alveolar bone around the roots of several maxillary cheek teeth, as well as root resorption. The diseased teeth were extracted and antibiotic and analgesic treatment commenced. Follow-up radiographs taken at a routine weekly debridement (Fig. 12.28b) showed a retained root fragment, which was subsequently extracted. The kangaroo responded well to treatment; no further lysis was noted on follow-up radiographs after 2 months and treatment ceased. However, 5 months later, the kangaroo was again inappetent with a left-sided facial swelling, partly closed left eye and ptyalism. Radiographs at this time (Fig. 12.28c) showed periodontitis of the third molar tooth, which was extracted, followed by 2 months of medical treatment together with weekly debridement under anaesthesia. The kangaroo was euthanased when it developed further disease of the left mandibular cheek teeth 5 months later.

Red-necked wallaby, female, 18 months old

This red-necked wallaby presented with a firm bony swelling midway along the right body of the mandible. On oral examination there was gingivitis and gingival recession at the level of the right mandibular incisor tooth. Radiographs showed marked bony proliferation and soft tissue swelling around the apex of the tooth (Fig. 12.29). Biopsies of the proliferative bone were submitted for culture and histopathology and the affected tooth was extracted. Histopathology of the bone revealed locally extensive, suppurative osteomyelitis. Culture revealed a mixture of Gram-negative obligate anaerobes. Soft tissue swelling resolved with antibiotic treatment and the bony lesion decreased in size.

Case submitted by: Dr Kathryn C. Gamble, Dr Lester E. Fisher and Dr Lauren M. Kao, Lincoln Park Zoo, IL, USA and Dr Janis Raines, Dallas Zoo and Children's Aquarium at Fair Park, TX, USA.

Red-necked wallaby, male, 6 years old

This red-necked wallaby (*Macropus rufogriseus*) was presented for loss of body condition and dysphagia. On oral examination there was an abscess involving the right mandibular first molar tooth, which was extracted. *Prevotella bivia* and *Fusobacterium* sp. were cultured from the abscess and after 6 weeks of antibiotic therapy repeat radiographs revealed no evidence of osteomyelitis. The animal developed a swollen jaw 1 month later and another tooth was extracted, together with a fragment of necrotic mandible. A CT scan of the head was undertaken (Fig. 12.30) to evaluate the extent of the bony changes and a further large mandibular sequestrum and a pathological fracture of the mandible were identified. A hemi-mandibulectomy was performed at the level of the caudal molar tooth and the animal was able to live without medication for the next 8 months before it died.

Case submitted by: Dr Nancy Carpenter and Dr Erika Crook, Utah's Hogle Zoo, UT, USA.

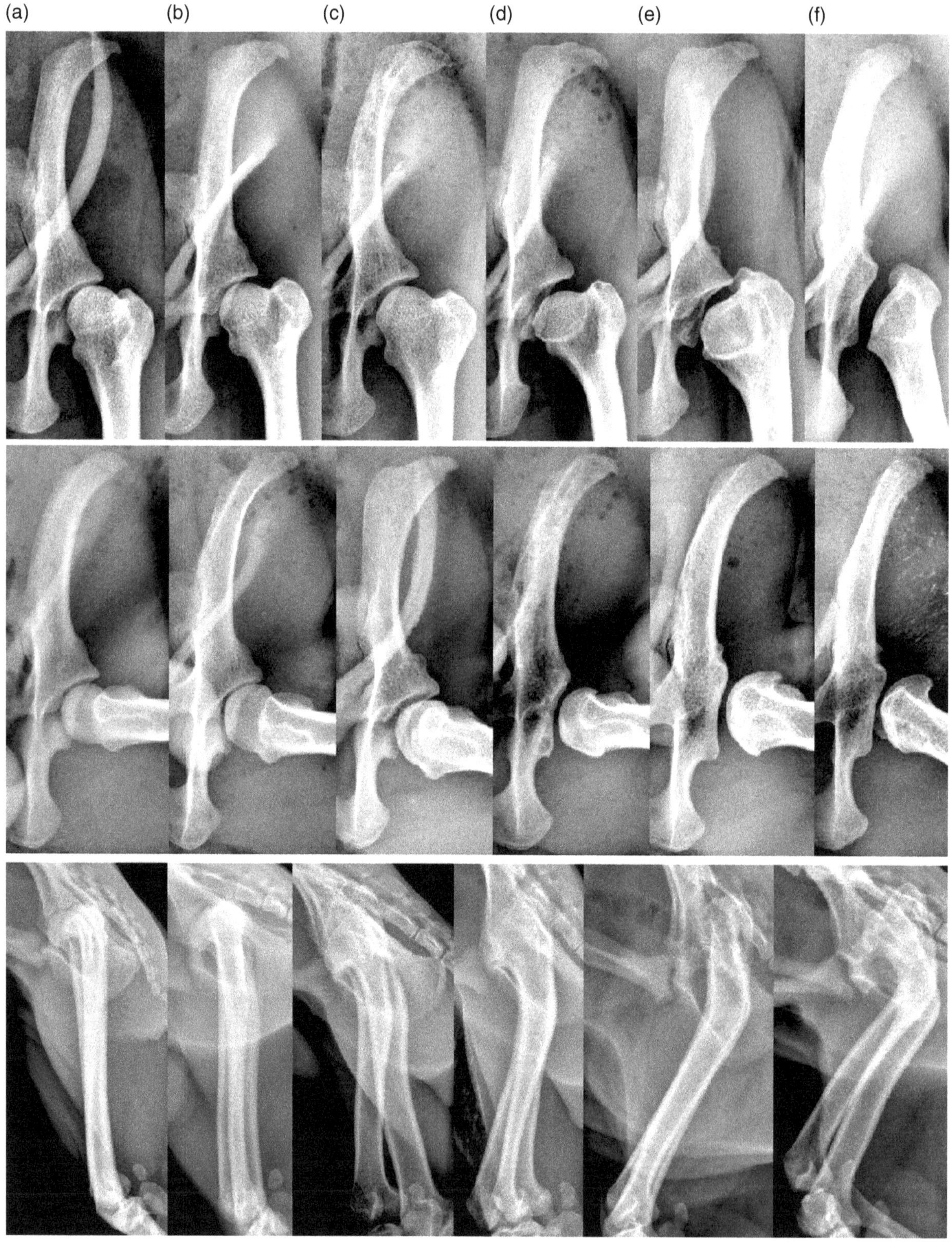

Figure 12.25: Coxofemoral joint, koala (*Phascolarctos cinereus*): (top) extended ventrodorsal view, (middle) frogleg ventrodorsal view, (bottom) lateral view. The images in column (a) represent normal joints, through to advanced dysplasia in column (f).

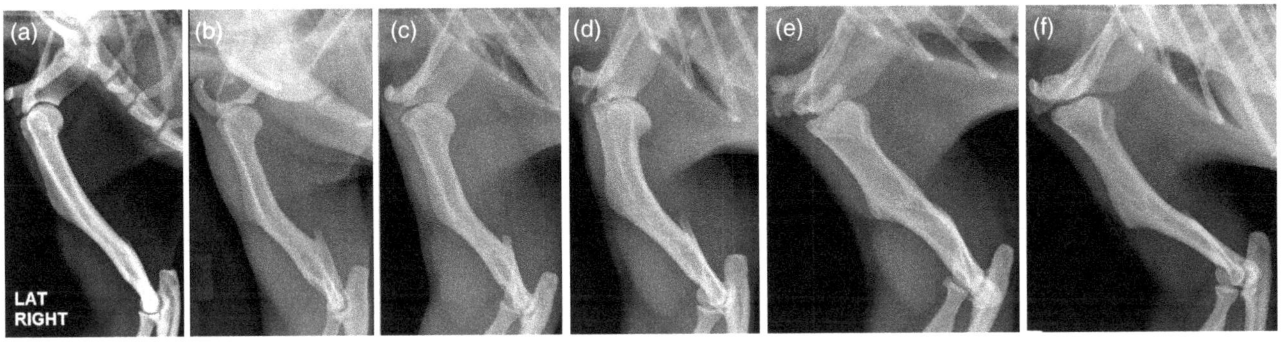

Figure 12.26: Shoulder and brachium, koala (*Phascolarctos cinereus*), lateral views: column (a) normal, through to column (f) advanced dysplasia.

Endodontic disease

Dental trauma may result in exposure of the pulp with subsequent apical abscess formation. Identifying the extent of infection, planning for endodontic treatment and monitoring of effectiveness of therapy relies on radiographic evaluation of the tooth.

Common wallaroo, unknown sex, adult

These intraoral dental radiographs shown in Figure 12.31 are of a cadaver specimen of a common wallaroo (*Macropus robustus*). The animal had a complicated crown fracture of the left mandibular incisor tooth. The radiograph shows the four radiographic signs of endodontic disease: loss of crown integrity, failure of the pulp canal to narrow, periapical lucency and root resorption.

Koala, male, 3 years old

This koala presented with persistent, mild unilateral ocular discharge because of obstruction of the corresponding nasolacrimal duct. There was facial swelling rostroventral to the affected eye. Oral examination showed that the maxillary first incisor tooth on this side was worn, but did not appear to have exposed pulp. Nasal discharge was absent. Skull radiographs taken at this time did not show any bony lysis around the tooth. CT reconstruction (Fig. 12.32) revealed lysis of the bone around the worn maxillary incisor tooth, consistent with tooth root abscessation. The affected tooth was removed, but the koala showed some ongoing dysphagia and deteriorating health.

Case submitted by: Dr Kate Bodley, Melbourne Zoo, VIC, Australia.

In the koala, the incisor teeth are used to manipulate branches and leaves, but are not used for cutting or mastication of leaves. Thus they may accumulate deposits of plaque and calculus that may lead to severe periodontitis and eventually abscess formation. In this koala, the exposed pulp was likely the route of entry of infection, although the cause of the dental wear was unknown.

Dental trauma leading to endodontic disease is common in carnivorous marsupials and may be seen in both free-ranging animals and those in permanent care.

Tasmanian devil, female, 4.5 years old

A Tasmanian devil housed in a zoo had sustained a complicated (exposed pulp) crown fracture of the right maxillary canine tooth at least 3 years prior to presentation. On oral examination, the tooth was greyish in colour and there was a purulent periodontal pocket on the mesiobuccal aspect of the tooth. Radiographs (Fig. 12.33) were taken before extraction of the tooth. The periapical infection had led to severe periodontitis and the formation of an oronasal fistula. The infection and fistula resolved after antibiotic treatment and extraction of the tooth.

Spot-tailed quoll, male, adult

This wild spot-tailed quoll (*Dasyurus maculatus*) was presented in poor body condition, lame on one

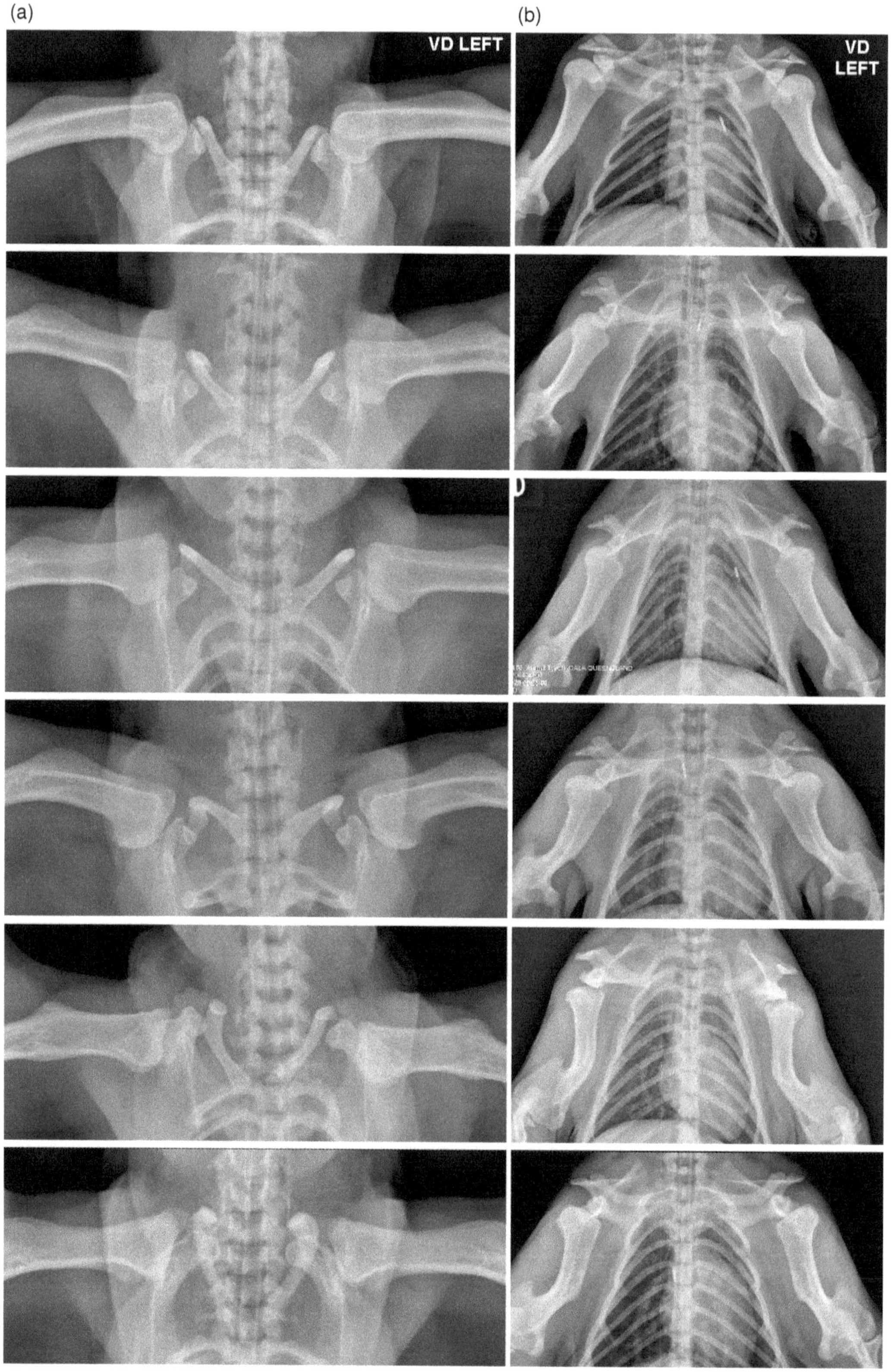

Figure 12.27: Shoulder and brachium, koala (*Phascolarctos cinereus*): column (a) ventrodorsal laterally extended views, column (b) ventrodorsal caudally extended views. (Top) Normal shoulders through to (bottom) advanced dysplasia.

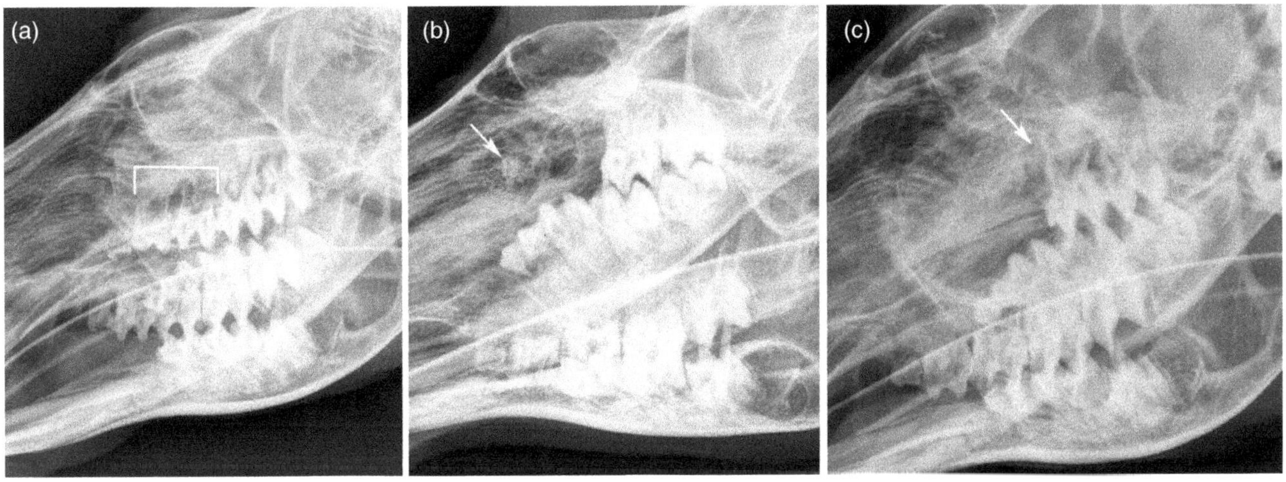

Figure 12.28: Skull, red kangaroo (*Macropus rufus*), dorsolateral-ventromedial oblique view. (a) Lysis of the alveolar bone around the roots of the left maxillary premolar and the 1st and 2nd molar tooth (square bracket). (b) Radiograph taken 2 weeks following the extractions shows the vacated alveoli of the 1st and 2nd molar teeth; a retained root fragment from the 1st molar tooth persists (arrow). (c) Radiograph obtained 5 months following initial extractions showing further extension of the lytic process around the roots of the left 3rd molar tooth (arrow).

hindlimb and with a large, purulent wound at the tail base. The mandibular canine teeth had complicated crown fractures and the exposed pulp was necrotic. The caudal maxillary and mandibular molar teeth were mobile and there was purulent discharge in the gingival sulci. Skull radiographs were taken to evaluate the extent of the dental disease (Fig. 12.34). The mandibular and maxillary canine teeth were misaligned because of abaxial rotation of both mandibular canines subsequent to periapical bone necrosis. There was periodontal disease of all teeth. The severity of the dental disease, together with leg and tail wounds, made the quoll unsuitable for rehabilitation and it was euthanased.

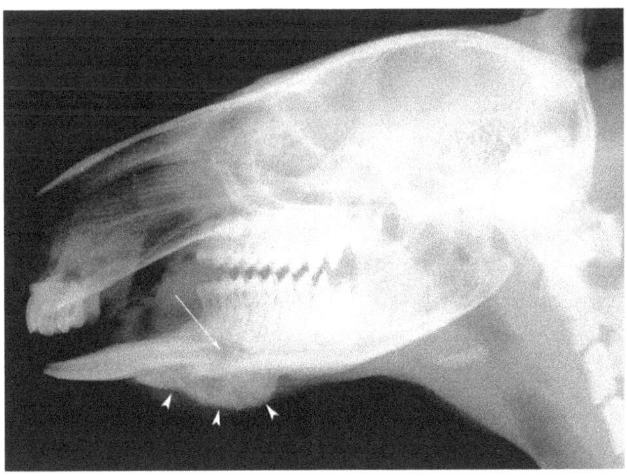

Figure 12.29: Skull, red-necked wallaby (*Macropus rufogriseus*), lateral view. There is a periapical lucency associated with the right mandibular incisor tooth (arrow), with expansive new bone formation of the surrounding mandibular bone (arrowheads). There is mild soft tissue swelling adjacent to the bony lesion.

Malocclusion in wombats

Chronic dental problems are common in wombats in permanent care (Bryant and Reiss 2008) and are primarily associated with their elodont (continually growing) dentition. Any tooth loss or malocclusion can result in abnormal tooth wear and associated sequelae. This often necessitates euthanasia in the case of rescued wild animals undergoing treatment and rehabilitation, or frequent correctional trimming for the life of the animal when in permanent care. All orphaned wombat joeys should undergo careful oral examination to exclude malocclusion.

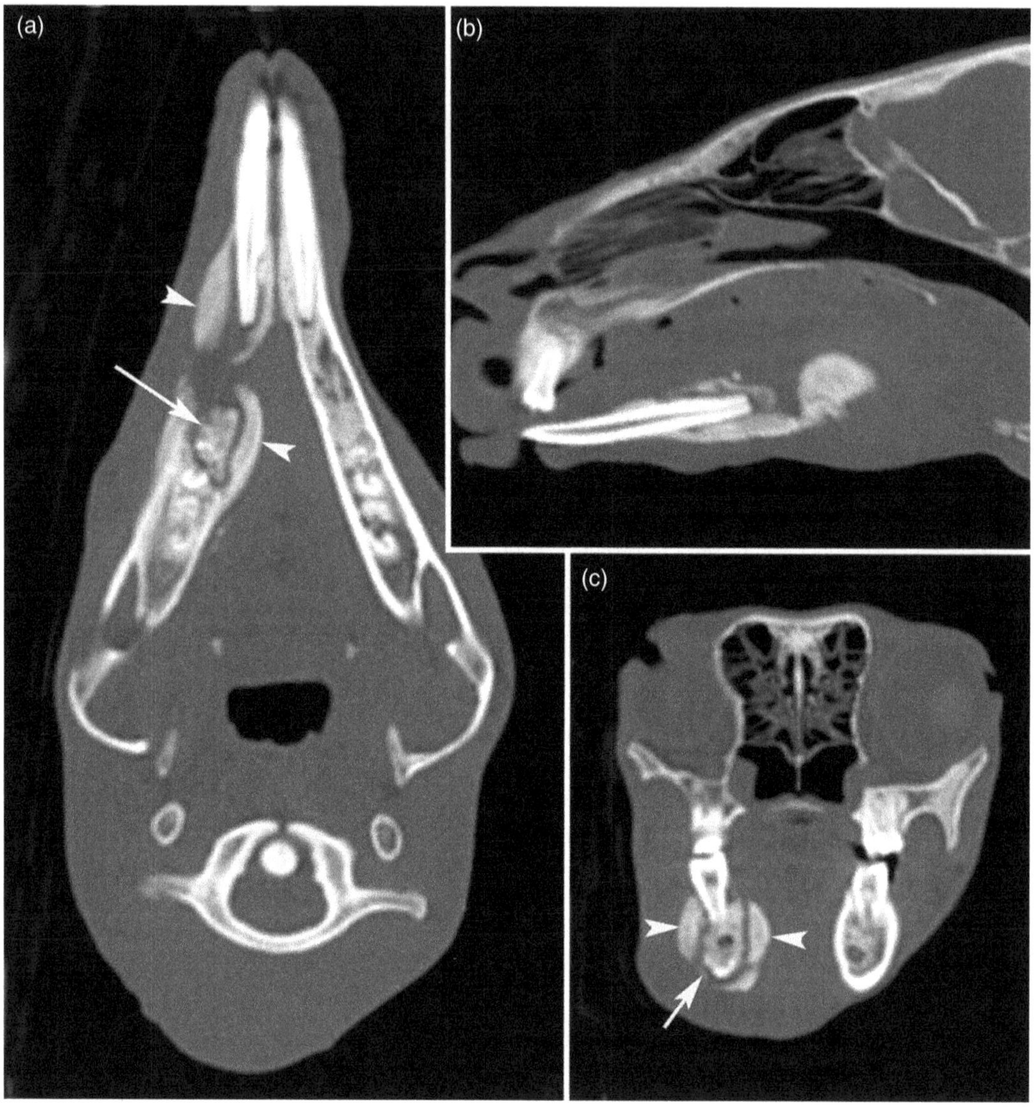

Figure 12.30: Head, red-necked wallaby (*Macropus rufogriseus*), CT reconstruction: (a) dorsal plane, (b) sagittal plane, (c) transverse plane. A sequestered piece of bone (arrow) is surrounded by an involucrum (arrowheads). The cavity between the involucrum and the sequestrum and the ipsilateral incisor root will contain infectious debris. The mandibular ramus is fractured.

Common wombat, female, 2 years old

This orphaned common wombat (*Vombatus ursinus*) joey had been hand-reared from approximately 6 months of age. It showed marked growth retardation, lethargy and inappetence, which was apparent only after it was introduced to solid foods. Oral examination and skull radiography (Fig. 12.35) under anaesthesia revealed elongated cheek teeth crowns with a corresponding malocclusion. It is likely the dental changes were present at the time of weaning and malocclusion prevented the joey from adjusting to a normal diet of fibre-rich grasses and the resultant lack of wear exacerbated the problem.

Case submitted by: Dr Michelle Campbell-Ward, Taronga Western Plains Zoo, NSW, Australia.

Common wombat, male, ~15 months old

This orphaned joey had been hand-reared from approximately 6 months of age. Despite its stage of

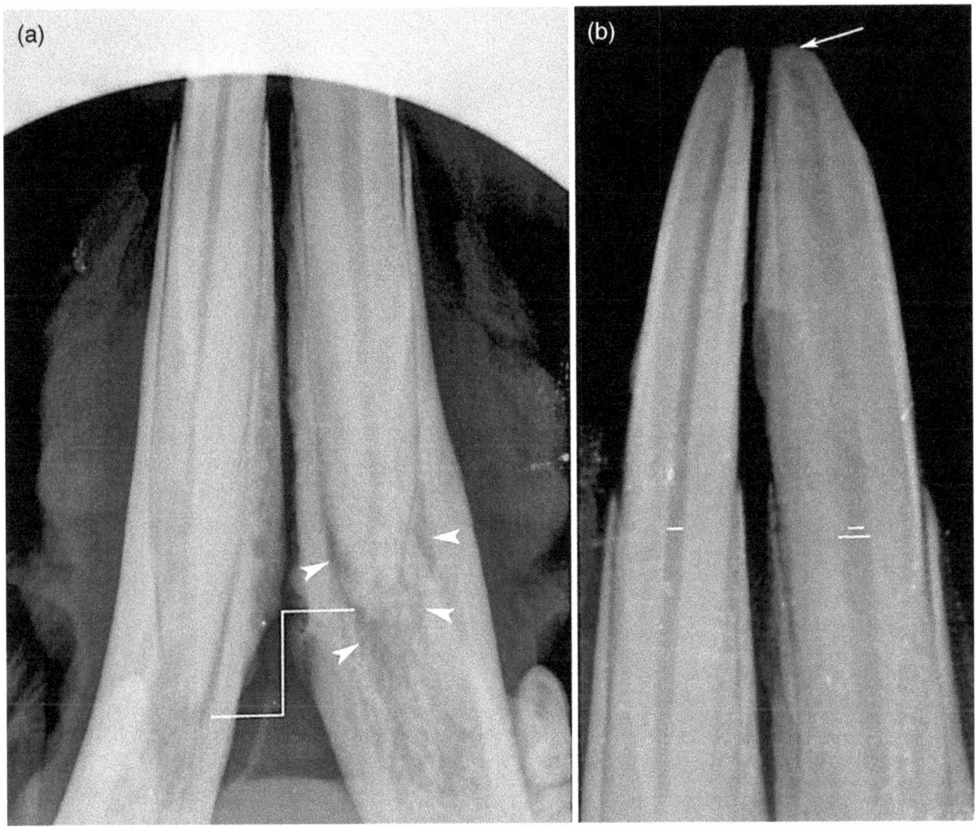

Figure 12.31: Mandibular incisors, common wallaroo (*Macropus robustus*), intra-oral view. The images show loss of crown integrity (arrow in b) and failure of the pulp canal to narrow (white lines in b). Pulp ceases to make dentin when a tooth is dead; hence, the pulp canal will be wider than that of its counterpart. There is periapical lucency because of infection and inflammation of the periapical tissues (arrowheads in a) and root resorption because of periapical inflammation, resulting in the apex of the affected tooth being shorter than its counterpart (white stepped line in a).

development at the time of presentation (15 months), it was still dependent on bottle feeds and did not consume much solid food. The wombat was unable to close its mouth, because of elongated cheek teeth, and the mandible appeared short and deviated to the right, resulting in malocclusion of all teeth (Fig. 12.36).

Common wombat, female, ~16 months old

This orphaned joey had also been hand-reared from approximately 6 months of age. It had weaned successfully onto a solid diet. Uneven wear and labial spurring of the cheek teeth was noted at a routine health assessment and radiographs were taken to evaluate the malocclusion (Fig. 12.37). Dental filing at 6-monthly intervals was needed to maintain the teeth in apposition.

The nasal passages

As is the case in domestic animals, nasal cavity disease in Australian mammals may result from infection, foreign bodies, neoplasia or extension of dental disease. Nasal masses are not uncommon in the koala, with cryptococcal granulomas being the most common cause (Bercier *et al.* 2012).

Cryptococcal granuloma

Koala, female, 4 years old

This zoo koala returned a positive latex cryptococcal antigen test at an otherwise unremarkable pre-shipment examination. Plain radiographs of the skull were normal at that time, but 5 weeks later, the koala became weak and lethargic and CT images of the skull (Fig. 12.38) showed a mass in

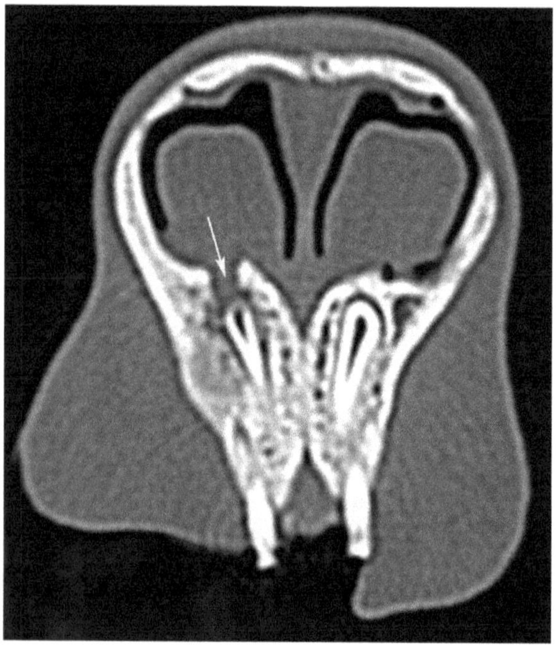

Figure 12.32: Rostral nasal cavity, koala (*Phascolarctos cinereus*), CT reconstruction, transverse plane. Bone surrounding the root of the right maxillary 1st incisor tooth has eroded, so the tooth root communicates with the nasal cavity (arrow). The contralateral incisor root appears normal.

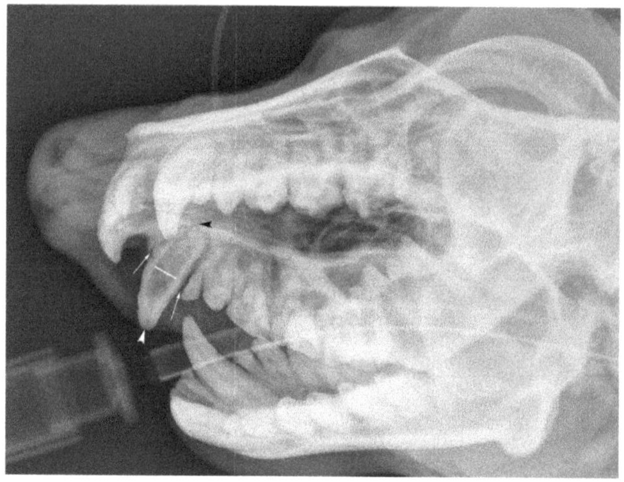

Figure 12.33: Skull, Tasmanian devil (*Sarcophilus harrisii*), dorsolateral–ventromedial oblique view. The right maxillary canine tooth has a fractured crown (arrowhead) exposing the pulp canal, which has failed to narrow (white line). There is periapical lucency, complete apical resorption and a fistula (black arrowhead) extending into the nasal cavity. The infection has extended into the periodontal ligament space, resulting in periodontitis (arrows).

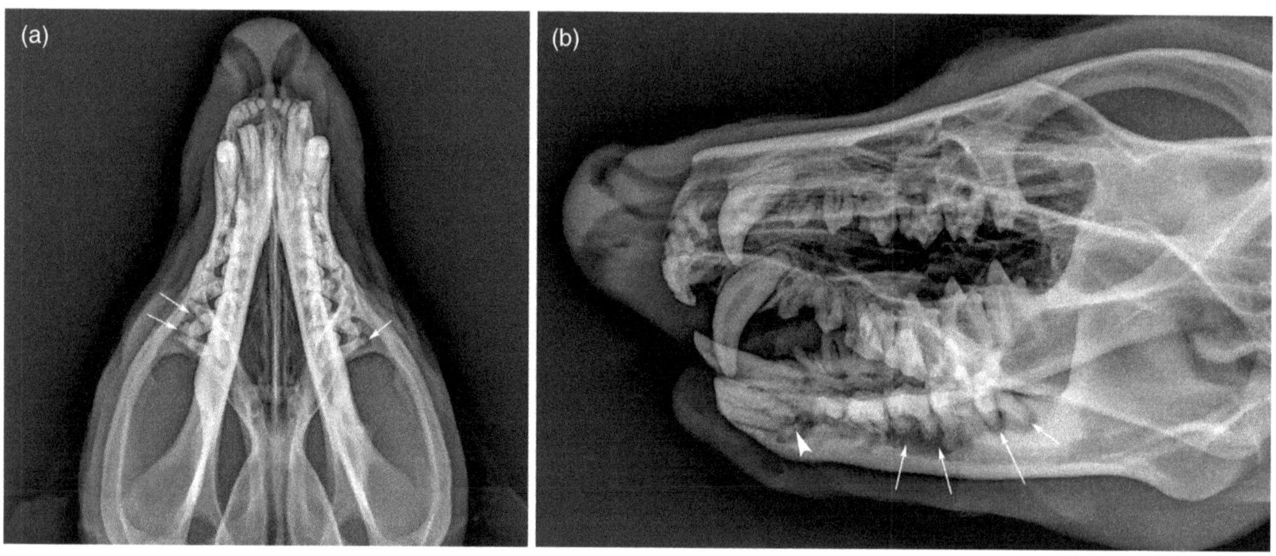

Figure 12.34: Skull, spot-tailed quoll (*Dasyurus maculatus*): (a) ventrodorsal view, (b) dorsolateral–ventromedial oblique view. There is severe periodontitis affecting several of the mandibular and maxillary premolars and molars (arrows). The mandibular canines show severe endodontic disease secondary to pulp exposure (arrowhead).

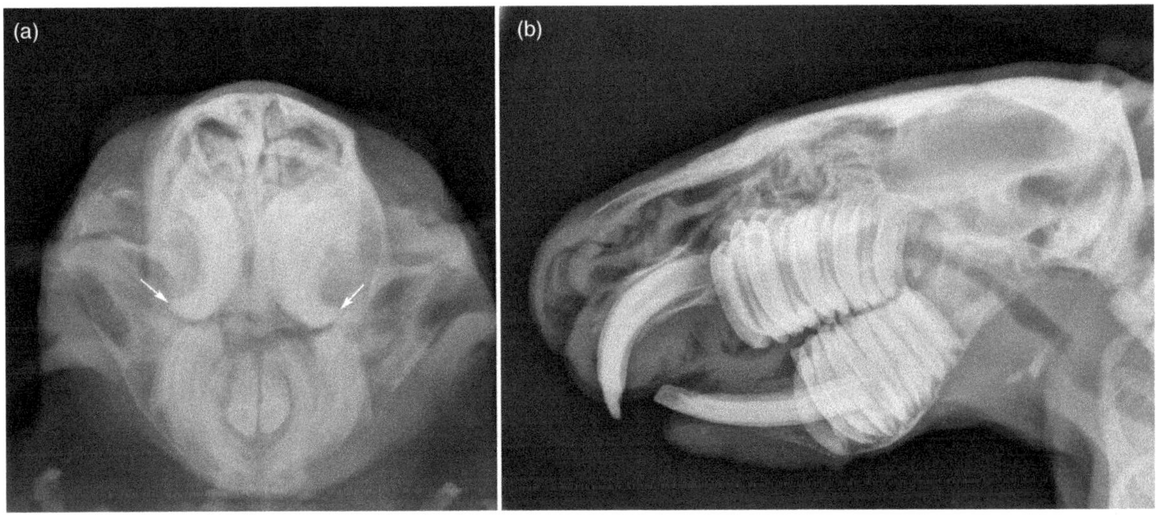

Figure 12.35: Skull, common wombat (*Vombatus ursinus*): (a) rostrocaudal view, (b) lateral view. There is elongation of the apices and crowns of all cheek teeth, curvature of the maxillary cheek teeth resulting in buccal spurs (arrows), an irregular occlusal plane and lysis of the ventral mandible.

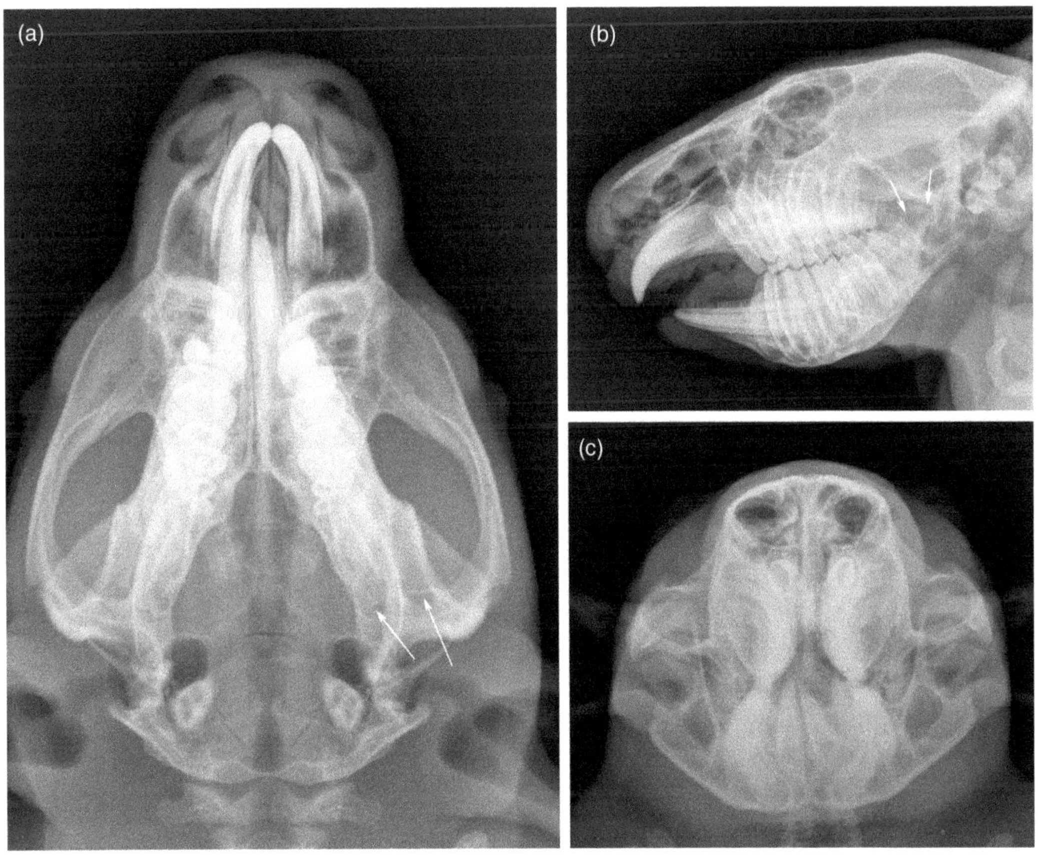

Figure 12.36: Skull, common wombat (*Vombatus ursinus*): (a) ventrodorsal view, (b) lateral view, (c) rostrocaudal view. There is relative mandibular brachygnathia and deviation to the right. The condyloid process of one mandible appears lucent (arrows), suggesting a destructive process in a temporomandibular joint, which may account for the jaw misalignment.

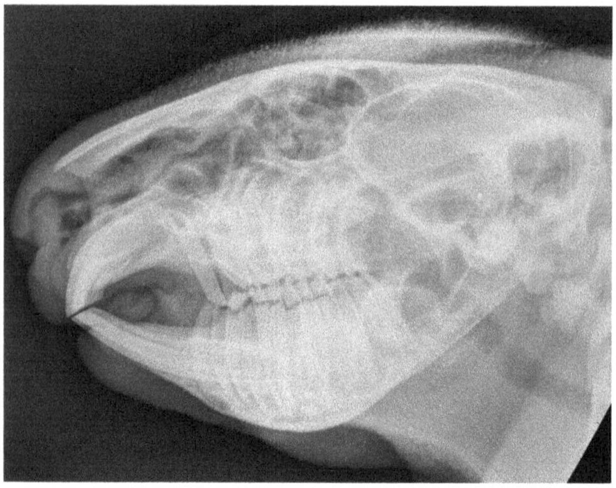

Figure 12.37: Skull, common wombat (*Vombatus ursinus*), lateral view. There is uneven wear and step formation of the cheek teeth, although the teeth are in apposition and the animal was able to eat solid food.

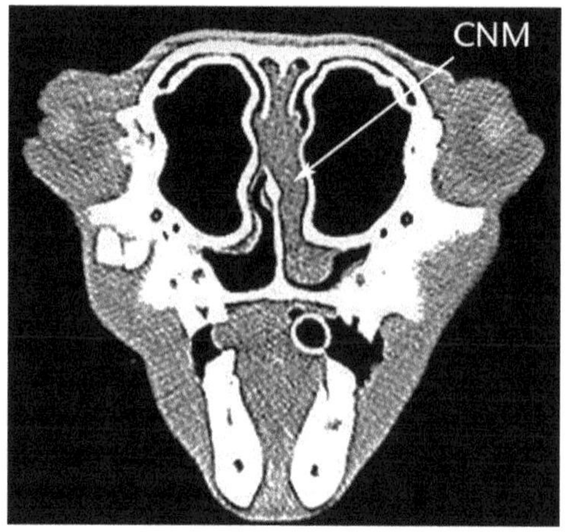

Figure 12.39: Skull, koala (*Phascolarctos cinereus*), CT reconstruction, transverse image. There is increased opacity in the left common nasal meatus (CNM), conforming to the shape of the conchae.

the nasopharynx. The mass was surgically excised and was confirmed as a cryptococcal granuloma on histopathology and culture (Bercier *et al.* 2012).

Case submitted by: Dr Geoff Pye, Disney's Animal Kingdom, FL, USA and images reproduced with permission from: Bercier M, Wynne J, Klause S, Stadler CK, Gorow A, Pye GW (2012) Nasal mass removal in the koala (*Phascolarctos cinereus*). *Journal of Zoo and Wildlife Medicine* 43, 898–908.

Koala, male, 18 years old

An aged zoo koala developed a left nasal discharge and upper respiratory noise. Plain skull

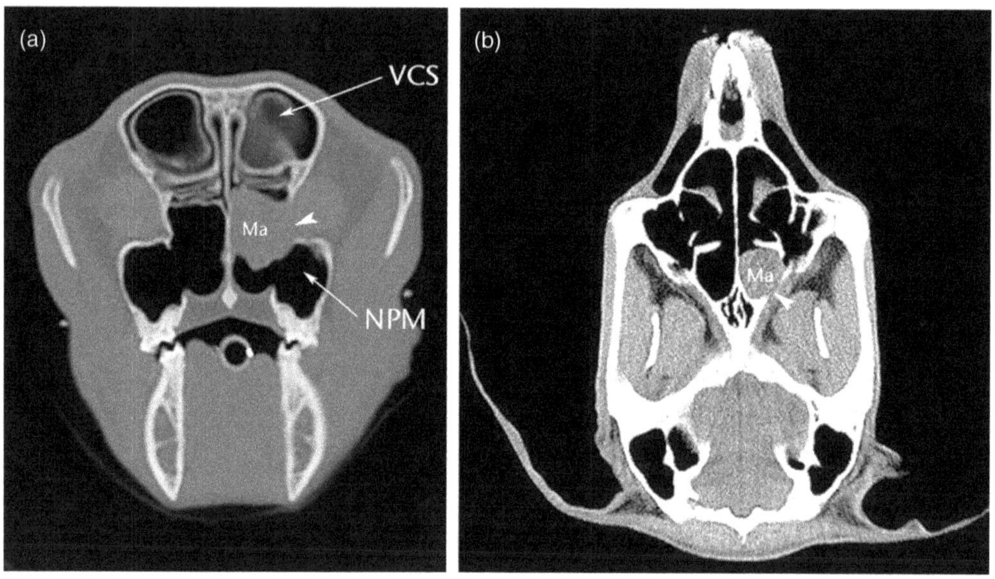

Figure 12.38: Skull, koala (*Phascolarctos cinereus*), CT reconstruction: (a) transverse and (b) dorsal plane images. There is a mass (Ma) in the nasopharyngeal meatus (NPM). It has destroyed adjacent bone (arrowhead) and extends into the post-orbital soft tissue. There is also increased opacity in the ventral conchal sinus (VCS), which may represent fluid accumulation, tissue inflammation or a mass extending into the sinus.

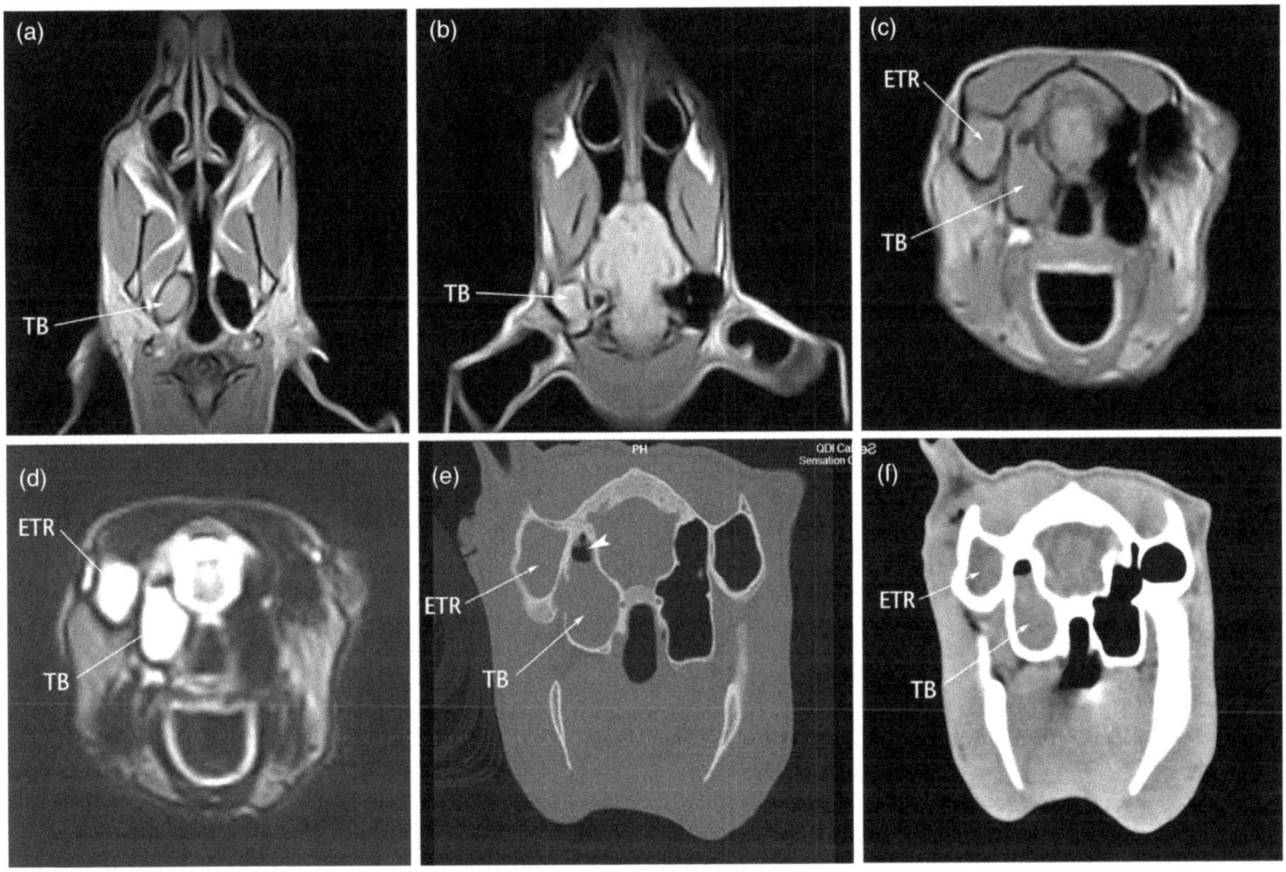

Figure 12.40: Skull, koala (*Phascolarctos cinereus*), comparative MRI and CT views centred on the middle ear structures. The CT scan was performed 11 days after the MRI. (a, b) Dorsal plane T1-weighted (T1W) MRI images demonstrating fluid in the right tympanic bulla (TB). (c) Transverse plane T1W MRI image showing fluid in the right tympanic bulla and epitympanic recess (ETR). (d) Transverse plane T2W MRI image in which the fluid signal is hyperintense, compared with the isointense fluid signal in the T1W images. (e) CT reconstruction, transverse plane optimised for bone. (f) CT reconstruction, transverse plane optimised for soft tissue. In both (e) and (f) fluid is visible in the TB and ETR. A small bubble of gas is also present in the TB (arrowhead).

radiographs did not reveal any nasal lesions, although a purple-grey mass was seen on rhinoscopy of the left nasal cavity. Both MRI with contrast enhancement and CT scan of the skull (Fig. 12.39) were performed and a mass conforming to the shape of the conchae and extending along the left nasal passage was identified. The mass was surgically excised through a dorsal rhinotomy. Histopathology of the excised tissue showed a spindle cell tumour or nasal polyp (Bercier *et al.* 2012).

Case submitted by: Dr Geoff Pye, Disney's Animal Kingdom, FL, USA and image reproduced with permission from: Bercier M, Wynne J, Klause S, Stadler CK, Gorow A, Pye GW (2012) Nasal mass removal in the koala (*Phascolarctos cinereus*). *Journal of Zoo and Wildlife Medicine* 43, 898–908.

The ear

Otitis media has been reported in dunnarts and quolls (Holz 2008) and possums and gliders (Johnson and Hemsley 2008) but appears to be a relatively rare finding in most Australian mammals. Otitis externa has been reported in koalas and in males may originate from pinna wounds inflicted during fighting (Blanshard and Bodley 2008). Extension of infection through the tympanic membrane, resulting in otitis media, is a possible sequela.

Otitis media

Koala, male, 3 years old

This wild koala was captured as part of a translocation project. It had a purulent right aural discharge from which the Gram-negative rod *Yokanella regensburgei* was cultured. Despite treatment, the otitis externa failed to resolve and intermittent mild ataxia developed. Skull MRI and CT scan (Fig. 12.40) revealed advanced otitis media. The koala was successfully returned to the wild following surgical external ear canal ablation and tympanic bulla osteotomy.

Case submitted by: Dr Amber Gillett, Australia Zoo Wildlife Hospital, QLD, Australia.

The thorax

The lungs

Many marsupials do not show overt signs of lung disease that can be recognised clinically, even in advanced cases, and other signs associated with respiratory disease such as haematological changes or pyrexia are often absent. Thoracic radiography is a useful modality for assessing the lower respiratory tract.

Pneumonia is relatively common in hand-reared marsupials, often as a result of aspiration or inappropriate husbandry in immunocompromised joeys. Affected animals may only show non-specific signs such as poor growth rates and lethargy.

Pulmonary oedema/pneumonia

Eastern grey kangaroo, female, pouch young

This hand-reared orphaned eastern grey kangaroo (*Macropus giganteus*) joey presented with a history of poor growth rate and a sudden onset of depression and inappetence. The joey was thin, undersized and tachypnoeic. Thoracic auscultation and body temperature were normal. Radiographs were taken to evaluate the lungs (Figs 12.41 and 12.42). The joey's condition deteriorated and it was euthanased. Histopathology confirmed marked pulmonary oedema and pneumonia.

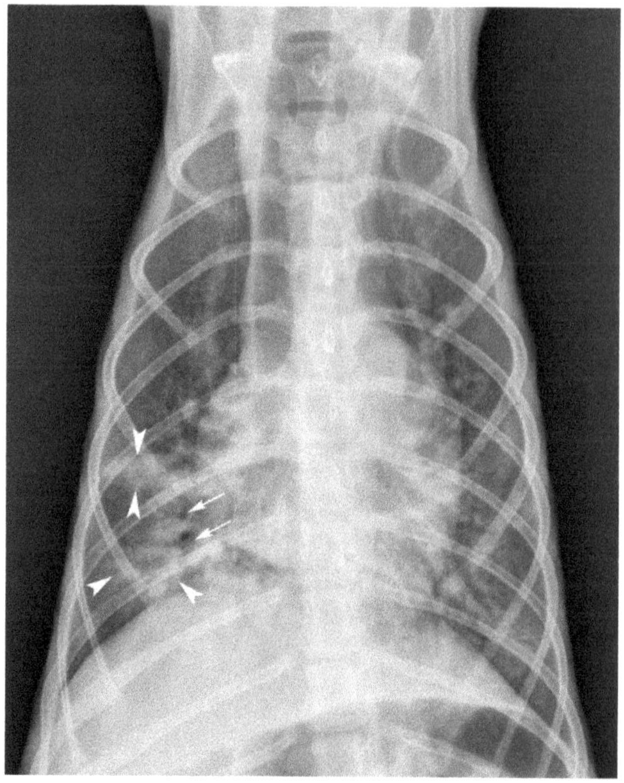

Figure 12.41: Thorax, eastern grey kangaroo (*Macropus giganteus*), ventrodorsal view. There is generalised increased pulmonary opacity with a mixed alveolar/interstitial pattern (arrowheads), most prevalent in the middle and caudal pulmonary regions. Air bronchograms are evident (arrows).

Case submitted by: Dr Benn Bryant, Taronga Western Plains Zoo, NSW, Australia.

Suppurative pneumonia

Swamp wallaby, female, 5 years old

This swamp wallaby had been treated successfully for mild periodontitis ('lumpy jaw') 1 year prior to presentation with anorexia, lethargy and poor body condition. It had increased abdominal respiratory effort with a soft audible grunt. Thoracic radiographs (Figs 12.43 and 12.44) showed consolidation of the ventral and caudal portion of both lungs. Tracheal wash cytology was consistent with a septic pneumonia. Despite treatment, the wallaby's condition further deteriorated and it was euthanased. At post-mortem examination the right

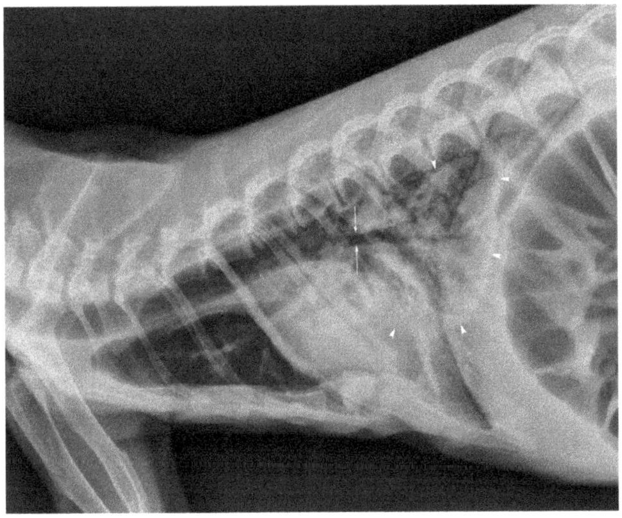

Figure 12.42: Thorax, eastern grey kangaroo (*Macropus giganteus*), lateral view. A caudal lung lobe is relatively opaque (arrowheads), rendering the bronchi more clearly defined (arrows). The pulmonary changes are consistent with inflammatory disease (infection, parasites), haemorrhage or pulmonary oedema. Note the fourth sternal element appears malformed and displaced in this view.

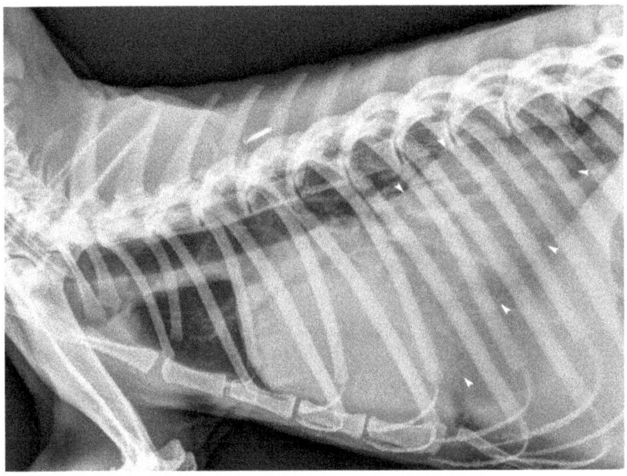

Figure 12.43: Thorax, swamp wallaby (*Wallabia bicolor*), right lateral view. The caudodorsal lungs appear airless (arrowheads), although superimposed blood vessels suggest that one caudal lobe is less consolidated than the other. Cranial lobar airways are visible. There is some peribronchial opacity in the cranial lobes, but the cranial lungs appear aerated. The size and shape of the cardiac silhouette is difficult to discern because of silhouetting of the caudal lobes and the heart. Note: a transponder can be seen.

middle lung lobe was necrotic and the remaining right lobes and caudal left lung had multifocal necrotic nodules. Bacterial culture of the lung tissue yielded heavy growth of *F. necrophorum*. Histopathology confirmed suppurative and necrotising pneumonia with bronchiectasis. *Fusobacterium necrophorum* is often associated with 'lumpy jaw' in macropods and its isolation suggests the previous oral disease may have been the inciting cause of this animal's pneumonia.

Pulmonary hydatid cysts

Tammar wallaby, female, 2 years old

A tammar wallaby (*Macropus eugenii*) was experimentally infected with *Echinococcus granulosus* eggs and subsequently developed two hydatid cysts in the lungs. The growth of the cysts was monitored radiographically for the next 14 months until the wallaby was euthanased because of substantial loss of lung volume (Fig. 12.45). The presence of two pulmonary hydatid cysts was confirmed at post-mortem examination. The wallaby showed no clinical signs even with substantially reduced lung volume.

Case submitted by: Dr Tamsin Barnes, School of Veterinary Science, The University of Queensland, QLD, Australia.

Naturally infected macropods primarily develop hydatid cysts in the lungs or thoracic cavity and typically have only one or two cysts (Barnes *et al.* 2007). Macropods are thought to play a key role in the epidemiology of hydatid disease in Australia. Caseation and calcification of degenerate cysts is not a common feature in wallabies (Barnes *et al.* 2008; Barnes *et al.* 2011), but have been reported in kangaroos (Jenkins and Morris 2003).

Mycobacterial lymphadenitis and pneumonia

Long-nosed potoroo, female, 21 months old

Bilateral firm swellings were noticed in the subcutaneous tissue ventral to the angle of the jaw of an anaesthetised long-nosed potoroo. Radiographically, the masses had soft tissue opacity with

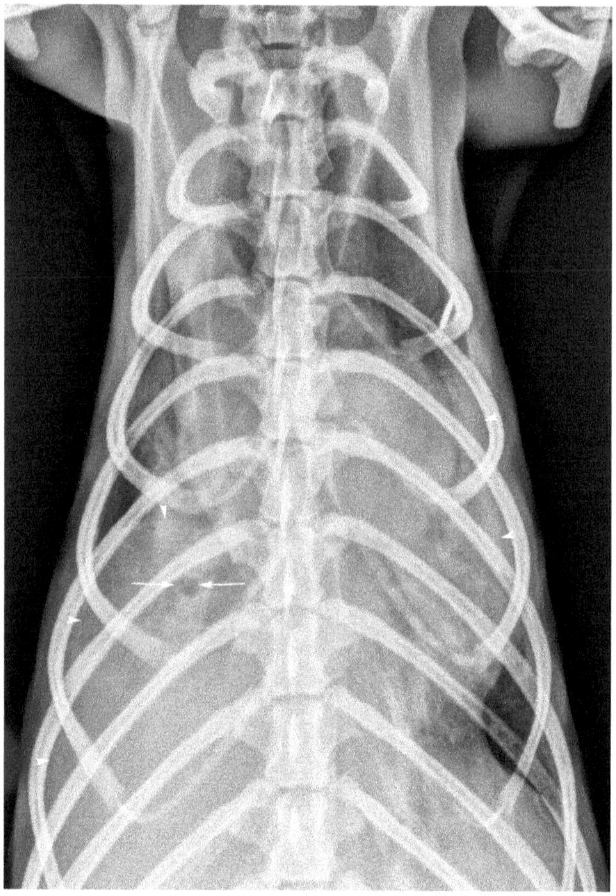

Figure 12.44: Thorax, swamp wallaby (*Wallabia bicolor*), ventrodorsal view. The middle and caudal lung fields appear airless (arrowheads), although end-on aerated bronchi (arrows) are visible in places. Well-defined air bronchograms are visible. The right scapula is superimposed on the right cranial lobe. The cardiac silhouette and diaphragm are camouflaged by the surrounding solid pulmonary parenchyma. Note: a transponder can be seen.

central foci of bone (mineral) opacity (Fig. 12.46a, b). Fine needle aspirate cytology suggested the swelling may be salivary gland and the mineralised foci were thought to represent sialoliths. No acid-fast organisms were seen with Ziehl-Neelsen stain. Eight months later, the potoroo had lost weight, was dyspnoeic and the submandibular masses had enlarged. On radiographs, there were pulmonary opacities, a dorsal mediastinal mass and narrowing of the cervical trachea (Fig. 12.46c). The potoroo was euthanased. On post-mortem examination the masses were identified as submandibular and retropharyngeal lymph nodes. Histopathology confirmed granulomatous and necrotising lymphadenitis, pneumonia and splenitis. Bacteria from the *Mycobacterium fortuitum* complex were cultured from the affected tissues. Mycobacteriosis is a common opportunistic environmental pathogen in Australian mammals in permanent care and repeated sampling may be needed to identify infection. Mineralisation of mycobacterial granulomas is an inconsistent finding.

Metastatic lung disease

Long-nosed bandicoot, male, 3 years old

This long-nosed bandicoot (*Perameles nasuta*) was found hunched in its enclosure, reluctant to move and dribbling urine. A firm mass was palpable in the caudoventral abdominal area and appeared to surround the epipubic bones. Radiographs taken of the abdomen and the thorax (Fig. 12.47) showed the abdominal mass and pulmonary changes consistent with metastatic lung disease. The bandicoot was euthanased. Post-mortem examination confirmed the presence of a multinodular caudal abdominal sarcoma (either histiocytic or rhabdomyosarcoma) involving the abdominal muscle wall and attached to one of the epipubic bones. The bladder was displaced dorsally by the tumour and distended with urine. The lymph nodes dorsal to the colon were enlarged. The heart was enlarged, with a dilated left ventricle. Histopathology confirmed metastases throughout the lungs and heart.

The oesophagus

Megaoesophagus

A high incidence of megaoesophagus has been described in captive-bred parma wallabies (*Macropus parma*) in a North American zoo. Those with more advanced disease showed varying degrees of weight loss, swelling of the cervical region, regurgitation, short flight distance, depression, collapse, dyspnoea or sudden death. The cause of megaoesophagus in parma wallabies has not been determined (Burgdorf-Moisuk *et al.* 2012).

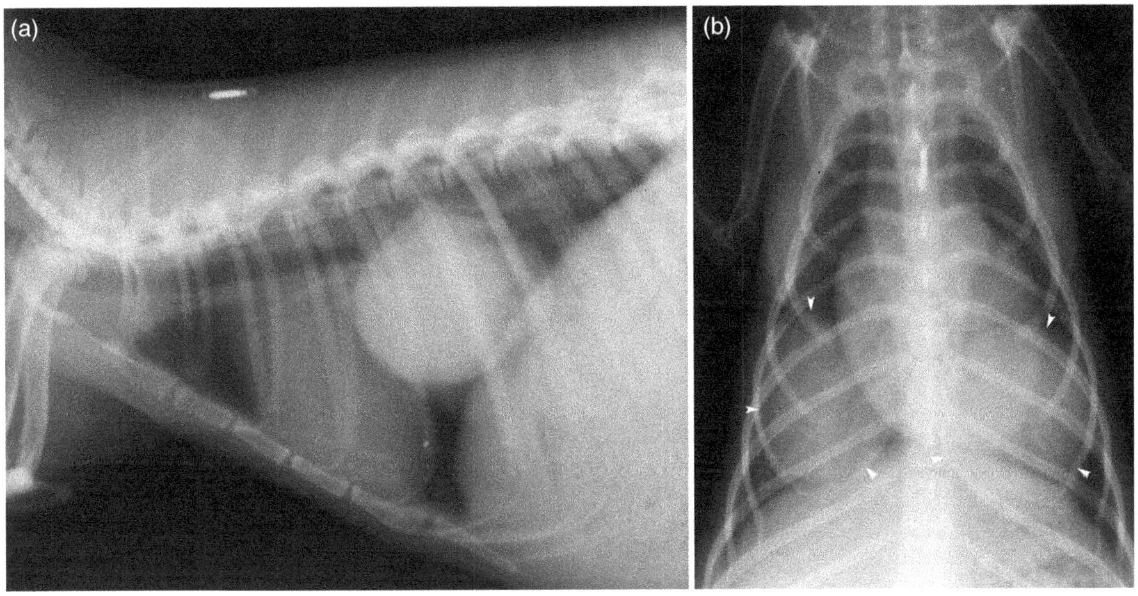

Figure 12.45: Thorax, tammar wallaby (*Macropus eugenii*): (a) lateral view, (b) ventrodorsal view. There are two ovoid opacities, one in each of the left and the right caudal lung fields (arrowheads). They have homogeneous soft tissue opacity and well-defined margins and are surrounded by well aerated pulmonary parenchyma that appears normal. Multiple pulmonary nodules are more likely to represent metastatic neoplasia, cysts or granulomas. Note: a transponder can be seen.

Radiographically, the oesophagus in all affected wallabies is dilated and contains gas. The dilation is more clearly demonstrated with contrast studies (Fig. 12.48).

Case submitted by: Dr Geoff Pye, Disney's Animal Kingdom, FL, USA and images reproduced with permission from: Burgdorf-Moisuk A, Pye GW, Smith JA, Papendick R, Ivy JA, Hamlin-Andrus C (2012) A retrospective and prospective study of megaesophagus in the parma wallaby (*Macropus parma*) at the San Diego Zoo, California, USA. *Journal of Zoo and Wildlife Medicine* **43**, 89–94.

The cardiovascular system

Primary cardiovascular disease appears to be uncommon in Australian mammals, with the exception of macropods. Sporadic cases have been described in wombats and Tasmanian devils. Endocarditis, myocardial infarction and primary or secondary cardiomyopathy have been reported in several species of kangaroo (Chineme *et al.* 1978; Bennet *et al.* 1993; Vogelnest and Portas 2008) and wallaby (Stern *et al.* 2009), with the kangaroo suggested as a model for the study of human hypertrophic cardiomyopathy (O'Rourke *et al.* 1986).

Cardiomyopathy

Red kangaroo, female, 12 years old

This red kangaroo was diagnosed 5 years previously with dilated cardiomyopathy based on radiography and ultrasound examination. The animal was managed medically for 5 years before becoming increasingly lethargic and inappetent. Radiographic (Fig. 12.49) and sonographic findings and abnormal electrocardiographic results led to a decision to euthanase the kangaroo. Dilated cardiomyopathy with liver damage was confirmed on gross post-mortem examination.

Case submitted by: Dr Genevieve Dumonceaux, Palm Beach Zoo, FL, USA.

Common ring-tailed possum, female, 8 years old

This common ring-tailed possum (*Pseudocheirus peregrinus*) was examined for a routine aged-animal

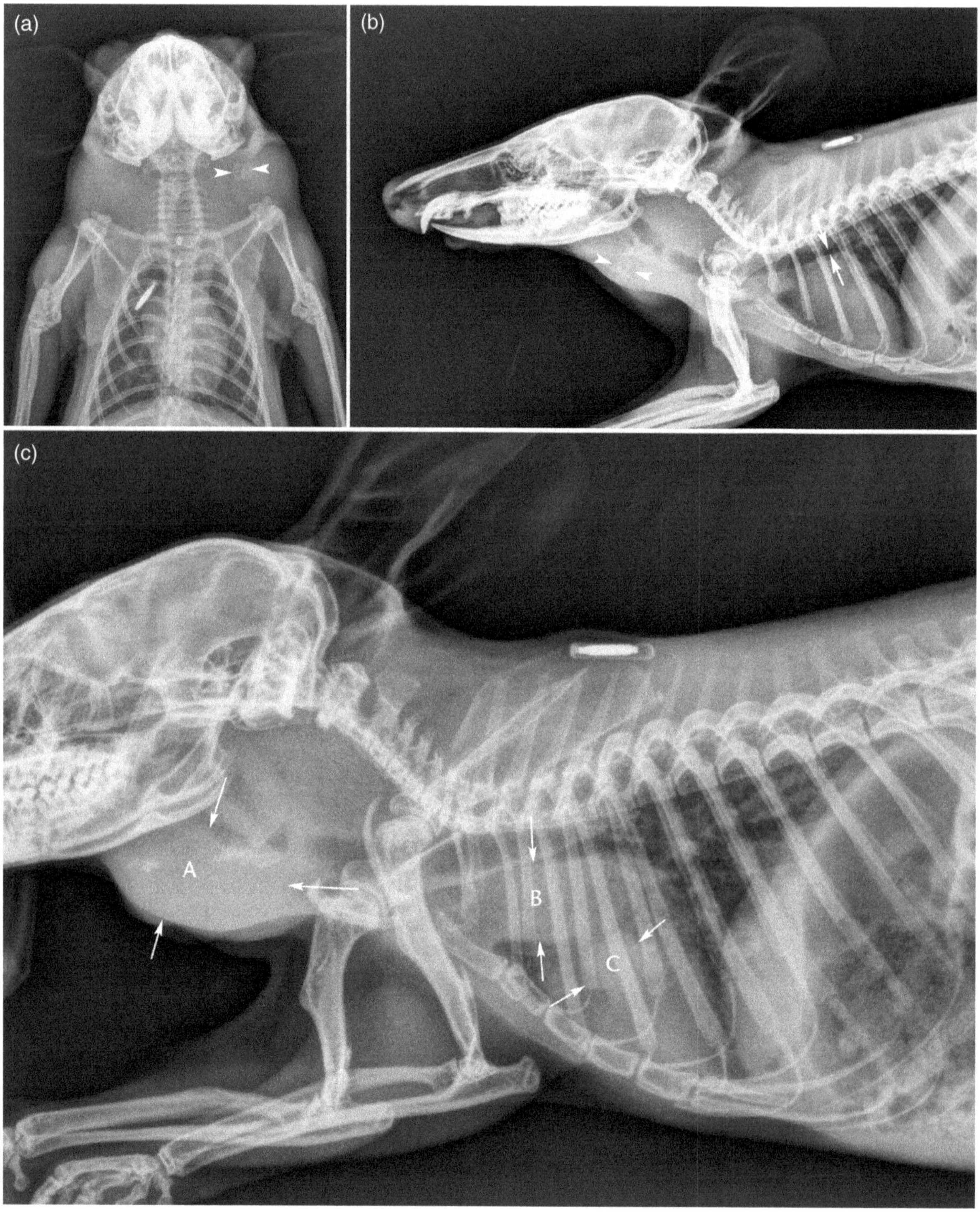

Figure 12.46: Head and thorax, long-nosed potoroo (*Potorous tridactylus*): (a) ventrodorsal view, (b) lateral view, (c) lateral view taken 8 months after (b). (a) Generalised swelling in the neck. A focal region of mineralisation (arrowheads) is present within this tissue on the left side. (b) The mineralised tissue appears ventral to the larynx. In the thorax the trachea appears narrow over the base of the heart (arrows). (c) The ventral neck swelling (A) has enlarged and appears to be constricting the laryngeal airway. The cranial mediastinum (B) is enlarged and a soft tissue nodule is present in the lungs (C). Note: transponders can be seen.

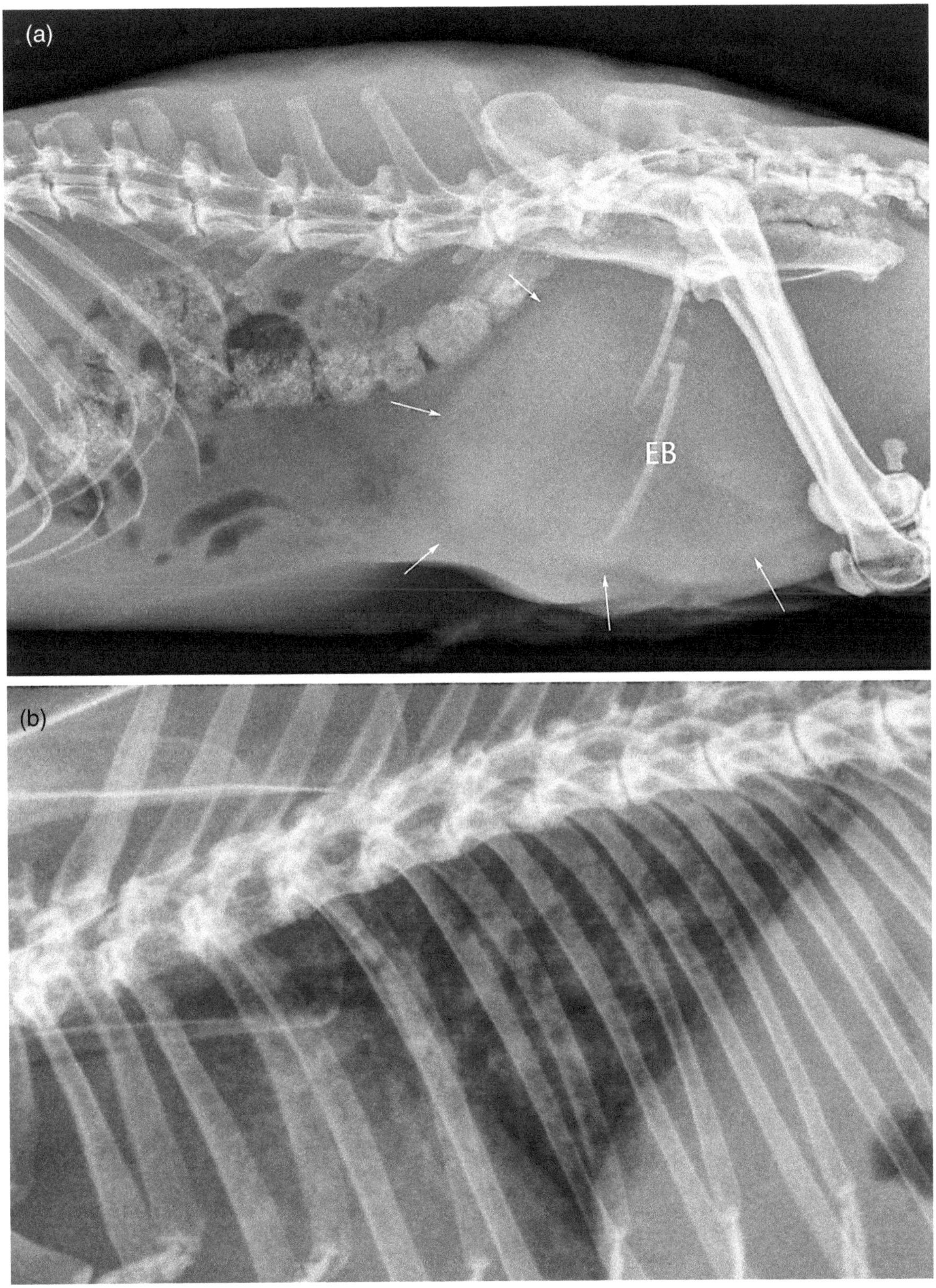

Figure 12.47: Long-nosed bandicoot (*Perameles nasuta*): (a) abdomen, lateral view, (b) thorax, lateral view. (a) The large soft tissue mass (arrows) has displaced an epipubic bone (EB). (b) Unstructured, mixed pattern of increased pulmonary opacity, histologically identified as metastatic sarcoma throughout the lung.

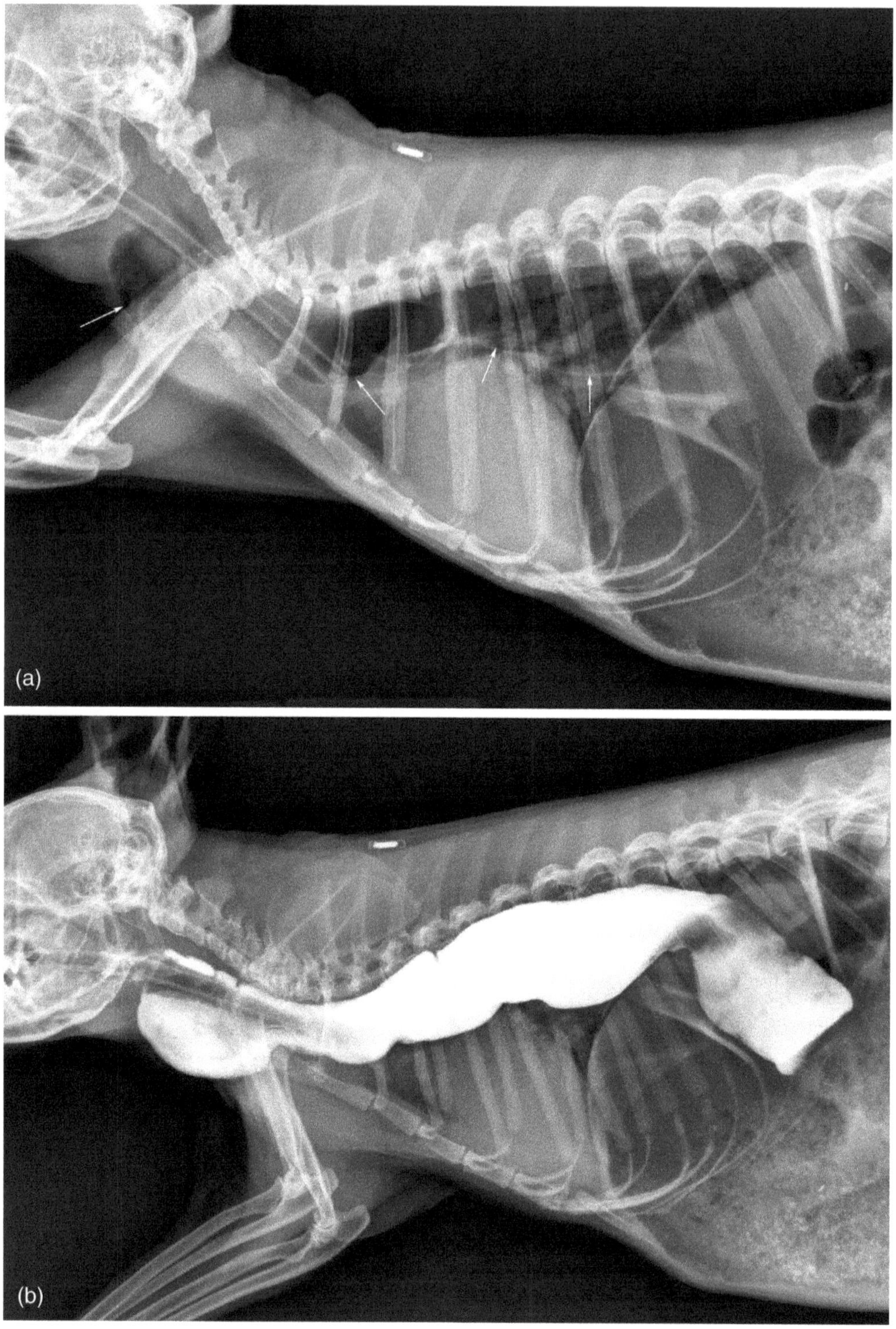

Figure 12.48: Neck and thorax, parma wallaby (*Macropus parma*), lateral view: (a) survey radiograph, (b) contrast radiograph. (a) The entire oesophagus is dilated and contains gas, which enables the thin oesophageal wall to be defined (arrows). (b) Following administration of barium, the entire oesophagus is dilated and manifests as a tube of irregular shape. The appearance indicates the underlying pathological change, which is loss of muscle tone in the oesophagus. Oesophageal dilation with barium persisted for more than 30 minutes, indicating oesophageal hypomotility. Note: transponders can be seen.

assessment. Thoracic radiographs (Fig. 12.50) revealed a cranial mediastinal soft tissue mass. Echocardiography showed mild aortic insufficiency and aortic ectasia, with a mass of varying echogenicity surrounding the ascending aorta. Ultrasound-guided fine needle aspiration (FNA) of the mass was attempted, but the possum died shortly after. Pneumothorax and pulmonary atelectasis as a result of the FNA attempt was identified at post-mortem examination. The cranial mediastinal mass was a focal aneurysm of the aorta at its origin, still within the pericardium. Cardiovascular disease is seldom reported in possums and gliders. In humans, aortic ectasia is associated with hypertension and ageing or may be familial (Isselbacher 2005) and can be an incidental finding on thoracic radiographs.

The abdomen

The gastrointestinal tract

Gastrointestinal pathology is common in Australian mammals held in permanent care. Wombats and koalas are hindgut fermenters with voluminous large bowels and correspondingly long gastrointestinal transit times. Gastrointestinal stasis, ileus and gas build-up are common features of gastrointestinal disturbance. If acute obstructive disease is not suspected, contrast provided by 5–10 mL of barium given orally can be useful for defining the gastrointestinal tract on radiographs.

Gaseous distension

Common wombat, female, ~2 years old

This juvenile common wombat had dental malocclusion with associated inability to wean onto a normal diet, which resulted in malnutrition and gastrointestinal ileus characterised by abnormal faecal passage and abdominal distension. Abdominal radiographs revealed uniform gaseous distension of the small and large intestines, occupying the entire abdomen (Fig. 12.51).

Case submitted by: Dr Michelle Campbell-Ward and Dr Benn Bryant, Taronga Western Plains Zoo, NSW, Australia.

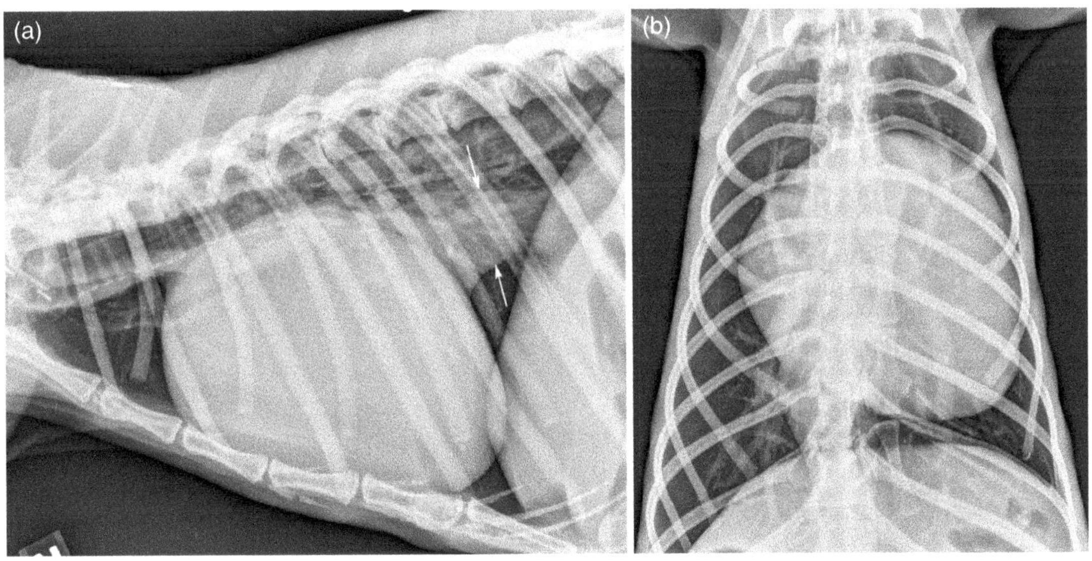

Figure 12.49: Thorax, red kangaroo (*Macropus rufus*): (a) lateral view, (b) ventrodorsal view. The cardiac silhouette is enlarged and has a spheroidal shape. The trachea is displaced dorsally and the caudal *vena cava* is dilated (arrows). The pulmonary parenchyma appears relatively hypovascular. These changes suggest right heart failure with pulmonary underperfusion and venous congestion of the caudal *vena cava*. Pericardial effusion was ruled out using cardiac ultrasonography.

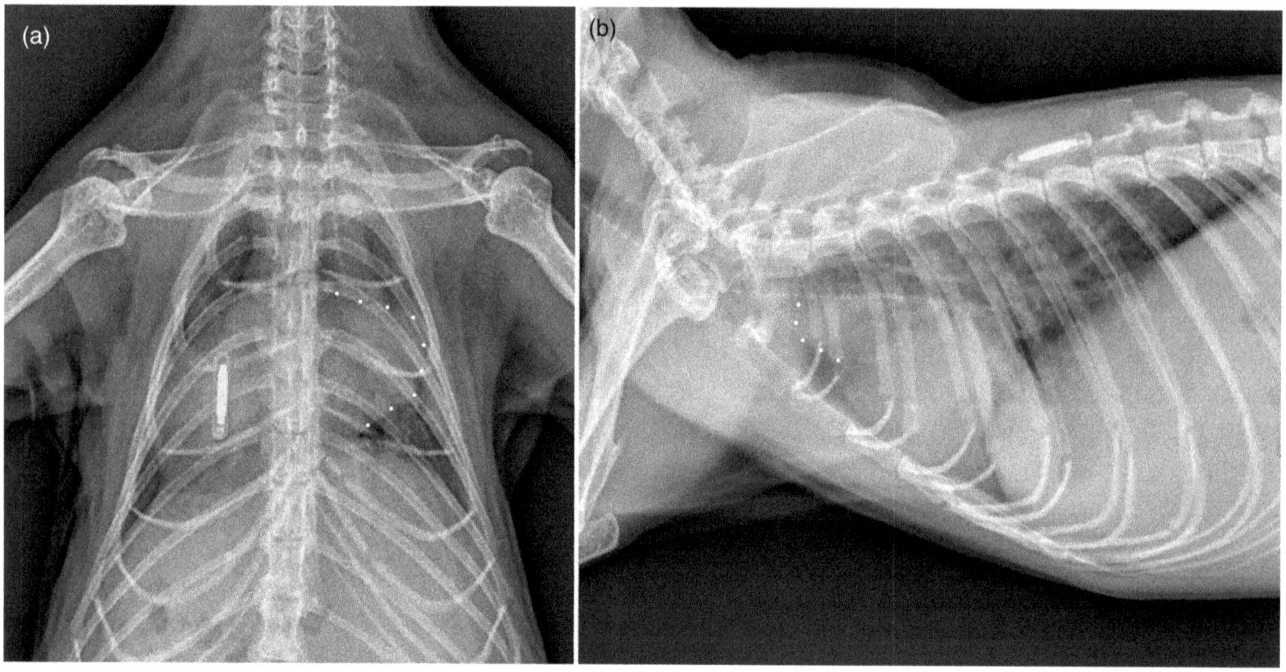

Figure 12.50: Thorax, common ring-tailed possum (*Pseudocheirus peregrinus*): (a) ventrodorsal view, (b) lateral view. (a) A soft tissue swelling (dotted edge) is present on the left side of the cardiac silhouette, in the region of the aortic arch. The heart is displaced to the right. (b) A swelling on the cranial margin of the cardiac silhouette (dotted line) is in the location of the aortic arch. This swelling was identified as an aneurysmal dilation of the aorta. Note: transponders can be seen.

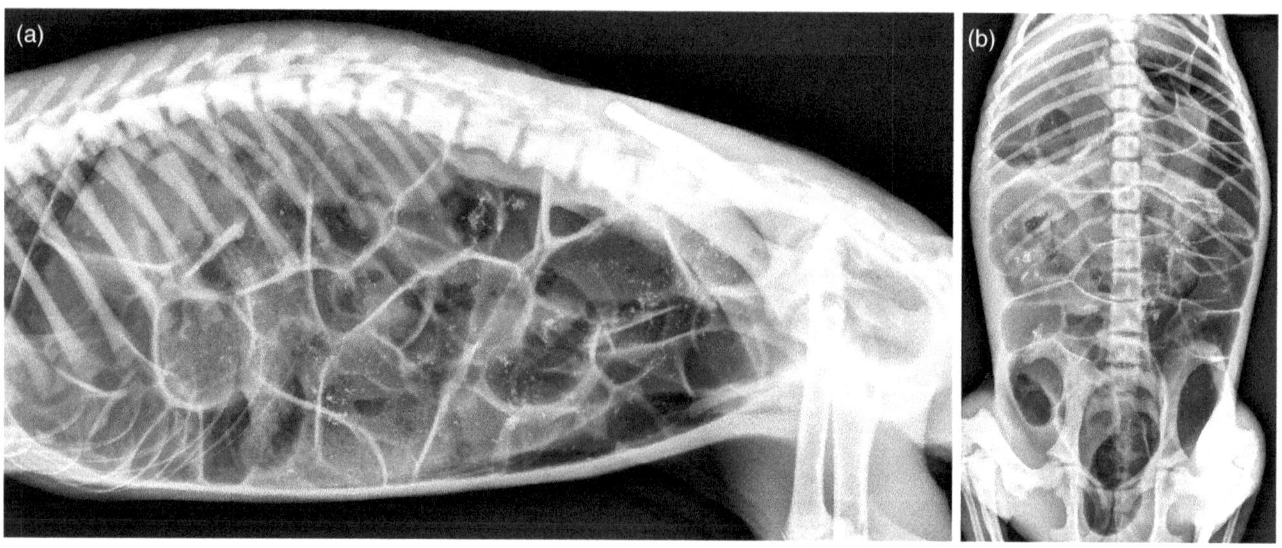

Figure 12.51: Abdomen, common wombat (*Vombatus ursinus*): (a) lateral view, (b) ventrodorsal view. There is moderate, uniform gaseous distension of the small and large intestines, occupying the entire abdomen. The haustrated proximal colon appears segmented. The radiographic diagnosis is ileus and most likely functional because the intestinal loops are uniformly dilated; however, intestinal obstruction or torsion cannot be ruled out.

Common wombat, male, ~13 months old

This juvenile common wombat was being hand-reared. It was inappetant and depressed, had abnormal faecal consistency and suffered progressive abdominal distension (Fig. 12.52). The animal subsequently underwent emergency surgery for correction of a large intestinal torsion.

Koalas

Gastrointestinal disease resulting in anorexia, lethargy, diarrhoea, reduced faecal output and dehydration is relatively common in koalas in permanent care. Koalas are highly specialised feeders. Any change in food intake or leaf quality can result in gastrointestinal dysfunction. Disease may upset the balance of gastrointestinal flora necessary for hindgut fermentation (see Chapter 5 for normal koala abdominal radiographic appearance). Koalas with gastrointestinal dysfunction may accumulate a varying amount of gas, which is readily apparent on abdominal radiographs. Figures 12.53–12.55 are examples of abnormal gas accumulation in ill koalas. Gastrointestinal torsion has been reported in koalas and should be also considered when intestinal gas accumulation is seen radiographically (Joyce-Zuniga *et al.* 2014).

The urogenital tract

The most common radiographic pathology affecting the urogenital tract in Australian mammals appears to be the formation of calculi within the upper or lower urinary tract. Uroliths have been seen in macropods (Vogelnest and Portas 2008), dasyurids (Holz 2008), koalas (Blanshard and Bodley 2008), and possums and gliders (Johnson and Hemsley 2008). Although the aetiopathogenesis of urolith formation in Australian mammals is not well understood, dietary factors may play a role.

Urolithiasis

Red kangaroo, vasectomised male, 6 years old

This red kangaroo presented inappetent, recumbent, hypothermic and dehydrated. It was azotaemic and abdominal ultrasound examination showed moderate, free abdominal fluid and an enlarged left kidney with pelvic dilation but no

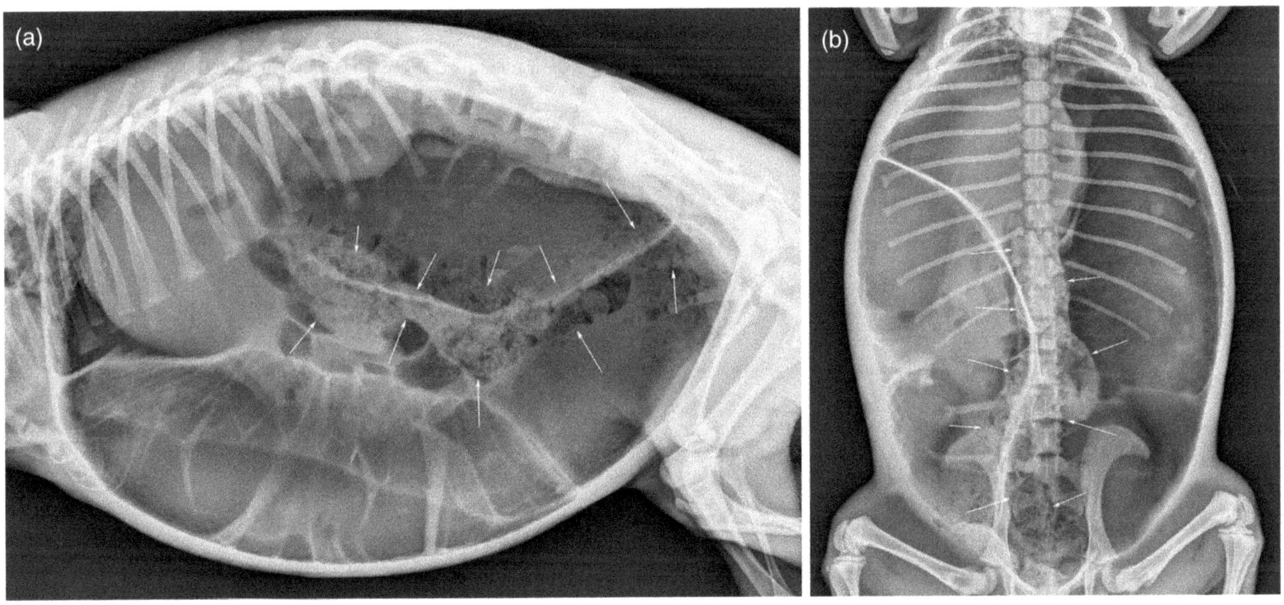

Figure 12.52: Abdomen, common wombat (*Vombatus ursinus*): (a) lateral view, (b) ventrodorsal view. There is massive dilation of the intestines, occupying most of the abdominal cavity. Haustration within the dilated viscera indicates that it is probably the proximal colon. The radiographic appearance is consistent with torsion of the colon, marked ileus or radiolucent foreign body impaction. The descending colon contains faeces and appears normal (arrows).

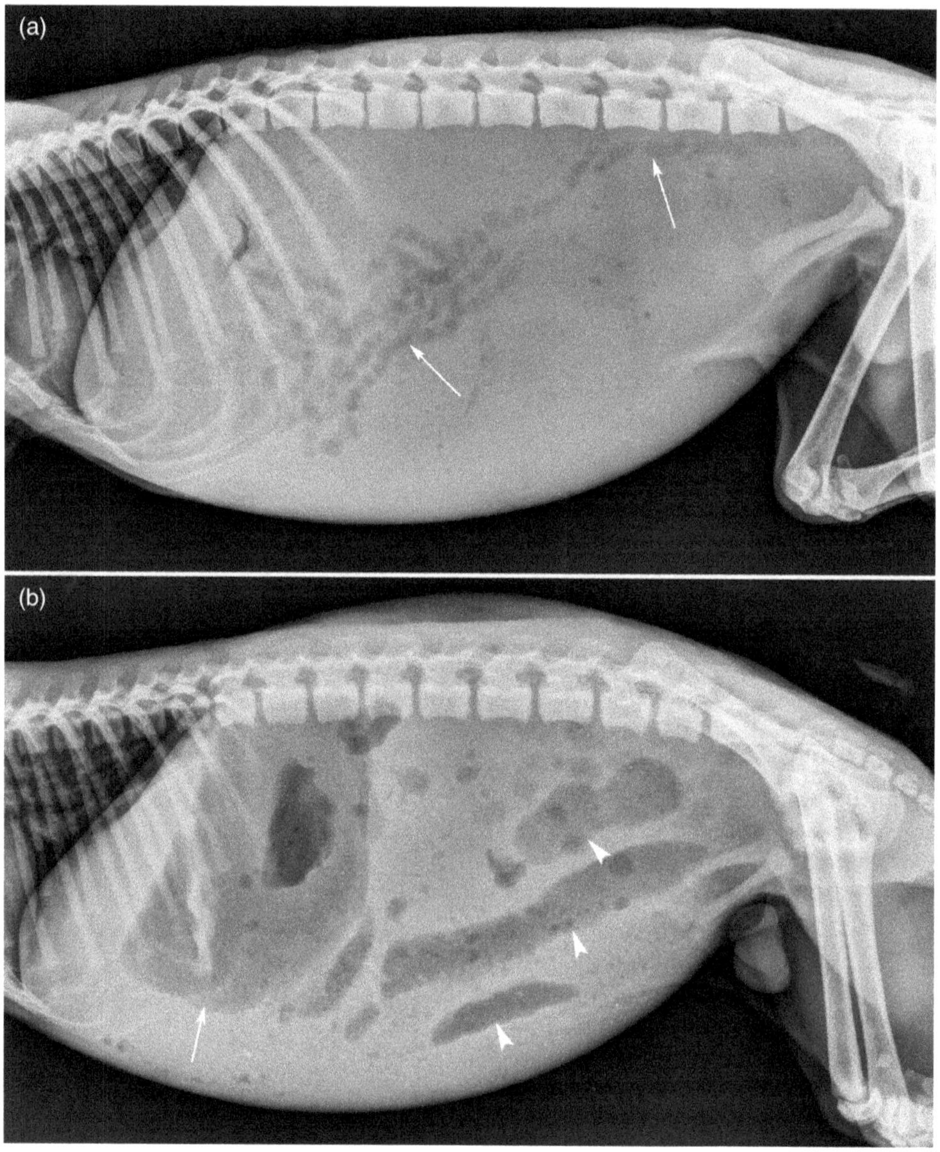

Figure 12.53: Abdomen, koala (*Phascolarctos cinereus*), right lateral view. (a) This adult male koala presented with diarrhoea and lethargy and died 3 days later with ileocaecal lymphoid depletion and mild mesenteric lymphadenitis but no intestinal abnormalities. There is a small amount of gas confined to the distal colon and rectum (arrows). (b) A different adult male koala had decreased appetite and diarrhoea and passed an intact gastropod in its faeces 2 days later, resulting in rapid return to normal. There is gas accumulation in the stomach (arrow) and to a lesser degree in the caecum and proximal colon (arrowheads). Larger particles of ingested leaf pass through the tract more quickly than finer particles (Blanshard and Bodley 2008) and presumably the gastropod was also passed fairly quickly after accidental ingestion.

acoustic shadowing. The bladder appeared intact; however, fluid consistent with urine was recovered from abdominocentesis and cystocentesis. Abdominal radiographs suggested intestinal ileus (Fig. 12.56). Urinary tract obstruction, hydronephrosis and uroabdomen were suspected and the kangaroo was euthanased. Post-mortem examination identified urinary bladder rupture, urethral obstruction by a single calculus and bilateral hydronephrosis with a single calculus in each ureter. All three calculi comprised 100% calcium carbonate. Review of the abdominal radiographs showed moderately

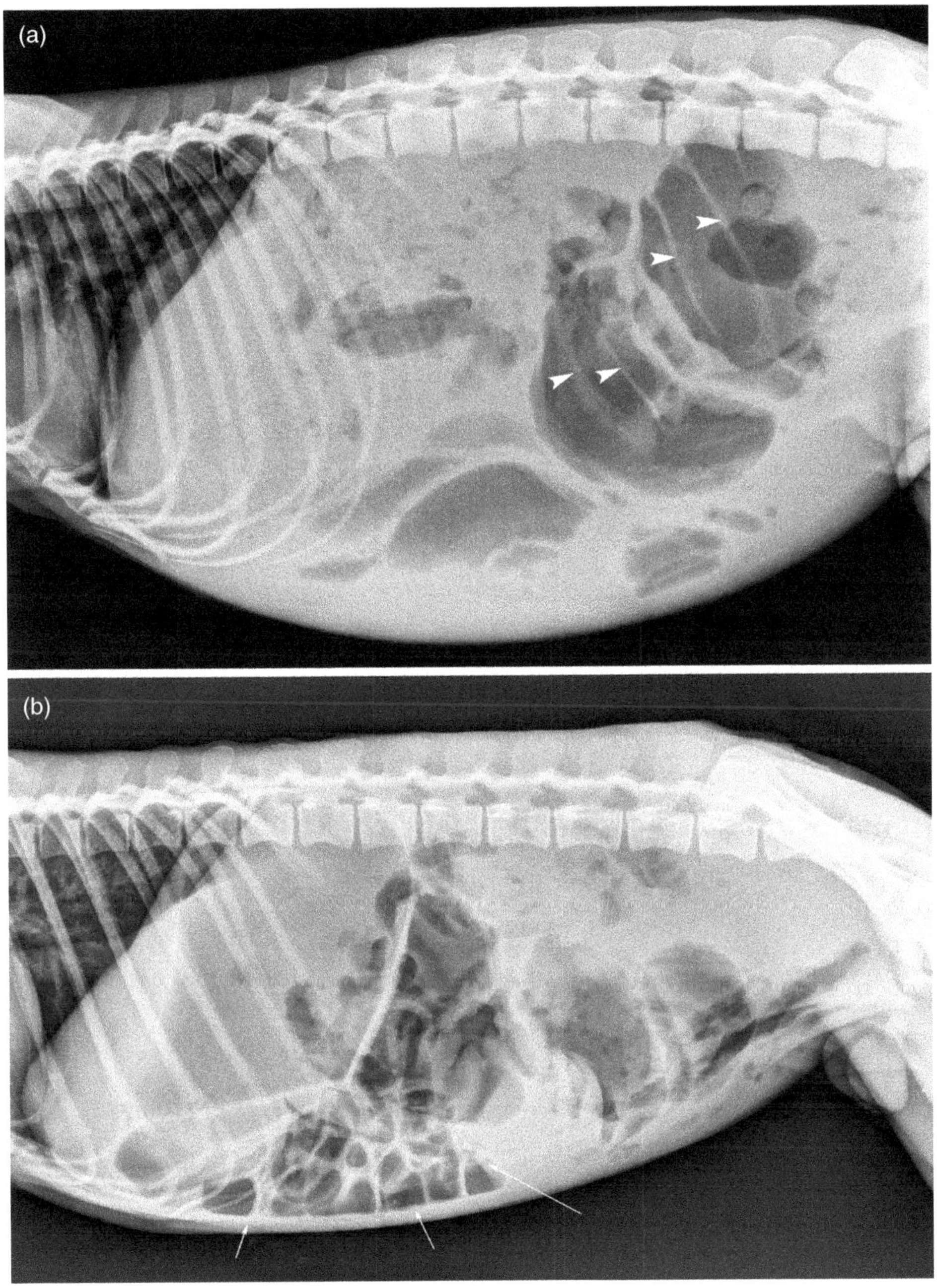

Figure 12.54: Abdomen, koala (*Phascolarctos cinereus*), lateral view. (a) This adult female koala showed decreased body condition, diarrhoea and anorexia and had a reversed neutrophil to lymphocyte ratio on haematology. The koala recovered with supportive treatment. There is a moderate amount of gas in the caecum and proximal colon, which in this case can be distinguished by the longitudinal folds (arrowheads). (b) This adult male koala was lethargic, dehydrated and anorexic and was euthanased because of a lack of response to treatment. At post-mortem examination there was moderate suppurative jejunitis. There is moderately extensive gas accumulation throughout the gastrointestinal tract, but particularly in the small intestine (arrows).

radiopaque calculi in the region of the ureters. Urolithiasis in this case was considered diet-related (Lindemann *et al.* 2013).

Case submitted by: Dr Kathryn Gamble, Lincoln Park Zoo, IL, USA and images reproduced with permission from: Lindemann DM, Gamble KC,

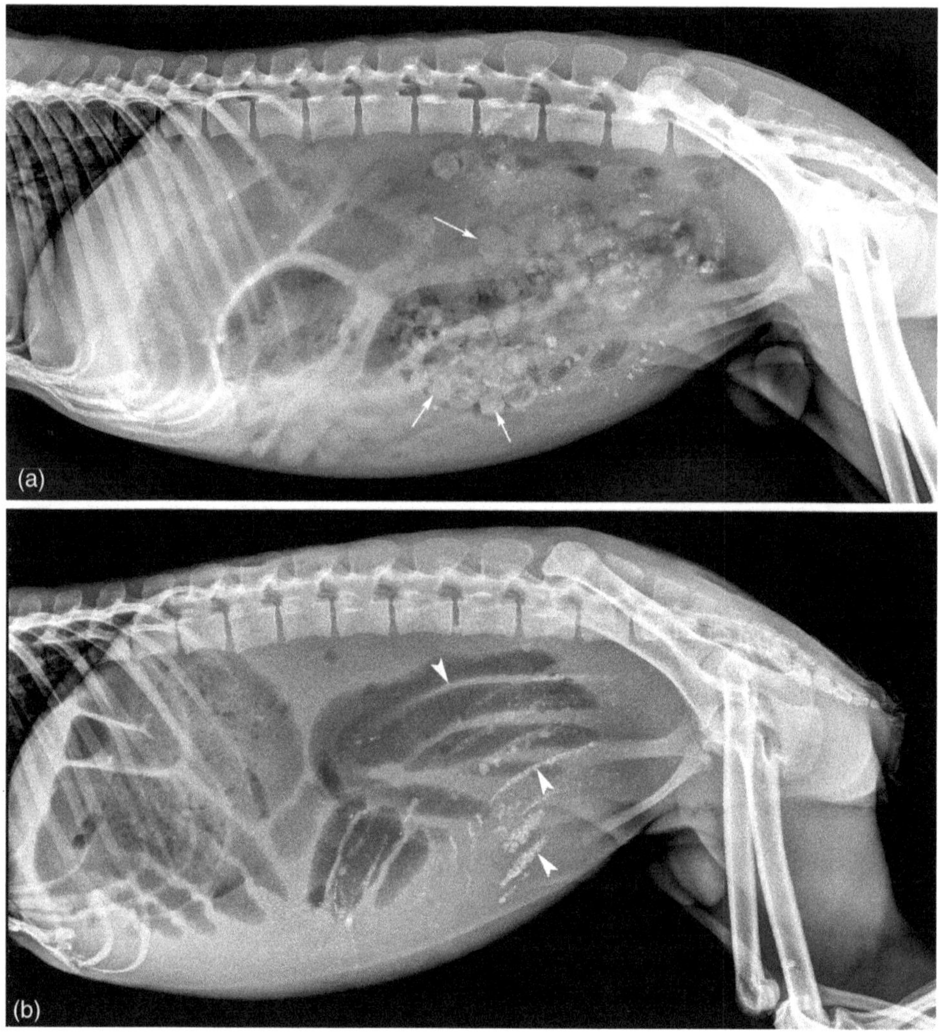

Figure 12.55: Abdomen, koala (*Phascolarctos cinereus*), lateral view: (a) 3 days after barium, (b) 9 days after barium. This adult male koala was lethargic, with a gradual loss of appetite and development of diarrhoea over a period of 2 weeks. It was euthanased after failing to respond to intensive supportive treatment. Post-mortem examination and histopathology showed caecal oedema with loss of luminal bacteria. One 5-mL dose of barium was administered orally prior to radiography. (a) Faecal pellets, which are formed in the distal colon, are outlined by barium (arrows). (b) The longitudinal folds of the caecum and proximal colon are outlined by barium and gas (arrowheads).

Corner S (2013) Calcium carbonate obstructive urolithiasis in a Red kangaroo (*Macropus rufus*). *Journal of Zoo and Wildlife Medicine* **44**, 196–199.

Spot-tailed quoll, female, 3.5 years old

This spot-tailed quoll was presented with sudden onset of stranguria and pollakiuria. Abdominal radiographs were normal, but the urinary bladder wall was thickened on ultrasound examination. No urinary tract calculi were noted; however, a small number of magnesium ammonium phosphate crystals were present in the urine. The quoll was treated for a lower urinary tract infection based on urine culture results. Clinical signs resolved, but 2 weeks later a firm mass was palpated in the bladder. The mass was radiopaque

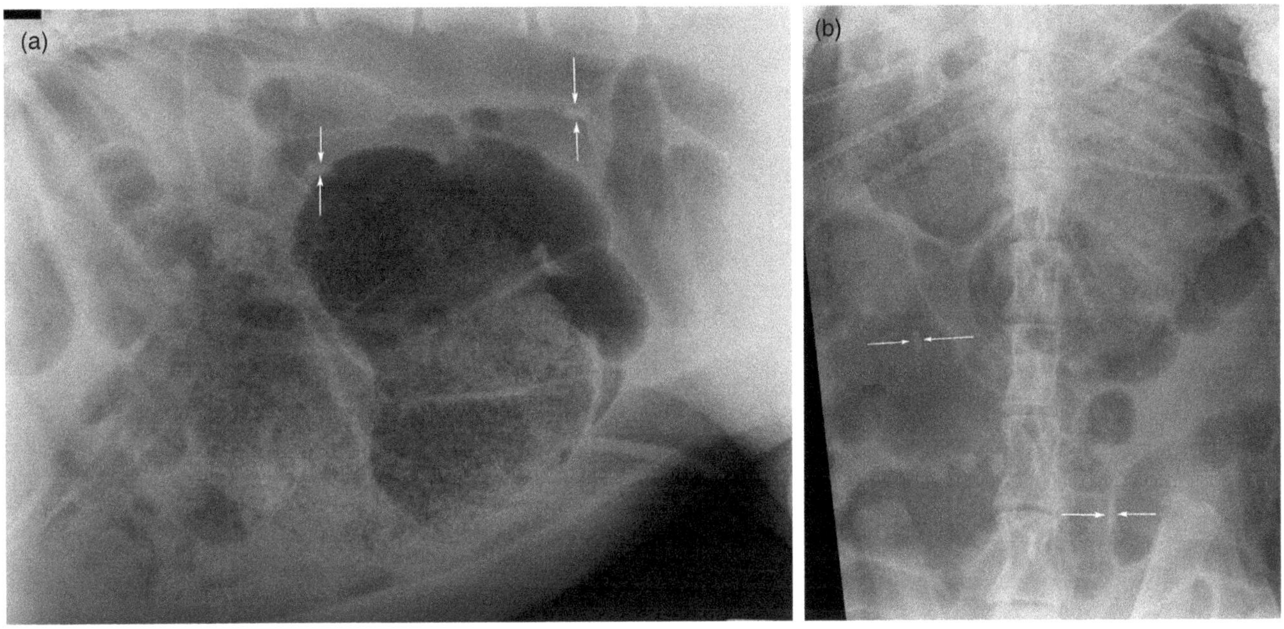

Figure 12.56: Abdomen, red kangaroo (*Macropus rufus*): (a) lateral view, (b) ventrodorsal view. (a) Focal, oblong opacities superimposed on the retroperitoneal space (arrows) are consistent with ureteric calculi, although care must be taken to exclude an intestinal or a peritoneal opacity. (b) Two calcific opacities (arrows) are visible superimposed on abdominal viscera. When available, sonography is useful to search for secondary signs of ureteric obstruction, such as renal pelvic or ureteric dilation.

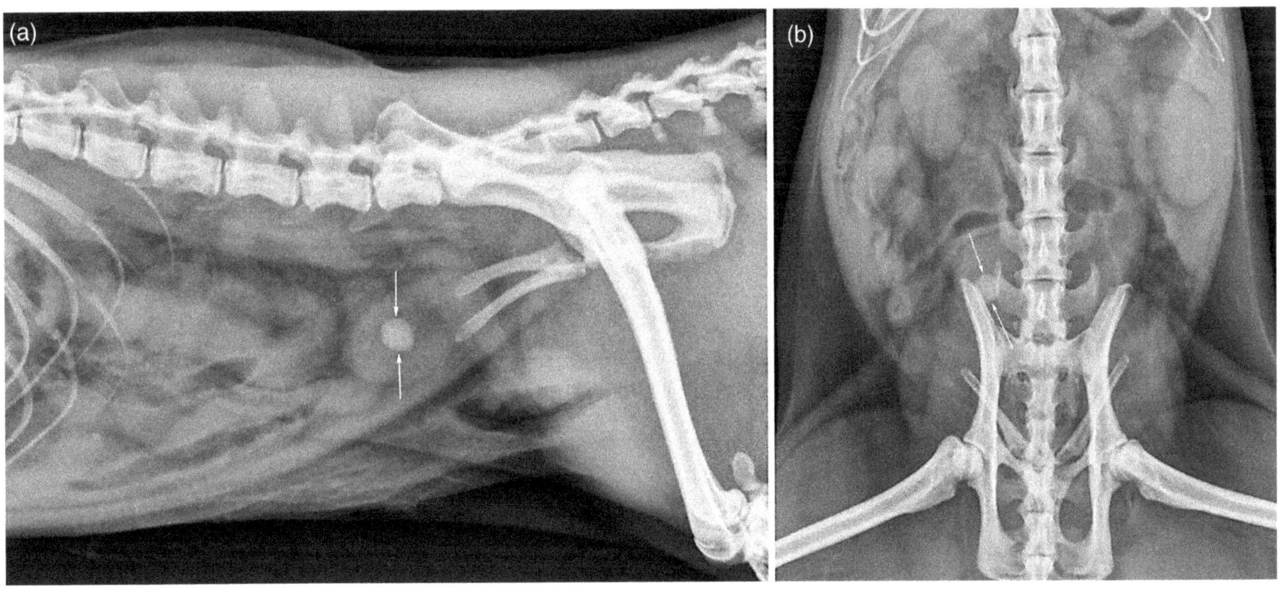

Figure 12.57: Abdomen, spot-tailed quoll (*Dasyurus maculatus*): (a) lateral view, (b) ventrodorsal view. There is a 6-mm diameter opaque, spherical nodule in the urinary bladder (arrows). The opacity is similar to bone, consistent with a cystic calculus.

(Fig. 12.57) and visible sonographically. A bladder calculus was surgically removed. It consisted of an aggregate of 1-mm bead-like concretions composed of 66% calcium oxalate and 34% magnesium ammonium phosphate. The calculus most likely formed around a nidus of bacteria and inflammatory cells during acute cystitis. No dietary modification was undertaken and there was no recurrence of calculus formation.

Prostatic calculi/mineralisation

Common ring-tailed possum, male, 9 years old

This common ring-tailed possum presented with a swollen hindfoot, but was otherwise considered healthy. Radiographs taken to evaluate the limb swelling revealed a radiopaque mass in the caudal abdomen, considered to be the urinary tract. Urinalysis of freely voided urine was normal and the mass could not be palpated in the bladder. Follow-up radiographs 3 months later (Fig. 12.58) confirmed the presence of a radiopaque mass caudal to the neck of the bladder and on this occasion the mass was palpable. Ultrasound examination showed parenchymal mineralisation in the cranial prostate gland with urinary sediment (amorphous phosphates) in the bladder. Serial urine cultures showed no bacterial growth and bacterial prostatitis was considered unlikely. Based on these finding, a diagnosis of prostatic calculi/mineralisation was made. In many species, prostatic calculi/mineralisation is asymptomatic. Corpora amylacea are small hyaline concretions of prostatic secretions that become mineralised with calcium salts and remain in the glandular lumina forming prostatic calculi. Corpora amylacea and prostate calculi are common in humans, but rare in domestic mammals. In dogs, prostatic mineralisation may be associated with neoplastic and non-neoplastic prostate disease (Bradbury *et al.* 2009). Prostatic calculi have been reported in a free-ranging koala concurrent with a chronic active prostatitis and cystitis (Krockenburger and Canfield 1999).

The mesentery

Mycobacteriosis

Mycobacteriosis has been reported in various macropod species held in permanent care. Ingestion of environmental organisms is a likely route of infection. Non-specific signs of disease make diagnosis difficult. Calcification of mycobacterial granulomas is likely dependent on both host and bacterial factors; however, when it occurs it aids in the radiographic localisation of lesions. Gastrointestinal foreign body, enterolith formation and neoplasia should be considered as differential diagnoses, for the presence of calcified masses suspected to be in the mesentery.

Mycobacteriosis of mesenteric lymph nodes

Red-necked wallabies, both female, 3–4 years old

The red-necked wallaby shown in Fig. 12.59 was lethargic with occasional diarrhoea. Abdominal radiographs identified two 5-cm oval radiopaque opacities in the mid-cranial abdomen. After 1 month, the two masses had increased in size and appeared more mineralised. Ultrasound guided fine needle aspiration of one mass yielded a small amount of caseous material containing acid-fast bacteria belonging to the *Mycobacterium avium* complex. The wallaby was euthanased and on post-mortem examination the masses were identified as abscessed mesenteric lymph nodes.

The wallaby shown in Fig. 12.60 was housed in the same zoo enclosure as the animal in shown Fig. 12.59. A radiopaque spherical mass was discovered in the right mid-abdomen on routine radiographs taken during a pre-shipment examination. The mass was palpable, with a firm outer texture and softer centre. Exploratory laparotomy identified the mass as an abscessed mesenteric lymph node, with a thin, mineralised cortex and purulent contents. The mass was surgically removed. Mycobacteria were identified on histopathology and culture, but the organism was unable to be further characterised by PCR.

Cases submitted by: Dr Kathryn C. Gamble, Dr Lester E Fisher and Dr Lauren M Kao, Lincoln Park

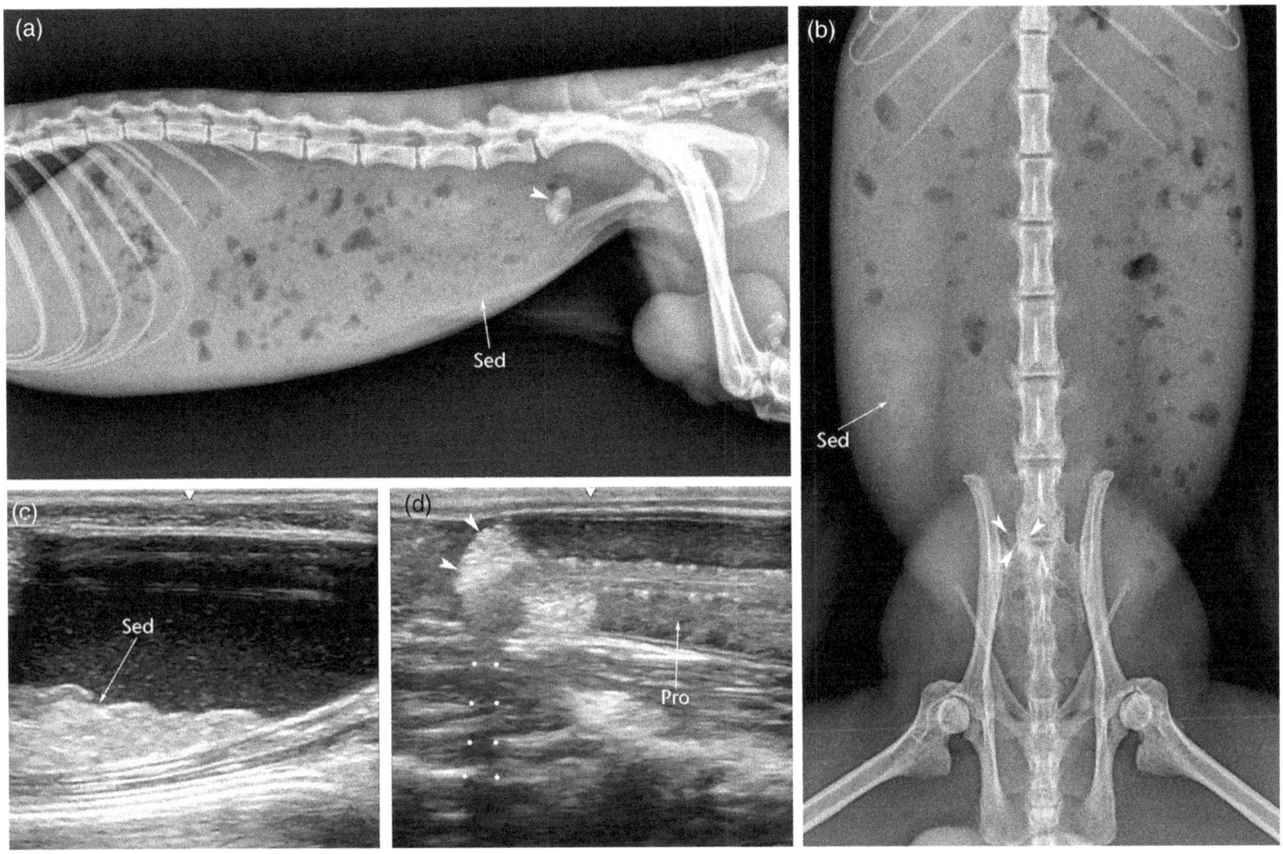

Figure 12.58: Abdomen, common ring-tailed possum (*Pseudocheirus peregrinus*): (a) lateral view, (b) ventrodorsal view, (c) urinary bladder, sagittal sonogram, (d) prostate, sagittal sonogram. There is an ovoid calcific nodule (arrowheads) located dorsal to the epipubic bones in (a) and in the midline (b). A calcific sediment (Sed) is present in the urinary bladder. The ultrasound images (c, d) were acquired using a 14-MHz transducer. They reveal a hyperechoic sediment in the urinary bladder (Sed) and calcific material (arrowheads) in the cranial end of the prostate gland (Pro), which produced a distant acoustic shadow (white dots). There are no visible urinary calculi. The ovoid calcific nodule represents mineralised prostatic tissue.

Zoo, IL, USA and Dr Janis Raines, Dallas Zoo and Children's Aquarium at Fair Park, TX, USA.

Neoplasia

Mesenteric and omental neoplasia

Greater bilby, male, 5 years old

This aged greater bilby (*Macrotis lagotis*) was examined because of weight loss. A firm irregular mass was palpated in the mid-abdomen. Whole-body radiographs (Figs 12.61 and 12.62) showed a mass with soft tissue opacity occupying much of the cranial abdomen. Sonography confirmed a multilobulated structure of varying echogenicity that was thought to be associated with the spleen. Exploratory laparotomy revealed a mass arising within the root of the mesentery involving the omentum and forming discrete and coalescing nodules, some containing large blood clots or free blood. The liver and spleen appeared grossly normal and the intestines were not involved. The mass could not be surgically resected and the bilby was euthanased. From the gross appearance of the tumour, haemangiosarcoma was highly suspected, but the cellular morphology was not typical and factor VIII IHC was negative. The cell line of origin of the tumour could not be determined on histopathology.

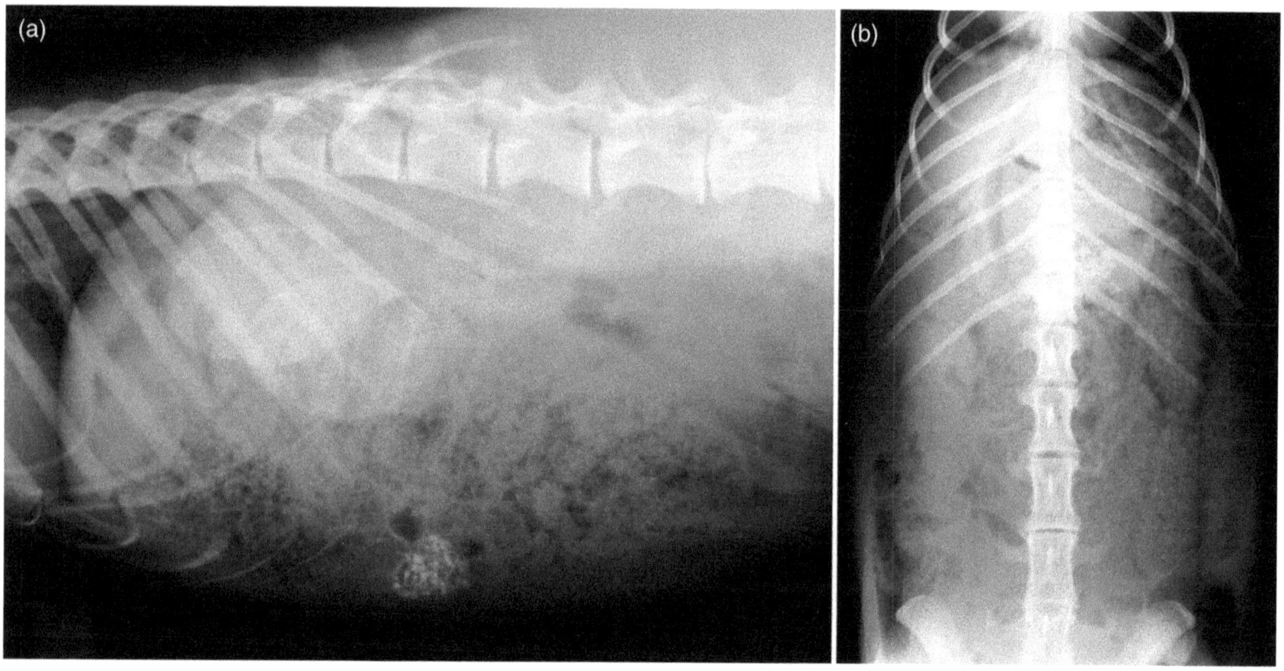

Figure 12.59: Abdomen, red-necked wallaby (*Macropus rufogriseus*): (a) lateral view, (b) ventrodorsal view. There are two egg-shaped opacities in the cranial abdomen. They appear to have a relatively lucent centre and slightly greater opacity of the outer 'shell'. There is an additional mottled ovoid opacity in the ventral abdomen. The opaque masses were identified as mineralised lymph nodes containing mycobacteria. The cause of the ventral abdominal mineralisation is unknown.

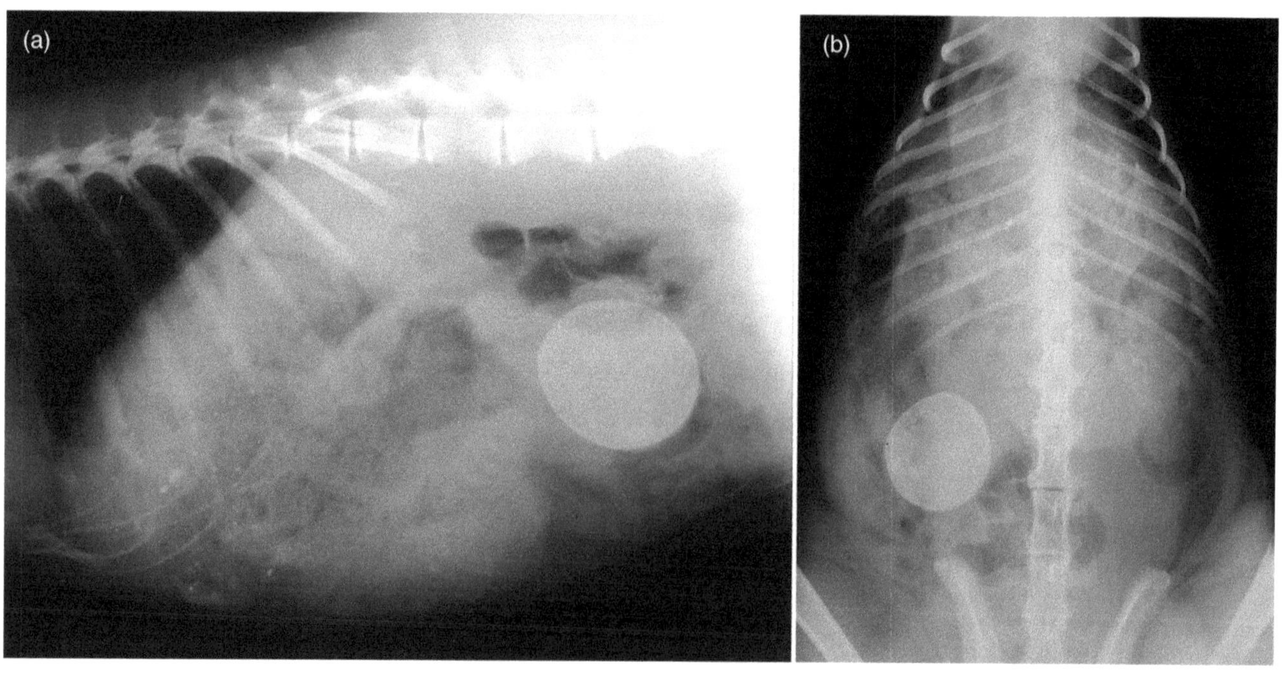

Figure 12.60: Abdomen, red-necked wallaby (*Macropus rufogriseus*): (a) lateral view, (b) ventrodorsal view. There is a single egg-shaped opacity, which is more uniformly opaque than the opacities seen in Fig. 12.59. It represents a mineralised lymph node containing mycobacteria.

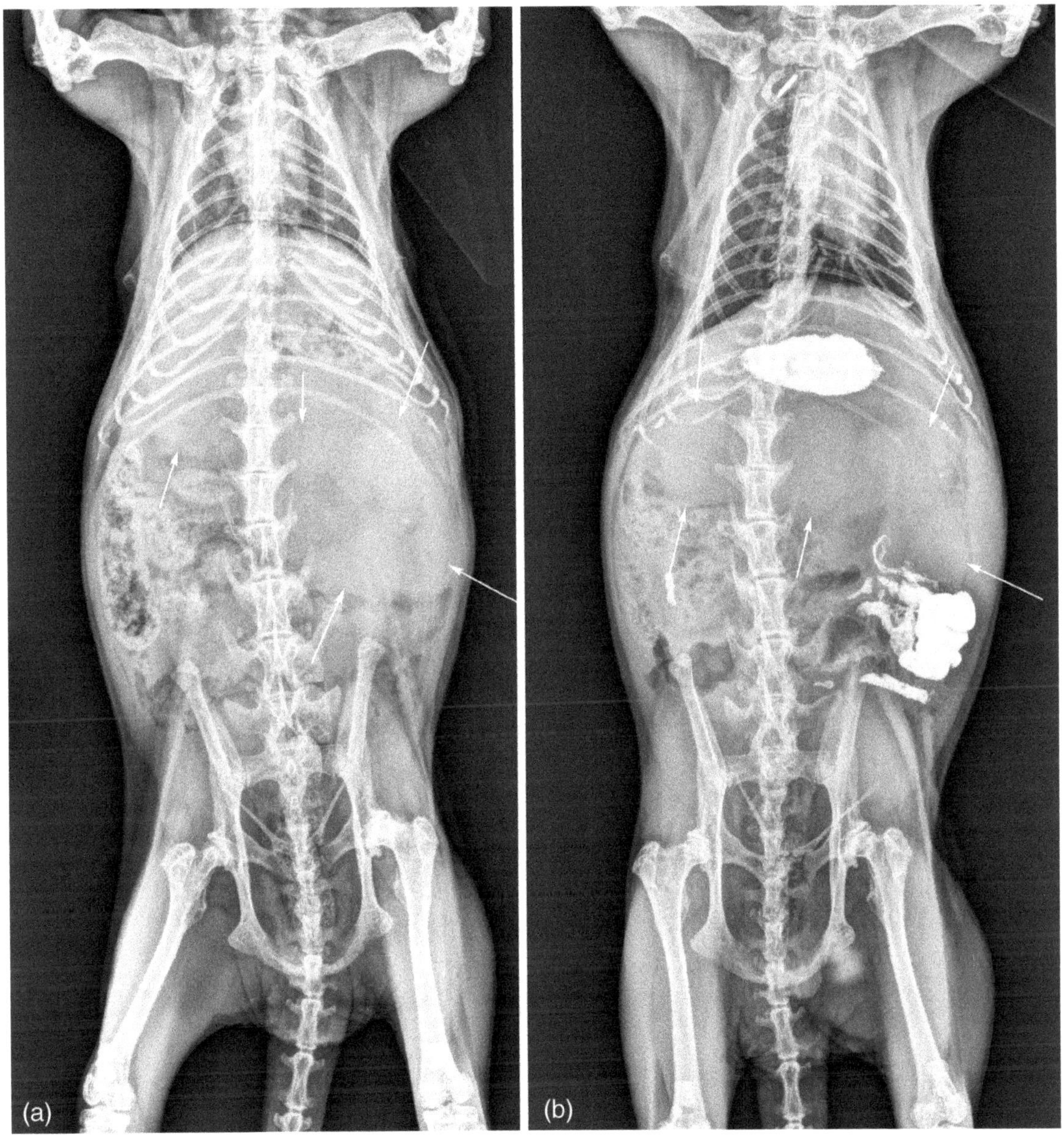

Figure 12.61: Thorax and abdomen, greater bilby (*Macrotis lagotis*), ventrodorsal view: (a) survey radiograph, (b) post oral barium administration. There is a nodular mass in the mid abdomen (margins indicated by arrows). The small and the large intestines are displaced caudally. Note: transponders can be seen.

Fat necrosis

Quokkas (*Setonix brachyurus*) appear to be predisposed to the development of mesenteric fat necrosis. This may be an incidental finding or may compromise the health of the animal. The cause is unknown, but may include genetic factors, altered fat metabolism because of obesity or a high intake of saturated fat (Hulst *et al.* 2010).

Mesenteric fat necrosis

Quokka, male, 9 years old

Mesenteric fat necrosis was suspected in a quokka based on survey radiographic findings. The quokka remained well for five years before developing weight loss, diarrhoea and polydipsia. The quokka failed to respond to treatment. Figure 12.63 shows abdominal radiographs taken shortly before

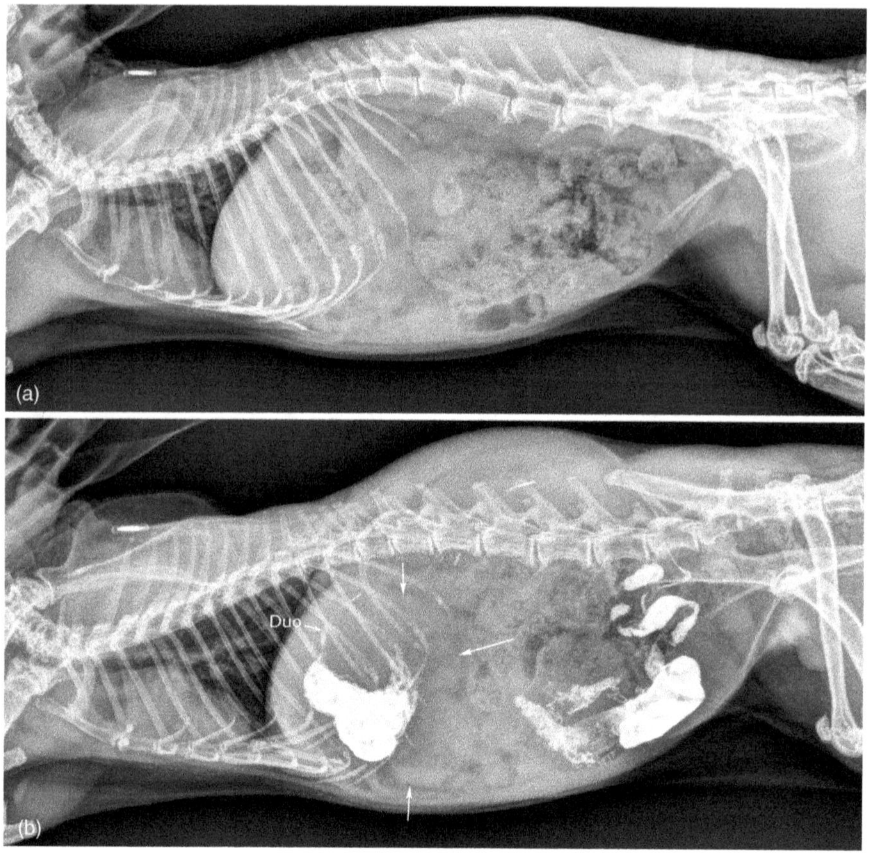

Figure 12.62: Thorax and abdomen, bilby (*Macrotis lagotis*), lateral view: (a) survey radiograph, (b) post oral barium administration. The duodenum (Duo) is displaced craniodorsally while the remaining intestines are displaced caudally. Arrows indicate the margins of the mass. Note: transponders can be seen.

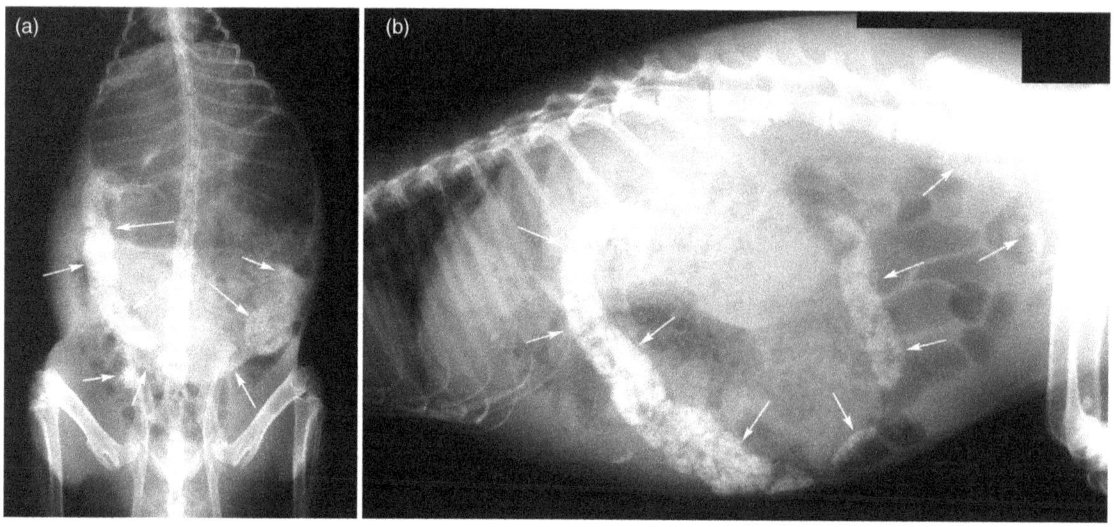

Figure 12.63: Abdomen, quokka (*Setonix brachyurus*): (a) ventrodorsal view, (b) lateral view. Mineralised margins of tissue (arrows) are visible forming elongated tubular structures in the peritoneal cavity. These structures consist of mineralised necrotic fat within the mesentery.

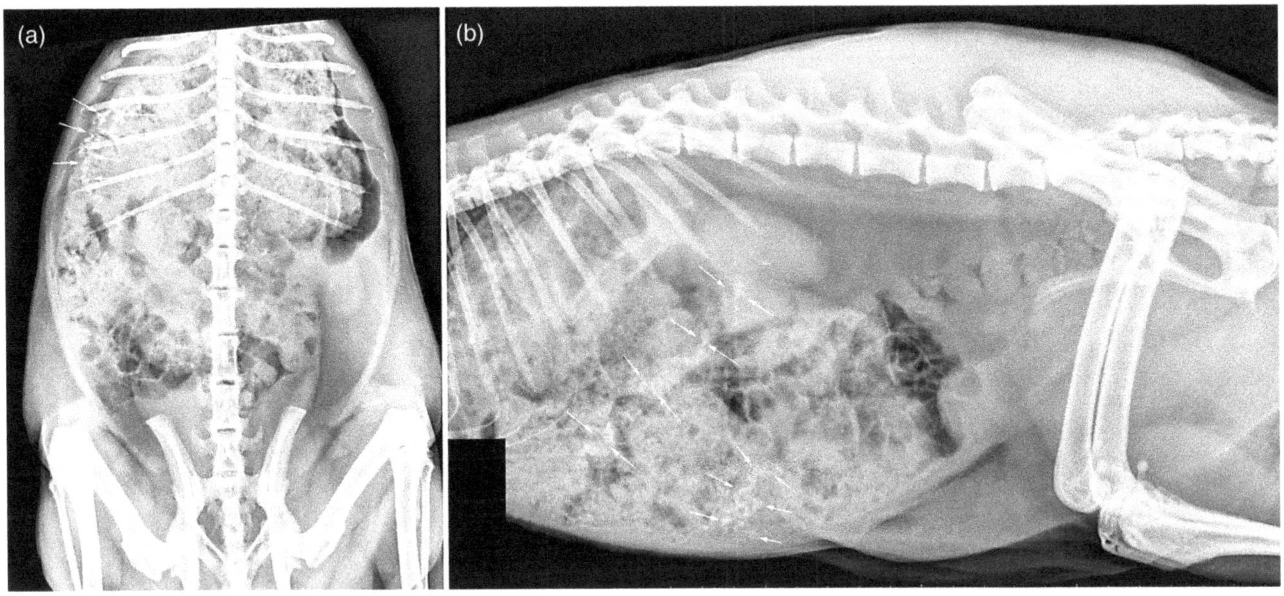

Figure 12.64: Abdomen, quokka (*Setonix* brachyurus): (a) ventrodorsal view, (b) lateral view. The arrows indicate mesenteric fat necrosis with less advanced mineralisation than is evident in Figure 12.63.

euthanasia. Post mortem examination revealed firm white linear masses in the mesentery that were confirmed on histopathology as fat calcification and necrosis.

Quokka, female, adult

A female quokka in the same collection as the animal in Figure 12.63 had a history of obesity with occasional episodes of lethargy and reduced appetite. The quokka was examined during one of these episodes when multiple linear, firm masses were palpable in the mid-abdomen. Radiographs (Fig. 12.64) were consistent with mesenteric fat necrosis. The quokka recovered without treatment and has remained otherwise healthy.

Miscellaneous soft tissue masses

A variety of soft tissue masses may occur in Australian mammals. Radiography can be useful in their detection and can aid diagnosis; however, other diagnostic modalities such as biopsy and microbiology are required to reach a definitive diagnosis.

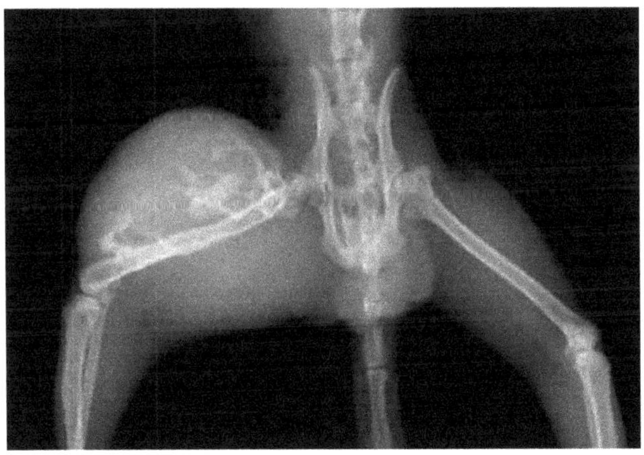

Figure 12.65: Pelvis and hind quarters, Leadbeater's possum (*Gymnobelideus leadbeateri*), ventrodorsal view. There is a large soft tissue mass arising adjacent to the right femur. It contains mineralised material and appears to intimately surround the femur without invading it. The mass is consistent with a neoplastic process arising within soft tissue and becoming mineralised. It is possible that the mineralised material is osseous, but this cannot be determined radiographically. It is interesting that the mass appears to engulf the femur without appearing to invade or destroy the adjacent cortices. Radiographic differential diagnoses could include soft tissue calcification subsequent to trauma or infection.

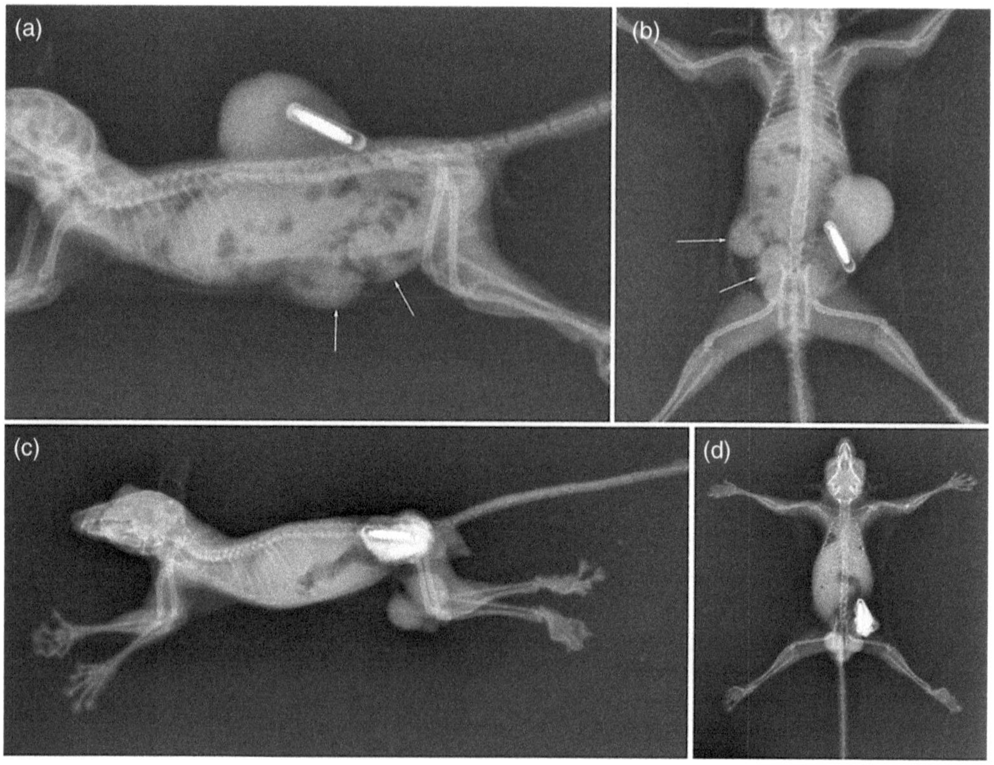

Figure 12.66: Thorax and abdomen, feather-tailed glider (*Acrobates pygmaeus*): (a) female, lateral view, (b) female, ventrodorsal view, (c) male lateral view, (d) male ventrodorsal view. (a, b) There is a large soft tissue mass on the dorsum, extending to the left side of the abdomen. It appears to incorporate a transponder. The mass is in a cutaneous or subcutaneous location and appears separate from the spine and abdominal viscera. There are two pouch young (arrows) faintly visible superimposed on the ventral abdomen. (c, d) There is a mineralised mass dorsolateral to the pelvis that appears to incorporate a transponder. The mass represents an unusual osteosarcoma.

Neoplasia

Soft tissue sarcoma

Leadbeater's possum, male, unknown age

This wild Leadbeater's possum (*Gymnobelideus leadbeateri*) was caught under permit for inclusion in a captive breeding program. The possum was thin with a firm mass palpable on the upper right hindlimb. Radiographs revealed a large, irregular partly mineralised mass associated with the femur (Fig. 12.65). An incisional biopsy was obtained and the mass was identified histologically as an undefined soft tissue sarcoma.

Case submitted by: Dr Franciscus Scheelings, Healesville Sanctuary, VIC, Australia.

Soft tissue sarcoma formation associated with subcutaneously implanted passive integrated transponders has been reported most commonly in feather-tailed gliders (*Acrobates pygmaeus*), as well as in other glider species. The tumours rarely metastasise to distant organs, but local tissue invasion is common (Vogelnest *et al.* 1997; Johnson and Hemsley 2008; Ladds 2009).

Feather-tailed gliders, female, 4 years old and male, 8 years old

A transponder was implanted subcutaneously over the dorsal thorax of a female feather-tailed glider for individual identification purposes and 4 years

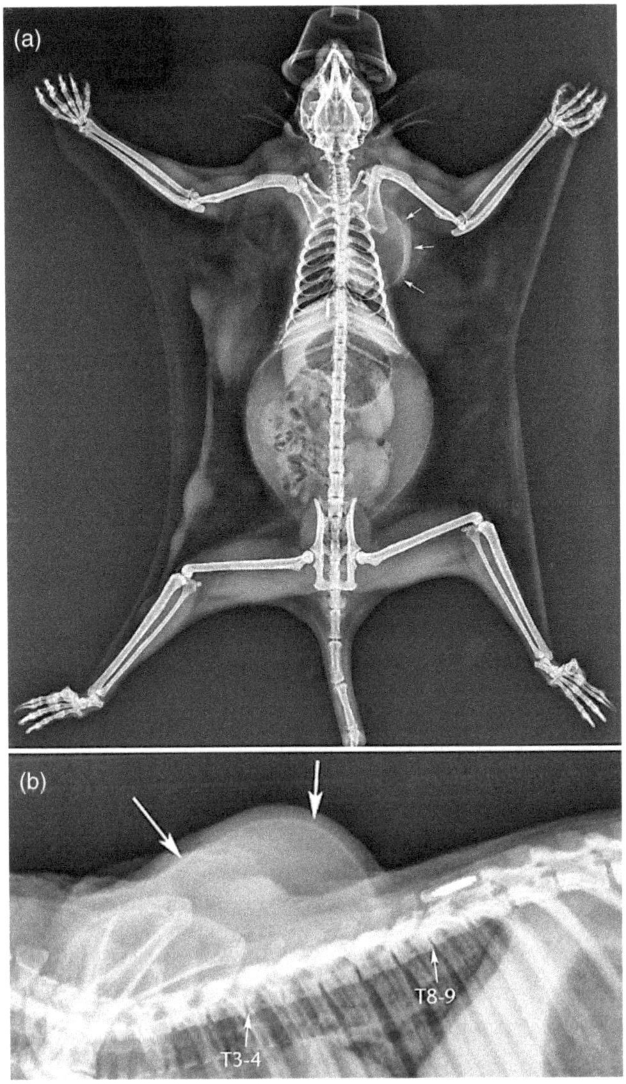

Figure 12.67: Yellow-bellied glider (*Petaurus australis*): (a) whole body, ventrodorsal view, (b) thoracic spine, lateral view. (a) A mass (arrows) is visible on the left side of the thorax adjacent to the scapula. Its opacity is less than that of muscle, indicating that it probably consists of adipose tissue. Smaller nodules are present in the patagium on the right side. There are large bilateral intra-abdominal fat pads of similar opacity to the thoracic mass. (b) A mass with well defined margins (arrows) is present on the dorsum of the thorax adjacent to the scapulae. The mass does not appear to encroach on the scapulae or the spine. The intervertebral disc spaces at T3–4 and T8–9 are narrower when compared with the adjacent disc spaces. Note: a transponder can be seen.

later, there was a large firm swelling that incorporated the transponder and had a peduncle that appeared to be attached to the ribcage. Whole-body radiographs (Fig. 12.66a, b) were taken before incisional biopsy. Biopsy results were characteristic of a soft tissue sarcoma.

A male feather-tailed glider developed a hard mass over the pelvis incorporating a transponder that had been implanted subcutaneously 8 years previously. Radiographs were taken (Fig. 12.66c, d) and the glider was euthanased. Histopathology revealed an osteosarcoma incorporating the transponder. The tumour was not attached to the underlying skeletal structures and was considered an unusual primary osteosarcoma rather than bony metaplasia of a soft tissue sarcoma.

Lipoma

Yellow-bellied glider, female, 16 years old

This yellow-bellied glider (*Petaurus australis*) presented with sudden onset of hindlimb paresis. A firm subcutaneous mass between the scapulae appeared to be attached to the cranial thoracic spine. Whole-body radiographs showed a soft tissue mass with fat opacity dorsal to the scapulae; no spinal abnormalities were noted. Fine needle aspirate cytology of the mass was consistent with a lipoma. The glider was treated on suspicion of spinal trauma or possible extension of the mass to involve the spinal cord. After 1 week, further radiographs were taken (Fig. 12.67) and thoracic intervertebral disc protrusion was suggested. The glider failed to respond to treatment and was euthanased. Histopathology confirmed an infiltrative lipoma external to the thoracic cavity, multiple cutaneous lipomas, thoracic intervertebral disc degeneration and prolapse, and spinal malacia, haemorrhage and Wallerian degeneration. The dorsal lipoma, although very infiltrative, did not appear to have penetrated the vertebral column, but had replaced adjacent musculature, which may have led to increased vertebral instability.

References

Ashwell KWS, Zhang LL (1997) Cyto- and myeloarchitectonic organization of the spinal cord of an echidna (*Tachyglossus aculeatus*). *Brain, Behavior and Evolution* **49**, 276–294. doi:10.1159/000112998

Barnes TS, Hinds LA, Jenkins DJ, Coleman GT (2007) Precocious development of hydatid cysts in a macropodid host. *International Journal for Parasitology* **37**, 1379–1389. doi:10.1016/j.ijpara.2007.04.012

Barnes TS, Goldizen AW, Morton JM, Coleman GT (2008) Cystic echinococcosis in a wild population of the brush-tailed rock-wallaby (Petrogale penicillata), a threatened macropodid. *Parasitology* **135**, 715–723. doi:10.1017/S0031182008004423

Barnes TS, Hinds LA, Jenkins DJ, Bielefeldt-Ohmann H, Lightowers MW, Coleman GT (2011) Comparative pathology of pulmonary hydatid cysts in macropods and sheep. *Journal of Comparative Pathology* **144**, 113–122. doi:10.1016/j.jcpa.2010.07.003

Bennet RA, Browning WN, Dunker FK (1993) Dilated cardiomyopathy in an antilopine kangaroo (*Macropus robustus*). In *Proceedings of the American Association of Zoo Veterinarians.* October 9–14, St Louis, Missouri. pp. 106–107. American Association of Zoo Veterinarians, Florida, USA.

Bercier M, Wynne J, Klause S, Stadler CK, Gorow A, Pye GW (2012) Nasal mass removal in the koala (*Phascolarctos cinereus*). *Journal of Zoo and Wildlife Medicine* **43**, 898–908. doi:10.1638/2011-0266R.1

Blanshard WH, Bodley KB (2008) Koalas. In *Medicine of Australian Mammals.* (Eds L Vogelnest and R Woods) pp. 227–327. CSIRO Publishing, Melbourne.

Bradbury CA, Westropp JL, Pollard RE (2009) Relationship between prostatomegaly, prostatic mineralization, and cytologic diagnosis. *Veterinary Radiology & Ultrasound* **50**, 167–171. doi:10.1111/j.1740-8261.2009.01510.x

Bryant B, Reiss A (2008) Wombats. In *Medicine of Australian Mammals.* (Eds L Vogelnest and R Woods) pp. 329–358. CSIRO Publishing, Melbourne.

Burgdorf-Moisuk A, Pye GW, Smith JA, Papendick R, Ivy JA, Hamlin-Andrus C (2012) A retrospective and prospective study of megaesophagus in the parma wallaby (*Macropus parma*) at the San Diego Zoo, California, USA. *Journal of Zoo and Wildlife Medicine* **43**, 89–94. doi:10.1638/2011-0202.1

Canfield P, Spencer A (1993) Secondary degenerative arthropathy (osteoarthrosis) of the hip joints in ageing, free-living koalas. *Australian Veterinary Journal* **70**, 394–395. doi:10.1111/j.1751-0813.1993.tb00827.x

Canfield PJ, Perry R, Brown AS, McKenzie RA (1987) Cranio-facial tumours of mixed cartilage and bone in koalas (*Phascolarctos cinereus*). *Australian Veterinary Journal* **64**, 20–22. doi:10.1111/j.1751-0813.1987.tb06051.x

Chineme CN, Njoku CO, Evbuoma N (1978) Degenerative cardiomyopathy in a captive kangaroo (*Macropus giganteus*). *Journal of Wildlife Diseases* **14**, 52–53. doi:10.7589/0090-3558-14.1.52

Connolly JH, Obendorf DL, Whittington RJ, Muir DB (1998) Causes of morbidity and mortality in platypus (*Ornithorhynchus anatinus*) from Tasmania with particular reference to *Mucor amphibiorum* infection. *Australian Mammalogy* **20**, 177–187.

Hanger JJ, McKee JJ, Tarlinton R, Yates A (2003) Cancer and haematological disease in koalas: a clinical and virological update. In *Proceedings of the 2003 Conference of the Australian Association of Veterinary Conservation Biologists,* Cairns. (Eds A Tribe and R Booth). pp. 19–30. Australian Association of Veterinary Conservation Biologists, Gatton, Queensland.

Holz P (2008) Dasyurids. In *Medicine of Australian Mammals.* (Eds L Vogelnest and R Woods) pp. 359–382. CSIRO Publishing, Melbourne.

Hulst F, Sangster C, Vogelnest L (2010) Mesenteric fat necrosis in the Quokka (*Setonix brachyurus*) In: *Proceedings of the Wildlife Disease Association,*

(*Australasian Section*), *Annual Conference*, Dover, Tasmania. pp 53–54. Wildlife Disease Association, (Australasian Section).

Isselbacher EM (2005) Contemporary reviews in cardiovascular medicine; thoracic and abdominal aortic aneurysms. *Circulation* **111**, 816–828. doi:10.1161/01.CIR.0000154569.08857.7A

Jackson S (2007) *Australian Mammals: Biology and Captive Management*. CSIRO Publishing, Melbourne.

Jenkins DJ, Morris B (2003) Echinococcus granulosus in wildlife in and around the Kosciuszko National Park, south-eastern Australia. *Australian Veterinary Journal* **81**, 81–85. doi:10.1111/j.1751-0813.2003.tb11440.x

Johnson RJ, Hemsley S (2008) Possums and gliders. In *Medicine of Australian Mammals*. (Eds L Vogelnest and R Woods) pp. 395–437. CSIRO Publishing, Melbourne.

Joyce-Zuniga NM, Roesler J, Hamlin Andrus C, Sutherland-Smith M, Ridcout BA, Pye GW (2014) Gastrointestinal torsions and intussusception in northern koalas (*Phascolarctos cinereus*) at San Diego Zoo (1976–2012). *Journal of Zoo and Wildlife Medicine* **45**, 118–126. doi:10.1638/2012-0182R.1

Krockenberger MB, Canfield PJ, Malik R (2003) *Cryptococcus neoformans* var. *gattii* in the koala (*Phascolarctos cinereus*): a review of 43 cases of cryptococcosis. *Medical Mycology* **41**, 225–234. doi:10.1080/369378031000137242

Krockenburger MB, Canfield P (1999) Prostatic calculi in a free-ranging koala. In *Proceedings of the Wildlife Disease Association (Australasian Section) Annual Conference*, Jervis Bay, NSW. p. 38. Wildlife Disease Association (Australasian section).

Ladds P (2009) *Pathology of Australian Native Wildlife*. CSIRO Publishing, Melbourne.

Lindemann DM, Gamble KC, Corner S (2013) Calcium carbonate obstructive urolithiasis in a Red kangaroo (*Macropus rufus*). *Journal of Zoo and Wildlife Medicine* **44**, 196–199. doi:10.1638/1042-7260-44.1.196

McCracken H (2008) Veterinary aspects of hand-rearing orphaned marsupials. In *Medicine of Australian mammals*. (Eds L Vogelnest and R Woods) pp. 13–37. CSIRO Publishing, Melbourne.

Middleton D (2008) Echidnas. In *Medicine of Australian Mammals*. (Eds L Vogelnest and R Woods) pp. 77–101. CSIRO Publishing, Melbourne.

O'Rourke MF, Avolio AP, Nichols WW (1986) The kangaroo as a model for the study of hypertrophic cardiomyopathy in man. *Cardiovascular Research* **20**, 398–402. doi:10.1093/cvr/20.6.398

Olsson A, Woods R (2008) Bats. In *Medicine of Australian Mammals*. (Eds L Vogelnest and R Woods) pp. 465–502. CSIRO Publishing, Melbourne.

Pye GW (2009) Shoulder dysplasia in koalas (*Phascolarctos cinereus*) at San Diego Zoo. *Journal of Zoo and Wildlife Medicine* **40**, 453–457. doi:10.1638/2008-0107.1

Pye GW, Hamlin-Andrus C, Moll J (2008) Hip dysplasia in koalas (*Phascolarctos cinereus*) at the San Diego Zoo. *Journal of Zoo and Wildlife Medicine* **39**, 61–68. doi:10.1638/2007-0106.1

Pye GW, Gait SC, Mulot B, Real de Asu MD, Martinez-Nevado E, Bonar CJ, Baines SJ, Baines EA (2013) Metabolic bone disease in juvenile koalas (*Phascolartcos cinereus*). *Journal of Zoo and Wildlife Medicine* **44**, 273–279. doi:10.1638/2012-0031R1.1

Stern AW, Smith S, Snider TA (2009) Hypertrophic cardiomyopathy in two captive Bennett's wallabies (*Macropus rufogriseus rufogriseus*). *Journal of Veterinary Diagnostic Investigation* **21**, 889–892. doi:10.1177/104063870902100623

Thrall DE (2013) Introduction to radiographic interpretation. In *Textbook of Veterinary Diagnostic Radiology*. (Ed DE Thrall) pp. 74–86. Elsevier-Saunders, Missouri.

Vogelnest L, Portas T (2008) Macropods. In *Medicine of Australian Mammals*. (Eds L Vogelnest and R Woods) pp. 133–225. CSIRO Publishing, Melbourne.

Vogelnest L, Woods R (Eds) (2008) *Medicine of Australian Mammals*. CSIRO Publishing, Melbourne.

Vogelnest L, Canfield P, Hartley JW, Reece RL (1997) Suspected transponder implant associated mesenchymal tumourigenesis in two feather-tailed gliders (*Acrobates pygmaeus*) and a koala (*Phascolarctos cinereus*). In *Proceedings of Wildlife Disease Association (Australasian Section) Annual Conference.* Flinders Island, Tasmania. pp. 50–54. Wildlife Disease Association (Australasian Section).

APPENDIX 1

A checklist of the mammals of Australia and its territories

Paul Andrew

This list includes extinct mammals that survived to European arrival and the few mammals that occur on Australian territories, such as Christmas Island and Lord Howe Island. Pelagic species have been included if they are known to have, or are likely to have, occurred in Australian waters.

Taxonomy

Beyond the traditional Linnaean ranks, Subclasses, Suborders and, for dasyurids and murids where Families are large, Subfamilies have been added. It should be noted that this reference list contains no taxonomic novelties and that no attempt has been made to resolve the taxonomic turmoil in bats and rodents.

The sequence of higher taxa follows Wilson and Reeder (2005). Species taxa are in alphabetical order within family (subfamily in dasyurids and murids). Walton (1988) is followed for Geoffroy (not É. Geoffroy), Blainville (not de Blainville), and Tarragon (not de Tarragon). Species limits follow *The Action Plan for Australian Mammals 2012* (Woinarski *et al.* 2014), excepting *Mormopterus*, which follows Reardon *et al.* (2014), and the *Petrogale brachyotis* group, which follows Potter *et al.* (2010); and for *Antechinus argentus* see Baker *et al.* (2013) and for *Antechinus arktos* see Baker *et al.* (2014). *Taxonomy of Australian Mammals* by Jackson and Groves (2015) is expected to provide a benchmark for Australian mammals. For the prehensile-tailed rat, which is technically without a name because the holotype has been re-identified (see Wilson and Reeder 2005: 1440), *Pogonomys mollipilosus* is retained for the purposes of this list.

Common names

There are no novel common names in this list. The primary sources are Strahan (1995) and Wilson and Reeder (2005), with occasional names from Menkhorst and Knight (2004).

There is general favour in using names that are unique, descriptive, short and pronounceable. Whenever possible, names that are parochial and convey little or no information (e.g. 'Monjon' or 'Narblek') should be avoided unless familiarised with a group-name. The use of scientific names as common names is also problematic when species change genus. More importantly, common names and scientific names have different purposes; common names should reflect the phenotype, whereas scientific names reflect the genotype. Following this principle, marsupial mole is used as a group; although not in the same order as the Eurasian moles, they are 'moles' by virtue of being subterranean.

There is a second issue with common names that has yet to attract much attention among Australian mammal names: their construction.

Duckworth and Pine (2003) discuss mammal common names at length, including construction etiquette, using Indochinese mammals as examples. However, their guidelines are drawn from ornithology where there is a long tradition of treating common names as proper names, including capitalisation and the rules of English grammar.

Perhaps the most trenchant issue is how to deal with two-part group names. For example, within the recent Australian mammal literature we find freetail-bat (Strahan 1995), freetail bat (Menkhorst and Knight 2004), and free-tailed bat (Wilson and Reeder 2005).

The protocol of compounding group names is well established in some Australian taxa, for example, rock-wallaby, tree-kangaroo. This works because the two groups are well known. Familiarity might one day make acceptable the groups ringtail and brushtail, as used by Groves in Wilson and Reeder (2005), but for the time being they are known as ring-tailed possum and brush-tailed possum. Elsewhere, retaining the adjectival form, where possible, keeps groups large for unfamiliar taxa, for example, bent-winged bat (rather than bentwing-bat, but horseshoe-bat because the adjectival form is too clumsy, as in horseshoe-marked bat). Thus the generalised rule adopted here is to compound group names comprising two nouns, but to retain two words when the first part of a group name is adjectival. This has the added advantage of being grammatical.

The application of a set of rules always highlights inconsistencies. The groups elephant-seal, pilot-whale and sperm-whale will be small, and while pebble-mound and stick-nest are well-established names, they are clumsily turned in to adjectives. Any committee established to review Australasian mammal names should include consideration of how the name is to be written, as well as spoken.

Checklist of the mammals of Australia and its territories

In this list the usual protocols have been followed, wherein the scientific name at each rank is followed by the name's author and the date of publication. The author and date are in parentheses if the taxon was first placed in a different genus.

Class Mammalia Linnaeus, 1758
Subclass Prototheria Gill, 1872
ORDER MONOTREMATA Bonaparte, 1832
Family Tachyglossidae Gill, 1872
Tachyglossus aculeatus (Shaw, 1792) Short-beaked echidna
Zaglossus bruijni (Peters & Doria, 1876) Long-beaked echidna
Family Ornithorhynchidae Gray, 1825
Ornithorhynchus anatinus (Shaw, 1799) Platypus
Subclass Marsupialia Iliger, 1811
ORDER NOTORYCTEMORPHIA Kirsch, 1977
Family Notoryctidae Ogilby, 1892
Notoryctes caurinus Thomas, 1920 Northern marsupial mole
Notoryctes typhlops (Stirling, 1889) Southern marsupial mole
ORDER DASYUROMORPHIA Gill, 1872
Family Thylacinidae Bonaparte, 1838
Thylacinus cynocephalus (Harris, 1808) Thylacine
Family Myrmecobiidae Waterhouse, 1841
Myrmecobius fasciatus Waterhouse, 1836 Numbat
Family Dasyuridae Goldfuss, 1820
Subfamily Dasyurinae Goldfuss, 1820
Antechinus adustus (Thomas, 1923) Rusty antechinus [tropical antechinus]
Antechinus agilis Dickman, Parnaby, Crowther & King, 1998 Agile antechinus
Antechinus argentus Baker, Mutton & Hines, 2013 Silver-headed antechinus
Antechinus arktos Baker, Mutton, Hines & Van Dyck, 2014 Black-tailed antechinus
Antechinus bellus (Thomas, 1904) Fawn antechinus
Antechinus flavipes (Waterhouse, 1838) Yellow-footed antechinus
Antechinus godmani (Thomas, 1923) Atherton antechinus
Antechinus leo Van Dyck, 1980 Cinnamon antechinus

Antechinus minimus (Geoffroy, 1803) Swamp antechinus
Antechinus mysticus Baker, Mutton & Van Dyck, 2012 Buff-footed antechinus
Antechinus stuartii Macleay, 1841 Brown antechinus
Antechinus subtropicus Van Dyck & Crowther, 2000 Subtropical antechinus
Antechinus swainsonii (Waterhouse, 1840) Dusky antechinus
Dasycercus blythi (Waite, 1904) Brush-tailed mulgara [ampurta]
Dasycercus cristicauda (Krefft, 1867) Crest-tailed mulgara
Dasykaluta rosamondae (Ride, 1964) Kaluta
Dasyuroides byrnei Spencer, 1896 Kowari
Dasyurus geoffroii Gould, 1841 Western quoll
Dasyurus hallucatus Gould, 1842 Northern quoll
Dasyurus maculatus (Kerr, 1792) Spot-tailed quoll [tiger quoll]
Dasyurus viverrinus (Shaw, 1800) Eastern quoll
Parantechinus apicalis (Gray, 1842) Dibbler
Phascogale calura Gould, 1844 Red-tailed phascogale
Phascogale pirata Thomas, 1904 Northern phascogale
Phascogale tapoatafa (Meyer, 1793) Brush-tailed phascogale
Pseudantechinus bilarni (Johnson, 1954) Sandstone false antechinus [sandstone dibbler]
Pseudantechinus macdonnellensis (Spencer, 1895) Fat-tailed false antechinus
Pseudantechinus mimulus (Thomas, 1906) Alexandria false antechinus
Pseudantechinus ningbing Kitchener, 1988 Ningbing false antechinus
Pseudantechinus woolleyae Kitchener & Caputi, 1988 Woolley's false antechinus
Sarcophilus harrisii (Boitard, 1841) Tasmanian devil

Subfamily Sminthopsinae Archer, 1982
Antechinomys laniger (Gould, 1856) Kultarr
Ningaui ridei Archer, 1975 Wongai ningaui
Ningaui timealeyi Archer, 1975 Pilbara ningaui
Ningaui yvonnae Kitchener, Stoddart, & Henry, 1983 Southern ningaui
Planigale gilesi Aitken, 1972 Paucident planigale
Planigale ingrami (Thomas, 1906) Long-tailed planigale
Planigale maculata (Gould, 1851) Pygmy planigale [common planigale]
Planigale tenuirostris Troughton, 1928 Narrow-nosed planigale
Sminthopsis archeri Van Dyck, 1986 Chestnut dunnart
Sminthopsis bindi Van Dyck, Woinarski & Press, 1994 Kakadu dunnart
Sminthopsis butleri Archer, 1979 Carpentarian dunnart [Butler's dunnart]
Sminthopsis crassicaudata (Gould, 1844) Fat-tailed dunnart
Sminthopsis dolichura Kitchener, Stoddart & Henry, 1984 Little long-tailed dunnart
Sminthopsis douglasi Archer, 1979 Julia Creek dunnart
Sminthopsis gilberti Kitchener, Stoddart, & Henry, 1984 Gilbert's dunnart
Sminthopsis granulipes Troughton, 1932 White-tailed dunnart
Sminthopsis griseoventer Kitchener, Stoddart, & Henry, 1984 Grey-bellied dunnart
Sminthopsis hirtipes Thomas, 1898 Greater hairy-footed dunnart
Sminthopsis leucopus (Gray, 1842) White-footed dunnart
Sminthopsis longicaudata Spencer, 1909 Large Long-tailed dunnart
Sminthopsis macroura (Gould, 1845) Stripe-faced dunnart
Sminthopsis murina (Waterhouse, 1838) Slender-tailed dunnart [common dunnart]
Sminthopsis ooldea Troughton, 1965 Ooldea Dunnart
Sminthopsis psammophila Spencer, 1895 Sandhill Dunnart
Sminthopsis virginiae (Tarragon, 1847) Red-cheeked Dunnart
Sminthopsis youngsoni McKenzie & Archer, 1982 Lesser hairy-footed Dunnart

ORDER PERAMELEMORPHIA Ameghino, 1889
Family Thylacomyidae Bensley, 1903
Macrotis lagotis (Reid, 1837) Greater bilby
Macrotis leucura (Thomas, 1887) Lesser bilby
Family Chaeropodidae Gill, 1872
Chaeropus ecaudatus (Ogilby, 1838) Pig-footed bandicoot
Family Peramelidae Gray, 1825
Echymipera rufescens (Peters & Doria, 1875) Rufous spiny bandicoot
Isoodon auratus (Ramsay, 1887) Golden bandicoot
Isoodon macrourus (Gould, 1842) Northern brown bandicoot
Isoodon obesulus (Shaw, 1797) Southern brown bandicoot
Perameles bougainville Quoy & Gaimard, 1824 Western barred bandicoot
Perameles eremiana Spencer, 1897 Desert bandicoot
Perameles gunnii Gray, 1838 Eastern barred bandicoot
Perameles nasuta Geoffroy, 1804 Southern long-nosed bandicoot
Perameles pallescens Thomas, 1923 Northern long-nosed bandicoot
ORDER DIPROTODONTIA Owen, 1877
SUBORDER VOMBATIFORMES Burnett, 1830
Family Phascolarctidae Owen, 1839
Phascolarctos cinereus (Goldfuss, 1817) Koala
Family Vombatidae Burnett, 1830
Lasiorhinus krefftii (Owen, 1872) Northern hairy-nosed wombat
Lasiorhinus latifrons (Owen, 1845) Southern hairy-nosed wombat
Vombatus ursinus (Shaw, 1800) Common wombat
SUBORDER PHALANGERIFORMES Szalay, 1982
Family Burramyidae Broom, 1898
Burramys parvus Broom, 1895 Mountain pygmy possum
Cercartetus caudatus (Milne-Edwards, 1877) Long-tailed pygmy possum
Cercartetus concinnus (Gould, 1845) Western pygmy possum
Cercartetus lepidus (Thomas, 1888) Little pygmy possum [Tasmanian pygmy possum]
Cercartetus nanus (Desmarest, 1817) Eastern pygmy possum
Family Phalangeridae Thomas, 1888
Phalanger mimicus Thomas, 1922 Southern common cuscus
Spilocuscus maculatus (Geoffroy, 1803) Common spotted cuscus
Trichosurus caninus (Ogilby, 1836) Short-eared brush-tailed possum
Trichosurus cunninghami Lindenmayer, Dubach & Viggers, 2002 Mountain brush-tailed possum
Trichosurus vulpecula (Kerr, 1792) Common brush-tailed possum
Wyulda squamicaudata Alexander, 1919 Scaly-tailed possum
Family Pseudocheiridae Winge, 1893
Hemibelideus lemuroides (Collett, 1884) Lemuroid ring-tailed possum
Petauroides volans (Kerr, 1792) Greater glider
Petropseudes dahli (Collett, 1895) Rock ring-tailed possum
Pseudocheirus occidentalis (Thomas, 1888) Western ring-tailed possum
Pseudocheirus peregrinus (Boddaert, 1785) Common ring-tailed possum
Pseudochirops archeri (Collett, 1884) Green ring-tailed possum
Pseudochirulus cinereus (Tate, 1945) Cinereous ring-tailed possum
Pseudochirulus herbertensis (Collett, 1884) Herbert River ring-tailed possum
Family Petauridae Bonaparte, 1832
Dactylopsila trivirgata Gray, 1858 Striped possum
Gymnobelideus leadbeateri McCoy, 1867 Leadbeater's possum
Petaurus australis Shaw, 1791 Yellow-bellied glider
Petaurus breviceps Waterhouse, 1838 Sugar glider
Petaurus gracilis (De Vis, 1883) Mahogany glider [ebony glider]
Petaurus norfolcensis (Kerr, 1792) Squirrel glider
Family Tarsipedidae Gervais & Verreaux, 1842

Tarsipes rostratus Gervais & Verreaux, 1842 Honey possum

Family Acrobatidae Aplin, 1987

Acrobates pygmaeus (Shaw, 1794) Feather-tailed glider

SUBORDER MACROPODIFORMES Ameghino, 1889

Family Hypsiprymnodontidae Collett, 1877

Hypsiprymnodon moschatus Ramsay, 1876 Musky rat-kangaroo

Family Potoroidae Gray, 1821

Aepyprymnus rufescens (Gray, 1837) Rufous bettong

Bettongia anhydra Finlayson, 1957 Desert bettong

Bettongia gaimardi (Desmarest, 1822) Eastern bettong

Bettongia lesueur (Quoy & Gaimard, 1824) Burrowing bettong

Bettongia penicillata Gray, 1837 Brush-tailed bettong

Bettongia pusilla McNamara, 1997 Nullarbor dwarf bettong

Bettongia tropica Wakefield, 1967 Northern bettong

Caloprymnus campestris (Gould, 1843) Desert rat-kangaroo

Potorous gilbertii (Gould, 1841) Gilbert's potoroo

Potorous longipes Seebeck & Johnston, 1980 Long-footed potoroo

Potorous platyops (Gould, 1844) Broad-faced potoroo

Potorous tridactylus (Kerr, 1792) Long-nosed potoroo

Family Macropodidae Gray, 1821

Dendrolagus bennettianus De Vis, 1887 Bennett's tree-kangaroo

Dendrolagus lumholtzi Collett, 1884 Lumholtz's tree-kangaroo

Lagorchestes asomatus Finlayson, 1943 Central hare-wallaby [Lake Mackay hare-wallaby]

Lagorchestes conspicillatus Gould, 1842 Spectacled hare-wallaby

Lagorchestes hirsutus Gould, 1844 Rufous hare-wallaby

Lagorchestes leporides (Gould, 1841) Eastern hare-wallaby

Lagostrophus fasciatus (Péron & Lesueur, 1807) Banded hare-wallaby

Macropus agilis (Gould, 1842) Agile wallaby

Macropus antilopinus (Gould, 1842) Antilopine kangaroo [antilopine wallaroo]

Macropus bernardus Rothschild, 1904 Black wallaroo [Woodward's wallaroo]

Macropus dorsalis (Gray, 1837) Black-striped wallaby

Macropus eugenii (Desmarest, 1817) Tammar wallaby

Macropus fuliginosus (Desmarest, 1817) Western grey kangaroo

Macropus giganteus Shaw, 1790 Eastern grey kangaroo

Macropus greyi Waterhouse, 1846 Toolache wallaby

Macropus irma (Jourdan, 1837) Western brush wallaby

Macropus parma Waterhouse, 1846 Parma wallaby

Macropus parryi Bennett, 1835 Pretty-faced wallaby [whip-tailed wallaby]

Macropus robustus Gould, 1840 Common wallaroo

Macropus rufogriseus (Desmarest, 1817) Red-necked wallaby

Macropus rufus (Desmarest, 1822) Red kangaroo

Onychogalea fraenata (Gould, 1841) Bridled nail-tailed wallaby

Onychogalea lunata (Gould, 1840) Crescent nail-tailed wallaby

Onychogalea unguifera (Gould, 1840) Northern nail-tailed wallaby

Petrogale assimilis Ramsay, 1877 Allied rock-wallaby

Petrogale brachyotis (Gould, 1841) Short-eared rock-wallaby

Petrogale burbidgei Kitchener & Sanson, 1978 Monjon rock-wallaby

Petrogale coenensis Eldridge & Close, 1992 Cape York rock-wallaby

Petrogale concinna Gould, 1842 Nabarlek rock-wallaby

Petrogale godmani Thomas, 1923 Godman's rock-wallaby
Petrogale herberti Thomas, 1926 Herbert's rock-wallaby
Petrogale inornata Gould, 1842 Unadorned rock-wallaby
Petrogale lateralis Gould, 1842 Black-footed rock-wallaby [black-flanked rock-wallaby]
Petrogale mareeba Eldridge & Close, 1992 Mareeba rock-wallaby
Petrogale penicillata (Gray, 1827) Brush-tailed rock-wallaby
Petrogale persephone Maynes, 1982 Proserpine rock-wallaby
Petrogale purpureicollis Le Souef, 1924 Purple-necked rock-wallaby
Petrogale rothschildi Thomas, 1904 Rothschild's rock-wallaby
Petrogale sharmani Eldridge & Close, 1992 Mt Claro rock-wallaby
Petrogale wilkinsi Thomas, 1926 Wilkins' rock-wallaby
Petrogale xanthopus Gray, 1855 Yellow-footed rock-wallaby
Setonix brachyurus (Quoy & Gaimard, 1830) Quokka
Thylogale billardierii (Desmarest, 1822) Tasmanian pademelon
Thylogale stigmatica (Gould, 1860) Red-legged pademelon
Thylogale thetis (Lesson, 1827) Red-necked pademelon
Wallabia bicolor (Desmarest, 1804) Swamp wallaby

Subclass Eutheria Gill, 1872

ORDER SIRENIA Iliger, 1811

Family Dugongidae Gray, 1821
Dugong dugon (Müller, 1776) Dugong

ORDER SORICOMORPHIA Gregory, 1910

Family Soricidae Fischer von Waldheim, 1817
Crocidura trichura Dobson, 1889 Christmas Island shrew

ORDER CHIROPTERA Blumenbach, 1779

Family Pteropodidae Gray, 1821
Dobsonia magna Thomas, 1905 Bare-backed fruit-bat
Macroglossus minimus (Geoffroy, 1810) Northern blossom-bat
Nyctimene robinsoni Thomas, 1904 Eastern tube-nosed bat
Pteropus alecto Temminck, 1837 Black flying-fox
Pteropus brunneus Dobson, 1878 Percy Island flying-fox
Pteropus conspicillatus Gould, 1850 Spectacled flying-fox
Pteropus macrotis Peters, 1867 Big-eared flying-fox
Pteropus natalis Thomas, 1887 Christmas Island flying-fox
Pteropus poliocephalus Temminck, 1825 Grey-headed flying-fox
Pteropus scapulatus Peters, 1862 Little red flying-fox
Syconycteris australis (Peters, 1867) Eastern blossom-bat [common blossom-bat]

Family Rhinolophidae Gray, 1825
Rhinolophus megaphyllus Gray 1834 Eastern horseshoe-bat
Rhinolophus robertsi Tate, 1952 Greater large-eared horseshoe-bat

Family Hipposideridae Lydekker, 1891
Hipposideros ater Templeton, 1848 Dusky leaf-nosed bat
Hipposideros cervinus (Gould, 1854) Fawn leaf-nosed bat
Hipposideros diadema (Geoffroy, 1813) Diadem leaf-nosed bat
Hipposideros inornatus McKean, 1970 Arnhem leaf-nosed bat
Hipposideros semoni Matschie, 1903 Semon's leaf-nosed bat
Hipposideros stenotis Thomas 1913 Northern leaf-nosed bat
Rhinonicteris aurantia (Gray, 1845) Orange leaf-nosed bat

Family Megadermatidae Allen, 1864
Macroderma gigas (Dobson, 1880) Ghost bat

Family Emballonuridae Gervais, 1855
Saccolaimus flaviventris (Peters, 1867) Yellow-bellied sheath-tailed bat

Saccolaimus mixtus Troughton, 1925 Cape York sheath-tailed bat
Saccolaimus saccolaimus (Temminck, 1838) Bare-rumped sheath-tailed bat
Taphozous australis Gould, 1854 Coastal sheath-tailed bat [coastal tomb bat]
Taphozous georgianus Thomas, 1915 Common sheath-tailed bat [sharp-nosed tomb bat]
Taphozous hilli Kitchener, 1980 Hill's Sheath-tailed bat
Taphozous kapalgensis McKean & Friend, 1979 Arnhem sheath-tailed bat
Taphozous troughtoni Tate, 1952 Troughton's sheath-tailed bat

Family Miniopteridae Dobson, 1975
Miniopterus australis (Tomes, 1858) Little bent-winged bat
Miniopterus orianae Thomas, 1922 Large bent-winged bat

Family Molossidae Gervais, 1855
Austronomus australis (Gray, 1838) White-striped free-tailed bat
Chaerephon jobensis (Miller, 1902) Great Northern free-tailed bat
Micronomus norfolkensis (Gray, 1839) Eastern free-tailed bat
Mormopterus cobourgianus (Johnson, 1959) Mangrove free-tailed bat
Mormopterus eleryi Reardon & McKenxie, 2008 Bristle-faced free-tailed bat
Mormopterus halli Reardon, McKenzie & Adams, 2014 Cape York free-tailed bat
Mormopterus kitcheneri McKenzie, Reardon & Adams, 2014 South-western free-tailed bat
Mormopterus lumsdenae Reardon, McKenzie & Adams, 2014 Northern free-tailed bat
Mormopterus petersi (Leche, 1884) Inland free-tailed bat
Mormopterus planiceps (Peters, 1866) South-eastern free-tailed bat
Mormopterus ridei (Felten, 1964) Ride's free-tailed bat

Family Vespertilionidae Gray, 1821
Chalinolobus dwyeri Ryan, 1966 Large-eared wattled bat
Chalinolobus gouldii (Gray, 1841) Gould's wattled bat
Chalinolobus morio (Gray, 1841) Chocolate wattled bat
Chalinolobus nigrogriseus (Gould, 1856) Hoary wattled bat
Chalinolobus picatus (Gould, 1852) Little wattled bat
Falsistrellus mackenziei Kitchener, Caputi & Jones, 1986 Western false pipistrelle
Falsistrellus tasmaniensis (Gould, 1858) Eastern false pipistrelle
Murina florium Thomas, 1908 Flute-nosed bat
Myotis macropus (Gould, 1855) Southern myotis
Nyctophilus arnhemensis Johnson, 1959 Arnhem long-eared bat
Nyctophilus bifax Thomas, 1915 Eastern long-eared bat
Nyctophilus corbeni Parnaby, 2009 South-eastern long-eared bat
Nyctophilus daedalus Thomas, 1915 Pallid long-eared bat
Nyctophilus geoffroyi Leach, 1821 Lesser long-eared bat
Nyctophilus gouldi Tomes, 1858 Gould's long-eared bat
Nyctophilus howensis McKean, 1975 Lord Howe Island long-eared bat
Nyctophilus major Gray, 1844 Greater long-eared bat
Nyctophilus sherrini Thomas, 1915 Tasmanian long-eared bat
Nyctophilus walkeri Thomas, 1892 Pygmy long-eared bat
Phoniscus papuensis (Dobson, 1878) Golden-tipped bat
Pipistrellus adamsi Kitchener, Caputi & Jones, 1986 Cape York pipistrelle
Pipistrellus murrayi Andrews, 1900 Christmas Island pipistrelle
Pipistrellus westralis Koopman, 1984 Northern pipistrelle
Scoteanax rueppellii (Peters 1866) Greater broad-nosed bat

Scotorepens balstoni (Thomas, 1906) Inland broad-nosed bat
Scotorepens greyii (Gray, 1843) Little broad-nosed bat
Scotorepens orion (Troughton, 1937) Eastern broad-nosed bat
Scotorepens sanborni (Troughton, 1937) Northern broad-nosed bat
Vespadelus baverstocki (Kitchener, Jones & Caputi, 1987) Inland forest-bat
Vespadelus caurinus (Thomas, 1914) Northern cave-bat
Vespadelus darlingtoni (Allen, 1933) Large forest-bat
Vespadelus douglasorum Kitchener, 1976 Yellow-lipped bat
Vespadelus finalysoni Kitchener, Jones & Caputi, 1987 Finlayson's bat
Vespadelus pumilus (Gray, 1841) Eastern forest-bat
Vespadelus regulus (Thomas, 1906) Southern forest-bat
Vespadelus troughtoni (Kitchener, Jones & Caputi, 1987) Eastern cave-bat
Vespadelus vulturnus (Thomas, 1914) Little forest-bat

ORDER CARNIVORA Bowdich, 1821
SUBORDER CANIFORMIA Kretzoi, 1943
Family Otariidae Gray, 1825
Arctocephalus pusillus (Schreber, 1775) Brown fur-seal
Arctophoca australis (Lesson, 1828) Southern fur-seal
Arctophoca gazella Peters, 1875 Antarctic fur-seal
Arctophoca tropicalis (Gray, 1872) Subantarctic fur-seal
Neophoca cinerea (Péron, 1816) Australian sea-lion
Phocarctos hookeri (Gray, 1844) New Zealand sea-lion

Family Phocidae Gray, 1821
Hydrurga leptonyx (Blainville, 1820) Leopard seal
Leptonychotes weddellii (Lesson, 1826) Weddell seal
Lobodon carcinophaga (Hombron & Jacquinot, 1842) Crab-eater seal
Mirounga leonina (Linnaeus, 1758) Southern elephant seal
Ommatophoca rossii Gray, 1844 Ross seal

ORDER CETACEA Brisson, 1762
SUBORDER MYSTICETI Flower, 1864
Family Balaenidae Gray, 1821
Eubalaena australis (Desmoulins, 1822) Southern right whale

Family Balaenopteridae Gray, 1864
Balaenoptera acutorostrata Lacépède, 1804 Dwarf minke whale
Balaenoptera bonaerensis Burmeister, 1867 Antarctic minke whale
Balaenoptera borealis Lesson, 1828 Sei whale
Balaenoptera edeni Anderson, 1878 Bryde's whale
Balaenoptera musculus (Linnaeus, 1758) Blue whale
Balaenoptera omurai Wada, Olshi & Yamada, 2003 Omura's whale
Balaenoptera physalus (Linnaeus, 1758) Fin whale
Megaptera novaeangliae (Borowski, 1781) Hump-backed whale

Family Neobalaenidae Gray, 1873
Caperea marginata (Gray, 1846) Pygmy right whale

SUBORDER ODONTOCETI Flower, 1869
Family Delphinidae Gray, 1821
Delphinus capensis, Gray, 1828 Long-beaked common dolphin
Delphinus delphis Linnaeus, 1758 Short-beaked common dolphin
Feresa attenuata Gray, 1874 Pygmy killer whale
Globicephala macrorhynchus Gray, 1846 Short-finned pilot whale
Globicephala melas (Traill, 1809) Long-finned pilot whale
Grampus griseus (Cuvier, 1812) Risso's dolphin
Lagenodelphis hosei Fraser, 1956 Fraser's dolphin
Lagenorhynchus cruciger (Quoy & Gaimard, 1824) Hourglass dolphin
Lagenorhynchus obscurus (Gray, 1828) Dusky dolphin

Lissodelphis peronii (Lacépède, 1804) Southern right whale dolphin

Orcaella heinsohni Beasley, Robertson & Arnold, 2005 Australian snub-finned dolphin

Orcinus orca (Linnaeus, 1758) Common killer whale

Peponocephala electra (Gray, 1846) Melon-headed whale

Pseudorca crassidens (Owen, 1846) False killer whale

Sousa chinensis (Osbeck, 1765) Indo-Pacific hump-backed dolphin

Stenella attenuata (Gray, 1846) Pantropical spotted dolphin

Stenella coeruleoalba (Meyen, 1833) Striped dolphin

Stenella longirostris (Gray, 1828) Spinner dolphin

Steno bredanensis (Lesson, 1828) Rough-toothed dolphin

Tursiops aduncus (Ehrenberg, 1833) Indo-Pacific bottle-nosed dolphin

Tursiops truncatus (Montagu, 1821) Common bottle-nosed dolphin

Family Phocoenidae Gray, 1825

Phocoena dioptrica Lahille, 1912 Spectacled porpoise

Family Physeteridae Gray, 1821

Kogia breviceps (Blainville, 1838) Pygmy sperm whale

Kogia sima (Owen, 1866) Dwarf sperm whale

Physeter catodon Linnaeus, 1758 Giant sperm whale

Family Ziphiidae Gray, 1865

Berardius arnuxii Duvernoy, 1851 Arnoux's beaked whale

Hyperoodon planifrons Flower, 1882 Southern bottle-nosed whale

Indopacetus pacificus (Longman, 1926) Longman's bottle-nosed whale

Mesoplodon bowdoini Andrews, 1908 Andrews's beaked whale

Mesoplodon densirostris (Blainville, 1817) Blainville's beaked whale

Mesoplodon ginkgodens Nishiwaki & Kamiya, 1958 Ginkgo-toothed beaked whale

Mesoplodon grayi Von Haast, 1876 Gray's beaked whale

Mesoplodon hectori (Gray, 1871) Hector's beaked whale

Mesoplodon layardii (Gray, 1865) Strap-toothed beaked whale

Mesoplodon mirus True, 1913 True's beaked whale

Tasmacetus shepherdi Oliver, 1937 Shepherd's beaked whale

Ziphius cavirostris Cuvier, 1823 Cuvier's beaked whale

ORDER RODENTIA Bowdich, 1821

Family Muridae Gray, 1821

Subfamily Hydromyinae Alston, 1876

Conilurus albipes (Lichtenstein, 1829) White-footed rabbit-rat

Conilurus capricornensis Cramb & Hocknull, 2010 Capricorn rabbit-rat

Conilurus penicillatus (Gould, 1842) Brush-tailed rabbit-rat

Hydromys chrysogaster Geoffroy, 1804 Water rat

Leggadina forresti (Thomas, 1906) Forrest's mouse

Leggadina lakedownensis Watts, 1976 Lakeland Downs mouse

Leporillus apicalis (Gould, 1853) Lesser stick-nest rat

Leporillus conditor (Sturt, 1848) Greater stick-nest rat

Mastacomys fuscus Thomas, 1882 Broad-toothed rat

Melomys burtoni (Ramsay, 1887) Grassland melomys

Melomys capensis Tate, 1951 Cape York melomys

Melomys cervinipes (Gould, 1852) Fawn-footed melomys

Melomys rubicola Thomas, 1924 Bramble Cay melomys

Mesembriomys gouldii (Gray, 1843) Black-footed tree-rat

Mesembriomys macrurus (Peters, 1876) Golden-backed tree-rat

Notomys alexis Thomas, 1922 Spinifex hopping mouse

Notomys amplus Brazenor, 1936 Short-tailed hopping mouse

Notomys aquilo Thomas, 1921 Northern hopping mouse
Notomys cervinus (Gould, 1853) Fawn hopping mouse
Notomys fuscus (Jones, 1925) Dusky hopping mouse
Notomys longicaudatus (Gould, 1844) Long-tailed hopping mouse
Notomys macrotis Thomas, 1921 Big-eared hopping mouse
Notomys mitchellii (Ogilby, 1838) Mitchell's hopping mouse
Notomys mordax Thomas, 1922 Darling Downs hopping mouse
Notomys robustus Mahoney, Smith & Medlin, 2008 Broad-cheeked hopping mouse
Pogonomys mollipilosus (Peters & Doria, 1881) Prehensile-tailed rat
Pseudomys albocinereus (Gould, 1845) Ash-grey mouse
Pseudomys apodemoides Finlayson, 1932 Silky mouse
Pseudomys auritus Thomas, 1910 Long-eared mouse
Pseudomys australis Gray, 1832 Plains rat
Pseudomys bolami Troughton, 1932 Bolam's mouse
Pseudomys calabyi Kitchener & Humphreys, 1987 Kakadu pebble-mound mouse
Pseudomys chapmani Kitchener, 1980 Western pebble-mouse
Pseudomys delicatulus (Gould, 1842) Delicate mouse
Pseudomys desertor Troughton, 1932 Desert mouse
Pseudomys fieldi (Waite, 1896) Shark Bay mouse
Pseudomys fumeus Brazenor, 1934 Smoky mouse
Pseudomys glaucus Thomas, 1910 Blue-grey mouse
Pseudomys gouldii (Waterhouse, 1839) Gould's mouse
Pseudomys gracilicaudatus (Gould, 1845) Eastern chestnut mouse
Pseudomys hermannsburgensis (Waite, 1896) Sandy inland mouse
Pseudomys higginsi (Trouessart, 1897) Long-tailed mouse
Pseudomys johnsoni Kitchener, 1985 Central pebble-mound mouse
Pseudomys nanus (Gould, 1858) Western chestnut mouse
Pseudomys novaehollandiae (Waterhouse, 1843) New Holland mouse
Pseudomys occidentalis Tate, 1951 Western mouse
Pseudomys oralis Thomas, 1921 Hastings River mouse
Pseudomys patrius (Thomas & Dollman, 1909) Eastern pebble-mouse
Pseudomys shortridgei (Thomas, 1907) Heath mouse
Uromys caudimaculatus (Krefft, 1867) Giant white-tailed rat
Uromys hadrourus (Winter, 1984) Masked white-tailed rat
Xeromys myoides Thomas, 1889 Water mouse [false water-rat]
Zyzomys argurus (Thomas, 1889) Common rock-rat
Zyzomys maini Kitchener, 1989 Arnhem Land rock-rat
Zyzomys palatalis Kitchener, 1989 Carpentarian rock-rat
Zyzomys pedunculatus (Waite, 1896) Central rock-rat
Zyzomys woodwardi (Thomas, 1909) Large rock-rat

Subfamily Murinae Murray, 1866
Rattus colletti (Thomas, 1904) Dusky rat
Rattus fuscipes (Waterhouse, 1839) Bush rat
Rattus leucopus (Gray, 1867) Cape York rat
Rattus lutreolus (Gray, 1841) Swamp rat
Rattus macleari (Thomas, 1887) Maclear's rat
Rattus nativitatis (Thomas, 1889) Christmas Island rat
Rattus sordidus (Gould, 1858) Canefield rat
Rattus tunneyi (Thomas, 1904) Tunney's rat
Rattus villosissimus (Waite, 1898) Long-haired rat

References

Baker AM, Mutton TY, Hines HB (2013) A new dasyurid marsupial from Kroombit Tops, south-east Queensland, Australia: the Silver-headed Antechinus, *Antechinus argentus* sp. nov. (Marsupialia: Dasyuridae). *Zootaxa* **3746**, 201–239. doi:10.11646/zootaxa.3746.2.1

Baker AM, Mutton TY, Hines HB, Van Dyck S (2014) The Black-tailed Antechinus, Antechinus arktos sp. nov.: a new species of carnivorous marsupial from montane regions of the Tweed Volcano caldera, eastern Australia. *Zootaxa* **3765**, 101–133. doi:10.11646/zootaxa.3765.2.1

Duckworth JW, Pine RH (2003) English names for a world list of mammals, exemplified by species of Indochina. *Mammal Review* **33**, 151–173. doi:10.1046/j.1365-2907.2003.00012.x

Jackson SM, Groves CP (2015) *Taxonomy of Australian Mammals.* CSIRO Publishing, Melbourne.

Menkhorst P, Knight F (2004) *A Field Guide to the Mammals of Australia* (2nd edn). Oxford University Press, Melbourne.

Potter S, Eldridge MD, Cooper S, Taggart DA (2010) Phylogeography of the *brachyotis* group with recognition of two species within *Petrogale brachyotis*. In *Proceedings of the 56th Scientific Meeting of the Australian Mammal Society and Rock-wallaby Symposium*, 5–9 July 2010, Canberra. (Eds IR Wallis and MDB Eldridge) p. 56. The Australian Mammal Society, Canberra.

Reardon TB, McKenzie NL, Cooper SJB, Appleton B, Carthew S, Adams M (2014) A molecular and morphological investigation of species boundaries and phylogenetic relationships in Australian free-tailed bats *Mormopterus* (Chiroptera: Molossidae). *Australian Journal of Zoology* **62**, 109–136. doi:10.1071/ZO13082

Strahan R (1995) *The Australian Museum Complete Book of Australian Mammals.* Reed New Holland, Sydney.

Walton DW (1988) *Zoological Catalogue of Australia, Volume 5, Mammalia.* Australian Biological Resources Study, Canberra.

Wilson DE, Reeder DM (Eds) (2005) *Mammal Species of the World: A Taxonomic and Geographic Reference.* John Hopkins University Press, Baltimore.

Woinarski JCZ, Burbidge AA, Harrison PL (2014) *The Action Plan for Australian Mammals 2012.* CSIRO Publishing, Melbourne.

APPENDIX 2

Glossary of abbreviations and terms used for annotation of images

A	Acetabulum
Ac	Acromion
AcB	Accessory bulla
AcC	Accessory carpal
AcP	Accessory process
AdT	Adipose tissue
AF	Acetabular fossa
AM	Alveolar margin
AnP	Angular process
Ao	Aorta
AO	Auditory ossicle
AP	Articular process
Apx	Apex
B	Beak
BNA	Bony nasal aperture
Br	Bronchus
BZ	Bifid zygomatic
C	Canine
Ca(x)	Carpal bone
Cal	Calcaneus
Calc	Calcar
CAP	Caudal articular process
Cd(x)	Caudal vertebra
CdE	Caudal epicondyle
Ce(x)	Cervical vertebra
CeR	Cervical rib
CeV	Cephalic vein
CF	Coxofemoral
CH	Capitulum humeri
Ch	Choana
ChE	Chondral element
ChTC	Aerated chamber of the thyroid cartilage
Cl	Clavicle
CM	Cranial mediastinum
CMS	Caudal maxillary sinus
CNM	Common nasal meatus
Co	Coracoid
Coc	Cochlea
Col	Colon
CoP	Condyloid process
CP	Coronoid process
CPM	Coronoid process of the mandible
Cr	Crown
CrE	Cranial epicondyle
CrP	Cribriform plate
CT	Central tarsal
D	Dentine
DaP	Dactylopatagium
DC	Descending colon
De	Dentary
DFE	Distal femoral epiphysis
DFiE	Distal fibular epiphysis
DNC	Dorsal nasal concha
DNM	Dorsal nasal meatus
DP(x)	Deciduous premolar
DR	Deltoid ridge
DRC	Distal row of carpal bones
DRE	Distal radial epiphysis
DRT	Distal row of tarsal bones
DT	Deltoid tuberosity
DTiE	Distal tibial epiphysis

DUE	Distal ulna epiphysis
Duo	Duodenum
E	Enamel
EAC	External auricular canal
EAM	External acoustic meatus
EB	Epipubic bone
EC	Ethmoidal conchae
Ep	Epicoracoid
ET	Ethmoid turbinates
ETR	Epitympanic recess
ETRl	Epitympanic recess lateral
ETRm	Epitympanic recess medial
F	Femur
Fa	Fabella
FC	Femoral condyle
FCE	Femoral condylar epiphysis
FH	Femoral head
Fi	Fibula
FLS	Filtering lip serrations
FP	Fat pad
Frc	Furcation
FrS	Frontal shield
FS	Frontal sinus
GB	Gall bladder
Gl	Glenoid
GTr	Greater trochanter
GTu	Greater tubercle
H	Humerus
HA	Haemal arch
HC	Humeral condyle
HH	Humeral head
HPt	Hamulus pterygoid bone
Hrt	Heart
Hy	Hyoid
I	Ilium
I(x)	Incisor
IA	Iliac apophysis
IdS	Interdental space
In	Interclavicle
Inc	Incisor
InC	Intermediate carpal
IoF	Infraorbital foramen
IPP	Iliopectineal process
Isc	Ischium
IscF	Ischial foramen
IT	Ischial tuberosity
JP	Jugular process
JuP	Jugal process
K	Kidney
Ke	Keel
L(x)	Lumbar vertebra
La	Laryngeal lumen, larynx
LaTr	Lateral trochanter
LC	Lateral condyle
LE	Lateral epicondyle
Li	Liver
LK	Left kidney
LManMo	Left mandibular molars
LMaxMo	Left maxillary molars
LTr	Lesser trochanter
LTu	Lesser tubercle
Lu	Lung
M	Mandible
M(x)	Molar
Ma	Mass
MamP	Mammillary process
ManI	Mandibular incisor
ManM	Mandibular molars and premolars
Max	Maxilla
MaxI	Maxillary incisor
MaxM	Maxillary molars and premolars
MC	Medial condyle
Mc(x)	Metacarpal
MCP	Medial coronoid process
ME	Medial epicondyle
MeF	Mental foramen
Mes	Mesosternum
MeTr	Medial trochanter
MF	Masseteric fossa
MI	Mandibular incisor
MIE	Middle and inner ear
MNM	Middle nasal meatus
Mo	Molar
MS	Mandibular symphysis
MT	Maxillary turbinates
Mt(x)	Metatarsal bone
MTR	Molar tooth root
NC	Nasal cavity
NF	Nutrient foramen
NMT	Nasomaxillary turbinates

No	Nostril
NP	Nasopharynx
NPM	Nasopharyngeal meatus
NS	Nasal septum
NV	Nasal vestibule
OApx	Open apex
OB	Occipital bone
OC	Occipital condyle
OF	Obturator foramen
Ol	Olecranon
ONS	Ossified nasal septum
OsC	*Os calcaris*
OsP	*Os penis*
Ov	Ovary
P(x)	Phalange
Pa	Patella
PA	Pulmonary artery
Pat	Patagium
PB	Pneumatised bone
PcG	Paracloacal glands
PdL	Periodontal ligament
Pe	Penis
PEM(x)	Partially erupted molar
PFE	Proximal femoral epiphysis
PFiE	Proximal fibular epiphysis
Ph	Physis
Pha	Pharynx
PlP	Plagiopatagium
PM	Premolar
PM(x)	Premolar
Po	Pouch
PoPr	Postorbital process
Pr	Presternum
PRC	Proximal row of carpal bones
Prem	Premaxilla
Pro	Prostate
PrP	Propatagium
PS	Pubic symphysis
Pt	Pterygoid bone
PTB	Pneumaticised temporal bone
PtF	Pterygoid fossa
PTiE	Proximal tibial epiphysis
PTiM	Proximal tibial metaphysis
Pu	Pubis
PuV	Pulmonary vein
PV	Palatine vacuities
R	Radius
R(x)	Rib
RC	Radial carpal
RH	Radial head
RK	Right kidney
RMaxMo	Right maxillary molars
RoC	Root canal
RT	Radial tuberosity
S	Scapula
Sa(x)	Sacral vertebra
SaI	Sacroiliac
SaP	Sacral plate
Sc	Scrotum
SC	Supinator crest
Se	Sesamoid
Sed	Sediment
SF	Sacral foramen
SGTu	Supraglenoid tubercle
SI	Small intestine
SoP	Soft palate
SP	Spinous process of vertebrae
Sp	Spur
Spl	Spleen
SS	Sphenoidal sinus
SSc	Scapular spinous process
ST	Soft tissue
St(x)	Sternal element(s)
Sto	Stomach
Ta(x)	Tarsal bone
Tal	Talus
TB	Tympanic bulla
TC	Tympanic cavity
TCA	Tibial crest apophysis
Te	Testicle
TE	Tibial epiphysis
TH	Trochlea humeri
Th(x)	Thoracic vertebra
Ti	Tibia
TiC	Tibial crest
TMJ	Temporomandibular joint
TP	Transverse process
Tr	Trachea
TT	Teres tuberosity
Tu	Tubercle

U	Ulna
UB	Urinary bladder
UC	Ulna carpal
Upm(x)	Unerupted premolar
UrP	Uropatagium
UrV	Uropatagial vein
V	Vomer
VCS	Ventral conchal sinus
VNC	Ventral nasal concha
VNM	Ventral nasal meatus
Xi	Xiphisternum
ZA	Zygomatic arch

INDEX

www.ingramcontent.com/pod-product-compliance
Lightning Source LLC
LaVergne TN
LVHW071343210926
848486LV00003B/214
9780643108646